SYMPOSIUM ON THE FOUNDATIONS OF MODERN PHYSICS 1990

Other titles in this series

SYMPOSIUM ON THE FOUNDATIONS OF MODERN PHYSICS '85
50 Years of the Einstein-Podolsky-Rosen Gedankenexperiment
Joensuu, Finland 16–20 June 1985

edited by **P Lahti** (*Turku*) and **P Mittelstaedt** (*Cologne*)

Contents: Quantum Mechanics and Reality (*N Rosen*); Tests of Bell's Inequalities with Pairs of Low Energy Correlated Photons: An Experimental Realization of Einstein-Podolsky-Rosen-Type Correlations (*A Aspect & P Grangier*); Reality in Classical Physics and Reality in Quantum Theory (*H Barreau*); The Non-Unique Decomposition of Mixtures: Some Remarks (*E G Beltrametti*); Dirac Revisted (*D J Foulis & C H Randall*); EPR: When is a Correlation not a Mystery? (*B C van Fraassen*); The EPR Problem in its Historical Development (*M Jammer*); A New Interpretation of Quantum Mechanics (*S Kochen*); Observations in Quantum Mechanics and the "Collapse of the Wave Function" (*R Peierls*); Quantum Mechanics: Fifty Years Later (*C Piron*); On the Unification of Quantum Theory and Classical Physics (*H P Stapp*); Quantum Theory and Space-Time (*C F von Weizsäcker*); Wolfgang Pauli, Physicist and Philosopher (*C P Enz*); On the Non-Locality of Pre- and Post-Selected Quantum Ensembles (*J Bub*); Realization of a "Delayed-Choice", Mach-Zehnder Interferometer (*T Hellmuth et al.*); Quantum Theory, Causality, and EPR Experiments (*K Kraus*); EPR Non-Separability and Global Aspects of Quantum Mechanics (*Y Ne'eman*); Quantum Geometry and the EPR Gedankenexperiment (*E Prugovečki*); Entropy and Measurement in Stochastic Quantum Mechanics (*F E Schroeck, Jr.*); New Variable Detection Probability Model for EPR-Type Experiments (*F Selleri*) and others.

SYMPOSIUM ON THE FOUNDATIONS OF MODERN PHYSICS '87
The Copenhagen Interpretation 60 years after the Como Lecture
Joensuu, Finland 6–8 Aug 1987

edited by **P Lahti** (*Turku*) & **P Mittelstaedt** (*Cologne*)

Contents: Niels Bohr's Discussions with Albert Einstein, Werner Heisenberg and Erwin Schrödinger: The Origins of the Principles of Uncertainty and Complementarity (*J Mehra*); The Role of Modern Mathematics in the Formulation of Basic Laws in Physics (*K Bleuler*); An Ontological Foundation for the Quantum Theory (*D Bohm*); Unsharp Reality and the Question of Quantum Systems (*P Busch*); The Space, Time, and Field Concepts in Wolfgang Pauli's Work (*C P Enz*); Veiled Reality (*B d'Espagnat*); Niels Bohr's Concept of Reality (*H J Folse*); Wolfgang Pauli's Conception of Reality (*K V Laurikainen*); Contextual Quantum Objects and their Ontic Interpretation (*H Primas*); Heisenberg's Philosophy (*C von Weizsäcker*); A Quantitative Criterion of the Applicability of the Classical Description within the Quantum Theory (*L A Khalfin & B S Tsirelson*); Unsharp Joint Measurement of Complementary Observables in a Photon Split Beam Experiment (*P Mittelstaedt & I Ohba*); Many-Hilbert-Spaces Theory of Quantum Measurements (*M Namiki & O Ohba*); Quantum Chaos and Irreversibility (*A Peres*) and others.

SYMPOSIUM ON THE FOUNDATIONS OF MODERN PHYSICS 1990

Quantum Theory of Measurement and Related Philosophical Problems

Joensuu, Finland
13–17 August 1990

Editors
Pekka Lahti
University of Turku

Peter Mittelstaedt
University of Cologne

World Scientific
Singapore • New Jersey • London • Hong Kong

Published by

World Scientific Publishing Co. Pte. Ltd.
5 Toh Tuck Link, Singapore 596224
USA office: 27 Warren Street, Suite 401-402, Hackensack, NJ 07601
UK office: 57 Shelton Street, Covent Garden, London WC2H 9HE

British Library Cataloguing-in-Publication Data
A catalogue record for this book is available from the British Library.

SYMPOSIUM ON THE FOUNDATIONS OF MODERN PHYSICS 1990

ISBN-13 978-981-02-0387-0
ISBN-10 981-02-0387-X
ISBN-13 978-981-02-0417-4 (pbk)
ISBN-10 981-02-0417-5 (pbk)

FOREWORD

This volume contains the proceedings of the SYMPOSIUM ON THE FOUNDATIONS OF MODERN PHYSICS 1990 which was held in Joensuu, Finland, at and under the aegis of the Department of Physics of the University of Joensuu during August 13–17, 1990. This symposium was the third in a series of similar conferences held in Joensuu in 1985 and 1987.

The theory of quantum mechanical measuring process has been a subject of increasing research interest during the recent years. The revival of interest on it was encouraged during the 1980's due to the advances on the formal and conceptual structures of quantum mechanics, accompanied with new experimental possibilities and technological demands, as well as due to new ideas on the interpretation of the theory. The third Joensuu Symposium on the foundations of modern physics was devoted to review the present situation of quantum measurement theory and to discuss the related philosophical problems.

The programme of the Symposium consisted of 18 invited lectures, 22 contributed lectures and a poster session. The majority of the lectures were devoted to the quantum theory of measurement and the related philosophical problems. Two lectures were concerned with the philosophical ideas of Wolfgang Pauli. During the conference informal seminars and two panel discussions were also organized. The first panel was chaired by Professor Bas van Fraassen and it was pondering *the influence of the unconscious on our conception of reality*. The second panel was chaired by Professor Enrico Beltrametti, and it was dealing with *the problem of measurement in quantum mechanics*. This programme emerged from a close collaboration between the Organizing Committee and the Scientific Advisory Committee. The programme was designed to fit best the aims of the Symposium. In particular, the contributed lectures were selected from the 81 submitted papers on the basis of the recommendations of the Scientific Advisory Committee.

The Organizing Committee consisted of K. V. Laurikainen (University of Helsinki, Chairman), Rauno Hämäläinen (University of Joensuu), Pertti Ketolainen (University of Joensuu), Pekka Lahti (University of Turku), Simo Liukkonen (Helsinki University of Technology), Kauko Mansikka (University of Turku), Jussi Rastas (Helsinki University of Technology), and Jouni Viiri (University of Joensuu, Secretary). The members of the Scientific Advisory Committee were Enrico Beltrametti (University of Genoa), Charles P. Enz (University of Geneva), Bernard d'Espagnat (University of Paris), Bas van Fraassen

(Princeton University), Peter Mittelstaedt (University of Cologne), Helmut Rauch (Atominstitut, Wien), Henry P. Stapp (University of California), C. F. von Weizsäcker (Max-Planck-Gesellschaft, Starnberg), and Anton Zeilinger (University of Innsbruck). The Joensuu Symposium was host to approximately 120 participants, whose names are listed at the end of this volume.

These proceedings contain the invited and the contributed lectures of the Symposium. In addition, it contains the opening address of the Symposium, as well as the summary talks and the closing words held at the end of the Symposium. The papers presented at the poster session are mentioned at the end of the volume.

The panel discussions were recorded and they will be published as a separate booklet by the end of this year in the Report Series of the Department of Physical Sciences of the University of Turku.

On behalf of the organizers we wish to thank the following organizations for their valuable support to the Symposium: City of Joensuu, Finnish Ministry of Education, Neste Oy Foundation, Research Institute for High Energy Physics in Helsinki, Research Institute for Theoretical Physics in Helsinki, University of Joensuu, University of Turku, and Alexander von Humboldt Foundation.

Pekka J. Lahti
Peter Mittelstaedt

Turku, October 1990

SYMPOSIUM ON THE FOUNDATIONS OF MODERN PHYSICS

CONTENTS

Opening Address

K.V.Laurikainen

Rector of the University of Joensuu
Ladies and Gentlemen

We are witnessing the continuation of the 25-year-old tradition of holding international physics conferences in North Karelia. At a time when establishing a university in Joensuu was merely a dream, the North Karelian Summer University began organizing annual international summer schools in physics. When the University was founded, the favourable attitude of the City of Joensuu made it possible to continue the tradition of holding physics conferences here.

On this occasion I have reason once again to stress that the Lord Mayor of Joensuu, Mr. Aaro Heikkilä, has special merits in this respect. I am certain that without his personal interest in natural philosophy this series of physics conferences would not have been continued. Lord Mayor Heikkilä has made a major contribution by supporting an alternative natural philosophy in our country despite the very negative attitudes displayed by our academic circles.

This is probably the last conference in which I shall chair the organizing committee. Therefore, I wish to use this opportunity for presenting certain personal views about the state in the fundamental research in physics in general and about the situation in Finland.

The discrepancy between Bohr and Einstein concerning the completeness of quantum theory is well known. However, the general conception that Einstein was a realist while Bohr had accepted a positivistic or instrumentalistic philosophy does not describe their attitudes correctly. Arthur Fine has coined the term *motivational realism* to describe Einstein's attitude. This was an expression of Einstein's way of life and the motivation for his work. Einstein wrote that he was convinced that by using a pure mathematical construction we are able to find concepts and laws combining them, which give the key to the understanding of nature. This was Einstein's rationalistic creed expressing his unshaken belief in an objective reality and in the ability of human thought to reach this reality more and more accurately. However, the relation between knowledge and reality is very abstract and remains to be comprehended on the basis the best scientific methods we have at our disposal.

I feel that Bohr also represented "motivational realism" in the sense Fine describes it. Bohr certainly believed in an objective reality as Einstein did, but he had lost belief in our ability to describe reality in an illustrative way. There are limitations in our abilities in this respect, and therefore Bohr was not willing to describe reality in more detail than was possible by using quantum mechanics. He considered this result an "epistemological lesson" which nature has given us in atomic research.

In fact, Wolfgang Pauli understood the implications of this "epistemological lesson" even more clearly than Bohr himself. Pauli's special contribution was the introduction of psychological viewpoints into the discussion about physical reality. Empirical research is based on measurements which are fundamentally psychic actions. So are the creation of new concepts and the discovery of new laws of nature. Pauli used, especially, his knowledge of the functioning of our *unconscious,* for discussing the problems of atomic theory. This resulted in a new conception of reality which *included matter and spirit as complementary aspects of one world.*

These views deserve special attention at this time. What is characteristic of Pauli's conception of reality is that reality has both rational *and irrational* aspects. I have tried to describe the irrationality of reality in an article which is available at our secretarial office, and later today I shall make some additional remarks concerning the problem of free will.

The idea of the irrationality of reality seems to receive strong criticism. We must remember, however, that we are now speaking of reality which contains also psychic aspects. We are substituting Cartesian dualism with a *complementary conception of reality*, and this presupposes that we acknowledge the limits of rational science.

Bohr did not want to speak of reality at all, and he repeatedly discouraged against introducing any kind of mysticism into the discussion concerning quantum mechanics. Pauli, instead, considered science and mysticism to be complementary ways of comprehending reality. Pauli also often emphasized that the origins of science and religion are the same.

In this respect Pauli's letters forty years ago expressed views which are becoming very topical today. The relation between knowledge and belief deserves rethinking from the aspect of the epistemological lesson nature has given in atomic research. I hope that these problems will not be neglected when speaking of the basic problems of quantum mechanics.

I know very well that strong ciriticsm is to be expected from different sides if one dares to open a discussion about these views. Pauli himself became rather pessimistic due to the reactions concerning his first attempts to describe his ontological views. As far as Finland is concerned, it is my experience that mixing science and religion creates severe difficulties both in natural science and in theology - not to speak of philosophy which in Finland is dominated by traditions based on logical empiricism and on Wittgenstein's philosophy. If one attempts to speak of *one world containing both matter and spirit,* the result is absolute silence and a "closing of all doors". I do not know about the situation in other countries, I only have experience from Finland. Perhaps certain political circumstances have made the intellectual atmosphere particularly dogmatic here. Without the steadfast support from the City of Joensuu, a conference of this kind would not have been possible in Finland.

I suppose that even for most of our participants the views which I am now describing are very strange. I have come to accept them by studying Pauli's articles and letters again and again for almost 15 years. A new conception of reality changes our thinking in many ways. Such changes are not easy to realize. For three hundred years Cartesian dualism has influenced the thought of Western people, and to a physicist it has meant an absolute cutting off of all signs of spirit from science. Natural science has become purely materialistic. It is time to think carefully what the role of consciousness is in empirical science - and also what the *role of the unconscious* is. I hope that these questions will be discussed here although physicists are not accustomed to thinking of them. For me the revaluation of the empirical method on the basis of the complementary conception of reality has resulted in a criticism of the direction of research in many fields of natural science. I shall mention some examples in my lecture today.

* * * * *

In addition to the City of Joensuu, which is the main sponsor of this meeting, the following organizations have supported our Symposium:

- University of Joensuu
- Finnish Ministry of Education
- Neste Oy Foundation
- Research Institute for Theoretical Physics, Univ. of Helsinki
- Research Institute for High Energy Physics, Univ. of Helsinki

On behalf of the Organizing Committee I wish to thank all of these sponsors for their valuable support.

We are very glad that so many excellent specialists in the foundations of quatum theory are present here in Joensuu. Let us hope that these discussions will help a little in clarifying the quantum mysteries - which perhaps are still more profound than we physicists of today understand. I wish to welcome all of you to this Symposium.

Invited and Contributed Lectures

THE SPIN OF A QUANTUM-ENTITY AND PROBLEMS OF NON-LOCALITY.

D. AERTS and J. REIGNIER
Theoretische Natuurkunde, Vrije Universiteit Brussel, Pleinlaan 2, Brussels, Belgium

1. ABSTRACT

We introduce a possible definition for the concept of non-locality in the quantum world, which seems to us a minimal operational definition, taking into account the results of actually performed experiments and reasonings about possible 'gedanken' experiments. The definition is the following: An entity is "non local" if it is possible to prepare it in a state such that it can be influenced from macroscopically separated regions of space by (macroscopically) local apparatus acting only in one (or several) of these separated regions at one time. We discuss two examples of spin superposition experiments which clearly show that quantum entities are non-local. In particular, we show that the familiar Stern-Gerlach experiment allows a nice illustration of this non-locality.

2. INTRODUCTION.

From the very beginning of the quantum theory it was clear that the so-called "quantum world" is very different from our familiar or "classical world". For a long time however the study of the quantum properties remained limited to the domain of the stationary states of microscopical physics and this has certainly greatly contributed to a comfortable attitude of "sleeping on a gentle pillow" which is still so general among the physicists [1]. Recent experiments on spin correlation measurements in EPR situations have shown that the paradoxal aspect of the quantum formalism put forward in the original EPR paper can be realized (although not easily) in the laboratory. Many different analys s have meanwhile been made of these situations, and the idea of "non-locality" has been put forward for the two quantum particles in EPR situation. However even in the case of one single particle, one meets experimental situations which, when expressed in classical language, are to be called non-local. These non-local effects are usually considered as side-effects of the wave aspect of the particle. Against this rather "soft" position, we would like to suggest that non-local effects should be investigated apart from the other quantum peculiarities. In order to be definite, we shall use a minimal operational definition for the concept of non-locality, avoiding it to be connected in a too specific way to EPR or interference situations: *"**An entity is 'non local' if it is possible to prepare it in a state such that it can***

be influenced from macroscopically separated regions of space by (macroscopically) local apparatus acting only in one (or several) of these separated regions at one time."

We shall illustrate this definition by means of concrete experimental situations. In section 3 we recall one of the many neutron interferometric experiments performed by the group of Rauch. In section 4 we analyze the Stern-Gerlach experiment and notice that non-locality is already present in this situation. This point is made more explicit by a " gedanken " experiment presented in section 5. In section 6 we give some embryonic thoughts about the possible changes in the classical world image that could be envisaged to attempt to understand somewhat more about this non-locality of the quantum world.

3. THE EXPERIMENT OF RAUCH'S GROUP.

Beautiful interference experiments with thermal or ultra-cold neutrons have been carried out at the Laue-Langevin Institute in Grenoble by the group of H. Rauch[2]. All these neutron interferometric experiments pertain to the case of self-interference, where during a certain time interval, only one neutron is inside the interferometer, if at all. We are interested in the experiment about the so called 4π-symmetry of spinors, where it is shown that if one introduces a local constant magnetic field on one of the two paths inside the interferometer with the aim of causing a Larmor precession of the spin of a neutron that "occasionally" follows this path, one has to create a 4π rotation in order to recover the original interference pattern. Let us briefly describe this experiment in order to make our point [3]: A perfect crystal of silicon has been cut to give an interferometer of the Laue type

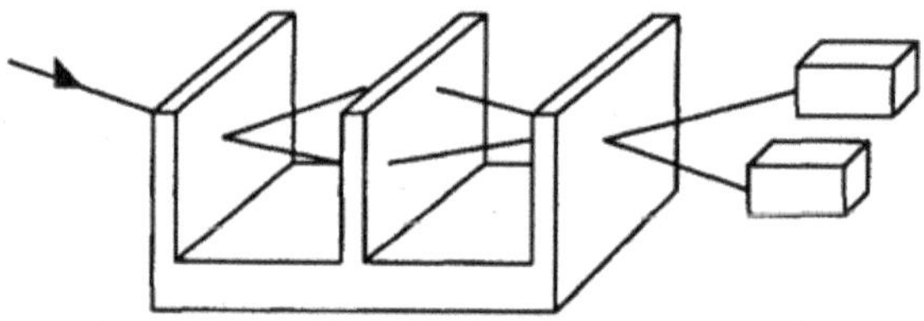

A perfect silicon crystal is cut in such a way to form an interferometer of the Laue type. The beam of neutrons comes in at the left, and counters measure the intensities of the outgoing beams (fig 1).

The thermal neutrons are dynamically Bragg-diffracted at the first crystal plate, then reflected at the middle plate, and finally superposed at the third plate (fig.1). Introducing a crystal plate of variable orientation (d on fig.2) one can change the relative phase of the two beams, and create an interference pattern in the detecting counters as a function of the orientation of the plate, as predicted by wave mechanics. One can modify this interference pattern by acting locally with a constant magnetic field **B** confined to a small region R on path 2, such that an "occasional" neutron flying on this path would pass for a short period of time Δt through **B**, and have its spin turned over an angle $\beta = \gamma \cdot B \cdot \Delta t$ around the direction of this field by the Larmor precession, γ being the gyromagnetic factor of the neutron. If **B** is taken to define the x_3-direction, a spin in direction $\mathbf{n}(\theta, \phi) = (\sin\theta \cdot \cos\phi, \sin\theta \cdot \sin\phi, \cos\theta)$ will be transformed into a spin in direction $\mathbf{n}(\theta, \phi+\beta)$. The complex vector (spinor) $S(\theta, \phi) = (\exp(-i\phi/2) \cdot \cos\theta/2, \exp(i\phi/2) \cdot \sin\theta/2)$ of C^2 representing the spin in the quantum formalism will be changed into S(θ, ϕ+β). We consider the case $\beta = 2\pi$; then $\mathbf{n}(\theta, \phi+2\pi) = \mathbf{n}(\theta, \phi)$; a Larmor precession of 2π does not change the spin direction, and this is logical since if we

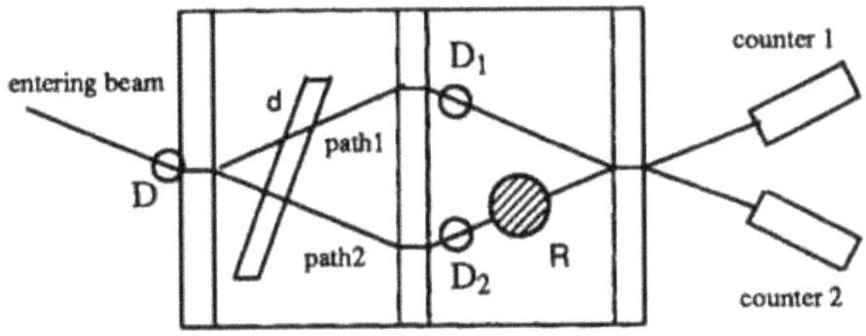

A drawing of a sight from above of the experiment with the neutron interferometer. On path 2 a magnetic field is applied confined to a small region R, such that an 'occasional' neutron flying on this path has to pass through the magnetic field (fig 2)

turn any space direction over an angle of 2π, we get the same space direction. However, since the vectors representing the spin states are vectors of C^2, and the rotation group of C^2 is SU(2), the action of such a 2π rotation is different for these vectors, in the sense that $S(\theta, \phi+2\pi) = -S(\theta, \phi)$. There is nothing mysterious in this fact, because in quantum mechanics the state is represented by the normalized vector, or by the ray of the Hilbert space. Therefore $S(\theta, \phi)$ and $-S(\theta, \phi)$ represent the same spin state. This point is often misunderstood although it is mathematically very simple and deeply rooted in the quantum formalism, and due to the fact that SO(3), the rotation group in three dimensional Euclidean space, is represented by SU(2) in the group of symmetries of C^2. This is so because one has to consider 'projective representations' of SO(3) [4)], i.e. to take into account explicitly that the state of the quantum entity is represented by the ray of the Hilbert space, and not by the vector. Rauch et al. found that if $\beta = 2\pi$, the interference pattern is changed completely. Regions where before a lot of neutrons were detected, become almost empty regions, and vice versa, regions that were almost empty, now show an abundance of detected neutrons. The original interference pattern is only recovered if $\beta = 4\pi$ (or a multiple of 4π), and this is the reason why they speak of the 4π- symmetry of spinors. From our point of view, this very fascinating result shows that the picture of a "localized" neutron following a definite path is wrong. Such a "localized" neutron would only explore a narrow neighbourhood of one of the two paths, with the following alternatives: either it travels on path 1 and is not affected by the magnetic field, or, it travels on path 2 and its spin is turned over an angle 2π. Hence, in both cases, nothing happens. This of course illustrates the necessity of a wave aspect of the quantum entity. But let us now consider a quantum mechanical description of this situation, using "localized" wave-packets to represent the individual neutron, travelling alone in the apparatus. Let us firstly define what we mean by an approximately localized particle state: *a normalized wave-packet $\Psi(x_1, x_2, x_3)$ is ε-localized in the space region D if the corresponding probability integral calculated on D is larger than 1-ε,*

$$\int_D dx_1 dx_2 dx_3 \, |\Psi(x_1, x_2, x_3)|^2 > 1 - \varepsilon \qquad (1)$$

which physically means that the probability of detecting the quantum entity in region D is larger than 1 - ε. Such a state will be represented by the symbol Ψ_D or alternatively by $\Psi_{D\,t}$ if we like to emphasize that this happens at time t. When it enters the interferometer, at time t=0, the neutron is ε-localized in a small region D just in front of the interferometer (fig2), with a definite spin in direction $\mathbf{n}(\theta, \phi)$:

$$\Psi'_{D\,0} = \Psi_{D\,0}(x_1,x_2,x_3) \otimes S(\theta,\phi) \qquad (2)$$

The experimental situation is a bit is too complicated to allow an exact treatment of the Schrödinger equation, but this is not really necessary in order to understand what is going on. The initial wave packet evolves into two separated wave packets the centres of which follow approximately the paths previously described. The state of the neutron at a certain time t, when it is inside the apparatus, is described by a wave function

$$\Psi_t = \Psi_{D_1\,t} + \Psi_{D_2\,t} \qquad (3)$$

$$\text{where } \Psi_{D_1\,t} = \Psi_{D_1}(x_1,x_2,x_3) \otimes S^1(\theta,\phi) \qquad (4)$$

$$\text{and } \quad \Psi_{D_2\,t} = \Psi_{D_2}(x_1,x_2,x_3) \otimes S^2(\theta,\phi)$$

$\Psi_{D1\,t}$ is a product state, the space part of which is represented by the wave packet Ψ_{D1}, ε–localized in a region D_1 on path 1(analogous $\Psi_{D2\,t}$). The precise form of the wave packets is rather irrelevant for our purpose, and therefore we forget about the deformation of the wave-packets ε-localized in regions D_1 and D_2, and consider instead these regions as evolving along the paths 1 and 2, respectively, when the time elapses (fig2). Rauch applies a magnetic field in a region R which is crossed by the region D_2 during some time interval $(t_0, t_0 + \Delta t)$ causing a Larmor precession of the spinor S^2, so that only one of the two parts of the wave function of the neutron is affected by this magnetic field. If the experimental arrangement is such that $\beta = 2\pi$, then the spinor S^2 changes its sign, which means that, comparing two experimental situations, the first one without the magnetic field (or B inactive) and the second one with the magnetic field as here above described (B active), one gets after the crossing of the region R:

$$\{\Psi_{D2}\}_{B\ \text{active}} = - \{\Psi_{D2}\}_{B\ \text{inactive}} \qquad (5)$$

But then the 'true' wave function Ψ_t representing the state of the neutron undergoes the following changes: Before the time t_0 where the movable region D_2 reaches the region R, the wave function remains the same in both cases (i.e. B active or inactive). After the time $t_0 + \Delta t$ where the movable region D_2 has fully crossed the region R, the wave function becomes when B is active :

$$\begin{aligned} \{\Psi\}_{B\ \text{active}} &= \{\Psi_{D1} + \Psi_{D2}\}_{B\ \text{active}} \qquad (6) \\ &= \{\Psi_{D1} - \Psi_{D2}\}_{B\ \text{inactive}} \end{aligned}$$

which is of course deeply different from $\{\Psi\}_{B}$ inactive at the same time. The change in the interference pattern that is observed by the experimenters is nothing else than the statistical realization of this elementary change, and it is indeed explained in that way in their publications. We only want to insist, on purpose, on the <u>local</u> character of the action which changes the state of <u>one single neutron</u> (even if the experimental observation of this fact is statistical). It is clear that one can make the same reasoning about an experiment placing the region R on the path of the wave-packet number 1.

This means that ***the single neutron can be influenced from macroscopically separated regions of space by (macroscopically) local apparatus acting only in one of these separated regions at one time***. This is exactly our criterion of non-locality.

To emphasize the amazing character of the effect of 'de-localization' of one neutron realized in the experiment, we would like to make a rough comparison based on a scaling which respects the relative orders of magnitude. The longitudinal coherence length of these neutrons, experimentally shown to be of the order of 20 Å, is a good measure of the size of a wave packet and hence in a certain sense an 'upper bound' for the 'size' of the neutron. Since the regions D, D_1, and D_2 are represented on the figures by means of spots of more or less 0.5 cm diameter, there is a scaling factor of $0.25 \cdot 10^7$. On the other hand, the incoming neutrons follow each other at an average distance of some 300 m, the real distance between the two paths is of the order of 5 cm, and the region R is a square with sides equal to 0.15 cm. Hence, in an "on scale" drawing, neutrons like little balls of 0.5 cm diameter would be shot having an average distance of 750.000 km between them (i.e. nearly 2 times the distance moon earth), the two paths would be separated by 125 km and the region R would be 4 km squared. At this scale, the experiment literally means that we can de-localize one single object for which we normally would imagine it to be enclosed in a sphere of 0.5 cm, over a distance of 125 km, in the sense that this 'one' object can be influenced from two space regions that are like small villages of a few km^2 some 125 km apart. It is clear that this 'non local Rauch effect', is not some kind of 'peculiar' property of 'neutrons' or 'magnetic fields' and the conclusion about the non-locality should be accepted also for other quantum-entities, and also for other force fields. Notice however that interference patterns are always the result of a statistical sampling of individual cases. Hence, to be able to draw more direct conclusions about single entities, it is probably better to discuss experiments where the effect of a local perturbation can be seen on each individual entity or equivalently, experiments where the statistical prediction turns out to be a certainty. We want to consider now such an experiment, the principle of which is quite analogue to the one of Rauch, in which exactly the same effect of de-localization of one quantum entity happens and can in principle be observed on each single entity.

4. THE STERN-GERLACH EXPERIMENT.

We now want to discuss the old Stern-Gerlach experiment, which in fact can be considered as the true discovery of the electron spin [5)]. Stern and Gerlach prepared a beam of silver atoms travelling with a rather well defined velocity, in a certain space direction **b** (beam). The beam enters a strong 'inhomogeneous' magnetic field, with a strong constant part along a space direction **a**, orthogonal to **b**, and also having a strong 'gradient'. When the atoms of the beam leave the magnet, they are detected on a screen, placed orthogonal to **b**. Stern and Gerlach found that the individual atoms are detected in two spots on the screen: one spot, 'upwards', in the **a** direction, and another spot 'downwards', in the **-a** direction.

In a time when modern quantum mechanics did not yet exist, but the idea of quantization was already very common, they interpreted their result as an indication that the silver atoms have a 'quantized magnetic moment'. They had in mind the following 'classical picture' of the mechanism taking place during the experiment : The silver atom is imagined to be a classical particle of mass M with a magnetic moment μ due to its orbiting valence electron. When it enters the region of the magnetic field, it interacts with this magnetic field in essentially a classical way: the magnetic moment vector μ precesses rapidly around the **a** direction and the whole atom undergoes a force due to the coupling of the magnetic moment to the gradient of the field. The classical calculations show that the field gradient causes a deflection of the impact of the beam on the screen which extends on a full interval whose extremities correspond to the spots detected in the experiment. This 'quantized' orientation of the angular momentum remains up to our days the basic concept that physicists use when they try 'to represent', 'to picture', or 'to talk about', the spin of a quantum entity. Let us now consider the modern quantum mechanical description of this situation. Unlike the case of the Rauch experiment, one can calculate in detail the quantum mechanical evolution of one single atom in the Stern-Gerlach experiment, by solving the time dependent Schrödinger-Pauli equation with an interaction potential energy equal to $-\mu\cdot(\sigma\cdot\mathbf{B})$, where μ is the magnetic moment of the atom, $\sigma = (\sigma_1, \sigma_2, \sigma_3)$ are the Pauli matrices, and $\mathbf{B}$ is the magnetic field (limited in practice to a small region along the magnet), with an appropriate initial condition.

$$i\hbar\frac{\partial}{\partial t}\Psi(t) = -\frac{\hbar^2}{2M}\Delta\Psi(t) - \mu\cdot\sigma\cdot B\cdot\Psi(t)$$

$$\Psi(0) = \Psi_0 \qquad (7)$$

We consider the situation where an atom with a well defined spin direction enters the Stern Gerlach magnet, at time t = 0. Its state is then represented by a normalized vector Ψ_0, of the form $\Psi_D(\mathbf{x}) \otimes (\alpha, \beta)$, where $\Psi_D(\mathbf{x})$ is a wave-packet ε-localized in region D, localized at the entrance of the Stern-Gerlach apparatus (fig3), and $(\alpha, \beta) = S(\theta, \phi) = (\cos(\theta/2)\exp(-i\phi/2), \sin(\theta/2)\exp i(\phi/2))$ representing the spin direction $\mathbf{n}(\theta, \phi)$ defined with respect to the directions $\mathbf{a} = (0,0,1)$ and $\mathbf{b} = (1,0,0)$. It is remarkable, taking into account the importance of the Stern-Gerlach experiment and the apparent simplicity of the problem, that <u>nowhere</u> in the literature a sound quantum mechanical treatment exists. Most treatments consider incorrect magnetic fields (not satisfying the Maxwell equations), or they do not give any satisfactory explanation for the approximations used to construct a 'solution' of the equation. Therefore we have investigated again the quantum mechanical treatment of the Stern-Gerlach experiment, trying to avoid the weakness of the earlier attempts [6].We consider the case of an the incoming atom represented by a small three dimensional gaussian wave packet with a mean velocity v_0 in the $\mathbf{b} = (1, 0, 0)$ direction. We follow the evolution of the wave packet, when it travels through the magnet and analyze in which way we can construct a solution, making the appropriate physically significant approximations.This allows us to compute the wave function at a time τ when the atom leaves the magnet and we find that it is then clearly separated into two wave packets. We then follow the further evolution of these free wave packets until they reach the screen. The result is rather complex to write down explicitly but for the present discussion a detailed expression is not necessary. Enough is to say that the wave function that leaves the magnetic field is of the form

$$\Psi_{out} = \alpha \cdot \Psi_A \otimes (1,0) + \beta \cdot \Psi_B \otimes (0,1) \qquad (8)$$

where A and B are two separated regions in space (fig3), and where we keep the same notations as previously to indicate the ε-localization of a wave packet in a region of space.

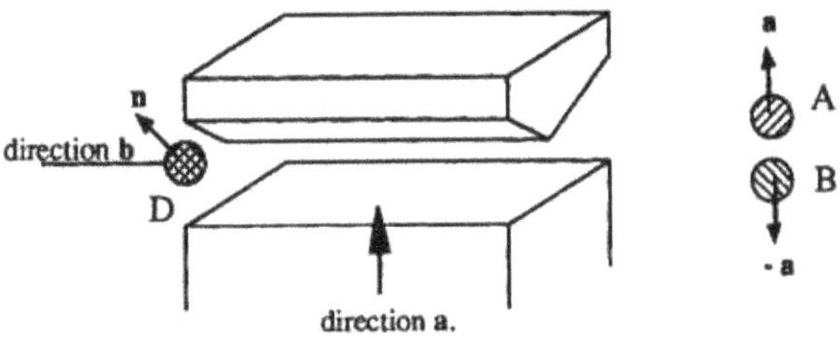

The region D is the space region where the incoming atom is ε–localized, and the vector **n** is the space direction of the spin of the incoming atom. The regions A and B are the space-regions where the component wave-functions of the superposition wave function representing the state of the outgoing atom, are ε–localized. These component wave functions appear in the superposition with spin-vectors corresponding to respectively space direction a and -a. The complete wave function representing the state of the outgoing atom does not have a spin corresponding to a fixed space direction (fig 3).

Remember that in all the fig. of this paper, we do not make a drawing of the quantum entities but only of the regions where they are ε-localized ; the arrows attached to these regions are mathematical directions which only occasionally indicate the spin direction of the atom (this is the case for region D, this is not the case for regions A and B). We want to analyse now the physical meaning of the state represented by this vector Ψ_{out} , the state of the single atom when it leaves the magnet before it is detected. Most physicists agree on the fact that the purely classical model of a particle having a 'quantized' rapidly precessing magnetic moment, is not a correct model for the 'spin' of a quantum-entity. Through the years some kind of 'semi-classical' picture has developed. The final state vector Ψ_{out} is often implicitly interpreted in the following way : When an atom enters the Stern-Gerlach apparatus, it either really 'flips up', and flies on the 'upper' beam, and it is then *de facto* in a state represented by the vector $\Psi_A \otimes (1, 0)$, or it 'flips down' and flies on the 'lower' beam, and is then *de facto* in a state represented by the vector $\Psi_B \otimes (0,1)$. The very fact that, according to the calculations of quantum mechanics, the final state is represented by the vector Ψ_{out} is interpreted as reflecting our lack of knowledge about which of the two possibilities the atom chooses. If this 'semi-classical' interpretation would be correct, then the atoms that fly through a Stern-Gerlach magnet have always a well defined spin direction, and are always ε-localized in one of the separated regions A or B. Of course, this is not true from first principles of quantum mechanics and also from the results of the Rauch experiment that one can translate *mutatis-mutandis* to the Stern-Gerlach case. But, if the 'semi-classical' interpretation is wrong, what else can we propose ? Let us try to describe a possible experiment of the Rauch type which sustains the thesis that atoms coming out the magnet are in a non-local state.

5. THE NON-LOCALITY OF AN ATOM LEAVING THE STERN GERLACH APPARATUS.

Experiments that consist in combining different Stern-Gerlach apparatus to illustrate the amazing properties of the superposition principle has been considered in the past by many authors and are described in various text-books [7)]. We consider such an experiment on purpose of showing that an atom in the state Ψ_{out} is non-local in the sense of our definition. It is true that the experiment we shall describe is probably impossible to realize now and perhaps for ever because one cannot control the inhomogeneous magnetic fields in the different magnets of the set-up with a sufficient accuracy [8)]. But we are confident in the results because of the great similarity with the work of Rauch et al. with neutrons. And this 'gedanken' experiment has the advantage of working directly on one single atom or equivalently with the statistical weight of certainty. One sends a beam of atoms flying in direction (1, 0, 0), with a definite spin direction along $\mathbf{n}(\theta, \phi)$ into a first Stern-Gerlach magnet and a magnetic field in direction $\mathbf{a}$ = (0, 0, 1). Then one puts right behind it a second Stern-Gerlach magnet identical to the first one except for the field direction which is opposite: $-\mathbf{a}$ = (0, 0, -1). This second magnet will converge the two beams (to describe the experiment we use the classical language) in such a way that they become parallel (fig4). Then one puts a third Stern-Gerlach magnet, identical to the second one, which will incline the beams

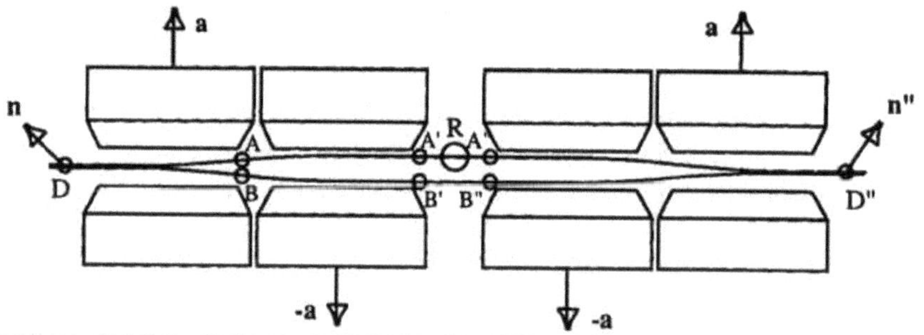

A representation of the experiment, combining four Stern-Gerlach magnets, such that the original beam is recombined. In region R we apply a constant magnetic field, and turn the spin over an angle of 2π. The spin of an outgoing atom has then been turned over an angle of π around the **a** direction (fig 4).

towards each other. Finally one puts a fourth Stern-Gerlach magnet, identical to the first one, which will have the effect of reconstituting the two beams into a single one which is in principle identical to the original incoming beam. In a region R one applies a constant magnetic field, such that the spin can be turned by the Larmor precession, over different angles. One makes the experimental arrangement in such a way that the spin will turn over an angle of 2π. Notice that because such a rotation is equivalent to the identity, the axis of rotation (i.e. the direction of the magnetic field) is irrelevant for the results of the experiment. Let us now analyze the quantum mechanical description, to see what will happen. As before, the state at time t = 0 is represented by a normalized vector Ψ_0 of the form $\Psi_D(x) \otimes (\alpha, \beta)$ (see section 3). Again we have to solve the Schrödinger-Pauli equation (7) for the evolution of this vector through the first Stern-Gerlach magnet, such that the state of the atom coming out of the first magnet is given by Ψ_{out} (see(8)), where A and B are separated regions (see section 3 and fig. 4). The second Stern-Gerlach magnet will not change anything to the spin components, since they are (1, 0) and (0, 1), but will influence the direction of flight of the two wave packets Ψ_A and Ψ_B, so that they become parallel at its end, so that the atom coming out of the second magnet will be in a state $\alpha \cdot \Psi_{A'} \otimes (1,0) + \beta \cdot \Psi_{B'} \otimes (0,1) = \Psi'_{out}$ (see fig. 4). If we now apply a constant magnetic field in the region R localized on the upper beam (fig4), then only the spin of the partial vector $\alpha \cdot \Psi_{A'} \otimes (1,0)$ will be influenced. If it is turned over an angle of 2π, then the spinor (1, 0) will be changed into the spinor (-1,0), no matter what the direction of the magnetic field applied in the region R can be. Of course, as we recall it in our reasonings about the Rauch experiment, these two spinors represent the same state. So by the effect of the Larmor rotation we find at the entry of the third Stern-Gerlach magnet a state of the atom represented by a vector $\alpha \cdot \Psi_{A''} \otimes (-1, 0) + \beta \cdot \Psi_{B''} \otimes (0, 1) = (-\alpha) \cdot \Psi_{A''} \otimes (1, 0) + \beta \cdot \Psi_{B''} \otimes (0, 1) = \Psi''_{out}$, where A" and B" are again separated regions in space (fig 4). To proceed and find the quantum mechanical effect of the two last Stern-Gerlach magnets, we notice that the Schrödinger dynamics is invariant for time reversal. So instead of trying to calculate the evolution up to the end of the fourth magnet of a state which is of the form Ψ''_{out} at the entrance of the third one, we can inverse the problem and consider the question : which state of the form $\Psi_{D''}(x) \otimes (\gamma, \delta)$ at the end of the fourth magnet will give rise to the state Ψ''_{out} at the entrance of the third magnet by a time reversed Schrödinger evolution? The answer to this question is immediately given by the first part of the calculation, and is of course $\Psi_{D''}(x) \otimes (-\alpha, \beta)$.

Hence if the spin of the incoming atom is in direction $\mathbf{n}(\theta, \phi)$, then using in the region R a constant magnetic field of appropriate magnitude but arbitrary direction, the spin state $\mathbf{n}''(\theta, \phi)$ of the outgoing atom has turned over an angle π around the direction $\mathbf{a}$ defined by the Stern-Gerlach magnets. In the classical picture of spinning particles this magnetic field in region R would have turned the spin over 2π, which means that each atom would leave the apparatus exactly in the same state as it enters. The same result would be found in the semi-classical picture. If we complete the experiment by a Stern-Gerlach measurement of the orientation of the spin of the final atom, there is no doubt that it will fully confirm the standard quantum mechanics prediction of a rotation of π ! In this sense it is clear that the experiment shows a realization of our definition of non-locality for quantum entities.

6. CONCLUSION.

We want to ask now again what the physical meaning is of the state of a quantum-entity of the form $\alpha \cdot \Psi_A \otimes (1,0) + \beta \cdot \Psi_B \otimes (0,1)$. In our opinion the following statement can be made : The state $\alpha \cdot \Psi_A \otimes (1,0) + \beta \cdot \Psi_B \otimes (0,1)$ represents one quantum entity, neither localized in region A with spin in the direction **n** = **a**, nor localized in region B with spin in direction **n** = - **a**. Moreover ***this one quantum-entity can be 'influenced' from two separated regions A and B by local apparatus acting only in one of these regions at one time.*** This does not allow to conclude that this one quantum-entity is at two places at the same time and not between. We seriously consider the possibility that our 'classical' conception of three dimensional Euclidean space and localized entities is just not convenient to picture what really happens. Probably "the human part of the construction of the reality" (see H. Poincaré, ref. 9) is to be reconsidered in order to begin to approach the quantum mechanical non-locality. In [10)] a similar proposal was made in relation with the problematic around the EPR situation , and on the hand of some example it was shown that it is not impossible, 'in principle', to imagine such an approach, that could lead to a human understanding of the quantum non-locality.

7. ACKNOWLEDGEMENT.

We want to thank J. Broekaert for presenting this paper at the conference.

8. REFERENCES.

1. This expression seems to have been used by Einstein in a letter to Schrödinger (See the article of Mermin, N.D., Physics Today , April , 9 (1989).
2. These experiments are described in great details in several publications of members of this group. One find a sufficient list of references in Rauch, H.,*'Neutron interferometric tests of quantum mechanics'* , Helv. Phys. Acta, 61, 589 (1988).
3. Rauch H., Zeilinger A., Badurek G., Wilfing A., Banspiess W., Bonse U.,*'Verification of Coherent Spinor Rotation of Fermions'*, Phys.Lett. 54A, 425 (1975).
4. Bargmann, V., Ann. Math., 59, 1, (1954), see also, Hammermesh, M.,*"Group theory and its applications to physical problems "*, Addison Wesley, Reading, Mass.,(1962), and Wigner, E.P.,*" Group theory and its applications to quantum mechanics and atomic spectra "*, Academic Press, New York, (1959).
5. Gerlach,W. and Stern,O.,*"Der experimentelle Nachweis der Richtingsquantelung im Magnetfeld"*, Zeitschrift für Physik, 9,349, (1922).
6. Aerts, D. and Reignier, J., *" A quantum mechanical treatment of the Stern Gerlach experiment "*, in preparation.
7. See for instance *'The Feynman Lectures on Physics'*, Vol. III, Ch. 5,6.
8. Schwinger J., Scully M.O., Englert B.-G., *'Is spin coherence like Humpty Dumpty ?'* Zeits. f. Phys. D., 10, 135 (1988).
9. About this question of the human part of the construction of space it is still very interesting to read again the small book of H. Poincaré *"La science et l'hypothèse "*
10. Aerts D.,*" An attempt to imagine parts of the reality of the micro-world"*, in the proceedings of the conference *'Problems in Quantum Physics ; Gdansk '89'* to be published by World Scientific Publishing Company, Singapore.

Two Attempts to Solve Schrodinger's Paradox

David Albert
Philosophy
Columbia University

Barry Loewer
Philosophy
Rutgers University

In his famous thought experiment Schrodinger imagined an arrangement like the following. A cat (whose initial state is $|R>$) interacts with an electron so that if the electron's x-spin is up then the cat is killed and if the electron's x-spin is down then the cat remains alive. That is, at the start of the interaction if the state of the electron+cat is $|\uparrow>|R>$ ($|\downarrow>|R>$) then at its conclusion it will be $|\uparrow>|Dead>$ ($|\downarrow>|Alive>$). Quantum theory assures us that there exist Hamiltonians which characterize such interactions as this one. Further, it follows from the linearity of Schrodinger's dynamical equation that at the conclusion of a measurement of an electron whose initial state is $c_1|\uparrow> + c_2|\downarrow>$ the final state will be

$$c_1|\uparrow>|Dead> + c_2|\downarrow>|Alive> \qquad \text{(CAT)}$$

This state superposes the possibilities of Dead and Alive. If one assumes, as standard interpretations of quantum theory do, that an observable has a well defined value for a system S when <u>and only when</u> S's quantum state is an eigenstate of that observable then the unfortunate cat who finds itself in CAT is niether dead nor alive. This struck Schrodinger and most others who have thought about it as intolerable.

Schrodinger's cat is a measuring device whose state of aliveness register's the spin of the electron and, a similar consequnce follows for any accurate measurement of an observable when the system being measured is not in an eigenstate of that observable. A perfectly accurate and non-destructive measurement of observable A of S by B of M is required to satisfy the condition

> If A is an observable pertaining to system S which has values a_i and B is an observable pertaining to a measuring device M with values b_i -these are the measuring devices pointer positions in which the outcome is recorded- then an

interaction between M and S is a measurement of A on S by B on M only if the Hamiltonian which characterizes the interaction guarantees that whenever S+M starts in state $|A=a_k>|R>$ it ends up in state $|a=A_k>|B=b_k>$ ($|R>$ is M's "ready to measure" state.)

The problem is that if S starts in a state which is not an eigenstate of A and if Schrodinger's equation characterizes the evolution of the combined system under the interaction then S+M will end up in the state

$$\mathrm{SUM}c_i|A=a_i>|B=b_i> \qquad \text{(MEAS)}$$

which is not an eigenstate of AxB. If observables possess values only in their eigenstates then at the end of the measurement niether A nor B will possess a definite value. The problem of avoiding this intolerable conclusion has come to be called "the measurement problem."

In this paper we will discuss two relatively recent suggestions for how to avoid the intolerable conclusion. The first (due to Kochen, Healey, Dieks, and van Frassen; 5,7,8), which we call KHD, suggests modifying the standard prescription for connecting values of observables with quantum states. It allows for certain observables to have values even when the system is not in an eignenstate of those observables. The second (due to Ghirhardi, Rimini, and Weber; 6,4), which we call GRW, says that Scrodinger's equation is not the right fundamental law of evolution of systems. While the two suggestions are rather different they share a need to resort to similar fudging at certain crucial points. If either account is to be made explicit then the fudging must be avoided. Here we will argue that KHD appears plausible only as long as it fudges. Once it is made explicit it is clearly seen to be inadequate. On the other hand it looks like the GRW can be made explicit. The result is a theory which has a number of curious feautres which we leave the reader to consider. Before describing these two accounts it would do well to remind the reader of how the "standard interpretation" handles the problem.

According to the standard interpretation (which was explicitly formulated by von Neuman; 9) Schrodinger's equation is not always right; it does nto govern every interaction. In the cat interaction the final state of cat+e is not CAT but is one of the states $|Dead>|\uparrow>$ or $|Alive>|>$ with probabilities c_1^2 and c_2^2. More generally, the final state of S+M in a genuine measurement is not MEAS but one of the states $|A=a_k>|B=b_k$. This is called the "projection" or "collapse" postulate and it, according to von Neuman, and not Schrodinger's equation governs measurement interactions.

This is a very ingenous solution since it allows one to maintain that observables possess values only for systems in eigenstates. But it is very difficult to believe that the collapse described by von Neumann can be a real physical process. The main problem is that there are in principle experimentally detectable differences between the states which result from Schrodinger's equation and those which result from collapses and for all those cases which have been experimentally investigated the final state has been found to be the one predicted by Schrodinger's equation. It is true that these experiments have always involved interactions between micro-systems (systems containing a few particles). In view of this, a proponent of the standard view can say (and many have) that genuine measurements must involve macroscopic devices. Their idea is that as long as the interaction is microscopic Schrodinger's equation obtains but when it involves a macroscopic system the collapse postulate obtains. But there seems to be no non-arbitrary way of drawing a line between micro and macro systems. At what point of complexity does an interaction which establishes a correlation between the observables of two systems become a measurement? And while the transistion from microscopic to macroscopic is gradual the differences between two kinds of dynamics are dramatic. The Schrodinger evolution is continuous, deterministic, and time reversible while the collapse dynamics is discontinuous, probabalistic, and time irreversible. It is simply incredible that at some point of complexity the linear dynamics are suspended and the collapse dynamics takes over. So those who endeavor to understand quantum theory as a true description of reality have looked for ways of avoiding von Neuman type collapses.

The KHD Interpretation

In describing KHD it will be useful to imagine a measurement of x-spin in which the result is recorded in the position of a pointer. The pointer pointing to UP if the spin is up and to DOWN if the spin is down (we will use |U> and |D> for the associated pointer states.) Recall that the measurement problem arises because the linear Schrodinger equation predicts that the post-measurement state of M+S is CAT (with the |U> replacing |Alive> etc.) and that state is not an eigenstate of the pointer's position. If we assume, as the standard interpretation of quantum theory does, that the pointer possesss a position only when its state is an eigenstate of position, then CAT is a state in which the pointer fails to have a position.

As we mentioned above, one way of responding to the measurement problem is to drop the assumption that an observable has a value only when a system is in one of its eigenstates. This is what is done by certain so called "hidden variable" theories. A hidden variable theory assumes that every genuine dynamical variable possesses a precise value in every state. It turns out that if such theories are to reproduce the statistical predications of quantum theory then they must exhibit some odd features. Specifically they must be non local and contextual. The first means, roughly, that changes in an dynamical quantity that refers to one component of a system may instantaneously affect the value of a dynamical quantitiy that refers to another component even if these components are spatially separated. The second means that precisely how a quantity is measured may affect the outcome of the measurement. These odd features have been taken as reasons for rejecting such theories though these so called "no hidden variables proofs" are, in our view, far from conclusive (see 2).

KHD is more sophisticated than more old-fashioned hidden-variable theories. It allows only some of the observables which pertain to a subsystem of a larger system to possess values when the system is not in an eigenstate of that observable. This is done in such a way, it is hoped, as will supply outcomes to measurements and definite positions to pointers (and definite states of aliveness to cats etc.). Here is how it works. The state CAT can be represented in various bases (as the sum of various complete sets of orthogonal vectors) but the particular representation used above has a special feature. As long as $c_1 \neq c_2$ it is the <u>unique</u> representation of the form SUM $c_i |a\rangle_i |b\rangle_i$ where the $|a\rangle_i$ are orthoganol and the $|b\rangle_i$ are orthogonal. Such representations are called "bi-orthonormal." The KHD interpretation is summarized by the following rules:

R1) if S is composed of S_1 andS_2 and the state of S is \$ with biorthonormal representation = SUM$c_i |a_i\rangle_{S1} |b_i\rangle_{S2}$ and the $|a\rangle_i$ are eigestates of observable A and the $|b\rangle_i$ are eigenstates of observable B then A is defined (has a value) on system S_1 and B is defined on S_2.

R2) If S is in the state referred to in R1 the probability that A has the value a_i and B has value b_i is c_i^2

At first sight it looks as though this proposal solves the measurement problem very nicely. In consequence of R1 the pointer position observable will possess a definite value at the conclusion of a measurement and the cat will be either dead or alive. Furthermore, this is done without there being any collapse of state (the Schrdoinger equation is the only description of state evolution). We might also add that this account does not run afoul of no-hidden variable thoerems (since not all observables obtain values via R1 and R2) and, while it has an inevitable non-locality, according to its advocates, this non-locality is no more objectionable than it is in standard quantum theory. There is a problem about what to say when there does not exist a unique bi-orthonormal decomposition (amusingly the usual presentations of Schrodinger's paradox invovle just such a state) but perhaps this shouldn't be a big worry since almost every state (i.e. the measure on such states in the Hilbert space is 1) possesses a unique bi-orthonormal decomposition.

There are two respects in which KHD account as so far described is incomplete. The first is that the prescription for determining the values of defined observables on a subsystem is underspecified. A many particle system can be decomposed many ways into subsystems with each way possessing a biorthonormal state representation. Rules need to be introduced to guarantee that the values of observables which are defined according to one such decomposition do not conflict with values defined in another. It is likely that this can be done in more than one way and so whatever rules are chosen may appear to be arbitrary.[5] The second, and more serious, lacuane concerns the dynamics of the physical quantities defined in accord with (R1). As the state \$(t) of a composite system evolves in accordance with Schrodinger's equation the vectors in the biorthonormal representation of \$(t) will typically change. In consequence the observables defined on the system's subsystems in accord with (R1) will change. The question that the dynamics needs to answer is how is the value of an observable defined at t+ related to the value defined at t? If there is no relation then R1 will have the effect that values will jump about around. For example if position is well defined at t and equals x, and is not well defined at t+, but defined again at t++ the position at t++ can be far from x. We don't consider the lack of dynamics to be a an irremediable defect but the theory cannot really be taken seriously as a physical theory until a dynmaics is supplied.

We now come to an objection which is, in our view, fatal to KHD. The problem is that measurements which are actually performed almost never satisfy the condition we placed on measurements. In a real measurement there is always some probability of the measuring device making an error. For this reason in a realistic measurement the post-measurement state of M+e will be

$$c_3 |\uparrow\rangle|U\rangle + c_4 |\uparrow\rangle|D\rangle + c_5 |\downarrow\rangle|U\rangle + c_6 |\downarrow\rangle|D\rangle \quad \text{(CAT*)}$$

The components $|\,|\rangle|U\rangle$ and $|\,|\rangle|D\rangle$ represent errors. If the measurement is a good one then the amplitudes of these error components will be small but, as we have noted, in realistic measurements they will inevitably be non-zero. CAT* is obviously not a biorthonormal representation. The biorthonormal representation of this state is

$$c_7 |\uparrow_@\rangle|G\rangle + c_8 |\downarrow_@\rangle|H\rangle \qquad \text{(BCAT*)}$$

where $|\,|_@\rangle$ represents the state in which spin in direction @ (not x-spin) is up and $|G\rangle$ represents a state of the pointer. (R1) does not assign a definite x-spin to the electron in BCAT* although it does assign a definite value to the spin in the @ direction. This is a bit disconcerting since M is supposed to be measuring x-spin. But the real problem is that $|G\rangle$ and $|H\rangle$ are not eigenstates of the observable corresponding to the pointer's position. That is, (R1) fails to assign a definite position to the pointer. If all M's observables which possess values are determined by (R1) then, according to KHD, there simply isn't going to be a matter of fact when CAT* obtains about whether the pointer is pointing to 'up' or to 'down'! (and similarly, if there is some small chance of the cat remaining alive when the x-spin is up then at the conclusion of the measurement the state of the cat and the electron will not be one to which KHD assigns a value to the cat's aliveness.

There is a response to our objection which has been suggested by proponents of KHD. It is that, while position pointer is not well defined in BCAT* the observable which is well defined, call it Q*, will be so near to the pointer's position that we may consider the pointer as having a position "for all practical purposes." The closeness can be measured by the degree to which Q* commutes with position. But this is a fudge. We want the cat dead or alive, not in an eigenstate of an observable which almost commutes with the aliveness observable. In any case, it turns out that even fudging won't save the account. There are Hamiltonians describing almost accurate measurements which result in final states in which the observable which is well defined is almost <u>maximally</u> non-commuting with the observable corresponding to the pointer's position (for example, in a state like CAT* in which c_1

is almost equal to c_2). So one cannot say that even if pointer position is not well defined an observable which is "close" to position will always be defined even for quite accurate measurements. And, of course, there are interactions in which a cat may find itself which are not measurements at all and in which Schrodinger evolution gets it into into states to which the KHD principle fails to assign a definite value to the Aliveness observable.

We conclude that, contrary to the intial appearances, KHD (and variants of it) fails to guarantee that measurements always have outcomes-indeed, if it were true then the kinds of measurements that are actually made would almost never have outcomes. Schrodinger's cat has not yet been delivered-dead or alive.

GRW

The second proposal we want to examine is an objective theory of the collpase of the wave function due to Ghirardi, Rimini, and Weber (see 6). It has recently been refined and advocated by John Bell (see 4). Its basic idea is that the Schrodinger equation is not exactly correct. The wave function of an N-particle system $\$(r_1, r_2, ..., t)$ usually evolves in accordance with Schrodinger's equation; but every so often (once in $1/N \times 10^{15}$ seconds), at random, but with a fixed probability per unit time, the wave function is multiplied by a gaussian (and normalized). The form of the Gaussian is

$$K \exp(-[x-r_k]^2/2D^2 \qquad (G)$$

where r_k is chosen at random from the arguments and the width of the Gaussian D is of the order of 10^{-5} cm. The probability of this Gaussian's being centered at any particular point x is stipulated to be proportional to the absolute square of the inner product of $\$(x,...,t)$ (evaluated at the instant just prior to the jump)) with G. The effect of the mulitplication is a "collapse" of the wave function to a more localized wave function. The probability of a jump per particle-once in 10^{15} seconds) and the width of the multiplying Gaussian (10^{-5} cm) are taken to be new fundamental constants.

What immediately strikes one about thie GRW theory (and what is so nice about it) is that unlike von Neuman's collapse theory it proposes a uniform dynamical law for all interactions and one in which the notion of "measurement" does not appear. Although the theory has in principle empirical differences with any non-collapse view (like the KHD interpretation previously considered) it is so designed that these differences will be very difficult to detect and certainly have not been detected up until now. For isolated microscopic systems (i.e. systems containing a samll number of particles) 'jumps' will be so infrequent as to be unoobservable in practice; and D (the width of he gaussian) has been chosen suffciently large so that violations of the conservation of energy (which jumps produce since they localize wave functions) will be sufficiently small as to be unobservable (over reasonable time intervals) even in macroscopic systems.

Of course the crucial question for the measurement problem is whther GRW collpases produce states in which observables like the pointer's position, the cat's aliveness, and so forth are well defined. Here is GRW's story about that. Consider, for example, the spin measurement previously discussed in which the x-spin of an electron is recorded by the position of a pointer. At the conclusion of the measurement the state of the pointer is CAT. But on GRW CAT is very unstable. In a very short time (about $1/N10^{15}$ seconds where N is the number of particles which is on the order of 10^{20})the wave function for the pointer + electron will get multiplied by a GRW gaussian and a GRW jump will occur and one of the terms in CAT will all but dissappear. We can picture the effect of a collapse as follows:

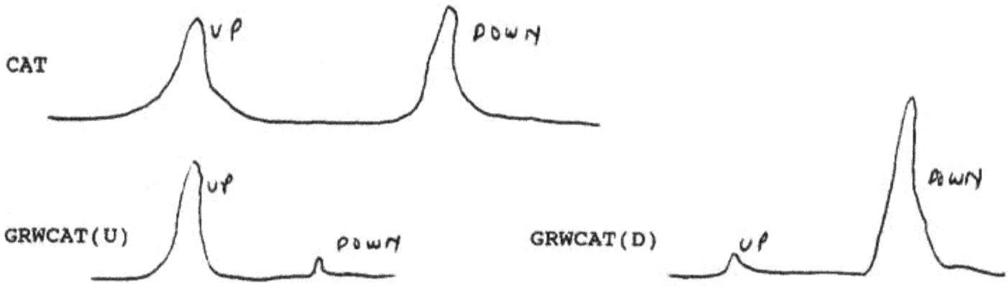

After the jump M+e will be either in GRWCAT(U) or GRWCAT(D) in which almost all the amplitude of the pointer+electron will locate the pointer at either the U or the D position. This seems, at least initially-we will question it later-, to justify the claim that after the jump the pointer possess a definite position.

Notice how the fact that the measurement apparatus is macroscopic enters this account. Since the macroscopic pointer contains many particles the probability that there will be a jump to either GRW(D) or GRW(U) in a small time is very great. Whereas for microscopic systems the probability of a jump is very small. Notice also that, unlike the standard view, there is a smooth transition from the microscopic to the macroscopic and the laws predict the interactions for measurment devices rather than assume them. And unlike the KHD interpretation no special difficulty is caused by measurements which are not completely accurate. GRW predicts that the state CAT* will in a short amount of time collapse to a state in which almost all the amplitude is concentrated in one of its four components. If the collpase is to either ¦¦>¦U> or ¦¦>¦D> the measurement is an accurate one. Otherwise it is inaccurate and subsequent measurements may reveal the error. And, of course, Schrodinger's cat, being a macroscopic cat, will occupy the no cat's land been life and death for only an instant.

Given all the above can we now say that the measurement problem has been solved? We are afraid not. One difficulty has already been discussed by Albert and Vaidman see 3). It is that there are certain interactions which are considered to be measurements by physicists but which do not involve the correlation of the position of something with the observable being measured. Even worse, there are certain observables which we all take to have defininite values (e.g. exactly what image appears on a TV screen) which do not involve the positions of particles (or sufficiently many particles) in a way that allows the GRW collapses (which are always onto position) to produce definite outcomes. Since this point has already been discussed we will not pursue it further here. Instead we want to reveal the way GRW 's solution is fudging and look at a way of making it explicit.

Our worry is that GRW collapses almost *never* produce definite outcomes even when outcomes are recorded in distinct positions of macroscopically many particles. The reason is that a GRW jump does not literally produce a collapse into an eigenstate of position. It does not collapse a state like CAT into one of its component states. A GRW collapse of CAT yields one of the states like GRWCAT(D) with tails in which *almost* all the amplitude is concentrated in the region around one of the two components but there is non-zero, though very small, amplitude associated with other regions. GRW collapses do not, and cannot, make the other component of the wave function vanish entirely. A GRW collapse is represented mathematically by multiplication of CAT by a gaussian which is centered at one of the possible positions of the elementary particles in CAT. Since this gaussian, although very narrow (recall its width is 10^{-5} cm.) has non zero value everywhere and although it produces a concentration of amplitude around its mean, the resulting state still has no zero amplitude at every point which it previously had non-zero amplitude. And it is very important to the theory that the multiplying Gaussian has non-zero amplitude everywhere. If it didn't then the theory would make predictions which we know are contrary to theose of quantum mechanics (in particular, it would lead not noticeable conflicts with conservation of energy). This means that the post collapse state is not an eigenstate of position and so does not actually assign a definite position to the pointer.

A proponent of GRW might respond to this point by claiming that although GRWCAT(D) is not in an eigenstate of position it is in an eigenstate of an operator which is very close (almost commutes) with the operator representing pointer position and that this is sufficient to allow us to consider the pointer as having a definite position for all practical purposes. This response is similar to Dieks' attempt to save KHD from our objection to it. It is also fudging. Position being <u>almost defined</u> is not the same as position being defined and a cat's almost being dead or almost being alive is not the same as its being dead or alive.

But here (as opposed to the KHD theory) there may well be aceptable ways of avoiding the fudging. One way (which emerged in the course of some conversations we had with John Bell in Erice in 1989) is to hold that what's being asserted when somebody asserts that (say) "the pointer is pointing at S" is just that <u>almost</u> all of the pointer's probability amplitude (that is: some specified very large percentage of that amplitude, but not necessarily all of it) is presently concentrated in the "pointing-at-D" sector of the pointer's configuration space. This suggestion will involve a not quite standard (but nonetheless <u>nearly</u> standard and perfectly explicit) mapping from quantum states onto ordinary English statements like "the pointer is poining at D" or "the cat is alive", which specifies precisely which quantum states they supervene on. If this suggestion were adopted, of course, we would have to take care to distinguish between descriptions of systems in terms of hermetian operators and descriptions of them in English, since (for example) an object which is not in an eigenstate of the hermetian position operator may nontheless have a definie "English position."

Suppose that we decide to adopt one of the rules of the sort that "the pointer is pointing at X" is true just in case at least k% (i.e. 99% or 99.9% or whatever) of the pointer's probability amplitude is concentrated in the pointing-at-X sector. It will turn out, then, for example, that if GRWCAT(D) is the staste of the world, the the pointer is pointing to D. But this proposal has a number of unsettling consequences. To begin with, there will be states of any particle (and of pointers and of everything else) wherein the particle is located at X, and wherein there will nontheless be a finite probability (albeit a very small one) of a collapse occurring which leaves the particle, an instant later, at Y, no matter how distantly separated X and Y may be, and without the particle's ever having occupied any region between X and Y. Further, these proposals will entail that a particle which is at a located at X (i.e. it is correct to say in English that it is located at X) to be affected by physical processes which occur, simultaneously at Y, no matter how distantly separated X and Y may be. And even more curiously, the proposals entail that a particle being in region R and being in region R* does not entail that the particle is within the intersection of R and R*. Finally, if any of the proposals of this type are true then there will apparently be no empiricial way to ascertain which one of these proposals-that is whether it is 99% or 99.9% or...) is the true one.

While the above features of a GRW theory without fudging are peculiar and deserve scrutiny they do not seem to rule out GRW. Whether we can learn to live (or must learn to live) with them will presumably depend on what other ways of resolving Schrodinger's paradox we can come up with (See 1,2).

References

1. Albert,D., and Loewer,B., "Intepreting the Many Worlds Interpretation" Synthese, 12, 1988.

2. Albert,D., and Loewer,B., "Two No-collapse Interpretations of Quantum Mechanics" NOUS 1989.

3. Albert,D., and Vaidman, L., "On a Theory of the Collpase of the Wave-Function" Proceedings of George Mason conference on the foundations of quantum mechanics.

4. Bell,John, The Speakable and the Unspeakable in Quantum Theory, Cambridge 1987.

5. Dieks, Dennis, "On Some Alleged Difficulties in Quantum Theory" Synthese forthcoming and "Resolution of the Measurement Problem Through Decoherence of the Quantum State", unpublished ms.

6. Ghirardi, Rimini, and Weber, Phys. Rev. D 34 1986.

7. Healey, Richard, The Philosophy of Quantum Mechanics Cambridge 1989.

8. Kochen, Simon, "A New Interpretation of Quantum Mechanics," Symposium on the Foundations of Modern Physics, Lahti and Middlestadt eds. New World.

9. von Neuman, John, The Mathematical Foundations of Quantum Mechanics Princeton 1955

QUANTUM THEORY OF SINGLE EVENTS

A. O. BARUT
Physics Department, University of Colorado
Boulder, Colorado, 80309

ABSTRACT

A clear distiction is made between a single individual event denoted by $\psi(x, a)$ depending on the parameters a, and a typical averaged event denoted by $\Psi(x)$. This avoids the notion of the collapse of the wave function, among others. Explicit localized solutions of the wave equations are constructed which move like relativistic particles and obey quantum relations $E = \lambda\omega$ and $P = \lambda k$. Quantum spin correlations are evaluated on the basis of individual events with classical "hidden" parameters.

1. A Single Event versus a Typical Event

1.1. In an electron interference experiment electrons are sent one after another, are collected as small dots on a screen and the pictures are stored in a memory device[1]. When the experiment begins we see a few isolated dots which gets denser and as the number of collected dots become very large we see the emergence of interference patterns as a regularity in repeated events.

I shall assume that a single tiny dot represents a single individual electron.This and the regularity in repeated events is shown schematically in Fig. 1. Also shown is the probabilty of a typical event by the distribution $|\Psi|^2$. I represent the single dot in Fig.1 by a localized wave lump $\psi(x, t; a)$, where a is a set of parameters labelling the position,velocity and the shape of the individual event. The blackening caused by a single event may be proportional to $|\psi|^2$. Clearly ψ and Ψ are entirely different types of entities, hence the different symbols used. One is an objective material individual event; the other is a derived measure of regularity of repeated events and does not exist if the experiment is not repeated. This distinction is fundamental for the present theory, hence we formalize it more precisely:

Def.1. A <u>single event</u> is an experimental outcome which can be attributed to an isolated objective individual entity, e.g. a track in a bubble chamber.

Def.2. A <u>typical event</u> is the expected averaged behaviour of a single entity in repeated experiments.

Both are observed,though in different ways, and there must be a theory for both. The corresponding wave functions are:

(i) $\psi(\boldsymbol{x},t;\boldsymbol{a})$ = wave function of the individuum with "name" $\boldsymbol{a}$, where $\boldsymbol{a}$ = a set of classical "hidden" parameters. The parameters are not really hidden, they are , as we mentioned above, the position, velocity and the shape parameters of a localized wave lump. I prefer, instead of the historical adjective "hidden", the terminology of classical (nonquantized) parameters. This individual wave function is deterministic, objective and highly localized. It is nonstationary, but it is not a wave-packet : we want it to be highly localized and nonspreading as seen in experiments. We shall say more about the spreading a little later. The superposition principle does not hold in the usual sense. The sum of two lumps with different parameters, even if it is a solution of the wave equation represents two individual particles.

(ii) $\Psi(\boldsymbol{x},t)$ = the wave function for a typical event without any hidden or classical parameters. The word typical is used here in the same way as we speak for example of a typical American.Thus this wave function is by definition probabilistic. It is not localized, usually represented by a plane wave or by a large wave packet in scattering experiments (of the order of the aperture of the instruments). It obeys the superposition principle : the linear combination of two such wave function can still represent another state of the same typical event. Finally Ψ can have stationary states, with $|\Psi|^2$ independent of time, unlike ψ.

1.2. Let us now see what we gain by clearly distinguishing these two different wave functions.

First of all the introduction of the individual wave function $\psi(\boldsymbol{x},t;\boldsymbol{a})$ removes completely the need for the notion "collapse of the wave function" and all the attempts to find mechanisms of how the collapse can arise: The localized wave lump ψ is the already collapsed single event, and Ψ never collapses, in fact must not collapse. In the practical use of quantum theory for repeated experiments one needs the uncollapsed form of Ψ and there is never any need to collapse or to reduce the wave function. In the theory of measurement this notion arises, but now we must use ψ. This notion arose historically because one wanted, or one thought, the wave function Ψ to represent **both** the regularity of repeated events and also the individual single event. That is the reason one talks about the discontinuity in the measurement theory : Quantum theory has two kinds of evolutions, we are often told, one the unitary deterministic evolution of Ψ given by the Schrödinger equation; the other the discontinuous probabilistic evolution during the measurement where Ψ is supposed to collapse to an individual event. I think this dilemma towards whose resolution heroic efforts have been spent since the beginning of quantum theory is unnecessary. We cannot describe two entirely different entities by the same mathematical quantity and hope for consistency. What we want to show here is that the evolutions of both ψ and Ψ are continuous and deterministic, but they have different physical meanings. If we are interested in the behaviour of an individual particle in a measuring device we must calculate the motion of the localized wave lump everywhere in the instrument, e.g. the motion of the spinning individual atom in the Stern-Gerlach magnet. We cannot represent the measuring device by a black box from which different eigenstates of an operator emerge. The latter is only true for Ψ. However we can pass from the description $\psi(\boldsymbol{x}, t; \boldsymbol{a})$ over to the description Ψ by averaging over the parameters $\boldsymbol{a}$ of ψ as the experimentalist in fact does when he wants to test Ψ. Furthermore in the development of quantum theory several other topics should be formulated in terms of individual wave functions Ψ rather then in terms of Ψ in order to avoid inconsistencies. For example, the problem of the classical limit of quantum theory is directly and easily obtained as the limit of the individual wavelets whose crests almost follow the classical trajectories. Classical mechanics is a theory of individual events. On the other hand, as is well known, it is very difficult to obtain classical trajectories from Ψ; it is not just letting $\hbar \to 0$. Similarly, the problems of quantum chaos or quantum statistical mechanics should be formulated in terms of $\psi(\boldsymbol{x}, t; \boldsymbol{a})$ where we can compare these phenomena in a direct one-to-one way with their corresponding classical counterparts.

2. Completeness of Quantum Mechanics and Brief History of the Theory of Single Events

Standard quantum theory deals only with Ψ i.e. with repeated events, hence with the typical behaviour of a single particle. It is of course very successful as such, and we all appreciate it. It cannot make any statements about individual events,—but we wish to describe an individual event,—hence for this simple reason it is incomplete. The experiment is here ahead of theory: experiments with individual particles abound. It is therefore legitimate, nay a duty, to try to complete the standard quantum theory by adding to it a "Theory of Single Events" in such a way that the standard theory emerges after averaging over the parameters of all the single events observed. Clearly this does not change anything about the standard interpretation, but it adds to it a new dimension so that we can keep quite apart statements made on single events versus on typical events. As we shall see this will be crucial in the discussion of spin correlation experiments. Furthermore, the theory of single events provides us, as mentioned above, a direct approach to classical theory in the limit of vanishing wave properties of the individual wavelets.

It may illuminate to observe that there is a similar situation to the above dichotomy (individual vs. typical) in sociology. Social scientists observe the behaviour of an individuum. But since one does not know much what to do with this single experiment, one sends qüestionnaires to a large number of individuals with the hope of deriving the laws for the typical behaviour. A typical person does not exists as such materially as an individual does. Moreover, we may get a different typical person if we perform a weighted average. One can well imagine to develop an equation for the behaviour of a typical person to predict its behaviour as well as for each individual.

Max Born[2], when he first applied the wave mechanics to scattering and introduced the statistical interpretation, was well aware of the distinction between individual and typical particles, which later has often been forgotten. He used a plane incoming wave and an outgoing spherical wave which he expanded into plane waves going into different directions and interpreted (in the second paper) the coefficients as the probability amplitudes of scattering from the initial momentom to the final momentum in the direction θ. He asserted that there were no internal properties of atomic systems which would fix and determine a definite result of any collision in a single event. But he added ,"but it is of course allowed to someone, who is not satisfied with this, to assume that there are additional parameters that are not in the theory which would determine the single event" . The additional parameters $\boldsymbol{a}$ that we envisage are indeed not in the theory but they are in the localized solutions as we shall show it explicitly.

L. de Broglie[3], however, from the very beginning "tried to magine a real physical wave which transported minute and localized objects through space in the course of time", and came back to this vision in the later period of his life.[5]

It is not a question of which interpretation is the exclusively correct one, Born's or de Broglie's, but there are , and there should be, two different entities ψ and Ψ, depending what question we ask. De Broglie's ψ is the more elementary, microscopic entity, and Born's Ψ is the derived,"macroscopic" quantity, much more under control.

Having introduced the wave function for single events and its distinction from the usual wave function, what is now the theory of single events, what are the laws or wave equations for $\psi(\boldsymbol{x}, t, \boldsymbol{a})$? It is generally believed that a theory of single events is *in principle* impossible. The best way to answer this belief is to go ahead and construct an explicit theory of single events. We shall use conservatively the usual wave equations also for ψ, but we should be open minded to modifications in the wave equations since the laws of single events must yet be developed inductively.

3. Localized Solutions of Wave Equations moving like Particles

3.1. We begin with the ordinary massless scalar wave equation $\Box\varphi = 0$.We look for a solution which is localized at a point $\boldsymbol{x}_0$ of the form

$$\varphi(\boldsymbol{x}, t) = F(\boldsymbol{x} - \boldsymbol{x}_0)e^{-i\Omega(t-t_0)} \tag{1}$$

Thus a point $\boldsymbol{x}_0$ is singled out- this is an individual solution. This localized wave is not static, but oscillates with an internal frequency Ω; it has the parameters $t_0, \boldsymbol{x}_0$ and Ω. Six more parameters will be introduced when we construct a moving solution by a Lorentz transformation, three velocity components and three angles of rotation. Actually, because the massless wave equation is also conformally invariant we could introduce five more parameters corresponding to dilatations and special conformal transformations making a total of 16 parameters. Note that the equation has no parameters. For simplicity, I consider here a spherically symmetric solution; the general case is given elsewhere[4]. In this case the moving solution has the form

$$\varphi(\boldsymbol{x}, t; \boldsymbol{a}) = F\left(r_{\perp}(t)\right) e^{(\boldsymbol{k}\cdot\boldsymbol{x} - \omega t)} \tag{2}$$

where $r_\perp$ is the transformed relativistic radial coordinate

$$r_\perp^2 = (r_\mu u^\mu)^2 - r^2$$

with $r^\mu = (t - t_0, \boldsymbol{x} - \boldsymbol{x}_0)$ and $u^\mu = \gamma(1, \boldsymbol{\beta})$ being the relativistic velocity in the direction of the Lorentz boost : $u^2 = 1, \gamma = 1/\sqrt{1-\beta^2}, \boldsymbol{\beta} = \boldsymbol{v}/c$. Again for simplicity we consider a pure boost without rotations and dilatations. For the frequency and the wave vector of the moving solution we find

$$\omega = \gamma\Omega \quad , \quad \boldsymbol{k} = \frac{\Omega}{c}\gamma\boldsymbol{\beta} \tag{3}$$

so that we have the dispersion relation

$$(\omega/c)^2 - \boldsymbol{k}^2 = (\Omega/c)^2 \tag{4}$$

which is that of a massive particle, although we have started with a massless equation. The reason is the localizing factor F in the solution instead of the pure plane wave solution. In order to see this in detail, we insert (1) into $\Box\varphi = 0$, and obtain the Helmholtz equation for F

$$\Delta F + (\Omega/c)^2 F = 0 \tag{5}$$

whose general solution can be expressed in terms of Bessel's functions[4]. The simplest spherically symmetric solution in the rest frame that we decided to consider here is

$$F = C\frac{\sin\left(\frac{\Omega}{c}r\right)}{r} \tag{6}$$

This is a wave concentrated inside a radius $R \sim c/\Omega$ but with a long decreasing oscillatory tail (Fig.2a). The moving solution, say in x-direction is

$$\begin{aligned} \varphi(\boldsymbol{x}, t; \boldsymbol{a}) &= C\frac{1}{r_\perp}\sin\left(\frac{\Omega}{c}r_\perp\right)e^{i(\boldsymbol{k}\cdot\boldsymbol{x} - \omega t)} \\ r_\perp &= \left[\gamma^2(x - vt) + y^2 + z^2\right]^{1/2} \end{aligned} \tag{7}$$

and is shown in Fig.2b for $v/c = .8$. We can associate a group velocity v to this moving lump; the phase velocity is $u = \omega/k = \frac{c^2}{v}$. Hence $uv = c^2$.

3.2. Mass and Localization.

After wave properties of electrons have been hypothesized by de Broglie and experimentally verified, one looked for an appropriate wave equation for these wave properties. Several physicists independently have added by hand a mass term to the wave equation

$$\left(\Box + \frac{m^2c^2}{\hbar^2}\right)\phi = 0 \tag{8}$$

The plane wave solutions of this Klein-Gordon equation satisfy the dispersion relation

$$(\omega/c)^2 - k^2 = m^2c^2/\hbar^2 \tag{9}$$

which is exactly the same as our dispersion relation (4) for the massless wave equation, but for its localized solutions. We can therefore relate the notion of mass of a quantum particle to the frequency Ω of internal oscillations of a localized wave by

$$\Omega = mc^2/\hbar \tag{10}$$

The Lorentz transformation of both sides of this equation gives

$$\gamma\Omega = \gamma mc^2/\hbar \quad , \text{or} \quad \hbar\omega = E \tag{11}$$

which is the basic quantum principle which we have now modelled with our localized solutions.

3.3. Energy and Momentum

We can reinforce and extend eq.(11) by actually evaluating the total field energy contained in our localized solution as an integral over the energy density. This is how the energy and momentum of an objective localized entity should be calculated. The field energy is given by

$$E = N_E \int \left\{ \left| \frac{1}{c}\frac{\partial\varphi}{\partial t} \right|^2 + |\Delta\varphi|^2 \right\} d\boldsymbol{r} \tag{12}$$

where N_E is a normalization constant, also for dimensional reasons. For a plane wave this energy is three-fold infinite ∞^3. For our localized solution (1) or (3) it is still one-fold infinite ∞^1. A method to obtain a finite energy even for a free particle is to consider a superposition of solutions over a narrow range of frequencies around some basic frequency Ω_0. Equation (1) becomes

$$\varphi(\boldsymbol{x}, t; \boldsymbol{a}) = \int d\boldsymbol{r}\Omega F(\boldsymbol{x} - \boldsymbol{x}_0, \Omega) e^{-i\Omega t} f(\Omega) \tag{13}$$

where $f(\Omega)$ is a distribution function. When (13) is inserted into (12) and use is made of the orthogonality relations of Bessel's functions

$$\int_0^\infty J_{\ell+1/2}\left(\frac{\Omega}{c}r\right) J_{\ell+1/2}\left(\frac{\Omega'}{c}r\right) r dr = \frac{c}{\Omega}\delta(\Omega - \Omega') \tag{14}$$

we obtain the finite result[4]

$$E = \lambda \int \Omega |f(\omega)|^2 d\Omega = \lambda \bar{\Omega} = \lambda \Omega_0 \tag{15}$$

with $\int |f(\omega)|^2 d\Omega = 1, \bar{\Omega}$ is the expectation value of the frequency and λ is a proportionality constant. In the moving frame eq.(16) becomes $E = \lambda\omega$ as in (11). The field momentum

$$\boldsymbol{P} \sim \int \frac{\partial \phi^*}{\partial t} \Delta \phi d\boldsymbol{r} \tag{16}$$

evaluated in a similar manner gives

$$\boldsymbol{P} = \lambda \boldsymbol{k} \tag{17}$$

with the same proportionality constant λ so that we get from (4) and (11) the relativistic energy momentum relation

$$(E/c)^2 - \boldsymbol{P}^2 = \lambda^2 (\Omega/c)^2 = (\lambda/\hbar)^2 m^2 c^2 \tag{18}$$

We cannot determine,of course, within this simple linear wave equation model that the value of λ is the Planck's constant $\hbar$, but perhaps after the interactions are introduced. Note that the standard quantum theory neither calculates $\hbar$, nor the relations (15)–(17); we have at least modelled the latter equations.

According to our discussion in Section 1., the localized solution is an individuum, and the plane wave solution characterizes the typical behaviour, i.e. Ψ, so that we can write our solution in a more suggestive way to be used later as

$$\psi(\boldsymbol{x}, t; \boldsymbol{a}) = F(\boldsymbol{x}, t; \boldsymbol{a}) \Psi(\boldsymbol{x}, t) e^{-i\Omega t} \tag{19}$$

where ψ is a solution of the massless wave equation and Ψ is a plane wave solution of the massive equation. Equation (19) realizes the early idea of de Broglie that a localized particle is guided by the (plane) wave Ψ.[3,5]

3.4 Similar solutions have been obtained for Maxwell, Dirac and other wave equations[6]. The general conclusions are the same. For our purpose in this talk it is sufficioent to have the simplest case of the scalar wave equation.

3.5. Localized Solutions of the Schrödinger Equation

The Schrödinger equation is obtained as a limit from the Klein-Gordon equation after separating a fast oscillating term $\exp(-i\Omega t)$. The massless wave equation has of course no unrelativistic limit, but the localized solution (2) has a limit for small velocities. We can proceed in two equivalent ways. Either we take the usual Schrödinger equation and look for localized solutions, or add a mass term to the Schrodinger equation. The difference is just the factor $\exp(-i\Omega t)$. At any rate the result is that the Schrödinger equation

$$i\hbar\frac{\partial\psi}{\partial t} = -\frac{\hbar^2}{2m}\Delta\psi \tag{20}$$

has a localized solution of the form

$$\psi(\boldsymbol{x},t;\boldsymbol{a}) = F(\boldsymbol{x}-\boldsymbol{x}_0-\boldsymbol{v}t)e^{\frac{i}{\hbar}(m\boldsymbol{v}\cdot\boldsymbol{x}-\frac{1}{2}mv^2t)}e^{-i\Omega t} \tag{21}$$

where F satisfies the same equation as (5)

$$\Delta F + \frac{2m}{\hbar}\Omega F = 0 \tag{22}$$

and

$$\boldsymbol{k} = \frac{m}{\hbar}\boldsymbol{v}\ ,\ \omega = \Omega + \frac{1}{2\hbar}mv^2 = \Omega + \frac{\hbar k^2}{2m} \tag{23}$$

Eq.(21) is the Galilei boosted rest frame solution $Fe^{-i\Omega t}$ as in eq. (1), and can also be written in the form (19) as

$$\psi(\boldsymbol{x},t;\boldsymbol{a}) = F(\boldsymbol{x}-\boldsymbol{x}_0-\boldsymbol{v}t)\Psi(\boldsymbol{x},t)e^{-i\Omega t} \tag{24}$$

where Ψ is the usual plane wave solution of the Schrödinger equation for free particles. In the calculation of phase differences the phase Ωt drops out.

The field energy and momentum calculated as before with

$$\begin{aligned} E &= \frac{i\hbar}{2}\int d\boldsymbol{r}\left[\psi^*\frac{\partial\psi}{\partial t} - \psi^*\frac{\partial\psi^*}{\partial t}\right] \\ \boldsymbol{P} &= \frac{\hbar}{2}\int d\boldsymbol{r}\left[\psi^*\frac{\Delta}{i}\psi - \psi\frac{\Delta}{i}\psi^*\right] \end{aligned} \tag{25}$$

give, using the normalization condition $\int\psi^*\psi d\boldsymbol{r} = 1$,

$$E = \hbar\omega \quad \text{and} \quad \boldsymbol{P} = \hbar\boldsymbol{k} \tag{26}$$

Now, because $\hbar$ is already in the wave equation, we get the correct proportionality constant $\hbar$.

We wish now to generalize the solution (24) to the Schroödinger equation with a potential

$$i\hbar\ \psi_{,t} = -\frac{\hbar^2}{2m}\Delta\psi + V(\boldsymbol{x})\psi \tag{27}$$

We thus look for a solution of the form[7]

$$\psi(\boldsymbol{x}, t; \boldsymbol{a}) = F\left(\boldsymbol{x} - \boldsymbol{g}(t; \boldsymbol{a})\right) \Psi(\boldsymbol{x}, t) e^{-i\Omega t} \tag{28}$$

where Ψ is the standard solution of the Schrödinger equation in the potential V, e.g. sum of stationary solutions, and the localization function F again satisfies (22). We obtain the following condition

$$i\dot{g} \cdot \nabla F(x - g) = \frac{\hbar}{m}\nabla F(x - g) \cdot \frac{\nabla\Psi}{\Psi} \tag{29}$$

For $V = 0, \Psi \sim e^{i(\boldsymbol{k}\cdot\boldsymbol{x} - \omega t)}, \frac{\nabla\Psi}{\Psi} = i\boldsymbol{k}$ and (29) is solved for $\dot{\boldsymbol{g}} = \boldsymbol{v}$ and (28) reduces to (21). In order to have more freedom we may introduce an additional phase in (28) which in the calculation of $|\psi|^2$ does not contribute:

$$\psi(\boldsymbol{x}, t; \boldsymbol{a}) = F(\boldsymbol{x} - \boldsymbol{g})\Psi(\boldsymbol{x}, t) e^{-1\Omega t} e^{\frac{i}{\hbar}H(\boldsymbol{x}, t)} \tag{30}$$

then the conditions are

$$\begin{aligned} H,t &= \frac{\hbar}{m}\left(\frac{\nabla F}{F} \cdot \frac{\nabla\Psi}{\Psi} - \frac{1}{2}(\nabla H)^2\right) \\ m\dot{g} \cdot \frac{\nabla F}{F} &= \left(\frac{\nabla F}{F} + \frac{\nabla\Psi}{\Psi}\right) + \Delta H \end{aligned} \tag{31}$$

the phase H may be even non-integrable.

3.6. It is interesting to compare the nonspreading solution (30) with Schrödinger's nonspreading coherent state for the harmonic oscillator. Due to the special form of the harmonic potential we can write

$$V = \frac{1}{2}m\omega^2\boldsymbol{x} = \frac{1}{2}m\omega^2\left[(\boldsymbol{x} - \boldsymbol{g})^2 - \boldsymbol{g}^2 + 2\boldsymbol{x}\cdot\boldsymbol{g}\right] \tag{32}$$

so that the first term $(\boldsymbol{x} - \boldsymbol{g})^2$ can be put into the localization function. Writing

$$\psi = G(\boldsymbol{x} - \boldsymbol{g}) e^{\frac{i}{\hbar}H(\boldsymbol{x}, t)} \tag{33}$$

where G satisfies

$$-\frac{\hbar^2}{2m}\Delta G(x - g) + \frac{1}{2}m\omega^2(\boldsymbol{x} - \boldsymbol{g})^2 G(x - g) = EG(x - g) \tag{34}$$

we obtain with $\nabla H = m\dot{g}$

$$H_{,t} = \frac{1}{2}m\dot{g}^2 + E + m\omega^2 \boldsymbol{x}\cdot g - \frac{1}{2}m\omega^2 g^2$$

A solution is found to this equation if $g(t)$ is the classical orbit satisfying $\ddot{g} = -\omega^2 g$. And if we take E to be the classical energy $E = \frac{1}{2}m\dot{g}^2 + \frac{1}{2}m\omega^2 g^2$ the solution ψ of the Schrödinger equation (33) becomes

$$\psi = G(\boldsymbol{x} - g)e^{\frac{i}{\hbar}(m\dot{g}\cdot\boldsymbol{x} - m\int^t \dot{g}^2 dt)} \tag{35}$$

which is the generalization of the Schrödinger's coherent state for an arbitrary solution G of the three-dimensional time-independent Schrödinger equation. The coherent states are localized within a size of the order of Bohr radius, whereas our solutions are localized within a size of the order of the Compton wavelength of the electron.

3.7. For the interpretation of the localized solutions (28) it is important important to discuss two limits:

(i) If $\Omega \to \infty, \frac{1}{r}\sin\left(\frac{\Omega}{c}\right) \to \delta(r)$, hence eq.(28) becomes

$$\psi(\boldsymbol{x}, t; g) \to \delta\left(\boldsymbol{x} - g(t)\right)\Psi e^{-i\Omega t} \tag{36}$$

This is the limit of localization to a point particle, and we obtain the classical trajectory $\boldsymbol{x} = g(t)$

(ii) We now average (28) over the parameters a. We integrate over the parameters

$$\int da\bar{F}\left(\boldsymbol{x} - g(t, \boldsymbol{a})\right)\Psi e^{-i\Omega t} \tag{37}$$

The F's at different positions add up to a constant, possible of infinite value, which we renormalize so that the result is

$$\Psi e^{-i\Omega t},$$

up to an indetermined constant, which is then determined by Born's interpretation.

Therefore the single quantum entity ψ is something inbetween these two limits, it has both particle and wave properties taken from the two limiting cases. It describes now more faithfully the experimental situation, for a single electron is neither a classical point particle, nor an infinitely extended plane wave, but something inbetween. And that is what we wanted to model.

3.8. Wave packets and nonspreading solutions.

It is true that Born's scattering theory can be formulated in terms of wave packets rather than plane waves and by averaging over the parameters of the wave packets. But there is a condition on the size of the wave packets; they have to be large compared to the characteristic size of the scattering potential. They are of the order of the size of the aperture of the beams, $\sim 10^{-3}$cm in atomic experiments compared to the Bohr radius $\sim 10^{-8}$cm. The reason is that if we would take a wave packet of the order of the radius of an atom, let alone the size of the electron, it would spread according to the formula

$$(\Delta x_t^2) = \frac{t^2\hbar^2}{4m^2(\Delta x_0^2)} \tag{38}$$

so that after $t \sim 10^{-5}$ sec. it would reach a size $\Delta x_t \sim 1$m! Thus very large wave packets are essentially equivalent to plane waves. But we wish to describe a single electron by a wavelength of very small size which does not spread. For a fixed frequency Ω our solutions do not spread. However, in order to obtain in addition a finite field energy we have taken a superposition of frequencies which may cause a small spreading. On the other hand, we have considered sofar linear wave equations. Now charged particles have selfenergies which add nonlinear terms to the wave equation. It is possible that these nonlinear terms which must be present would imply further localization without spreading.

4. Single Spin and the Calculation of Quantum Correlations

4.1. The idea of developing a practical quantum theory based on the wave functions $\psi(\boldsymbol{x}, t, \boldsymbol{a})$ for a single event is the following. Calculate the value of an observable A for a single event with parameters $\boldsymbol{a}$, $A(a)$, and then average over $\boldsymbol{a}$. For two observables A and B the correlation function is

$$E(A,B) = \frac{< AB > - < A >< B >}{[< A^2 >< B^2 >]^{1/2}} = \frac{\int d\boldsymbol{a} A(a)B(a) - \int d\boldsymbol{a} A(a) \int d\boldsymbol{a} B(b)}{[\int d\boldsymbol{a} A^2(a) \int d\boldsymbol{a} B^2(a)]^{1/2}} \tag{39}$$

This formula is of course valid in classical physics as well as in the standard quantum theory where $< |A| >$ means the quantum mechanical average of the operator A in a state $| >$. In this expression the values of the observables $A(a), B(a)$ are deterministic. Probability comes in only in the distribution of the parameters $\boldsymbol{a}$ which is determined by the experimental setup. Usually for similarly repeated events the values of a are equally distributed.

We shall next evaluate these correlations for the much discussed spin variables.

4.2. What is an Individual Spin?

Ever since Pauli introduced his spin formalism with σ-matrices as a "classically not explainable two-valuedness" it has been tacitly assumed that not only on the average but also individually the measurement of any spin component can yield only two sharp values. The belief that a single atom can only have two spin values is so ingrained that it is almost universally never even stated that this is a new assumption about single events. It is of course true in standard quantum theory, that is at the level of repeated events, i.e. at the level of Ψ, that the measurement of spin components yields on the average two values. But as we have seen standard quantum theory cannot make any statements about single events, i.e about the wave function $\psi(\boldsymbol{x},t;\boldsymbol{a})$ of an individual with parameters $\boldsymbol{a}$, which remains to be explored. The basis for the above assumption is the Stern-Gerlach experiment which however refers to repeated experiments and results in two broad maxima. We will argue now that this assumption of two-valuedness cannot be true for a single event and shall modify it without changing anything in the standard quantum theory for repeated events. The legitimate question has been posed whether we can reproduce observed quantum correlations by averaging over the parameters of individual events. The crucial point is what are the parameters for an individual spin. If we can completely reproduce observed correlations by a model of spin we have solved our problem.

In spite of Pauli's remark a single spin can actually be accurately modeled, like we modeled above the coordinates and velocities; the spin should be a dynamical variable on the same footing. The best relativistic spin model is perhaps that of a point charge performing a helical motion as its natural motion in such a way that the relative orbital angular momentum of the helical motion (called zitterbewegung) gives the spin magnetic moment with correct g-factor. This motion when quantized gives precisely the Dirac theory of the electron[8]. For our purpose here we shall use a nonrelativistic limit and model spin as a direction in space ; a vector $\boldsymbol{S}$ in the direction parametrized by the angles θ,φ : $\boldsymbol{S}=\boldsymbol{S}(\theta,\varphi)$. The angles (θ,φ) are now the parameters $\boldsymbol{a}$ of the individual spin wave function ψ. This is in fact how we intuitively picture the spin.

Let us apply now this model to the EPR-spin correlations of two spins in a total spin singlet state. We measure the spin component of $\boldsymbol{S}_1$ in some direction $\hat{a}$ and the spin component of $\boldsymbol{S}_2$ in some direction $\hat{b}$ and evaluate the correlation (39) by integrating over all the angles. With $A=\vec{S}_1\cdot\hat{a}$ and $B=\vec{S}_2\cdot\hat{b}$, eq. (39) becomes

$$E(A,B)=\frac{\int d\theta\sin\theta d\varphi\, S_1(\theta,\varphi)\cdot\hat{a}S_2(\theta,\varphi)\cdot\hat{b}}{[\int d\theta\sin\theta d\varphi(S_1\cdot\hat{a})^2\int d\theta\sin\theta d\varphi(S_2\cdot\hat{b})^2]^{1/2}} \tag{40}$$

In the singlet state, for example, the spins are correlated back to back

$$\boldsymbol{S}_2=-\boldsymbol{S}_1 \tag{41}$$

and assuming all single events to be equally probable, we can evaluate the integrals.The result is[9]

$$E(A,B)=-\hat{a}\cdot\hat{b} \tag{42}$$

which is exactly the quantum mechanical correlation calculated by the same formula (39) as

$$E(A,B) = \frac{\left\langle \sigma_1 \cdot \hat{a} \sigma_2 \cdot \hat{b} \right\rangle}{\left[\langle |(\sigma_1 \cdot \hat{a})^2| \rangle \left\langle \left| (\sigma_2 \cdot \hat{b})^2 \right| \right\rangle \right]^{1/2}} = -\hat{a} \cdot \hat{b} \tag{43}$$

The quantities $A(a), B(b)$ can also be interpreted as the expectation values of the operators in "localized" spin coherent states: $S(\theta, \varphi) = {}_c < \theta, \varphi | S | \theta, \varphi >_c$

The result (42) seems to contradict the well known discussions leading to Bell's inequalities. The difference comes from what kind of assumptions we make about individual events. In deriving Bell's inequalities one assumes that the values of the observables $A(a)$ and $B(b)$ for each individuum are discrete and equal to $\pm 1/2$ no matter what direction $\hat{a}$ we choose and this simultaneously. In our calculation these observables have continuous values, yet we obtain the correct quantum correlations, and the discrete case gives the wrong answer. So the whole issue really boils down to the question of what a single spin is.

This question can be perhaps answered by carefully monitoring the Stern-Gerlach experiment for individual atoms, by eliminating all effects of broadening of the two peaks to see if a residual broadening remains, and by taking into account the effect of the fringe fields. We have at the present time no complete theory of the Stern-Gerlach apparatus; it is very difficult to solve the nonlinear equations of the motion of spins through the apparatus.

We must be careful, also for spin, to distinguish between the wave function Ψ for repeated events and ψ for single events. The standard theory deals with the former for which the superposition principle holds. Thus superposition states like $\frac{1}{\sqrt{2}}(|\uparrow> + |\downarrow>)$ always mean statistical states, 50% up and 50% down in repeated experiments and it is used as such in practice. A single spin can never be in a superposition state. I think this distinction again eliminates a number of dilemmas in thinking about spin problems.

Similar correlations can be evaluated on the basis of eq. (40) for triplet and other states[10] as well as for the recently discussed correlations between three or four spins[11]. There is always a classical correlation of individual spin vectors $S_i(\theta_i, \varphi_i)$ which when averaged over angles reproduces the quantum correlations.

5. Conclusions

We have seen that for many interpretational questions in quantum theory it is not only necessary to attempt a theory of single events, but it is also possible despite some existing impossibility "theorems". These theorems have all some assumptions whose physical validity is challenged here and counterexamples are given explicitely.

Single events have classical (unquantized) parameters in space-time and in spin space and for each value of these parameters we get a single individual localized event.

The probabilistic wave function Ψ arises from the individual $\psi(a)$ by averaging over the parameters a and answers to questions concerning repeated events. This view overcomes the problem of the collapse of the wave function which becomes unnecessary.

Problems concerning the classical limit of quantum theory, quantum chaos and quantum statistical mechanics are best treated in terms of the wave function $\psi(a)$ rather than Ψ.

In particular we account for spin correlations by evaluating them classically for individual spins and integrating over their parameters, i.e. not assuming that each individual spin is already quantized.

Finally, the Schrödinger's cat being an individual with name Kedi, with a wave function $\psi(\text{Ked}i)$ and not Ψ, like an individual atom or an individual DNA molecule, is never in a superposition state!

6. References

1. A. Tonomura *et al. Amer. J. Phys.* **57** (1989) 117.
2. M. Born, *Z. f. Physik*,**37** (1926) 863; **38** (1926) 803.
3. L. de Broglie , C.R. *Acad. Sci.* **B180** (1925)4 98; *Ann.Phys.* (Paris) **3** (1925) 22.
4. A. O. Barut, *Phys.Lett.* **A143** (1990) 349.
5. L. de Broglie, C.R. *Acad. Sci.* **B 277** (1973) 71 .L.Mackinnon, *Lett.Nuovo Cim.* **32** (1985) 125.
6. A.O.Barut and A. Grant, *Found. Phys. Lett.* **3** (1990) 303.
7. A.O.Barut, *Found. Phys.* **20** (1990),No.10.
8. A.O.Barut and N.Zanghi ,*Phys. Rev. Lett.* **52** (1984) 2009.
9. A. O. Barut and P. Meystre, *Phys. Lett.* **A105** (1984) 1021.
10. A.O.Barut,in *Symposium on the Foundations of Modern Physics*,eds. P.Lahti and P.Mittelstaed (World Scientific,1985), p.321.
11. A.Zeilinger, These Proceedings.

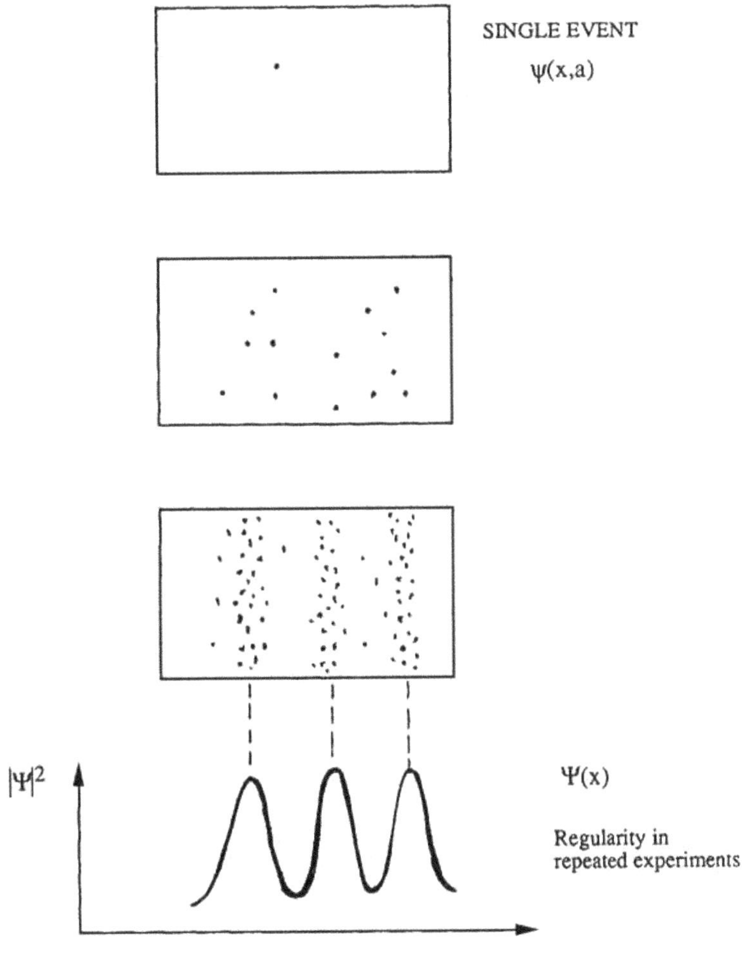

Fig.1

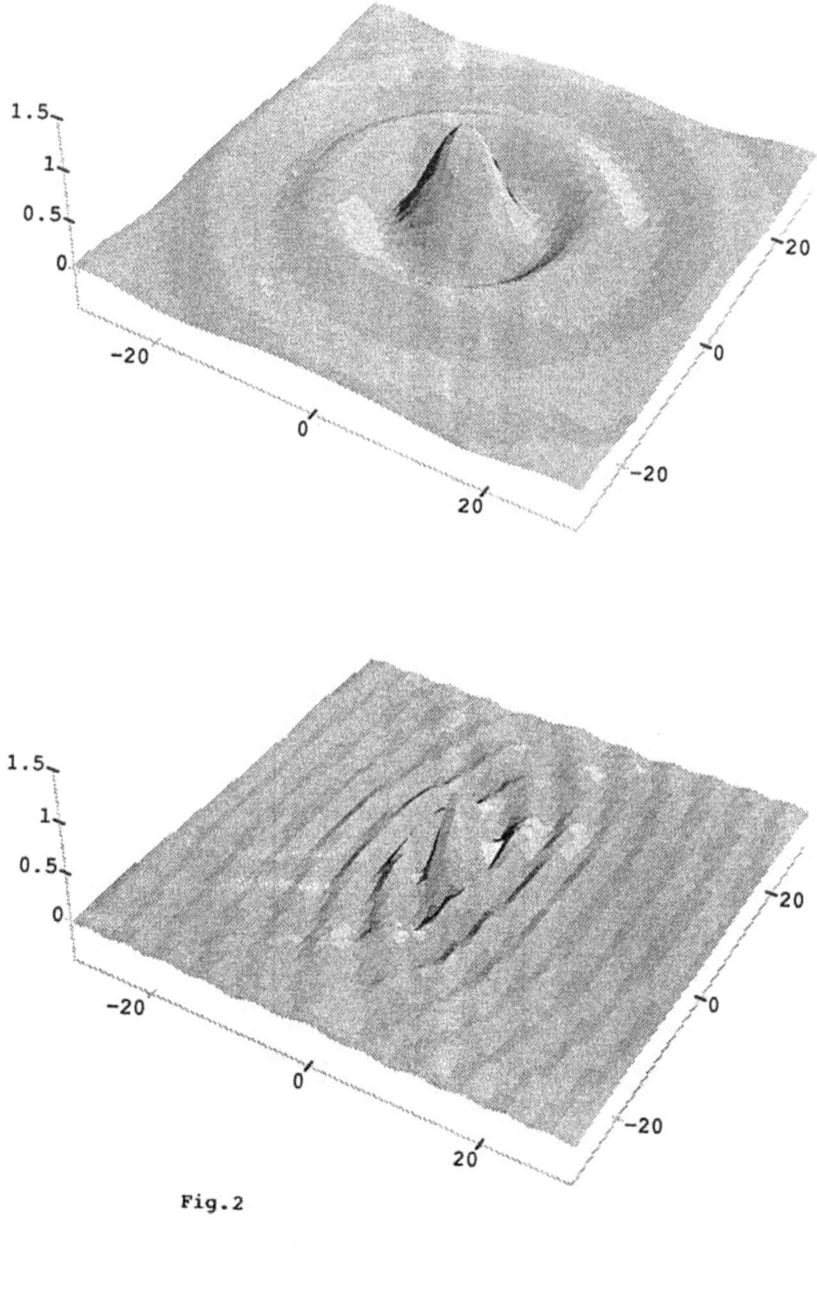

Fig.2

PERSPECTIVAL REALISM AND QUANTUM MECHANICS

MICHEL BITBOL
C.N.R.S., Institut d'histoire et philosophie des sciences et des techniques,
13, rue du four, 75006, Paris

ABSTRACT

A complete reappraisal of the philosophical meaning of Everett's interpretation of quantum mechanics is carried out, by analysing carefully the role of the concept of "observer" in physics. It is shown that Everett's interpretation is the limiting case of a series of conceptions of the measurement problem which leave less and less of the observer out of the quantum description of the measuring interaction. This limiting case, however, should not be considered as one wherein nothing is left outside the description. Something is still needed besides this description: pure cognitive capacity, the subject, or, in a very abstract sense: "mind". The set of branches which arise, according to Everett, from a measuring interaction, gain a renewed signification. They do not refer to distinct "worlds", but to the points of view "mind" can identify itself to. This idea is compared and contrasted with Squires' "selection" of a branch by the mind (without quotation marks). Finally, the notion of indeterminism in quantum mechanics gains an unexpected and new light from a strict application of the previous ideas.

1. Perspectival reality

The central idea of perspectival realism is the following: a "real object" is but the class of all its aspects, seen from different points of view. Some refinements must be added to this definition, if one is to reach a consistent conception of reality. First of all, one must avoid any reference to the object when dealing with the "aspects" which are its constitutive building blocks. Saying that an object is defined by the class of "its" aspects, i.e. by the class of aspects *of this object* obviously departs from the previous rule. To prevent circularity, one should rather use an internal criterion allowing one to link up the "aspects" in a single bundle called "the object". The said internal criterion can be, for instance, the existence of a transformation T (usually considered as smooth, continuous and reversible) which allows one to infer any "aspect" of the bundle from the knowledge of some other aspects. Secondly, to be general, the definition of the "real objects" must go beyond spatial metaphors. The "aspects" constituting a real objects should include experimental results. The concept of point of view should accordingly be extended so as to encompass that of preparation of a measurement, and/or that of the particular context on the background of which an instrument reading becomes an experimental result. Thirdly, we cannot limit the definition of a "real object" to a set of *actual* aspects, lest we are content with a flat version of Machian positivism. We must then supplement the actual aspects with all the (visual, auditive, tactile, or experimental) *possible* aspects which can also be inferred from actual aspects by using the transformation T.

As it stand, this definition of real objects was developed by several thinkers during our century. Let me first recall Russell:

"The 'thing' of common sense may in fact be identified with the whole class of its appearances, where, however, we must include among appearences not only those which are actual sense-data, but also (other) sensibilia(...)"[1] Russell here insists on the third of the above condition, since "sensibilia" refers both to actual and potential sense-data: "I shall give the name sensibilia to those objects which have the same metaphysical and physical status as sense-data, without necessarily being data to any mind."

Schrödinger agreed with Russell about the necessity of supplementing actual appearances by potential ones in the definition of "real objects". He also stressed the second condition: supplementing the spatial points of view by the *experimental* ones.

"(a material object) is composed, not only of my own sensations of it, but also of the imagined sensations I would have in turning it round, looking at it from different angles; in addition, it is composed of the perceptions I imagine you to have of it, and also, if you think of it purely scientifically, of all you could verify and would actually find, if you took it and dissected it to convince yourself of its intrinsic nature and composition, and so on. There is no end to enumerating all the potential percepts and sensations (...) that are included in my speaking of its (object) as of an objective feature of the 'real world around us' "[2].

More recently, P. Heelan reformulated, within a phenomenological framework of thought, the basic idea of perspectival realism:

"A profile of a particular object is a particular manifestation of that object in and through perceptions (...). Systematically associated with any profile is a manifold of different possible profiles of the object which exhibit all the various facets that the object can manifest under a certain system of variations."[3]

It is interesting to notice that, in Heelan's formulation, the first of the above three conditions is stressed, under the name of "essence" which replaces that of "transformation". "An essence is the generative law of a system of profiles", or "The essence of the object is the set of invariant structures that generates the manifold of its profiles."

The metaphysical originality of perspectival realism appears to be that this doctrine is just as little ontologically committed as positivism, while it retains an epistemological version of transcendance. Let me develop this apparent contradiction. To begin with, fulfilling the first above condition allows one to *state* the rules of construction of a "real object" using pre-objective building blocks, rather than merely *pointing* at the object. Such a construction removes from the object the kind of ontological primacy and opacity involved by purely ostensive definitions. Moreover, it leaves no ground for the concept of and "external" world. The objects being but bundles of percepts, it is indeed absurd to contend that a percept is "caused" by an object. This implies that, conversely, inferring the existence of some "external" object from a given actual percept is not a valid move.

Russell claimed in 1914 that abandoning this kind of inference should be construed as a philosophical ideal:

"Wherever possible, logical constructions are to be substituted for inferred entities."[1]

Now, despite its radical lack of ontological committment, perspectival realism leaves enough room for a minimal, purely epistemological, variety of transcendance. Indeed, a bundle of "sensibilia" goes clearly beyond any set of *actual* percepts, in so far as it both relies on a generative transformation, and includes all the potential percepts which are mutually connected through this transformation. This epistemological transcendance may explain the common illusion of ontological transcendence:

"Since the thing cannot without indefensible partiality be identified with any single one of its appearances, it comes to be thought of as something distinct from all of them and underlying them."[1]

The questions on which I wish to focus my attention at this stage, are the following: can the perspectival definition of reality be applied at the quantum level? In particular, what is the status of *potential* percepts in this case?[4].

2. Context and beyond

The answer to the first of the previous questions must apparently be negative. The notion of result of a potential experiment, stated by means of a counterfactual proposition, cannot be extended to the case of atomic or subatomic objects without generating, in many cases, a contradiction with quantum predictions. For instance, it suffices to assume both a certain version of locality condition, and the hypothesis of counterfactual definiteness, to derive Bell's inequalities[5].

The features which underly this incompatibility between the predictions of quantum mechanics and a systematic use of the concept of potential experiments are well-known. First of all, once an observable has been measured on a microscopic object, the result of subsequent measurements of other observables depend on the result of the first measurement. This renders irreversible the transformation T which associates the "aspect" of a given bundle to one another. Secondly, the measuring apparatuses of non-commuting observables A and B are mutually exclusive. Any conjunction of statements bearing on the outcomes of measurements of two non-commuting observables is thus operationally meaningless, even if one of these statements is counterfactual. This idea can be restated as follows[6]: each context (in this case each incompatible experimental arrangement) gives rise to a particular event-language, in which the results of measurements of any set of commuting observables can be expressed. A conjunction (or disjunction) of (actual or counterfactual) statements bearing on non-commuting observables merely lacks the one event-language in which it is meaningful. Now, the structure, at a meta-linguistic level, of the set of these event-languages have been shown to be that of an orthocomplemented modular lattice[7]. It is thus isomorphic to Birkhoff's and Von Neumann quantum logic[8]. This, and other related results[9] constitute the basis for a certain class of perspectival interpretations of quantum mechanics[10]. According to these interpretations, any experimental statement is relative to a certain Boolean sublattice, considered as a frame of observation.

We must examine at this point what sounds like a verbal paradox. Quantum mechanics fits quite well with a perspectival interpretation, whereas the same theory seems reluctant to be comprehended within the framework of perspectival realism. In fact, this "paradox" only brings out in an illustrative manner what we are already aware of, namely that for perspectival realism to be possible, the perspectives on which the definition of the objects is based should not be strictly exclusive of one another. In perspectival realism, the concept of an object arises from the idea of a kind of universal context wherein each point of view is adopted by an (actual or potential) observer and wherein the whole set of significant dynamical variables is (potentially or actually) measured. The perspectival definition of an object as a conjunction of counterfactual experimental statements involving a large (or even infinite) number of possible contexts is then transformed into a conjunction of counterfactual statements involving a *single* context: the conjunction of every possible context.

Now, the very possibility of the latter conjunction amounts to saying that every experimental proposition can be expressed within a single event-language or, equivalently, within a single Boolean frame. It has indeed been demonstrated by Watanabe[11] that a set of experimental propositions, stating both a context (or experimental procedure) and the associated result, has the structure of a Boolean lattice *if the context is unique*. On the other hand, when several contexts are involved, the (complemented) lattice of the experimental propositions is in general not Boolean. This lattice becomes Boolean in a particular case: when it is possible to define a product (or conjunction) of contexts. The field of applicability of perspectival realism as we stated it previously thus corresponds to the conditions which transform the set of the experimental propositions into a Boolean lattice. This correspondance seems to be but another way of asserting the incompatibility between perspectival realism and quantum mechanics as they stand. However, if the said correspondance is granted, the study of lattices of experimental propositions can become a powerful tool allowing one to determine what one should modify in both perspectival realism and (the interpretation of) quantum mechanics, if they are to be rendered compatible.

To sum up, a set of experimental propositions is a Boolean lattice provided one of the three following conditions is fulfilled:

1. The propositions involve no context at all.

(Indeed, in this case, the logical operations between the propositions are the set-theoretic operations among subsets of results)

2. The propositions involve a single context.

(When the context of experimental propositions is unique, the conjunction or disjunction of experimental propositions can be reduced, as in the previous case, to the corresponding set theoretic operations among subsets of results.[11])

3. There are several contexts, but it is possible to define a conjunction of these contexts.

(The set of contexts here reduces to a single composite context, and one is thus brought back to the second situation.)

The third condition is obviously incompatible with experiments at the atomic or subatomic level, but what about the first and the second?

In so far as the traditional cut between the object and the experimental device is maintained, each of the (possibly incompatible) experimental devices represents a well-definite context. In this case, no possibility is left to perform a shift from a many-context language to the one-context or the no-context language which are required by conditions 1. or 2. But as it is well known, this cut (which is almost exactly identical to Von Neumann's cut) has no reason to be frozen at a definite position in the description of any quantum measurement. It is thus conceivable that part or the totality of what we call in the traditional view "the context" be encompassed by an overall object or objective process. This displacement of the boundaries between "the context" and the "object" could lead to fulfillment of conditions 1. or 2.

Let us take as an instance the most extreme position of this boundary. This extreme position is such that every *thing* is described as an object of the theory, whereas no *thing* remains outside its domain of description to play the role of the context. Thus, condition 1 is fulfilled. Now, this corresponds quite closely to Everett's view of quantum mechanics. Everett indeed suggested that the status of 'real object' in quantum mechanics must be ascribed to the multidimensional wave describing not only the set of those particles which are submitted to a measurement, but also those which belong to the whole apparatus and to the observer's body.

Such a conception of the object as being every *thing* which partakes in the measurement process (or even every *thing* in the universe) however raises considerable problems as soon as one tries to apply to it the general perspectival construction of everyday-life and classical real objects. How could we transpose, to begin with, the idea that an object is made of all the possible sensations I would have "looking at it from different angles"? This expression amounts to supposing that the "observer" (myself), being completely independent from the observed object, he can change his point of view on the object at will. But Everett's 'real object' includes the observer's body. Even what we would call classically "the (experimental or spatial) point of view from which an observer looks at an object (i.e. at the object in the restricted sense)", is encompassed within the definition of the overall 'real object'. Similarly, how could we understand, in Everett's frame, that an object is also made of all the results of possible experiments, of "all you could verify and would actually find" on it? Indeed, the whole experimental set-up is part of the object itself, and choosing a new experiment would amount to change the object.

These difficulties clearly arise because the construction of real objects, which we are now trying to transpose to Everett's interpretation of quantum mechanics, relies heavily on a common-sense conception of the observer. In this conception, the observer is implicitly a "person" with a spatial position and bodily features, and the very definition of the possible situations whose combination gives rise to a 'real' entity depends in a crucial way on this assumption. The observer has to *move* around the object to see it from different places, or he has to *change* its experimental set up without changing the object itself.

There is a way to sort out the difficulty. To summarize, it amounts to saying that, in Everett's interpretation, every *thing* is within the field of quantum description, but not *everything*. There is *something* which remains definitely outside this description, in so far as it remains outside any description of the objective world. This *something* is the mind in a very abstract sense: the pure knowing subject. As Schrödinger would say, to explain the absence of mind in the objective world: "Mind is eminently subject, and it thus eludes any objective analysis."[12]. Our task is now to substitute this abstract mind to the human observer, in the perspectival construction of a real object.

Carrying out this substitution analogically, we would fancy that the complex 'real' system (object in the restricted sense+apparatus +observer's body) is the set of all the aspects of it the mind would perceive by "turning it round, moving it, looking at it from all different angles". This seems absurd. If mind is placeless, how could it change its *point* of view? If it has no eyes, how could it look at all?

But Everett's formalism offers a very straightforward generalization of the concept of point of view. Once the wave function of the overall 'real system' has been defined, it displays, at any given time, the set of memory states which result from the interaction between the object in the restricted sense, and the observer's brain. Rather than saying that this interaction results in a branching off of the world, I shall rather consider that it confronts Mind with all its own *possible states* in a given situation. Take then the concept of "point of view" in its most figurative sense, the one you use when you notice, for instance, that your point of view on life is determined by your past experiences. This kind of "point of view" represents in some way no less than your personal identity. Now, in this figurative sense, the set of memory contents displayed in Everett's state vector of the 'real system' under consideration is identical with the set of all the 'points of view' Mind can possibly have on the system. Reciprocally, the 'real system' can be defined as the set of (figurative) points of view Mind can possibly have on it, in good (formal) agreement with the perspectival conception of reality.

3. One mind, many points of view.

The idea that Everett's formalism can be interpreted as a technique for displaying the whole set of figurative points of view on a 'real system' Mind can adopt, is not devoid of difficulty.

It is at least clear that these difficulties are definitely distinct from those faced by the many-worlds conception of the same formalism. Whereas the many-worlds interpretation sounds like an imaginative fantasy whose challenge is to justify the necessity of generating many unobservable situations, the possible-points-of-view interpretation threatens our most intimately rooted beliefs concerning personal identity, and its relations to space and time. These are the problems I must now address.

To begin with, there seems to be something odd in the contention that Everett's writing of the state of a 'real system' displays a set of 'possibilities'. After all, a debate took place between physicists who contended that a measurement yields transition from several possibilities to a single actuality, and Everett, according to whom all the elements of the superposition are 'actual'. But to understand Everett's position, one must recognize that his primary concern was not to eliminate the concept of 'possibility' from the interpretation of quantum mechanics. It was rather to avoid considering any physical *transition* from the possible to the actual. And, since 'actuality' is usually considered as the most obvious characteristic of an experimental result, Everett had but one solution to eliminate the necessity of considering a transition: it was to spread out 'actuality' onto every term of the superposition. In the possible-points-of-view interpretation, the last stage of an experiment, its cognitive level, amounts to the identification of Mind to one of the possible points of view it can adopt upon the 'real system'. This identification is definitely not a physical process, not any more than Bell's 'attribution' of a beable is. It does not involve anything which would *change* the state vector describing the real system. However, at this stage, the reasoning remains incomplete. For, if Mind identifies itself *irreversibly* to one of its possible points of view, what difference other than purely verbal is there between 'identification' and 'actualisation'? Don't they both ascribe a particular status to one term in the superposition? To clear up the non-trivial difference between the two concepts, we need a precise spatio-temporal analysis of 'identification'. This analysis is likely to eliminate radically the possibility of speaking of 'identification' as if it were a 'process' able to occur 'reversibly' or 'irreversibly'. But before performing this analysis, I would like to study another difficulty.

A problem in the present analysis is that the modified perspectival description of the world we propose lacks a meta-level from which the points of view can be described. There is no "point of view of points of view" which would be able, as the Leibnizian God, to encompass the whole series of monads. Indeed, as soon as Mind identifies itself with a point of view, it can but identify itself to a *particular* one. However, this being granted, the mere insistance on the "particularity" of a point of view sounds artificial. Since no point of view is available from which all the other points of view would be seen as equivalent, the point of view Mind adopts, when adopted, is not one among others; it is *the* point of view, self-referred to as *my* point of view.

Let me retain that Mind, having no point of view of its own, can but adopt *particular* points of view and identify itself completely with each of them. Mind is by itself point-of-view-less, just as it is placeless and timeless. The aporia is the following: Mind is not within the world since, even if it can identify itself to any available point of view, it is not *identical* to this point of view. Nor does Mind stand outside the world, since it has no point of view of its own, independent from the points of view the world can offer. Wittgenstein would say that Mind is the limit of the world.

More formally, Mind can be considered as an empty space in the triadic relation: "point of view of (   ) on a 'real universe'". This scheme provides another way of seeing why Mind retains its necessity, even though the "real universe" gathers all that falls under the categories of knowledge: Mind plays a key role in the very *constitutive relations* of this knowledge.

4. Changing world, Timeless mind.

Everett's wave function can now rightly be called, even in the perspectival scheme, "a continuous representation of 'real objects' in space and time". Indeed, the wave function displays the whole set of "aspects" and (figurative) points of view which define the real object, and it allows one, through its evolution ruled by the Schrödinger equation, to predict the *becoming* of the said points of view. Moreover, it enables one to calculate the set of possible points of view defining a particular real system at time t, given this set at time t' (generally, one takes t<t', but the reversibility of the Schrödinger equation does not preclude t'<t). Howether, the last step of an experiment, cognition, 'identification' as I have called it, is not included in this description.

Indeed, involving a timeless entity, identification of the mind to a given point of view is definitely not a temporal process. It is metaphorically appealing to fancy it as a beam of light (of heavenly, non-disturbing light, of course!) illuminating a particular point of view, whereas the others remain in the darkness. But this is again misleading. No *thing* can give any distinct character to a point of view Mind has adopted, since every relevant *thing* is already included in the spatio-temporal description afforded by the wave function. In fact, Mind is the only entity concerned by 'identification'.

The strangest point is that, in this conception, nothing prevents Mind from identifying itself to this and that point of view, here and there, before and after. Mind has no spatio-temporal location, and it cannot be aware of any 'change' occurring while it performs the odd trip we have just sketched. Indeed, if it identifies itself to a given point of view, this implies that it adopts the whole associated memory content. It can never remember a "previous" point of view, for all its available 'memories' are confined to the particular point of view it occupies "presently". Of course, the words "previous" and "presently" have no other significance than to help giving an analogical picture of 'identification'. Mind has no history, and present, past and future are meaningless for it. Having a history would mean holding *traces* of a past. But all the possible traces are included in the description of the objective world, where they appear as (figurative) points of view. The only history Mind can have pertains to the point of view it identifies itself to.

More generally, the "direction of time" is a concept which can be shown to rely on the structure of memory in any (figurative) point of view, including information about other points of view[13]. Mind has nothing to do with that.

The difference between actualisation and identification is thus considerable. Actualisation is an irreversible (physical) process, by which one of the possible (figurative) points of view is selected, or given a distinctive status, so that any further description of possible points of view is dramatically affected by it. 'Identification' has no bearing with physics at all, it does not modify the becoming of the set of possible (figurative) points of view. Again, it is pure cognition. 'Identification' provides a connection between the objective world unfolding in space-time, and the placeless and timeless Mind. But this connection is necessarily very peculiar.

For instance, it would be absurd to state that 'identification' (of Mind) with a given point of view has occurred, say, at ten o'clock in the morning. For temporal location pertains to the objective world. It would be no less absurd to infer from this impossibility that 'identification' *never* occurs, since cognition would then prove impossible. 'Identification' occurs *now*. A 'now' which, provided it is not given a restricted token-reflexive meaning, cannot be ascribed any proper location in time[14], even though it bears relations (past and future) with locations in time. To summarize, in Schrödinger's words, "(...)it has a particular time-table, namely Mind is always now."[15].

An interpretation of Everett's theory of measurement based on 'identification' occurring 'now', apparently gives some credit to Bell's contention that Everett only offers an account for correlations between *present* memories. However, I think Bell is wrong when he opposes Everett's attempt at associating "each particular branch at any past time in a tree-like structure" to his emphasis on "memory contents as the important thing"[16]. There is actually no contradiction between the two concepts: Everett's branched scheme is a representation of the 'real' objective world unfolding in time, whereas emphasis on the memory contents can be construed as expressing the point of contact between the objective world and cognition, through 'identification' occurring now.

Similarly, Bell dislikes Everett's theory, because he thinks that "Everett's replacement of the past by memories is a radical solipsism (...)"[16]. By radical solipsism, Bell here means that not only, in Everett's theory, "I" am alone, but also that "the present", considered as a particular time t, is alone. I think that Bell's fear is not justified. I should rather like to show that Everett's picture provides the most decisive departure from 'radical solipsism' one may wish. For that, it is useful to recall a quantum mechanical account of time, first reported by Page and Wootters[17] and then analysed in Everettian context by Deutsch[18]. Let us consider the state vector $|\Psi\rangle$ of a 'real system' made of: an observer, its immediate surroundings and a clock. $|t\rangle$ is an eigenstate of a clock reading observable t, and $|t\rangle|a(t)\rangle$ the simultaneous eigenstates of t and of an observable A representing the rest of the system. According to Page and Wootters, it is possible in quantum mechanics to replace entirely the usual time dependence involving a variable t, by steady correlation with a clock. Let then $|\Psi\rangle$ write:

$$|\Psi\rangle = \int dt\ c(t)\ |t\rangle|a(t)\rangle$$

In the possible-point-of-view interpretation of Everett's formalism, identification of mind with a particular a(t) takes place 'now', but since 'now' does not belong to time, 'identification' does not prevent Mind to identify itself 'now' to any other a(t'). The only firm statement of the theory is its account of the internal consistency of each (figurative) point of view, including any particular reading of the watch. If the difference between the timeless 'now' and a particular clock time t is fully appreciated, it then becomes clear that 'radical solipsism' does not pertain to Everett's interpretation.

5. Privileged points of view.

Bell's thorough criticism of Everett's interpretation also bears on another point. According to Bell, Everett kept something of the Copenhagen ideas, by making assumptions about the "complexity" of the observer's body or measuring apparatus. After all, a complete objective physical description of "reality" should avoid any reference to particular *extra-theoretical*, properties of observer's bodies. According to Bell, it should describe these bodies as a set of particles, interacting with other particles. Now, if any bias in describing the observer's body is thus avoided, some important characteristics of Everett's picture disappears. In particular, at the microscopical level, there are as much coalescences as there are splittings of branches, and one is brought back to Feynman's path integral concept. Bell is certainly right on this point. A theory of "real" objects should be able to encompass indifferently any kind of object, without particular stress on the "complexity" of some of them. But to understand Everett's bias, one has to consider again the relation between the theory of "real" objects and cognition.

This can easily be done through the notion of 'identification'. Let us confront Mind with a "real system" made of only two particles whose states are correlated, as a result of an interaction having taken place between them. Nothing precludes that Mind identifies itself to one of the points of view offered by this simple system. But this point of view does not offer to mind more than one item of memory. Moreover, this memory is likely to be lost through coalescence of the splitted branches. At any rate, it does not afford any possibility for Mind to put into practice the set of epistemological criteria which define objective knowledge, such as reproducibility of the experiment, and intersubjective agreement. Testing reproducibility requires at the very least two items of memory to enable *comparison* between the result of the first measurement and the results of subsequent measurements. It requires also intellectual capacity to evaluate similarities or differences. Testing intersubjective agreement needs capacity to communicate by *signs*. Identification of Mind to 'simple' points of view is thus not impossible, but in such a 'process', Mind has not gained anything like knowledge. It is only from points of view pertaining to sufficiently complex systems that self-conscious objective knowledge is reached by the Mind.

To summarize, Everett's postulate of complexity is not needed in the possible-points-of-view interpretation of the relative state formalism. One only need a selection criterion analogous to the anthropic principle. The points of view pertaining to sufficiently complex systems (human bodies) are the only ones to be considered, because only from them can Mind be aware of objective knowledge, theories, and the like. In simple points of view, the questions we have just dealt with do not even arise.

6. Indeterminism, choice and context.

A theory can rightly be said deterministic if two conditions are fulfilled. First, the theory must afford a set of laws allowing one to predict *univocally* the state of any object at time $t+\delta t$, from the state of the same object at time t. Along with this criterion, a version of quantum theory which only retains continuous development in time of the state vector according to the Schrödinger equation is obviously deterministic.

The second criterion is that the theory should allow one to predict univocally the result of an experiment, given the specification of the measuring apparatus, of the individual object on which the measurement bears, and of their initial *state*. Quantum mechanics is usually considered to be indeterministic according to this criterion.

The divergence between the two criteria is at first sight surprising. If any experiment could be given a complete description within the framework of the continuous approach of quantum mechanics, determinism according to the second criterion would be a mere consequence of determinism according to the first criterion. One is thus compelled to aknowledge that *something* in the experiments is out of reach of pure continuous wave mechanics. Many attempts at disclosing the nature of this foreign "something" have been performed in the past. They go from the usual idea according to which the object is disturbed by interaction with the measuring apparatus in a way which is impossible to control, to the direct intervention of the subconscious (See Pauli's ideas[19]). None of these attempts have proved fully satisfactory. For instance, the idea of a "disturbance", even though intellectually appealing, amounts to suppose that there was a property before the measurement and that it has been modified by the physical interaction with the apparatus. But nothing precludes that this interaction be accounted for within the framework of wave mechanics, and, when this is done, the system object+apparatus appears as a whole, in which properties no more pertain to such and such part of it. In this situation, an important advance towards identification of the above-mentioned "something" could be afforded by an interpretation of quantum mechanics submitted to a proper version of the principle of objectivation, such as the possible-points-of-view interpretation of Everett's formalism. Indeed, by following this kind of interpretation, I shall show that the said "something" is not only out of reach of a particular physical theory such as wave mechanics; it will rather be found identical to that which is necessarily excluded from *any* act of knowledge. As a consequence, it will arise more clearly than ever that indeterminism must not be looked for within the framework of the quantum formalism itself. This indeterminism is more likely to be the sign that the interpretation of quantum mechanics cannot avoid dealing with a very far-reaching epistemic boundary.

To begin with, I would like to make the provocative statement that, if the second criterion of determinism is adequately generalized in order to match the possible-points-of-view interpretation of the relative state formalism, then quantum theory turns out to be intrinsically just as completely deterministic as classical mechanics is. Let us then proceed to this generalization. A theory can be said to be deterministic according to the second criterion, provided it allows one to state a set of propositions of the following kind:

D1: "If you choose the experimental set-up E and the initial condition I, then you will certainly obtain the result R" (The concept of choice is not absolutely necessary, but it is interesting to make use of it in a first approach, because it helps revealing the relevant questions. Statements dispensing with the word "choice" will be discussed later.)

Now, using the spatial metaphor which proved so useful, the deterministic statement D1 can be changed into:

D2: "If you choose to adopt the point of view P, then you will certainly see the aspect A of the object you are looking at.".

Here, specification of P is equivalent to that of both E and I in D1. As for A, it is but the spatial transposition of R. This form D2 of the deterministic statement can easily be transposed to the case a rigorous version of the principle of objectivation is adopted. For this, we only need to substitute "Mind" for "you" (conceived as a "person"), "figurative point of view" for "spatial or experimental point of view", "identify itself" for "adopt", "is aware" for "sees", and "result R' " for "aspect A". The sentence then becomes:

S: "If Mind chooses to identify itself with the (figurative) point of view P', then it will certainly be aware of the result R' "

Now, let us suppose more precisely that P' specifies a particular memory content correlated to one eigenstate among a given set of eigenstates corresponding to some observable. We are then brought back to the possible-points-of-view interpretations of the relative state formalism. It thus arises that, in such an interpretation, the linear superposition which represents the state of an object in the restricted sense correlated with an apparatus and an observer's body, just corresponds to a set of statements S. This being granted, quantum theory appears to be fully deterministic, in an almost trivial fashion: if Mind chooses to identify itself with the (figurative) point of view (or memory content) which is partly defined by its containing the result R', it will obviously be aware of R'.

One must nevertheless realize that sentence S requires that a basic condition be fulfilled: "If Mind chooses...". But what can it mean for Mind to "choose" something? To answer this question, we must first examine the concept of choice in the usual situation where *you* perform it. You (person) are aware that you have made a choice, if you can compare the final outcome of your process of decision making, and the subsequent action, when the latter has been performed. Knowing the action alone is not enough, since this action can be the result of any other process than "choice". Remembering the process of decision making is thus a crucial requirement to give meaning to the concept of choice. However, it is not necessarily *your own* memory which is involved. If, for instance, you happen to be amnesic, a friend of yours, who was aware of your process of decision making, can remind it to you after you have performed the corresponding action. Your past decision can after all be made part of the objective world, just to the same extent as your action, and as such it can be *known* and remembered by somebody else.

But Mind has no memory of its own. It has no memory which does not pertain to one of the points of view it may "choose" to identify to. It thus cannot "remember" any process of decision logically prior to this identification. More generally, no entity can possibly know Mind's process of decision making, if any. Mind, in the narrow acception we have adopted in this paper, is pure *knowing*. It is a pure pre-requisite for knowledge. It does not pertain to the domain of the *known*. Choice has thus definitely no meaning for Mind, since Mind cannot even be aware of having decided anything. Then, what about statement S, which involves comparison of a decision taken by the mind and the result R'? We could say that in a sense, it expresses a determinism that nobody can know, and especially not Mind itself.

The previous reasoning relies heavily on discussing the concept of choice. But it is by no means obvious that this concept cannot be dispensed with in the usual definition of experimental determinism. After all, the measuring apparatus could have been prepared through purely external circumstances, which owe nothing to any observer's volitions. Let us then examine another version of sentence D2, wherefrom any reference to "choice" have been removed:

D'2: "If you happen to adopt the point of view P, then you will certainly see aspect A"

The sentence may be analysed in the following way: the aspect A obtains on the background of a double context. A context which, at least in principle, can be modified (point of view P) and a quasi-invariant context ("You", your body, your personal identity, your memories etc...). For you can change your point of view without ceasing to be "you". This invariant context is crucial to give access to the full meaning of the sentence D'2. Indeed, it is only by remaining "you" that *you* can *compare* several outcomes and points of view, and consider P as a *particular* point of view, which *you* could leave. This is true even if you do not personnally leave the points of view P, and rather get information about aspects A_i from points of view P_i, by asking other observers or machines which have adopted P_i. It is this possibility you have of comparing directly or indirectly the aspects you get from many points of view, while remaining the invariant context on which the comparison may be performed, that enables you to test the deterministic content of sentences D'2 on *several* tokens. And this is necessary, for the fact that a *single* token of D'2 is true could after all be ascribed to pure chance. (The concept of choice, as used in the previous version of the reasoning, had no other role than its involving very intuitively such a possibility of comparison. If, before your decision, you were able to perform a choice, this meant implicitly that you could have taken another decision than the one you actually took.)

Let us now transform sentence S in the same way sentence D2 have been transformed into D'2:

S': "If Mind identifies itself with the (figurative) point of view P', it is certainly aware of the result R' ".

The fundamental difference between D'2 and S' is the following: there is no invariant context involved in S'. Mind can by no means adopt a point of view P', then a point of view P", and remember having been *itself* in P' when it is in P". Nor can Mind get indirectly the information concerning P" when it has adopted P': to whom could it ask this information? Mind, in the stronger acception of the word *identifies itself* with a particular point of view. It strictly adheres to a (figurative) point of view without any possibility to stand back.

Thus, there is no way by which Mind can test the deterministic content of a set of sentences S'. For Mind, the situation described by P', to which it has identified itself, is contingent and unique. Changing point of view, say to P", would only yield pure adhesion to it. P" would then be considered as contingent and unique as P' was.

To sum up, the incapacity of quantum mechanics to predict with certainty the result of an individual experiment does not result from an inbuilt indeterminism of the theory. It rather arises from the fact that the underlying determination cannot, even in principle, be mastered. It cannot be mastered by the only knower which may be left outside the quantum account of an experiment, namely Mind (in the severely restrictive sense we have progressively defined).

7. Fuzzy mind.

The possible-point-of view interpretation of quantum mechanics that I have advocated in the present paper bears some strong structural similarities with Squires' interpretation[20]. There are also major differences.

To deal with the fact that, in quantum mechanics, the outcome of a measuring interaction is a superposition of eigenstates, whereas only one of them is actually observed, E. Squires assumes that universal "consciousness" or "Mind" selects freely *one* of the terms of the superposition. But, by enabling the popular fuzzy components of the concept of "mind" to intrude into the analysis, the author is led to predict paranormal effects. To begin with, according to him, particular observers can consult the universal mind about other observers' states of "mind": this is "quantum telepathy". Secondly the "universal mind" can perform specific choices instead of "random selections" of eigenstates: this yields "quantum psychokinesis". My own paper can be construed as an attempt at reaching enough rigour in the definition of "mind" for any spurious justification of paranormal effects to become pointless. In the restrictive acception of the word "Mind" I used, namely that of pure *knowing*, or pure pre-requisite for knowledge, there is nothing to be *known* in Mind. Mind has no memory content of its own. Thus, in particular, no "observer" (or "point of view", in my own terms) can consult Squires' "universal consciousness" or "Mind" directly, if he wants to know another observer's memory content. No ground is therefore left for Squires' "quantum telepathy". For similar reasons, I noticed that the concept of choice is merely meaningless for Mind. Its "choices", if any, are bound to be blind. There is thus no way by which Mind could perform a *specific choice* among the available terms of the superposition of states. Hence, the absence of any ground for Squires' "quantum psychokinesis".

From a philosophical standpoint, Squires' conclusions follow from a very straightforward version of dualism. My view is rather related to a softened version of dualism, which I shall call "functional dualism". In such a doctrine, the notion of "knowing" is required in order to take into account the unavoidable incompleteness of any description of the known. The "knowing" has accordingly no other content than that of a mere name for incompleteness. It has no content, but it still retains a *function*: it is the "*to whom* " a result appear, or the "*by whom* " a physical structure of some brain becomes *experience*. At this point, it is worth discussing briefly another recent reading of Everett's interpretation, which is remarkably close to the one I am advocating although it rules out any version of dualism. According to M. Lockwood[21], the interaction which carries the observer's brain into a superposition of eigenstates yields a correlative differenciation of the streams of consciousness (or biographies) of the said observer. Along with this interpretation, Lockwood faces two main difficulties, that he fully acknowledges. The first difficulty is related to the self-referential aspects which undermines any identity theory. It is eschewed by a formal move towards dualism. The author of "Mind, Brain and the Quantum" calls such a move a "disclosure theory", whereby "awareness" is thought as "disclosing certain attributes of states of, or events within, our brains". The second difficulty is quite usual when dealing with Everett-like interpretations of quantum mechanics: why should the biography where I find myself privileged in any respect? This difficulty embodies the same kind of confusion, Lockwood argues, than the one which we face when asking "why is it now *now*?". I fully agree with the idea of conflating the two problems, but my solution to them is very different. Whereas, according to Lockwood, *now* is any time I am able to experience as present, I consider that *now* does not even belong to time[14]. According to Merleau-Ponty, *now* can be conceived metaphorically as a kind of point of view upon time. Similarly, I *defined* mind in such a limiting way that it does not belong to the physical world (it can be considered metaphorically as a kind of "point of view of points of view", even though some aspects of this metaphor are misleading, as discussed previously). Mind therefore lacks the continuity of a biography: it can only *experience* it, by identification to the corresponding brain state.

8. Conclusion.

The outstanding importance of quantum theory for modern thought is likely to lie in its having led us to make coïncide the particular limit of a physical description with the general limit of objective knowledge. Quantum theory has also put such a strain on the concept of reality, that it has only left a choice between abandoning this concept or attempting a radical generalization of it. The possible-points-of view interpretation of Everett's relative state formalism lies right at the intersection between these two generalizations. It relies on an extended version of the perspectival definition of "real objects", and also on a limiting concept of the observer, namely the pure knowing subject.

9. References

1. B. Russell, *The relations of sense-data to physics* (1914), in: *The collected papers of B. Russell*, (G. Allen and Unwin, London, 1986) Vol. 8.
2. E. Schrödinger, *Nature and the greeks*, (Cambridge University Press, Cambridge, 1954).
3. P. Heelan, *Space perception and the philosophy of science*, (University of California Press, Berkeley, 1983). See also M. Merleau-Ponty, *Phénoménologie de la perception*, (Gallimard, Paris, 1945)
4. E. Schrödinger, *Science, theory and man*, (Dover, New York, 1957).
5. B. d'Espagnat, *Phys. rep.* 110 (1982) 201.
6. P. Heelan, *Synthese*, 21 (1970) 2.
7. P. Heelan, *Found. Phys.* 1 (1970) 95.
8. G. Birkhoff and J. Von Neumann, *Ann. Math.* 37 (1936) 823.
9. M. Davis, *Int. J. theor. Phys.* 16 (1977) 867.
10. R. Healey, *Noûs*, 18 (1984) 591.
11. S. Watanabe, *Suppl. Prog. Theor. Phys.*, 37-38 (1966) 350.
12. E. Schrödinger, *Eranos Jahrbuch*, 14 (1946) 491.
13. M. Bitbol, *Philosophy of science*, 55 (1988) 349.
14. M. Bitbol, *Contextos*, VI/11 (1988) 7.
15. E. Schrödinger *Mind and mattèr* (Cambridge University Press, Cambridge, 1983), p. 145.
16. J. Bell, *Speakable and unspeakable in quantum mechanics*, (Cambridge University Press, Cambridge, 1988).
17. D.N. Page and W. K. Wootters, *Phys. Rev.* D.27 (1983) 2887.
18. D. Deutsch, in: *Quantum concepts in time and space*, eds R. Penrose and I.J. Isham, (Oxford University Press, Oxford, 1986).
19. K.V. Laurikainen, *Beyond the atom*, (Springer-Verlag, Berlin, 1988).
20. E. J. Squires, *Found. Phys. Lett.* 1 (1988) 13.
21. M. Lockwood, *Mind, Brain and the Quantum*, (Blackwell, Oxford, 1989)

Macroscopic Quantum Systems and the Objectification Problem

PAUL BUSCH
Institute for Theoretical Physics, University of Cologne
D-5000 Cologne 41, Germany

ABSTRACT

The quantum mechanical measurement problem is given a precise formulation as the problem of objectification. A systematic survey of the possible approaches towards its solution shows the present status of the quantum theory of measurement. A new proposal – unsharp objectification – is sketched which is based on the existence of quasi-classical states for macroscopic quantum systems.

1. Introduction.

The measurement process has been considered problematic ever since the early days of quantum mechanics. In spite of a great variety of proposals, up to now there is no generally accepted solution to the objectification problem, that is, the question of how a nonobjective physical quantity may give rise to definite measurement outcomes. This situation has led to serious doubts concerning the universal validity of Hilbert space quantum mechanics and to the consideration of more general theories. However, before accepting such a far-reaching conclusion, one should carefully investigate the limitations of the present theory. This is the task of a review on the quantum theory of measurement being under preparation.[1] The present modest contribution aims at a survey of parts of this work as well as an elaboration of some details.

We first try to develop a rigorous formulation of the objectification problem (Section 2). This enables us to give a systematic overview of the possible directions towards its solution and to briefly review and comment several recent approaches in light of the resulting scheme (Section 3). Special emphasis is put on the need for classical pointer observables and the relevance of the macroscopic nature of measuring devices. Conclusions are drawn in the final Section 4.

2. Nonobjectivity, Measurement, Objectification.

The phenomenon of the nonobjectivity of quantum mechanical observables raises the question about the meaning of the term measurement. The subsequent determination of the notion of measurement entails the *objectification requirement*, by which we are led to a precise formulation of the *objectification problem*.

Let S be a physical system with associated Hilbert space $\mathcal{H}$. The states of S are represented by and identified with the positive trace class operators of trace one, the state operators W. The observables of S are generally represented as positive operator valued measures on some Boolean σ-algebra of subsets of a set Ω. We shall restrict our attention to an ordinary discrete observable, represented by a selfadjoint operator A with spectral representation $A = \sum_{k \in K} a_k E_k$. Here the $a_k \in \Re$ are the different eigenvalues of A, with corresponding complete family of spectral projections E_k. Let E denote the spectral measure of A which is defined on the Borel algebra $(\Re, \mathcal{B}(\Re))$. The pair (E, W) defines a probability measure E_W such that $E_W(\{a_k\}) = tr[WE_k]$ is the probability for the eigenvalue a_k in the state W. It is the task of the quantum theory of measurement to establish the physical meaning of these probabilities, that is, their interpretation as relative frequencies of measurement outcomes (minimal interpretation) or as relative frequencies of realizing the values a_k in the object systems after measurement (realistic interpretation).

An observable A will be called *objective* in a state W if W represents a *Gemenge* $\{(p_k, W_k) | k \in K\}$ (also known as "proper mixture") of eigenstates W_k of A with weights p_k.[1,2] (Recall that a state W represents a Gemenge if it possesses a decomposition $W = \sum p_k W_k$ which admits an ignorance interpretation: the system S is actually in one of a collection of states W_k, but it is not known in which one.) Hence, $W = \sum_{k \in K} p_k W_k$. A state W is an eigenstate of A if it satisfies the relation $E_k W = W$ for some k, or equivalently, $E_W(\{a_k\}) = 1$. In general, A will be *nonobjective* in a pure states $W = P[\varphi]$, $\varphi \in \mathcal{H}_S$, unless W is an eigenstate of A. Yet, in spite of this nonobjectivity an A-measurement is supposed to lead to a definite outcome which can be predicted in a probabilistic sense. This fact is the root of the measurement problem. Indeed, it must be noted first of all that in view of the general phenomenon of nonobjectivity it is by no means clear what kind of information a measurement could yield. We tentatively take for granted the following intuitive description: a measurement is a physical interaction between an object system S and an apparatus $\mathcal{A}$ establishing a suitable correlation and finally leading to a permanent record of some outcome. More precisely, we adopt the following explication which at the same time denotes the problems to be solved by the quantum theory of measurement.

A measurement process consists of three stages. *Stage 1*: It is generally taken for granted that it is possible to prepare the systems S and $\mathcal{A}$ independently of each other, so that at the beginning of a measuring process the compound state is of product form $W \otimes W_{\mathcal{A}}$. One may employ repeatable measurements as *filters* for preparing the state of S.[1] *Stage 2*: The interaction between S and $\mathcal{A}$ produces correlations which are suitable for an information gain.[1,3] The nature of the "suitable" correlations is described by a minimal characterization of measurement in terms of the following *calibration requirement.*

($\mathcal{C}$) *If the system S is in an eigenstate of A, then an A-measurement should lead with certainty to an outcome corresponding to the respective eigenvalue.*

Due to the linearity of the measurement coupling this condition implies an apparently stronger one, the *probability reproducibility condition:*

($\mathcal{PR}$) *The probability for an eigenvalue a_k in the state W is the same as the probability for the corresponding pointer position after the measurement interaction.*

This latter requirement furnishes the minimal interpretation of the quantum mechanical probability measures and as such the most general notion of measurement. It can be extended in an obvious way to general observables to which the calibration condition usually cannot be applied.
Stage 3: While at the end of stage 2 a situation is reached in which all pointer positions are *possible* whose associated eigenvalues have a nonzero probability, any measurement process terminates with the appearance of a definite pointer position. This observation leads to the *objectification requirement:*

($\mathcal{O}$) *After the measurement interaction the pointer observable must have become objective.*

In the quantum theory of measurement the measuring apparatus $\mathcal{A}$ is represented by a Hilbert space $\mathcal{H}_\mathcal{A}$. The measurement coupling between S and $\mathcal{A}$ is effected by means of a unitary mapping U acting on the Hilbert space $\mathcal{H}_S \otimes \mathcal{H}_\mathcal{A}$ of the compound system $S + \mathcal{A}$. If the system $\mathcal{A}$ is a measuring device for the observable A, then it must possess a *pointer observable* Π the spectrum of which shall be taken to be in one-to-one correspondence with the spectrum of A. We denote the spectral representation of Π as $\Pi = \sum_{k\in K} \pi_k P_k$, with an obvious understanding of the terms. It may be assumed without loss of generality that the initial state of $\mathcal{A}$ is a vector state, that is, $W_\mathcal{A} = P[\Phi]$. The constituents of an A-measurement can be collected into a quadruple $\mathcal{M} =< \mathcal{H}_\mathcal{A}, \Pi, \Phi, U >$. Following the terminology of Ref. 1, we call such a quadruple a *normal unitary premeasurement* of the observable A if it satisfies the probability reproducibility condition ($\mathcal{PR}$). If, in addition, the objectification has been ensured, then $\mathcal{M}$ represents an A-measurement.

By virtue of the measurement coupling U the initial state $\mathcal{W} = W \otimes P[\Phi]$ of the system $S + \mathcal{A}$ is transformed into $\mathcal{W}' = U\mathcal{W}U^+$. The calibration requirement now reads:

($\mathcal{C}$) $$tr[\mathcal{W}' I \otimes P_k] = 1 \quad whenever \quad tr[W E_k] = 1.$$

It is a straightforward task to show that this condition is equivalent to the probability reproducibility property:

$$(\mathcal{PR}) \qquad p_W(a_k) \doteq E_W(\{a_k\}) = tr[WE_k] = tr[\mathcal{W}'I\otimes P_k] \doteq p_{\mathcal{W}'}(\pi_k).$$

The objectification requirement now assumes the following precise form:

($\mathcal{O}$) *The final state $\mathcal{W}'$ of $\mathcal{S}+\mathcal{A}$ represents a Gemenge of pointer eigenstates. In other words, the evolution $\mathcal{W} \longrightarrow \mathcal{W}'$ of $\mathcal{S}+\mathcal{A}$ corresponds to a stochastic process such that the initial state $\mathcal{W}$ is transformed into some state $\mathcal{W}_k$ satisfying $p_{\mathcal{W}_k}(\pi_k) = tr[\mathcal{W}_k I\otimes P_k] = 1$, with probability $p_k \doteq p_W(a_k) = p_{\mathcal{W}'}(\pi_k)$. Thus the resulting Gemenge is $\{(\mathcal{W}_k, p_k)|k \in K\}$.*

If this situation would be actually reached then *pointer objectification* would have been achieved. If, furthermore, one manages the measurement to be *repeatable*, then pointer objectification entails also *value objectification.* In fact, an A-measurement is repeatable if the states $\mathcal{W}_k$ satisfy $p_{\mathcal{W}_k}(a_k) = 1$. In that case the measurement forces the observable A to become objective. With this one would have established the realistic interpretation of the quantum mechanical probabilities.

There are no difficulties in principle for the quantum theory of measurement to explain the stages 1 and 2 of measuring processes. The hard part of the measurement problem is the objectification problem, that is, the task of ensuring on the basis of a unitary measurement coupling ($\mathcal{U}$) that the objectification – stage 3 – does indeed take place. The objectivity of Π in the state $\mathcal{W}'$ requires that this state is physically equivalent to the mixture $\mathcal{W}'' = \sum p_k \mathcal{W}_k$. In other words, these two states must be indistinguishable with respect to the observables of the compound system $\mathcal{S}+\mathcal{A}$. Assuming that $\mathcal{S}$ is a pure quantum system with no superselection rules ($\mathcal{QS}$), it then follows that the apparatus system $\mathcal{A}$ possesses an *effective pointer observable* $\tilde{\Pi} = \sum \pi_k \tilde{P}_k$, $\tilde{P}_k \leq P_k$, which necessarily is of a classical nature ($\mathcal{CP}$) in the sense described in the Appendix. In this way we obtain a theorem stating a first aspect of the measurement problem:

$$\text{(MP1)} \qquad (\mathcal{O})\ \&\ (\mathcal{QS})\ \&\ (\mathcal{U}) \implies (\mathcal{CP}).$$

The difficulty with this is, of course, the question how to *explain* the appearance of superselection rules without introducing an ad hoc restriction of the set of observables of macroscopic measuring devices.

A closely related aspect of the objectification problem is the incompatibility of the usual Hamiltonian description of the dynamics and the objectification. In fact, if a measurement is to be achieved by means of a unitary measurement coupling ($\mathcal{U}$), and if this dynamics is assumed to be generated by a Hamiltonian *observable* H (assumption ($\mathcal{HO}$)), then the requirement of a classical pointer observable ($\mathcal{CP}$) entails a severe violation of the probability reproducibility condition ($\mathcal{PR}$):[1,4]

$$\text{(MP2)} \qquad (\mathcal{CP})\ \&\ (\mathcal{U})\ \&\ (\mathcal{HO}) \implies \neg(\mathcal{PR}).$$

The proof of this verdict is straightforward: assumptions $(\mathcal{U})$, $(\mathcal{HO})$ and $(\mathcal{CP})$ entail that the unitary operator U commutes with the pointer observable Π; hence one obtains:

$$p_{W'}(\pi_k) = tr\big[UW \otimes P[\Phi]U^+ \, I \otimes P_k\big] = tr\big[P[\Phi]P_k\big] \neq p_W(a_k),$$

in contradiction to Equation $(\mathcal{PR})$ Statements (MP1) and (MP2) describe the logical situation encountered with the measurement problem. The very condition $(\mathcal{CP})$ necessary for the realization $(\mathcal{O})$ of a measurement precludes the realization of the condition $(\mathcal{PR})$ defining the term measurement. In other words, the concept of measurement excludes its realizability within ordinary quantum mechanics. Hence, to ensure the basic constituents of the notion of measurement, possibly on the basis of unitary dynamics, forces one into the strange idea of not considering the Hamiltonian as an observable of $\mathcal{A}$. This result may be taken as an indication that the conception of a unitary dynamics may be too narrow in order to deal with the irreversibility inherent in measurement.

3. Some Attempted Solutions to the Measurement Problem.

To begin with, there are two extreme (and extremal) options for escaping the object ification problem as summarized in the implications (MP1) and (MP2).

(1) *"There is no nonobjectivity."* – The denial of the phenomenon of nonobjectivity is characteristic of the hidden variable theories. Accordingly, in this view quantum mechanics is regarded *incomplete*, so that the measurement problem becomes irrelevant as there is no need for a quantum theory of measurement. However, there is no hidden variable theory comparable in generality to quantum mechancis; and the known hidden variable theories are plagued by strange features forced upon them by strong no-go theorems. I therefore do not recommend this attitude towards the measurement problem as a very attractive one.

(2) *"Objectification does not take place."* – This statement could be the motto of the many worlds interpretation which has been commented by P. Mittelstaedt on this meeting.[2] It seems that this or similar attempts[1] of circumventing (MP1) and (MP2) are the only solution tenable strictly within ordinary Hilbert space quantum mechanics.

Instead of sacrificing either quantum mechanics or the goal of objectification, one may consider possibilities of explaining classical (pointer) observables, if necessary by means of modifications of the axiomatics but staying *within* quantum mechanics, or even by means of going *beyond* quantum mechanics. A third way would be to try a weakening of the objectification requirement.

(3) *Giving up* $(\mathcal{QS})$: Sometimes it has been argued that the object system $\mathcal{S}$ itself as well as the apparatus system $\mathcal{A}$ have naturally restricted sets of observables so that not all selfadjoint operators correspond to observables. Such limitations on measurability may be due to fundamental conservation laws,[5] or due to the limited number of actually existing interactions. While it seems possible to avoid

the implication (MP1), it is open whether objectification can be achieved in this way. Also, such a solution would not have the status of a theoretically well-founded approach but rest upon accidental facts as long as there is no thorough theory of the fundamental interactions. This, admittedly vague, option is mentioned here just for completeness as it appears as one logical possibility of dealing with (MP1).

(4) *Giving up* ($\mathcal{U}$): A modification of the dynamics allows one to avoid the implication (MP1), but at the same time it may give rise to *dynamically induced superselection rules*. An example of this type of approach is the *unified dynamics* theory of Ghirardi, Rimini and Weber (GRW).[6] In this theory the von Neumann–Liouville equation for microsystems is modified by adding a linear term which models a spontaneous localization process taking place at random times. The GRW equation represents a quantum dynamical semigroup with a non-selfadjoint generator, so that the irreversibility needed for measurement is built in from the outset. The mean rate of the localization process can be chosen small enough so that microscopic systems pratically evolve according to the Schrödinger equation, while the localization events become noticeable only for maroscopic systems having a large number of constituents. In this way the localization of a macroscopic system is an observable which is *dynamically objectified* practically at any instant of time. Consequently, in the measurement context system $\mathcal{A}$ is a proper quantum system itself, without any need of an ad hoc introduction of classical pointer observables; on the contrary, postulate ($\mathcal{CP}$), and thus also ($\mathcal{O}$), is ensured in the sense of a dynamical restriction of the set of states of $\mathcal{A}$. It should be kept in mind that the GRW localization process corresponds to an *unsharp* position observable, so that the induced Gemenge is a continuous family of nonorthogonal and only unsharply localized states. It is an open question whether on the basis of the GRW dynamics one encounters a consistency problem of the form (MP2). Finally, while the theory provides an interesting step towards a unified quantum mechanical description of microscopic and macroscopic systems, it still would be desirable to find a fundamental theoretical basis for its new dynamical principle.

A related, but more general way of giving up ($\mathcal{U}$) is the introduction of nonlinear stochastic modifications of the Schrödinger equation. This again could be utilized to achieve dynamical objectification by means of taking literally measurements (as well as other quantum mechanical phenomena) as stochastic processes.[7] It should be noted that the recourse to nonlinearity in quantum mechanics bears somewhat dangerous consequences such as the possible existence of superluminal signals.[8]

Both of the above approaches – spontaneous localization as well as the more general stochastic nonlinear Schrödinger dynamics – are intended to regard the generalized dynamics as an autonomous process pertaining to isolated systems. On the other hand one may argue that due to the macroscopic nature of the apparatus $\mathcal{A}$ its interaction with the environment cannot be neglected but rather must be made responsible for the emergence of classical features and irreversibility as required for measurement. Incorporating the environment $\mathcal{E}$ into the description and reducing the unitary ($\mathcal{U}$), Hamiltonian ($\mathcal{HO}$) dynamics of $\mathcal{S}+\mathcal{A}+\mathcal{E}$ to that of $\mathcal{S}+\mathcal{A}$ leads, after certain approximations of the thermodynamic limit type (e.g., continuum limit), to Markovian Master equations similar to the GRW equation. The environment may be part of the apparatus system as in the Daneri–Loinger–Prosperi theory,[9,10] or a separate entity external to $\mathcal{S}+\mathcal{A}$, hopefully guaranteeing the emergence of classical properties[11] via environment–induced superselection rules.[12] A macroscopically large number of constituents of the environment is required for obtaining both the quasi-diagonality of the reduced $\mathcal{S}+\mathcal{A}$ state with respect to the desired pointer basis, as well as large Poincaré recurrence times. A suitable interaction between $\mathcal{A}$ and $\mathcal{E}$ ensures the robustness of the pointer states due to a continuous nondemolition-type monitoring of the pointer observable through the environment.[12]

It must be emphasized that these approaches do not reach the goal of objectification ($\mathcal{O}$) since the effective environment-induced superselection rules are obtained by voluntarily neglecting the – *practically* unobservable – degrees of freedom of the environment. Indeed, reading (MP1) and (MP2) as statements about the total system $\mathcal{S}+\mathcal{A}+\mathcal{E}$, these obstacles remain perfectly valid. Ignoring the environment then amounts to an apparent abandonning of the postulate ($\mathcal{U}$) of unitary dynamics. Nevertheless, the results obtained, particularly the explanation of a preferred pointer basis, provide an improved understanding of the robustness of macroscopic properties. As such they are important contributions to any approach to the measurement problem. The same remarks apply to many of the more recent measurement models (some of which can be found in these proceedings) aiming at an explanation of the (approximate) vanishing of nondiagonal terms of the final state of the apparatus

(5) *Taking seriously classical properties*: While superselection rules can be easily incorporated into the Hilbert space framework of quantum mechanics, they nevertheless represent an element which does not follow from the standard axiomatics. Thus the problem is to provide a convincing theoretical explanation for such restrictions of the sets of observables and states. The only known mechanism of producing superselection rules seems to be spontaneous symmetry breaking, which was discovered in theories for systems having infinitely many degrees of freedom. This amounts to going beyond ordinary Hilbert space quantum mechanics and adopting a more general theory. Having thus coped with implication (MP1), the goals of objectification and probability reproducibility still require to escape the dynamical consistency problem (MP2). In other words, the irreversibility inherent in measurement still demands a generalized notion of dynamics. Considering irreversibility as a spontaneous breaking of the fundamental time inversion symmetry, one is led again to invoking the theory of infinite systems. An eloquent argumentation in this spirit in favour of *algebraic quantum mechanics* has been given by H. Primas on this conference.[13] Some representative contributions along these lines, and mostly written within the framework of the C^*-algebra approach, are the papers of Hepp[14], Whitten-Wolfe and Emch[15], Machida and Namiki[16,17], Lockhart and Misra[18], and Primas[13].

The algebraic approach provides a powerful mathematical framework allowing one to study in a rigorous way the idealization of systems having infinitely many degrees of freedom. However, its contribution to the measurement problem does not yet allow a conclusive estimation since, due to enormous technical difficulties, the elaboration of measurement models still is in a rather preliminary stage. The necessity of going beyond quantum mechanics wa s also emphasized for similar reasons by other groups, such as the Geneva (Jauch, Piron[19]) or Marburg (Ludwig[20]) schools. We conclude that insisting upon objectification (involving classical observables and irreversibility) forces one – presumably with logical necessity – to consider theories more general than ordinary Hilbert space quantum mechanics. Yet, it would be desirable to have this result expressed as a formal no-go theorem.

Some simple hints towards the feasibility of this aim can be given at once. Assume the pointer observable Π (Section 2) of $\mathcal{A}$ to be a nondegenerate discrete quantity. Then (MP1) entails that Π itself is a classical quantity, so that $\mathcal{A}$ must be a discrete classical system. It follows that a realistic description of measuring devices has to be based on a Hilbert space $\mathcal{H}_{\mathcal{A}}$ which must not be "too small". In fact, the incorporation of continuous kinematical symmetries pertaining to the classical pointer "position" seems to require continuous superselection rules.

To summarize, the objectification problem seems to force one to make a decision between two options: either one tries to maintain the universal validity of ordinary quantum mechanics with no changes; then objectification seems impossible. Or one takes for granted that objectification actually occurs, then a more or less radical modification of the theory seems inevitable. Contrary to the usual textbook wisdom it has been argued as a result of careful analysis that macrophysical theories, like thermodynamics or classical mechanics, cannot be derived as approximate limiting cases of quantum mechanics. Such theories should rather be regarded as theories of their own right.[20] Yet, as macroscopic systems are aggregates of microscopic systems, it is important to understand the interrelation between quantum mechanics and macroscopic theories. This program has been carried out to quite a large extent by the Marburg[20] and Milano[21] groups in the form of a consistent embedding of macrotheories into a quantum mechanical many-body theory. The effect of such an embedding can be described, roughly speaking, as a restriction of the sets of states and observables of the many-body quantum theory so that the remaining structure allows for the emergence of a classical (deterministic) behaviour of the macroscopic observables. In the light of these investigations I now come to sketch a last possibility of dealing with the objectification problem based on a weakening of the objectification requirement ($\mathcal{O}$). With this option we have completed our discussion of the relevance of all four components of the implication (MP1). I must leave open the question whether this approach will ultimately save the universality of quantum mechanics, possibly without being merely a variant of the many-worlds option (2), or whether it is just an illustration of the limited validity of quantum mechanics.

(6) *Unsharp objectification?* This program is motivated by the following observations. In spite of the claim that classical structures cannot be *derived* within

quantum mechanics, it has become evident in a number of examples that quantum mechanics does possess a *quasi–classical* level of description. For instance, quantum mechanical phase space measurements behave like classical trajectory determinations provided the involved position and momentum unsharpnesses are macroscopically large.[22] Hence, classical trajectories can be approximately realized in terms of unsharp quantum observables. Further, the so-called AHA (et al) effect[23] displays a classical rotation behavior of quantum mechanical spin observables in the course of a sequence of weak measurements. This classical feature has been shown to be a consequence of an Einstein–Podolsky–Rosen type conspiracy between the involved spin N system and the weak measurement device.[24] Due to the huge unsharpness of the weak measurement the entanglement between object and apparatus is not completely destroyed but, on the contrary, guarantees the appearance of a surprising quantum effect: the occurrence of apparently forbidden readings lying far beyond the spectrum of the observables measured. The large number of constituents of the spin N system ensures that noncommuting spin components simultaneously have fairly well-defined values. Taking into account that such a weak measurement defines an unsharp observable, one finds a natural explanation of the miracle.[24] One lesson of the AHA effect is that the emergence of classical features may even be considered as a macroscopic quantum effect. As a third example it may be recalled that Bell's inequalities are satisfied *within quantum mechanics* if *coexistent* triples of spin observables are involved in the Einstein–Podolsky–Rosen experiment.[25,26]

The common point of these examples is the decisive role ascribed to unsharp observables. The measurement unsh arpness involved must not be confused with a mere inaccuracy; on the contrary, it is rooted in a genuine quantum indeterminacy inherent in the measuring device.[22,27,28] An important quasi–classical feature of unsharp quantum observables is the existence of *coexistent* (jointly measurable) sets of noncommuting observables. This joint measurability is due to a sufficiently large unsharpness and thus goes along with a blurring of possible interference effects. In this way a quasi–classical behavior of macroscopic observables emerges in a rather natural way without any ad hoc selection of a set of ordinary observables as the macroscopic observables. Indeed, I would propose to understand macroscopic observables as unsharp observables with a macroscopic degree of intrinsic unsharpness.[22] The magnitude of this unsharpness, which may be interpreted as inducing a kind of *natural coarse-graining*, fixes a scale on which the macroscopic observable assumes fairly well-defined values. In this context another observation is of great importance: the existence of quasi–classical states. In a model of a quantum mechanical amplifier Glauber[29] describes a very peculiar type of "macroscopic" oscillator (resp. photon field) states which have the property of being *quasi–diagonal* with respect to all kinds of observables: position, momentum, energy, or phase–space observables. The term quasi–diagonality refers to the exponential decay of nondiagonal elements with increasing separation from diagonal terms. It is plausible that a measurement of macroscopically unsharp observables performed on such states does not allow one to detect interference effects. Hence, the mentioned noncommuting observables may be assumed to be unsharply objective since this assumption presumably cannot be disproved.

Applying these ideas to the measurement context leads to the following scenario. Presumably any realistic pointer observable is an intrinsically and macroscopically unsharp observable, hence a macroscopic observable in the sense defined above. It may turn out that the interaction of the measuring device with the environment forces the apparatus to remain in some quasi–classical states in which, in particular, the pointer observable is quasi–diagonal. Then it becomes impossible to "see" (in the sense of macroscopic observation) the apparatus in a superposition of macroscopically distinguishable states. The "objectification" reached in this way is an essentially unsharp one and certainly does not require classical properties in the sense of superselection rules.

The description of macroscopic observables as macroscopically unsharp observables is not only plausible but has been shown to be a necessary consistency condition for the embedding of macroscopic deterministic theories into quantum mechanics (cf. Ref. 20, Chapter X.2.5). An interesting contribution which seems to fit with the spirit of our unsharp objectification proposal is the model of Khalfin and Tsirelson.[26] In this paper the quasi–classical behaviour of a macroscopic quantum system $\mathcal{A}$ is shown to result from a weak interaction with its environment $\mathcal{E}$. Under certain conditions which are typically satisfied for a large class of macrosystems the reduced dynamics of $\mathcal{A}$ is approximately given by a Markov process consisting of Gaussian (in a certain sense) operations (i.e., state transformations). I would expect that a closer analysis of their model shows that the resulting $\mathcal{A}$ states are quasi–classical states. It is interesting to note that other approaches to the measurement problem and to the question of macroscopic quantum systems make use of unsharp observables in an essential way.[6,21]

4. Conclusion.

In summary, the problem of measurement in quantum mechanics is, first of all, the question about the meaning of the term measurement. In view of the phenomenon of nonobjectivity the calibration requirement gives rise to an understanding of measurement as an interaction process leading first to probability reproducibility (via correlations) and then terminating irreversibly with pointer objectification. Objectification, in turn, requires classical pointer properties (MP1), but then it cannot be reconciled with probability reproducibility on the basis of Hamiltonian dynamics (MP2). This situation – the difficulty of explaining the presence of superselection rules and irreversibility within quantum mechanics – forces one to make a decision between the following alternatives from which the reader may make his choice.

If the universal validity of quantum mechanics is maintained, then at best unsharp objectification can be achieved, and (even macroscopic, classical) reality is fundamentally unsharp. On the other hand, insisting in strict objectification, quantum mechanics indeed needs to be generalized, either by means of introducing a modified dynamics (dynamical objectification) or by allowing for systems with infinitely many degrees of freedom entailing the required superselection rules in a fairly natural way. In view of the present state of affairs I would put some hope on the former option – unsharp objectification. Anyway, if it would turn out not to reach its goal, its failure would contribute to demonstrating the inescapability of the latter approach. To conclude, the discussion of the present work leads to the decision tree displayed in Table 1.

Table 1: Decision Tree: Interpretations of quantum mechanics and approaches to the objectification problem.

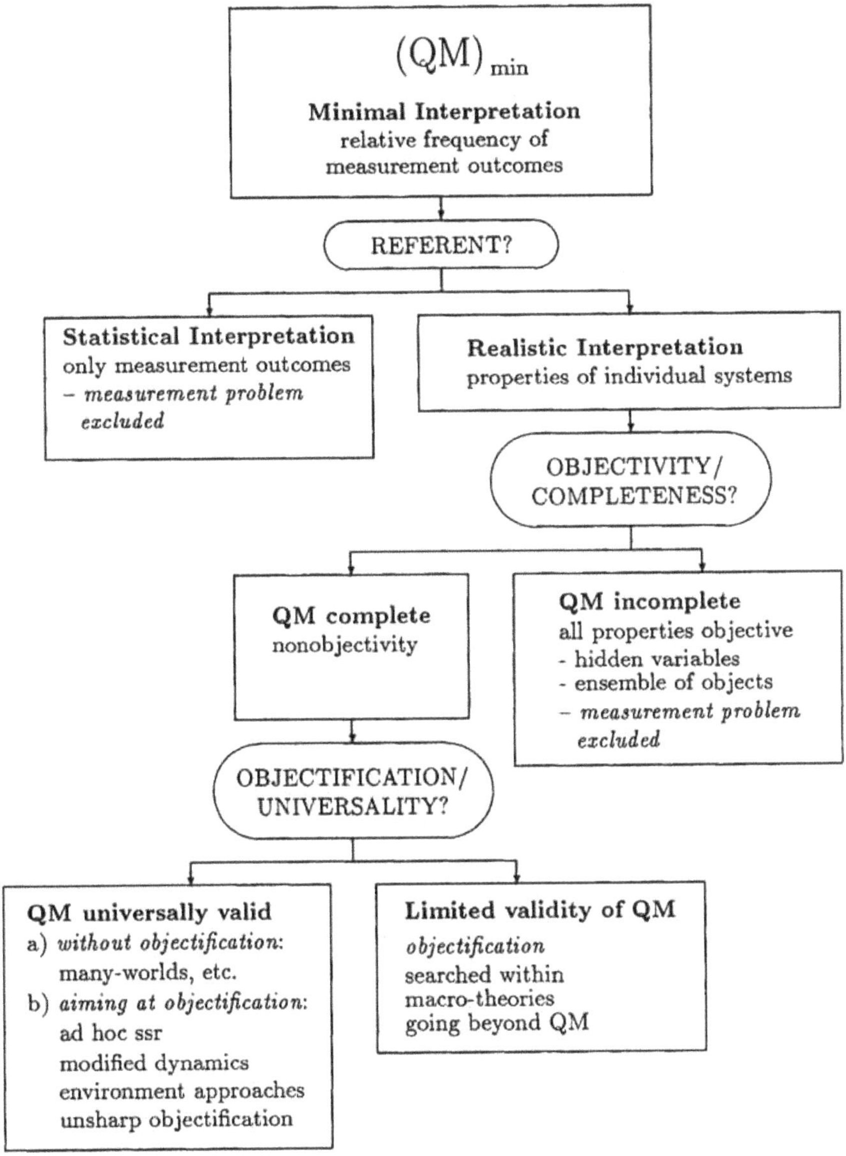

Appendix: Proof of (MP1).

The objectification requirement ($\mathcal{O}$) stipulates the equivalence of the states $\mathcal{W}' = UW \otimes P[\Phi]U^+$ and $\mathcal{W}'' = \sum pw(a_k)\mathcal{W}_k$ for all states $W = P[\varphi]$, $\varphi \in \mathcal{H}$, where $\mathcal{W}_k = I \otimes P_k \mathcal{W}' I \otimes P_k \cdot (\mathrm{tr}[\mathcal{W}' I \otimes P_k])^{-1}$ whenever $\mathrm{tr}[\mathcal{W}' I \otimes P_k] = pw(a_k) \neq 0$. Thus, the following equation must hold for all vectors $\varphi \in \mathcal{H}$, all observables B of $\mathcal{S}$ and all observables C of $\mathcal{A}$:

(A1) $$(U\varphi \otimes \Phi, B \otimes CU\varphi \otimes \Phi) = \sum_k (U\varphi \otimes \Phi, B \otimes P_k C P_k U\varphi \otimes \Phi)$$

Let Φ_{kj} be a complete orthonormal system of eigenvectors of Π such that $P_k \Phi_{kj} = \Phi_{kj}$. The unitary mapping U induces a family of bounded linear operators A_{kj} such that

$$U\varphi \otimes \Phi = \sum A_{kj}\varphi \otimes \Phi_{kj}.$$

Unitarity ($\mathcal{U}$) of U yields

(A2) $$\sum_{kj} A_{kj}^+ A_{kj} = I,$$

while the probability reproducibility is equivalent to the following:

(A3) $$\sum_j A_{kj}^+ A_{kj} = E_k \quad \text{for all } k.$$

Condition (A3) is equivalent to (A2) & (A4), where

(A4) $$A_{kj} E_l = \delta_{kl} A_{kj} \quad \text{for all } k, j, l.$$

Let U be such that it satisfies (A3). If $\mathcal{S}$ is a proper quantum system ($\mathcal{QS}$), then all bounded selfadjoint operators represent observables of $\mathcal{S}$. Now Equation (A1), if read as a condition selecting the observables C of $\mathcal{A}$, is equivalent to the following:

(A5) $$\left(\sum_i (\xi, A_{ki}\varphi)\Phi_{ki}, C \sum_j (\eta, A_{lj}\varphi)\Phi_{lj}\right) = 0 \quad \text{for all } \varphi, \xi, \eta \in \mathcal{H},$$

whenever $k \neq l$. By virtue of (A3) there exists, for each value of k, at least one value $i = i_k$ such that $A_{ki_k} \neq O$. Then (A5) implies that $(\Phi_{ki_k}, C\Phi_{lj_l}) = 0$ whenever $k \neq l$. It follows that there exists a family of projection operators $\tilde{P}_k$ such that $0 \neq \tilde{P}_k \leq P_k$ and $\tilde{P}_k C \tilde{P}_l = 0$ for all observables C of $\mathcal{A}$ whenever $k \neq l$. These projection operators determine the subspaces of $\mathcal{H}_\mathcal{A}$ which contain the (supports of the) possible component states of the reduced states of $\mathcal{A}$ after the measurement interaction.

Let $\tilde{P} = \sum \tilde{P}_k$. One concludes that in the subspace $\tilde{\mathcal{H}}_{\mathcal{A}} = \tilde{P}\mathcal{H}_{\mathcal{A}}$ of *accessible pointer states* the *effective pointer observable* $\tilde{\Pi} = \sum \pi_k \tilde{P}_k$ induces a superselection rule: only those operators C can be observables of $\mathcal{A}$ which satisfy the commutation relations $P_k(\tilde{P}C\tilde{P}) = (\tilde{P}C\tilde{P})P_k$ for all k.

References.

1. P. Busch, P. Lahti, P. Mittelstaedt, The Problem of Measurement in Quantum Mechanics, 1990, in preparation.
2. P. Mittelstaedt, The Objectification in the Measuring Process and the Many Worlds Interpretation, 1990, these Proceedings.
3. P. Lahti, Entropy and Measurement in Quantum Theory of Measurement, 1990, these Proceedings.
4. E. G. Beltrametti, G. Cassinelli, P. J. Lahti, Unitary Measurements of Discrete Quantities in Quantum Mechanics, *J. Math. Phys.* **31** (1990) 91–98.
5. E.P. Wigner, Die Messung quantenmechanischer Operatoren, *Z. Physik* **133** (1952) 101–108.
6. G.C. Ghirardi, A. Rimini, T. Weber, Unified Dynamics for Microscopic and Macroscopic Systems, *Phys. Rev.* **D34** (1986) 470–491.
7. N. Gisin, Quantum Measurements and Stochastic Processes, *Phys. Rev. Lett.* **52** (1984) 1657–1660.
8. N. Gisin, Weinberg's Non-linear Quantum Mechanics and Superluminal Communication, *Phys. Lett.* **A143** (1990) 1-2.
9. G.M. Prosperi, Macroscopic Physics and the Problem of Measurement in Quantum Mechanics, in *Foundations of Quantum Mechanics*, ed. B. d'Espagnat, (Academic Press, New York, 1971).
10. F. Haake, D.F. Walls, Overdamped and Amplifying Meters in the Quantum Theory of Measurement, *Phys. Rev.* **A36** (1987) 730–739.
11. E. Joos, H.D. Zeh, The Emergence of Classical Properties Through Interactions with the Environment, *Z. Physik* **B59** (1985) 223–243.
12. W.H. Zurek, Environment-induced Superselection Rules, *Phys. Rev.* **D26** (1982) 1862–1880.
13. H. Primas, Necessary and Sufficient Conditions for an Individual Description of the Measurement Process, 1990, these Proceedings.
14. K. Hepp, Quantum Theory of Measurement and Macroscopic Observables, *Helv. Phys. Acta* **45** (1972) 237–248.
15. B. Whitten-Wolfe, G.G. Emch, A Mechanical Quantum Measuring Process, *Helv. Phys. Acta* **49** (1976) 45–55.
16. S. Machida, M. Namiki, Theory of Measurement – A Mechanism of Reduction of Wave Packet.I and II, *Prog. Theor. Phys.* **63** (1980) 1457–1473 and 1833–1847.

17. M. NAMIKI, I. OHBA, Many-Hilbert-Spaces Theory of Quantum Measurements, in *Symposium on the Foundations of Modern Physics 1987*, eds. P. Lahti,P. Mittelstaedt, (World Scientific, Singapore, 1987).
18. C.M. LOCKHART, B. MISRA, Irreversibility and Measurement in Quantum Mechanics, *Physica* **136A** (1986) 47–76.
19. C. PIRON, Quantum Mechanics: Fifty Years Later, in *Symposium on the Foundations of Modern Physics 1985*, eds. P. Lahti, P. Mittelstaedt, (World Scientific, Singapore, 1985).
20. G. LUDWIG, *An Axiomatic Basis for Quantum Mechanics, Vols. 1 and 2*, (Springer, Berlin, 1985 and 1987).
21. A. BARCHIELLI, L. LANZ, G.M. PROSPERI, A Model for the Macroscopic Description and Continual Observation in Quantum Mechanics, *Nuovo Cim.* **72B** (1982) 79–121.
22. P. BUSCH, Indeterminacy Relations and Simultaneous Measurements in Quantum Theory, *Int. J. Theor. Phys.* **24** (1985) 63–92.
23. Y. AHARONOV, D.Z. ALBERT, A. CASHER, L. VAIDMAN, Surprising Quantum Effect, *Phys. Lett.* **A124** (1987) 199–203.
24. P. BUSCH, Surprising Features of Unsharp Quantum Measurements, *Phys. Lett.* **A130** (1988) 323–329.
25. P. BUSCH, Elements of Unsharp Reality in the EPR Experiment, in *Symposium on the Foundations of Modern Physics 1985*, eds. P. Lahti, P. Mittelstaedt, (World Scientific, Singapore, 1985).
26. L.A. KHALFIN, B.S. TSIRELSON, A Quantitative Criterion of the Applicability of the Classical Description within the Quantum Theory, in *Symposium on the Foundations of Modern Physics 1987*, eds. P. Lahti, P. Mittelstaedt, (World Scientific, Singapore, 1987).
27. M. GRABOWSKI, Quantum Measurement Scheme and new Examples of Generalized Observables, 1990, these Proceedings.
28. F.E. SCHROECK, JR., Unsharpness in Measurement Yields Informational Completeness, 1990, these Proceedings.
29. R.J. GLAUBER, Amplifiers, Attenuators and the Quantum Theory of Measurement, in *Frontiers in Quantum Optics*, eds. E.R. Pike, S. Sarkar, (Adam Hilger, Bristol, 1986).

NONINVASIVE MEASURABILITY, NEGATIVE-RESULT MEASUREMENTS, AND WATCHED-POTS: ANOTHER LOOK AT LEGGETT'S ARGUMENTS FOR THE INCOMPATIBILITY BETWEEN MACRO-REALISM AND QUANTUM MECHANICS

ROBERT K. CLIFTON

Dept. of History and Philosophy of Science, Cambridge University, Free School Lane, Cambridge UK CB2 3RH

ABSTRACT

A clarification is given of Leggett's proof that the macrorealist is committed to a Bell-type inequality (violated by quantum mechanical predictions) for a *single* SQUID system undergoing quantum flux tunnelling. The clarification helps to identify the flaw in Home's recent argument that the most controversial assumption in the derivation, NonInvasive Measurability, is unnecessary. The macrorealist's commitment to NonInvasive Measurability is then re-examined in light of two of Leggett's recent arguments in its favour. These are found to make tacit assumptions of reductionism and noncontextuality which are violated, for example, by de Broglie-Bohm's 'causal interpretation' of quantum mechanics. Finally, using a novel argument based on the 'chained' Bell inequality, the quantum watched-pot effect is deduced and shown to be related to, but independent of, the failure of NonInvasive Measurability in quantum mechanics for superposed states. Nevertheless, this effect poses no further problems for the macro-realist beyond those already engendered by Leggett's proof and subsequent arguments. And it is concluded that these fall short of excluding macrorealism *per se*. At best, they put an interesting constraint, albeit physically counterintuitive, on the *macro*-realist's ability to explain the failure of NonInvasive Measurability for *timelike*-separated measurements on *single* systems; much like Bell's theorem puts a nonlocality constraint on the *micro*-realist's explanation of quantum correlations for *spacelike*-separated measurements on *multiple* systems.

1. The Structure of Leggett's Proof

Leggett[1] has proposed a proof that the predictions of quantum mechanics (QM) pertaining to rf SQUID rings are incompatible with the two assumptions of: Macroscopic Realism (MR) - a macroscopic system with two or more macroscopically distinct states available to it will at all times *be* in one or the other of these states; and NonInvasive Measurability (NIM) at the macro-level - it is possible, in principle, to determine the state of a *macroscopic* system with arbitrarily small perturbation on its subsequent dynamics. In this first section, I aim to clarify the rôle played by the various assumptions required in Leggett's proof, paralleling a similar clarification of the Stapp-Eberhard proof of nonlocality given by Redhead[2]. In the next section, I pinpoint why Leggett's NIM assumption is contentious; and then criticize Home's[3] recent argument that NIM can be dropped, leaving a proof that MR (alone) conflicts with QM.

1.1 *Stage 1*

Following the idealizations adopted in refs. 1, and 3-6, consider an ensemble of nondissipative rf SQUID rings each totally decoupled from its environment and able to pass from a 'clockwise' to an 'anti-clockwise' flux state by tunnelling. These two states (which are 'macroscopically distinct' because the associated current flow in either direction is ≈ a few μA) can be characterized by a variable Q(t) which takes on the value +1 (clockwise) or -1 (anti-clockwise) after a measurement of the current direction at time t. For the correlation between the measured values of Q(t) at two different instants t_i and t_f (> t_i) *with no intervening measurements*, QM predicts that $<Q(t_i)Q(t_f)> = \cos[\omega(t_f - t_i)]$ (where ω = the characteristic resonance frequency of the SQUIDs). MR attributes to each system a *possessed* value for the observable Q(t) independent of whether this value is actually measured. I shall label such values using the square bracket notation $[Q(t_j)]_n^{t_k}$ which denotes the nth system's possessed value (=±1) corresponding to the observable $Q(t_j)$, *given* that a flux measurement on the system has been (in the case $t_k < t_j$), or will be (in the case $t_k > t_j$), undertaken at t_k. When no measurement at t_k is undertaken, no superscript shall appear. For measured values, I henceforth use the same notation but drop the square brackets.

To prove a Bell-type inequality for this SQUID ensemble, define:

$$\gamma_n \equiv [Q(t_1)]_n^{t_2}[Q(t_2)]_n^{t_1}+[Q(t_2)]_n^{t_3}[Q(t_3)]_n^{t_2}+[Q(t_3)]_n^{t_4}[Q(t_4)]_n^{t_3} - [Q(t_1)]_n^{t_4}[Q(t_4)]_n^{t_1}$$

where $t_4 > t_3 > t_2 > t_1$. Each factor in γ_n will be assigned a determinate value in any theory satisfying MR, although I shall assume, following Leggett[1], that for each n (i.e. for each member of the ensemble) Q(t) is only actually *measured* at either t_1 and t_2, or t_2 and t_3, or t_3 and t_4, or t_1 and t_4. Henceforth, these will be denoted as measurements m(1,2), m(2,3), m(3,4), and m(1,4), respectively. Now assuming the following four pairs of equations, which I shall collectively call the 'Matching Condition', hold ∀n:

$$\text{(i) } [Q(t_1)]_n^{t_2} = [Q(t_1)]_n = [Q(t_1)]_n^{t_4} \quad \text{(ii) } [Q(t_2)]_n^{t_1} = [Q(t_2)]_n = [Q(t_2)]_n^{t_3}$$

$$\text{(iii) } [Q(t_3)]_n^{t_2} = [Q(t_3)]_n = [Q(t_3)]_n^{t_4} \quad \text{(iv) } [Q(t_4)]_n^{t_1} = [Q(t_4)]_n = [Q(t_4)]_n^{t_3}$$

it is easy to verify that $\gamma_n = \pm 2$, ∀n. Given this, correlation coefficients for the various possessed values can be computed (in the usual way) from γ_n, with the result that these coefficients must satisfy the Bell-type inequality:

$$| <[Q(t_1)][Q(t_2)]>+<[Q(t_2)][Q(t_3)]>+<[Q(t_3)][Q(t_4)]>-<[Q(t_1)][Q(t_4)]> | \le 2 \quad (1.1)$$

Two further assumptions are needed in order to use this inequality to generate a contradiction with QM's predictions, since those predictions apply only to measured, not possessed, values for Q(t). The first is an assumption of 'Faithful Measurement' - For each i,f and all n, if m(i,f) is undertaken, then the values $Q(t_i)_n^{t_f}$ and $Q(t_f)_n^{t_i}$ revealed by the measurement will, respectively, be $[Q(t_i)]_n^{t_f}$ and $[Q(t_f)]_n^{t_i}$. An additional 'Random Selection' assumption is also needed to identify the actually measured correlation coefficients with those computed from possessed values in deriving the above inequality, since only a subset of the total ensemble is subjected to each of m(1,2) through m(1,4). Random Selection amounts to assuming that the sequences of possessed values used to compute each of the four correlations in (1.1) are random in the Von Mises - Wald - Church sense, i.e. their limiting relative frequencies are insensitive to computable place selections (cf. refs. 2,7, and 8 for further elaboration); and that the choice of which ensemble member to subject to which measurement out of m(1,2) through m(1,4) is governed by some effectively computable rule. With these assumptions we can regard the 4 subensembles subjected to each measurement type as being a statisfically faithful sample of the whole ensemble, and so $<[Q(t_i)][Q(t_f)]>_{(possessed)} = <Q(t_i)Q(t_f)>_{(measured)}$ for each i, f in (1.1), which then violates QM's predictions; for instance when $t_4-t_3 = t_3-t_2 = t_2-t_1 = \pi/(4\omega)$. So the argument thus far establishes that:

Stage 1 MR & Matching Condition & Faithful Measurement & Random Selection
$\Rightarrow$ Bell-type Inequality $\Rightarrow$ ~QM

1.2 *Stage 2*

In Leggett's[1] treatment NIM (alone) is portrayed as implying the Matching Condition. However in the context of the Stapp-Eberhard nonlocality proof, where polarizer settings play the rôle of the various measurement times here and locality helps to justify the Matching Condition, it has been cogently argued that determinism, in the sense of deterministic evolution of possessed values in time, is also required for this condition's justification. I shall only briefly show how the same argument applies to it's justification in the Leggett proof (cf. refs. 2,7,9, and 10 for more elaborate presentations of this argument in the Stapp-Eberhard context).

Clearly for each n, either the flux *is* or *is not* actually measured at t_1 (and similarly for t_2-t_4). So each of the Matching Condition's eight equations, which have the form $[Q(t_j)]_n^{t_k} = [Q(t_j)]_n$, must (assuming a measurement at t_k is actually performed on the nth SQUID) be interpreted as a counterfactual conditional of the form: '*If* a measurement at t_k had *not* been performed on the nth system, *then* the possessed value of Q at t_j *would* have been the same as it actually was'. But this assertion, with its claim that a particular value *would* have been possessed by Q at t_j (viz. the value Q *actually* possessed at t_j), conflicts with the possibility that these values might evolve stochastically. For such evolution means (by definition!) that either +1 or -1 *could* be possessed by Q at any given moment irrespective of whether or not any other measurements earlier (or, for that matter, later) than that moment are undertaken on the system; in short, it means the Matching Condition's equations *need not* hold if determinism fails. Thus we must add an assumption of determinism so that it becomes true that either +1 or -1 *would* be Q's possessed value at t_j, as necessitated by the state(s) of the nth SQUID prior to t_j. Only then can each of the Matching Condition's counterfactuals be implied by the condition that possessed values at t_j be causally independent of earlier (and, of course, later) measurements undertaken on the system. And to make this implication good, we must also invoke an additional 'No Conspiracy' assumption, viz. that the experimenter's n^{th} choice of measurement out of m(1,2) through m(1,4) and the nth SQUID state, itself, are not jointly determined by a common cause (just as in the Stapp-Eberhard proof the possibility of a 'backwards light cone conspiracy' between any measurement in one wing of an EPR-type experiment and the particles' evolution in the other must be ruled out).

In sum, for the second stage of Leggett's proof we have:

Stage 2 Determinism & NIM & No Backwards Causation & No Conspiracy
$\Rightarrow$ Matching Condition

(It is not clear whether Leggett, himself, would agree with the need for determinism in these assumptions. But elsewhere[11] he clearly distinguishes MR from the Matching Condition's counterfactuals, and appears to recognize that, even assuming the 3 'no influence' assumptions above, neither necessarily implies the other without extra assumptions.) Stages 1 and 2 combine to give:

Leggett's Proof
MR & NIM & No Backwards Causation & No Conspiracy & Determinism
& Faithful Measurement & Random Selection $\Rightarrow$ Bell-type Inequality $\Rightarrow$ ~QM

2. Home's Proposal

2.1 *Motivation*

Home aims to resolve a recent debate[4,5] concerning the legitimacy of assuming NIM. The problem is that, quite generically, QM violates 'StatNIM' - which asserts that the *statistics* for a SQUID's measured values are causally independent of earlier measurements. This violation occurs, not just because measurements collapse the state vector (for the projection postulate is controversial, and strictly not part of the theory), but because earlier measurements on superposed states destroy later interference effects.

To see this in general, let an ensemble of quantum systems each be prepared at t in the composite 'system + apparatuses' state:

$$\Sigma_i B_i(t)|q_i\rangle|A^1{}_j(t)\rangle|A^2{}_k(t)\rangle \tag{2.1}$$

where $\{|q_i\rangle\}$ are eigenstates of the measured observable Q (which we assume, for simplicity, is maximal), while $|A^1{}_j(t)\rangle$ and $|A^2{}_k(t)\rangle$) are the initial states of each apparatus, the first used to measure each system at t, and the second at t'>t. In the absence of system-apparatus interactions between t and t', the following evolutions occur:

$$\forall i,\ |q_i\rangle \rightarrow \Sigma_l\, S_{il}(t' - t)|q_l\rangle\ ;$$
$$\forall j,\ |A^1{}_j(t)\rangle \rightarrow |A^1{}_j(t')\rangle\ ;\ \forall k,\ |A^2{}_k(t)\rangle \rightarrow |A^2{}_k(t')\rangle \tag{2.2}$$

where the S_{il} are unspecified time-dependent coefficients. (For our idealized 'two-state' SQUIDs, these would be $\cos(\omega(t' - t)/2)$ and $-i\sin(\omega(t' - t)/2)$, if $|q_i\rangle = |Q(t)=+1\rangle$, etc.) At any time t where a system-apparatus interaction takes place we assume:

$$\forall i,j\ |q_i\rangle|A^1{}_j(t)\rangle \rightarrow (\Sigma_l\, I^1{}_{f(i,j),l}|q_l\rangle)|A^1{}_i(t)\rangle\ ;$$
$$\forall i,k\ |q_i\rangle|A^2{}_k(t)\rangle \rightarrow (\Sigma_l\, I^2{}_{g(i,k),l}|q_l\rangle)|A^2{}_i(t)\rangle \tag{2.3}$$

where f and g are integer functions which allow the possibility of disturbing measurements when $I^1{}_{f(i,j),l}$ and $I^2{}_{g(i,k),l} \neq \delta_{i,l}$. (Note, however, that both interactions in (2.3) still qualify as measurements since they preserve a 1:1 correspondence between final apparatus state and initial object state.) I have intentionally denoted the superposition coefficients the system 'begins' with at t by 'B', the 'spontaneous' isolated evolution coefficients by 'S', and the two measurement 'induced' evolution coefficients by 'I', in order to keep track of their various contributions to the probabilities computed below, and hence trace the origin of QM's violation of StatNIM.

Consider now 2 cases: (I) The joint system freely evolves from t until t' where it interacts with apparatus **2**; and (II) The same as in I, but each system first interacts with apparatus **1** at t. For case I, eqns. (2.2) and (2.3) imply that just after t' the joint system (2.1) will be in the state:

$$|\Psi_I(t')\rangle = \Sigma_{ilm}\, B_i(t)S_{il}(t'-t)I^2{}_{g(l,k),m}\, |q_m\rangle|A^1{}_j(t')\rangle|A^2{}_l(t')\rangle \tag{2.4}$$

whence: $\quad \mathrm{Prob}\{Q(t') = q_a\} = \Sigma_{uv}\, |\langle q_u|\langle A^1{}_v(t')|\langle A^2{}_a(t')|\, |\Psi_I(t')\rangle|^2$

$$= \Sigma_i\, |B_i(t)|^2|S_{ia}(t'-t)|^2 + \Sigma_{i,j\neq i}\, B_i(t)B_j(t)^* S_{ia}(t'-t)S_{ja}(t'-t)^* \tag{2.5}$$

But in case II, state (2.1) evolves to:

$$|\Psi_{II}(t')\rangle = \Sigma_{ilmn}\, B_i(t)I^1{}_{f(i,j),l}S_{lm}(t'-t)I^2{}_{g(m,k),n}\, |q_n\rangle|A^1{}_i(t')\rangle|A^2{}_m(t')\rangle \tag{2.6}$$

just after t', whence: $\mathrm{Prob}\{Q(t') = q_a\} = \Sigma_{uv}\, |\langle q_u|\langle A^1{}_v(t')|\langle A^2{}_a(t')|\, |\Psi_{II}(t')\rangle|^2$

$$= \Sigma_i\, |B_i(t)|^2|S_{ia}(t'-t)|^2 + \text{ terms} \neq 0 \textit{ only if } \text{measurement } \mathbf{1} \text{ is disturbing} \tag{2.7}$$

Comparing (2.5) with (2.7) shows that probabilities for results at t' depend on whether or not a prior t-measurement has been undertaken, even if that measurement was 'nondisturbing'! The crucial difference is in the second nonclassical interference term in (2.5) which is nonzero only if the system begins at t in a nontrivial superposition of states, rather than an eigenstate (in which case $B_i(t) = \delta_{is}$ for some s, and that interference term goes to zero).

Now it is easy to see how QM's violation of StatNIM, in particular its violation for SQUIDS, calls NIM into question. Suppose that in the actual world, two successive measurements at t and t' are undertaken on each member of an ensemble of SQUIDS. As we saw in deriving the Matching Condition in section **1.2**, determinism and NIM (plus the other two 'no influence' assumptions) makes true the counterfactual: 'Had no first measurement been undertaken on each SQUID, *then* prior to each second measurement the SQUID's possessed value *would* have been the same as it actually was'. Assuming Faithful Measurement, this means that under the hypothetical state of affairs one is led to imagine by this counterfactual, the sequence of t'-measurement results *would* have been the same as in the actual world, and hence the statistics would be the same, making StatNIM true. So using an even simpler argument than Leggett's we already have:

Simple Proof MR & NIM & No Backwards Causation & No Conspiracy

& Determinism & Faithful Measurement & Random Selection ⇒ StatNIM ⇒ ~QM

(Here Random Selection is again required. For to empirically check whether StatNIM is *in fact* violated, as QM predicts, the ensemble must be randomly partitioned into two subensembles - those SQUIDs on which t and t'-measurements are made, and those only subjected to the second measurement - and their statistics must be compared.) The above proof exposes an easy escape route for the macro-realist. Her most plausible response would be to deny the macrorealist analogue of StatNIM, viz. NIM! And in contrast to the Stapp-Eberhard proof (where 'locality' plays the role of NIM), it would not be physically problematic to deny NIM from the point of view of relativity (or any other physical theory). Hence the debate surrounding NIM.

It is worth noting the simplicity of the above proof as against Leggett's, which goes via a Bell-type inequality and uses a more complicated set of measurements. The motivation behind Leggett's choice of proof is his desire to generate 'the analog of what has been done in the EPR context: namely, to show that *irrespective of the interpretational framework* the experiments exclude a clearly formulated conjunction of "common-sense" hypotheses'[11]. However the simple proof operates just like Leggett's in not depending on QM's formalism or its interpretation, and in only invoking its statistical predictions at the final stage for comparison with macro-realist theories. So Leggett uses a sledgehammer to crack a nut. But one should not think that section **1**'s clarification of his proof was itself ill-motivated. For it shall now provide a framework from within which Home's proposal to avoid using NIM, which involves modifying the conceptual basis for Leggett's proof, can be analyzed and its essential difficulty clearly seen. After exposing this difficulty, I shall go on in section **3** to give a constructive response to the debate over NIM, with particular reference to two counterarguments of Leggett which sharpen the issues. There, I shall return to the simple proof, since neither of his counterarguments turn on using his more complex one.

2.2 *The Proposal*

Home[3] claims that "by appropriately formulating the notion of macro-realism" it is possible to derive a Bell-type inequality without invoking NIM. In fact, Home's formulation of MR simply includes within it determinism of possessed values (which the macro-realist, as usually understood, is not necessarily committed to). Rather, it is the procedure with which Home implements a condition analogous to the Matching Condition that is novel in his proof. Although Home is correct in claiming that his procedure successfully avoids the need for assuming NIM (and the other 'no influence' assumptions), I shall show that this procedure, itself, builds into his proof an inconsistency with the Random Selection assumption, thus forcing his proof that MR conflicts with QM to be invalid. But first I shall present and clarify Home's argument.

Each member of Home's ensemble involves 3 quantum mechanically identical SQUID rings A, B, and C initially prepared in the same superposition of '+1' and '-1' flux states. The first stage of Home's argument assumes that rings A - C are "chosen such that the initial state of B is the same as the final state of A; ...and the initial state of C is the same as the final state of B", i.e. they are *chosen* to satisfy the equations $Q_A(t_2)_n{}^{t_1} = Q_B(t_2)_n{}^{t_3}$ and $Q_B(t_3)_n{}^{t_2} = Q_C(t_3)_n{}^{t_4}$ which are analogous to parts (ii) and (iii) of the Matching Condition but pertain to different SQUID rings and measured, rather than possessed, values. Home is not totally explicit about these being *measured* values, but it follows from his further assertion that the choices can be checked "operationally" without additional assumptions - and clearly the only way to carry this out is via measurements.

In the second stage of his argument, Home appeals to what he (rather misleadingly) calls a "corollary" to MR (which, for him, includes determinism) that asserts:

Corollary - "it is always possible to envisage a macroscopic system which will evolve from a given initial state to a preassigned final state"

He then adds a fourth SQUID ring, D, to each '3-SQUID' member of his ensemble whose possessed values satisfy the equations $[Q_D(t_1)]_n{}^{t_4} = Q_A(t_1)_n{}^{t_2}$ and $[Q_D(t_4)]_n{}^{t_1} = Q_C(t_4)_n{}^{t_3}$. These equations are analogous to the rest of the Matching Condition, i.e. parts (i) and (iv), and constitute "the key input due to MR" in the argument, according to Home. As in the first stage, Home fails to explicitly state that D's *possessed* values are involved in these equations; but if not, it would be hard to see what rôle MR has to play in the argument. (In any case, we shall see below that MR has very little rôle to play anyway.)

Home's four equations (two from each stage of his argument) define a 'matching' condition, call it MC_H, which (assuming Faithful Measurement for the terms which involve possessed values) implies that an ensemble of sets of rings, like the chosen rings A - D, will have measured statistics satisfying a Bell-type inequality (cf. section **1.1**'s derivation from γ_n). On this basis, Home claims to generate a contradiction solely between QM and MR/determinism (which by the second stage of his argument, appear to commit one to a Bell-type inequality).

2.3 *Critique*

Home's corollary requires, more precisely, that:

Precise Corollary - $\forall$ pairs of states $(Q_A(t_1)_n{}^{t_2}, Q_C(t_4)_n{}^{t_3})$, $\exists$ a SQUID system, D, which can evolve in time from the first to the second state

Now if determinism obtains such that each possible initial state, +1 or -1, plus (possibly) the measurement of that state (since Home's proof is supposed to allow NIM to fail) uniquely determines a final state of either +1 or -1, at most two of the four possible pairs of states $(Q_A(t_1)_n{}^{t_2}, Q_C(t_4)_n{}^{t_3})$ will correspond to possible evolutions of a fourth D ring. However, this conflicts with Home's need to assume that *any* one of these four pairs corresponds to a possible evolution. So either Home must: (1) Drop his assumption of determinism and replace it with indeterminism, the assumption that all possible final states can be reached (though perhaps with small probability) from any given initial state; or (2) Assume that there is some other causally relevant factor, say F, of a SQUID's state *other than* the direction of circulation of its flux-induced current (e.g. possibly F = de Broglie-Bohm's 'quantum potential'[12], or some other less elusive physical factor) which contributes to determining the flux direction at later times so that a given direction at one time can still evolve into either +1 or -1 at any later time depending upon the value of F. Clearly Home would want to endorse (2) since his conception of macro-realism includes determinism. And endorsing (2) would not weaken his proof. For the existence of at least one additional causally relevant factor, F, is *already required* by a deterministic macrorealist theory if it is to be able to explain the fact that SQUIDs which start out at t_1 in, say, a flux '+1' eigenstate (and so, have a Q-possessed value of +1) can, according to QM, evolve into a state which predicts nonzero probabilities for obtaining both +1 and -1 as outcomes of a later flux measurement. (As yet, we do not know how such factors must operate to be consistent with the results of negative-result measurements, etc. - but cf. section 3!)

Now it is should be clear how Home avoids introducing assumptions like NIM or the other 'no influence' assumptions. In Leggett's proof, the need for these arises because one is considering, for a given SQUID system, what possessed value would have obtained had no earlier (or later) measurement been undertaken on that system. So Leggett's Matching Condition relates values (hypothetical to actual) on a *single* system. In contrast, Home's MC_H relates values for *different* systems, in a way justified by measurements and a corollary. Together they gaurantee that a particular choice of rings A-D can be made which satisfies a Bell-type inequality, while having no single SQUID measured more than twice so that the relevant QM predictions (specifically, $\cos[\omega(t_f - t_i)]$) can still apply.

But *do* they still apply after deliberately "choosing" A-D to satisfy MC_H? I submit that they do not, and that Home's 'derivation' of the Bell inequality simply begs the question. Take first the choice of A-C in the first stage. What about the rings that don't satisfy MC_H - why aren't they included in Home's ensemble, and why don't they figure in his statistics? Similarly with Home's second stage D ring choice. Yes, the corollary gaurantees the existence of a D ring with the required matching property. But it also gaurantees that rings will exist that won't 'match' with A and C in the required way - why aren't these included in Home's ensemble from which statistics satisfying the Bell inequality are computed?

The problem is that Home fails to (and indeed cannot) establish that the statistical predictions of QM apply to an ensemble of systems which has been deliberately constructed, in the light of its particular properties, in order to violate these predictions. For it is well-known that if we take an ensemble of systems whose member's properties (e.g. possessed or measured Q-values represented by ± 1) are recorded in a sequence (or sequences) with some limiting relative frequency (e.g. given by QM's predictions), and then select privileged members out of this sequence(s) *using the knowledge of those properties* (e.g. whether or not quadruples of systems satisfy MC_H) for inclusion in a new sequence(s), this new sequence(s) will not in general have the same limiting frequencies as the original sequence(s) with respect to the properties at issue[13]. In short, Home's particular way of selecting rings of type A - D biases the statistics of his ensemble forcing it to inaccurately reflect the statistics of the original unselected ensemble - in effect, forcing the Random Selection assumption to fail. But it is only the original unbiased ensemble, and any *randomly selected* subensembles, to which QM's predictions apply. So Home fails to provide any contradiction with QM's predictions, and hence any incompatibility between the latter and MR/determinism.

3. Leggett's Arguments for the Macro-Realist's Committment to Statistical NonInvasive Measurability

3.1 *Argument 1 - Substituting a Mixture for a Superposition*

Briefly, Leggett's first argument[1] starts by noting that QM rejects StatNIM precisely because it rejects MR; and then questions the ability of theories satisfying MR to explain StatNIM's violation in a natural/meaningful way.

The argument can be amplified along the following lines. The aim is to avoid scepticism about NIM engendered by section 2's simple proof, by establishing that StatNIM virtually follows from MR without even using NIM! Assume that it is empirically verified that StatNIM holds for any ensemble, E_n, of systems starting at t in an eigenstate $|q_n>$ of the observable measured, in agreement with QM's predictions when the first measurement interaction in case II is ideal (cf. eqns. (2.5) and (2.7)). It then follows that measurements undertaken on an ensemble composed of a mixture of subensembles $E_1,..E_n,..E_N$ starting at t in eigenstates $|q_1>,..|q_n>,..|q_N>$, respectively, will also satisfy StatNIM. And since the measured value at t is predictable with certainty in each of these subensembles, at t a definite possessed value can be said to exist, making MR true at t. But consider the case of an ensemble in a QM superposition at t. From the macrorealist perspective, this ensemble should be ontologically (not just "for all practical purposes"!) indistinguishable from the aforementioned mixture, because MR is inconsistent with superposing mutually incompatible possibilities. So for the macrorealist, StatNIM should hold, even for QM superpositions.

This argument's obvious weak point is that it assumes that observables cannot be assigned definite values in superposed states - i.e. that the only way a macrorealist can avoid superposing possibilities is to substitute mixtures for superpositions at the conceptual level. But Leggett would respond by pointing out that, if this assumption were denied, StatNIM's failure would surely be mysterious. For it would then be the case that although "the system "really has" the value (say) Q(t)=+1, it is nevertheless physically affected by the fact that the amplitude for Q(t)=-1, as determined by the quantum formalism, is nonzero - a maneuver which seems contorted in the extreme and would seem to make the first part of the statement [viz. that the system "really has" a value] virtually meaningless" (ref. 1, p. 950). To put it rhetorically: How can something nonexistent for the macrorealist, such as the other possible but unactualized value for Q, be causally *relevant* to later measurement result statistics?

Rather than viewing this argument as impugning MR *per se*, I believe there is another more reasonable way to view it. Namely, it shows that, to maintain MR, one must: (a) Ascribe physical reality (in some form or other) to the other components of the wavefunction not pertaining to the actual value possessed by the system, in order to explain their causal relevance; but, in so doing, one must (b) Stop short of allowing these other components to 'include' the other quantum mechanically possible possessed values of the system, in order to maintain consistency with MR. de Broglie-Bohm's 'causal interpretation'[12], for example, does precisely this through their idea of 'empty pilot' waves which have the potential to be causally relevant to later measurements by interfering with the 'occupied' wave associated with, for example, the actual current circulating around the SQUID loop. Leggett's argument also shows that a macrorealist cannot regard superpositions as reducible to mixtures, but must distinguish them according to whether or not these extra causally relevant wavefunction components (or something similar) are present. So whether or not they are present cannot be determined merely by looking at the properties (e.g. flux values) of the individual members of an ensemble - they must exist over and above these properties; and, in particular, be causally efficacious independently of those properties. Again, this type of 'wholism' is a well-known feature of the causal interpretation[14]. But this counterexample to Leggett's argument does not undermine its significance. For its interest lies in the fact that it shows that features analogous to those just mentioned which arise in the causal interpretation must exist in *all* MR theories if they are to be compatible with QM.

3.2 *Argument 2 - Exploiting Negative-Result Measurements*

Leggett's other argument involves making the first measurement at t in case II of the simple proof a negative-result measurement in which the measuring apparatus is set up so that an interaction with the system takes place for all but one of the eigenstates of the measured observable[15]. In that case, the fact that a change of state of the apparatus does not occur on a given occasion can be used to infer that the eigenvalue corresponding to the eigenstate which doesn't interact with the apparatus is the measurement result. Here, there would seem to be no question of the measurement disturbing the system in any 'classical' sense in order to produce a result.

Let us compute the effect on the statistics of t'-measurement results using this measurement technique. We again adopt the general approach set out in **2.1**, which makes no specific reference to SQUIDs (for a particular physically realizable negative-result setup for these, cf. ref. 16). Since there is obviously no single apparatus which can negatively-measure all the different possible results at t, we are forced to introduce N different negative-result setups, one for each possible measurement result. In general, the pth such setup will involve disturbing interactions with all other N-1 eigenstates except the pth. So a pth negative-result interaction at t will be characterized by the following replacement for the first equation in (2.3):

$$\forall i\neq p \quad |q_i>|A^1{}_0(t)> \to (\Sigma_l\, I^1{}_{f(i,0),l}|q_l>)|A^1{}_i(t)> \; ; \; |q_p>|A^1{}_0(t)> \to |q_p>|A^1{}_0(t)> \qquad (3.1)$$

(Note that we can no longer allow apparatus **1** to start at t in one of the possible eigenstates it could be left in after the first interaction, as we did in (2.3). For if we do, and it was found on some occasion that the apparatus was left undisturbed in, say, state $|A^1{}_k(t)>$, then one could not infer that the measurement result was q_p - it could have been q_k, which could have *also* fixed the final apparatus state to be $|A^1{}_k(t)>$ via a *disturbing* interaction.)

Evolving the system (2.1) (with $|A^1{}_0(t)>$ replacing $|A^1{}_j(t)>$) from t to t' for case II, but assuming a pth negative-result measurement at t, again yields $|\Psi_{II}(t')>$ in eqn. (2.6), but subject to the changes in (3.1). Using this, one can compute:

$$\text{Prob}\{Q(t) = q_p, Q(t') = q_a\} =$$

$$\Sigma_u\, |<q_u|<A^1{}_0(t')|<A^2{}_a(t')|\,|\Psi_{II}(t')>|^2 = |B_p(t)|^2|S_{pa}(t'-t)|^2 \qquad (3.2)$$

This is the probability that the systems undisturbed by the pth negative result measurement yield q_a at t'. To get the probability for q_a at t' *simpliciter*, one must sum this result over p, which yields exactly the same result as in eqn. (2.7) but, as one might expect, the terms due to the first measurement's disturbance drop out. On the other hand, QM's violation of StatNIM remains - whether or not the first measurement exploits negative-results, it still destroys the interference term in (2.5).

Leggett and Garg see this as showing that MR commits one to StatNIM for negative-result measurements, for otherwise their influence on t'-measurements would be inexplicable: "Should anyone wish to interpret the results of our proposed experiment (assumed for present purposes to agree with QM) by saying that the macroscopic object is indeed in a definite macroscopic state, but is nevertheless physically affected by an interaction which we know could have occurred only if it had been in a *different* macroscopic state, he or she is of course free to do so; we leave it to the reader to judge whether such an interpretation in any way diminishes ...the significance of our proposed experiment"[5]. Again, to put it rhetorically: How can a 'nonevent' (i.e. the lack of apparatus **1** response) precipitate an event (i.e. the alteration of subsequent propensities/statistics)?

Again, I believe, this question cannot threaten MR *per se*, but only tell us more about the features that a macrorealist theory must have to be internally consistent, and of course consistent with QM. In particular, Leggett and Garg's "only if" assertion is too strong. For one could explain the effect by noting that, whatever form the causally relevant 'empty' components of the wavefunction might take (recall their existence, or something similar, is required by Leggett's first argument in **3.1**), their causal relevance must be susceptible to being rendered impotent depending on the measurement context in which they reside. (Again, this is exactly what happens in the causal interpretation[17].) So Leggett must tacitly assume "noncontextuality" to get his argument off the ground. On the other hand, one can still agree with Leggett and Garg that this doesn't diminish the significance of the experiment on which the argument is based - for an empirical confirmation of contextuality is far from uninteresting!

4. NonInvasive Measurability and the Watched-Pot Effect

Recall from **2.1** that Leggett need not have resorted to a Bell-type argument to establish his proof's result. I now turn to the quantum 'watched-pot' effect, where a Bell inequality *does* improve the argument - specifically, the 'chained' Bell inequality. This was first discovered independently by Nelson and Warrington[18], and by Pearle[19]; and more recently rediscovered by Braunstein and Caves[20] who note the 'close relation' with the watched-pot effect, but don't make the connection explicit, as I shall now do. I shall then point out this effect's relation to QM's violation of StatNIM, and ask whether it creates any further problems for the macro-realist.

Consider a sequence of N measurements undertaken on the ith system in an ensemble of M similarly prepared quantum systems, each measurement having possible results '+1' (signifying that the state of the system lies in some given subspace, S, of Hilbert space - e.g. the subspace corresponding to 'no decay') and '-1' (signifying that the state lies somewhere in Hilbert space *other than* in S - e.g. the subspace corresponding to 'decay'). Let the sequence of measurement results which *happens* to be recorded at successive times for the ith system be $Q_i(t_1)$,... $Q_i(t_N)$. Since the square of each is +1, pure algebra dictates that:

$$Q_i(t_1)Q_i(t_2)-Q_i(t_1)Q_i(t_N) = Q_i(t_1)Q_i(t_2)[1-Q_i(t_2)Q_i(t_3)] + Q_i(t_1)Q_i(t_3)[1-Q_i(t_3)Q_i(t_4)] + ... + Q_i(t_1)Q_i(t_{N-1})[1-Q_i(t_{N-1})Q_i(t_N)] \quad (4.1)$$

and using the fact that the product of any two results is ≤ 1, (4.1) implies:

$$Q_i(t_1)Q_i(t_2)+Q_i(t_2)Q_i(t_3)+Q_i(t_3)Q_i(t_4)+ ... +Q_i(t_{N-1})Q_i(t_N)-Q_i(t_1)Q_i(t_N) \leq N - 2 \quad (4.2)$$

Summing (4.2) over M, dividing by M, and then taking the limit $M \to \infty$, yields the following chained Bell inequality which *any* theory (including QM!) must satisfy to be consistent:

$$<Q(t_1)Q_i(t_2)> + <Q(t_2)Q(t_3)> + ... + <Q(t_{N-1})Q(t_N)> - <Q(t_1)Q(t_N)> \leq N - 2 \quad (4.3)$$

Still manipulating algebra, we have from (4.3):

$$<Q(t_1)Q(t_N)> \geq (N - 1)\tau(N) - N + 2 \quad (4.4)$$

where $\tau(N) = \mathrm{Min}\{<Q(t_i)Q(t_{i+1})>, \forall i\}$, for a given N. Now physical assumptions enter the argument! Suppose that the interval $t_N - t_1$ remains constant, but $N \to \infty$. Then assuming that each measurement is ideal and instantaneous, QM predicts that $\tau(N) \to 1$

and hence, by (4.4), QM must predict that $<Q(t_1)Q(t_N)> = +1$, which is just the usual 'watched-pot' conclusion: A measure-one set of states which start out in a given state, must remain in that state after any finite interval of continuous measurement.

The main advantage of this alternative proof over standard watched-pot proofs[21] is that it is independent of QM's formalism; in particular, it does not rely on the projection postulate (dispelling some authors' concerns that the 'ensemble' interpretation of QM can avoid the effect by denying that postulate[22]). True, QM's prediction of $\tau(N) \rightarrow 1$ arises out of the peculiar way quantum states evolve (in first order approximation, having a t^2 dependence for small times). But other theories might ascribe laws of evolution that differ in detail from QM, yet still predict $\tau(N) \rightarrow 1$. The above watched-pot theorem applies to these also; and does not depend on the specific nature of how that prediction is generated from a formalism. Clearly, it also works if $\tau(N) \rightarrow 1$ has been verified experimentally, independent of any theory, by performing a large number of sufficiently nondisturbing, sufficiently instantaneous, measurements and extrapolating the results.

Interestingly, the watched-pot result also involves a violation of StatNIM, but of a different sort than discussed in **2.1**. Consider any two times t_1 and t_N, between which there are no measurements, and at which statistics for measurement results *differ* (such times always exist because the coefficients of QM superpositions fluctuate with time). The watched-pot theorem says that if, counterfactually, we had continuously measured *between* these two times as well, the statistics at the end of the interval would have been the same as they were at the beginning - which implies that they would be *different* from what they actually were. So continuous measurement also forces StatNIM to fail, but in a way that is dependent on a continuous number of previous measurements, rather than just one. Another difference is that, while StatNIM can fail in the watched-pot case even if the system starts at t in a mixture of states and is then continuously measured, it cannot if only the first t-measurement (assuming it is ideal) is performed. So these effects are clearly independent.

Does the watched-pot failure of StatNIM lead to new troubles for the macro-realist? I believe that there is no more trouble here than already generated by Leggett's arguments. For these already tell us that 'empty' wave packets can have causal relevance for later results, and lose their relevance when earlier measurements are made. So if one then postulates a law of evolution of these waves approximating QM's, or at least consistent with experiment, then it is no further surprise that such waves will be allowed less and less time to develop and be causally relevant as more and more measurements render them causally impotent.

5. Conclusion

We have seen that Leggett's proof can be given a secure foundation once all its required assumptions, including determinism, are made explicit. Leggett's result also admits of a simpler proof, which then leads naturally to questioning his proof's reliance on NIM. Unfortunately, Home's proposal fails to produce a valid proof without invoking NIM. But both of Leggett's counterarguments, rather than impugning MR *per se*, sharpen the issues by broadly showing what features of the world the macrorealist must invoke to explain QM's violation of StatNIM. Finally, the watched-pot effect was freed from reliance on the projection postulate, but shown to pose no further problems for the macro-realist.

There are various proofs in the literature that show *micro*scopic interference effects to be incompatible with micro-realism, once certain locality, reduction, and/or noncontextuality assumptions are made (cf. e.g. ref. 10, Ch. 1). Leggett's genius has been to isolate a *macro*scopic case where similar arguments can be applied, and experiments involving negligible dissipation are feasible[23] (*pace* Van Zandt and Peres[24], who take 'nondissipative' to *mean* microscopic). In this respect, Leggett has achieved his aim: To equal the generality of the Bell arguments. But in neither the Leggett nor Bell cases should we think realism *per se* is excluded. Of course, the question of whether the requirements on realism these arguments impose make it sufficiently counterintuitive to be rejected, is another issue.

Acknowledgements

I wish to thank Mr. Andrew Elby, Ms. Sara Foster, Prof. Dipanker Home, Dr. Tony Sudbury, and especially Dr. Jeremy Butterfield for helpful criticisms on various points; and both the Royal Commission for the Exhibition of 1851, and the Governing Body of Peterhouse, Cambridge for financial support.

References

1. A. Leggett, *Found. Phys.* **18** (1988) 939.
2. M. Redhead, *Incompleteness, Nonlocality, and Realism* (Oxford Univ. Press, Oxford, 1987) Ch. 4.
3. D. Home, in R. Nair (ed.) *Quantum Queries (Proceedings of the Conference on the Philosophical Foundations of Quantum Theory*, New Delhi, March 24 - 26, 1988) (World Scientific, Singapore, 1990).
4. L. Ballentine, *Phys. Rev. Lett.* **59** (1987) 1493.
5. A. Leggett and A. Garg, *Phys. Rev. Lett.* **59** (1987) 1621.
6. A. Leggett and A. Garg, *Phys. Rev. Lett.* **54** (1985) 857.
7. R. Clifton, J. Butterfield, and M. Redhead, *Brit. J. Phil. Sci.* **41** (1990) 5.
8. G. Svetlichny, M. Redhead, H. Brown, and J. Butterfield, *Phil. Sci.* **55** (1988) 387.
9. G. Hellman, *Synthese* **53** (1982) 461.
10. H. Krips, *The Metaphysics of Quantum Theory* (Clarendon Press, Oxford, 1987) Ch. 8.
11. A. Leggett, in S. Kamefuchi *et al.* (eds.) *Proc. 2nd Int. Symp. Found. of Quantum Mechanics* (Physical Society of Japan, Tokyo, 1986) 287.
12. D. Bohm, *Phys. Rev.* **85** (1952) 166 and 180.
13. R. von Mises, *Mathematical Theory of Probability and Statistics* (Academic Press, New York, 1964) p. 10 (note 4).
14. F. J. Belinfante, *A Survey of Hidden-Variables Theories* (Pergamon Press, Oxford, 1973) 84.
15. M. Renninger, *Zeitschrift für Physik* **158** (1960) 417.
16. C. D. Tesche, *Phys. Rev. Lett.* **64** (1990) 2358.
17. D. Bohm and B. Hiley, *The Causal Interpretation of Quantum Mechanics* (London preprint, 1990) Ch. 6.
18. Cited in M. Jammer, *The Philosophy of Quantum Mechanics* (Wiley & Sons, New York, 1974) 336.
19. P. Pearle, *Phys. Rev.* **2D** (1970) 1422.
20. S. L. Braunstein and C. M. Caves, *Phys. Rev. Lett.* **61** (1988) 655; *Nucl. Phys.* (*Proc. Suppl.*) **6B** (1989) 211.
21. B. Misra and E. C. Sudarshan, *J. Math. Phys.* **18** (1976) 756; A. Sudbury, *Ann. Phys. N. Y.* **129** (1984) 512.
22. A. Sudbury, in R. Penrose and C. J. Isham (eds.) *Quantum Concepts in Space and Time* (Oxford, Oxford Univ. Press, 1986) Ch. 6; P. Pearle, "Continuous Measurements and the Rules of Quantum Theory", Hamilton College preprint (1980).
23. A. J. Leggett, in G. Grinstein and G. Mezenko (eds.) *Directions in Condensed Matter Physics* (World Scientific, Singapore, 1986).
24. L. L. Van Zandt, *Am. J. Phys.* **45** (1977) 52 ; A. Peres, *Phys. Rev. Lett.* **61** (1988) 2019.

COPENHAGEN HEGEMONY: *NEED* IT BE SO?*

James T. Cushing
Departments of Physics and of Philosophy
University of Notre Dame
Notre Dame, IN 46556
U.S.A.

ABSTRACT

A physical theory has two logically distinct components: a mathematical formalism and a physical interpretation. Since the formalism of quantum mechanics has viable interpretations other than and with ontological commitments radically different from the standard Copenhagen version, one can ask how it came about historically that this indeterminist/acausal interpretation has been accepted to the exclusion of all others. An examination of the historical record shows that the philosophical and epistemological commitments of the creators of quantum mechanics played a major role in the final outcome of the interpretation dispute. It was not simply a matter of logic and empirical adequacy. Personal and social factors were involved as well. The choice was *rational*, but not uniquely compelled by scientific criteria alone. A plausible scenario, alternative to the actually-occurring historically contingent one, is sketched.

The well-known logical point that no set of data uniquely determines a theory is sometimes referred to as the Duhem-Quine thesis. While that rather abstract logical observation may be relatively uncontroversial, the *practical* determination of a physical theory by empirical considerations and by logic is often taken to be quite another matter. Einstein was of the opinion that "in practice the world of phenomena uniquely determines the theoretical system." Now Einstein to the contrary notwithstanding, there *are* people who have looked carefully at the development of certain major episodes in the history of physics and who have concluded that factors other than just "the world of phenomena" have been essential for specific theory choice and that, but for contingency, the final choice might have turned out other than it did.[1] This is *not* to claim that just *any* theory can be made to work, but rather to emphasize that logic and physical phenomena *alone* are not sufficient to select uniquely *one* theory (although, in many cases, they *alone* are enough to rule out or reject some theories as viable candidates). This is not to deny Einstein's claim that "in practice" one theory *is* finally chosen as "decidedly superior to all the rest." The question, then, is *how* is such a choice made? In this paper we point out a known equivalence between two opposing interpretations of quantum mechanics and discuss the factors that were involved in accepting the so-called "Copenhagen" interpretation over its rival.

1. "Copenhagen" vs. Bohm's Causal Quantum Theory

It is not necessary to describe here in great detail either the Copenhagen interpretation of quantum mechanics or David Bohm's alternative causal quantum theory.[2] Therefore, we simply highlight a few features of each interpretation so that we may contrast them later. We are concerned mainly with the epistemological/ontological commitments of each.

The formalism of the Copenhagen version of quantum mechanics is essentially characterized by the Schrödinger equation (or some equivalent dynamical equation in a transformed representation) for a state vector or wave function ψ and a set of rules for calculating probabilities of observing various allowed outcomes of an observation or experiment. The interpretation or ontology read onto the formalism assumes an inherent and irreducible indeterminism at the most fundamental level of microprocesses. Bohr's complementarity in terms of dual, but mutually exclusive, descriptions is at the heart of this picture of the world. An instance of complementarity – one that will be central to our discussion below – is that of a causal description vs. a space-time description. Related to this is the notorious measurement problem.

An alternative to this standard interpretation is the so-called causal one according to which microprocesses follow (in principle) deterministic laws and are subject to event-by-event causality. For the purposes of this paper, it is sufficient to restrict our attention to the original Bohm quantum potential scenario. (A detailed discussion of several causal or realistic quantum theories has been given by Selleri.[3]) Even though most practicing physicists "know" that this Bohm interpretation is "wrong" (although they can't always say *specifically* why), a careful examination of that interpretation, which is based on the same *formalism* as is "Copenhagen", shows that it and the standard interpretation are empirically indistinguishable. Conceptually, in Bohm's interpretation, the wave function ψ (through his quantum potential) represents the effect (or influence) of the environment on the microsystem. Here microsystems ("particles") follow well-defined trajectories in space-time and the dynamics evolve causally. This provides an understandable (even a visualizable) view of microprocesses and there is no collapse of the wave function (i.e., no measurement problem). Yet, such an interpretation has never been seriously considered by any sizeable fraction of the theoretical physics community. We discuss the historical reasons for this below.

2. Opposing Commitments, Opposing Schools

While it is not a new insight that there were two essentially independent routes to quantum mechanics, one leading to wave mechanics and the other to matrix mechanics,[4] it is nevertheless important for our purposes here to see what were the basic philosophical or methodological commitments of these two opposing "schools". This difference is often characterized loosely as determinism vs. indeterminism. We should emphasize that the concept of indeterminism as an essential feature of nature did not for the first time become a seriously-considered option just in the early part of the twentieth century with the advent of quantum physics. The cultural/philosophical milieu within which science develops and fashions its theories does provide a background of ideas and concepts that can influence (but neither uniquely nor solely) the direction of science.

2.1 *Cultural Milieux*

By the late nineteenth century there were significant philosophical precedents for the concept of indeterminism in nature,[5] as opposed to the straightforward determinism often associated with classical physics. For example, Charles-Bernard Renouvier (1815-1903) questioned the causality principle for physical processes, challenged the (Kantian) doctrine that acceptance of causality was a precondition for human understanding and held that an object and its representation cannot be divorced even in principle. Émile Boutroux (1845-1921) stressed contingency and *inherent* chance in nature's actions. Their thinking opposed belief in a completely rational universe and proposed instead an element of irrationality in a nature having contingent laws. In a given situation, there would be equally possible alternatives to what does, *in fact*, occur. Henri Poincaré (1854-1912) read and was influenced by Renouvier, while Einstein, Louis de Broglie and other founders of the quantum theory studied the writings of Poincaré. It has been quite well documented that Søren Kierkegaard's (1813-1855) philosophy made an impression on Niels Bohr through the teachings of Harold Høffding (1843-1931). Bohr explicitly acknowledged the influence of Høffding's philosophy on his own formulation of complementarity. One of Høffding's tenets was that in life decisive events proceed through sudden "jerks" or discontinuities. The point of these brief comments is not that these philosophical currents alone determined the course of quantum theory in the early part of this century, but rather that these concepts were available to, and in the minds of, the creators of quantum theory and did exert *an* influence on the choice of the final, "accepted" form of quantum theory. And, of course, logical positivism was a part of this backdrop with its emphasis on the central role of the empirical determination or definition of a term.

Brush[6] sees the rise of indeterminism as having its origin in the nineteenth century and growing gradually. For him, the primary opposition of concepts in the evolution of quantum mechanics was realism vs. instrumentalism ("positivism"). In his reconstruction, the acceptance or consideration of indeterminism had its roots in thermodynamics. In the larger arena of Western intellectual history, Brush contrasts romanticism with realism and claims there has been an oscillation between these two poles and that this has gone through several cycles in modern times. These different philosophical commitments came to be focused into opposing positions split along a continuity/discontinuity, or determinism/indeterminism, divide.

2.2 *The Wave-Mechanics Route*

It has been argued that Einstein's general position on foundational questions in physics – relativity, his stance on quantum theory and his long-standing commitment to a unified field theory – can best be understood as a search for the God of Spinoza.[7] Einstein believed that this God shows himself in the rational structure of the external physical world. The physicist tries to represent this in a causal space-time theory.

Einstein's basic *Weltanschauung* was that of a rational, causal world that could be comprehended in terms of an objective reality. He also acknowledged explicitly the philosophical influence on him of Poincaré and of Mach. It is, of course, the intersection of general predilections like these with the puzzles presented by physical phenomena that results in a definite theory or research program.

To illustrate this, we now sketch how the opposing positions – wave- vs. matrix-mechanics – were arrived at and what the arguments were for each view. The one traces its roots back to the nature of electromagnetic phenomena, the other to a study of (the discrete) spectral lines. So, not only does it turn out that the general philosophical outlooks of these two groups of key players were quite different, but also it was with very different classes of physical phenomena that each group began. The discontinuity vs. continuity dichotomy is contingently rooted in philosophical commitments and in specific classes of physical phenomena.

In 1909 Einstein used the Planck blackbody radiation law and Planck's energy quantization condition to compute the mean square energy fluctuation for blackbody radiation in thermal equilibrium with another system. His final result consisted of two terms. Einstein pointed out that the first term would be expected if radiation were composed of independent particles (his "photons") and the second term if radiation consisted of waves that could interfere with each other.[4]

In a paper published in 1917 on the quantum theory of radiation, Einstein showed that, when molecules emit and absorb radiation under the influence of an external radiation field, then momentum and energy must each be conserved.[4] It is well-known that Compton's experimental work on the scattering of radiation by electrons gave support to the hypothesis of free electromagnetic quanta. In this same 1917 paper, Einstein stated that the recoil direction of the molecule, which has emitted radiation, is "... only determined by 'chance', *according to the present state of the theory*" (my emphasis) and that "the weakness of the theory lies ... in the fact ... that it leaves the duration and direction of the elementary process to 'chance'." Here, as later, Einstein took this to be a shortcoming of the theory, a provisional fault to be overcome (hopefully) in the future.

The next key figure in this "continuity" school is, of course, Louis de Broglie. Not only was there little interest in quantum physics in France after World War I (and France and Germany remained scientifically insulated from each other then), but even earlier in the century theoretical physics had fallen very much behind the times. There was certainly no existing tradition of research in quantum theory in France when Louis de Broglie was a student. What theoretical research there was was dominated by classical physics. In 1923 de Broglie initiated the theory of wave mechanics in attempting to understand the dual nature of Einstein's photon. In his early youth, Louis de Broglie had been impressed by the well-known formal (mathematical) analogy between wave optics and classical particle mechanics. Building on this analogy and on some of his own previous work on Einstein's light quanta, de Broglie constructed a model of a particle that follows the trajectory of its associated phase wave. There was a great affinity of views between Einstein and de Broglie. In 1925 Einstein stated his opinion that de Broglie's ideas "involve more than merely an analogy". This is not surprising since Einstein had previously, in the context of general relativity, attempted to treat "particles" as the singularities in an underlying field.

Schrödinger recalled that "My theory was stimulated by de Broglie's thesis and by short but infinitely far-seeing remarks by Einstein." In a paper on Einstein's gas theory, Schrödinger concluded that photons can be seen as the energy levels of the aether oscillators and that cavity radiation need not " ... correspond to the extreme light-quantum representation" and that, therefore, the de Broglie-Einstein theory should be taken seriously. In that same year, Schrödinger exploited Hamilton's analogy between mechanics and optics to obtain his wave equation. Quantization was implemented by boundary conditions imposed on a continuous wave function.

The last thread in this part of our story is de Broglie's proposal in 1927 of a principle of the double solution in which the basic entity is a wave, but having a singularity whose trajectory would correspond to a particle path. Because of mathematical difficulties with that proposal, at the 1927 Solvay Congress he presented a simplified pilot-wave model. We return to this subject later.

What we have sketched, only in outline above, is one (historical) route to *a* formulation of quantum mechanics – namely Schrödinger's wave mechanics. It should be evident that this small group (Einstein, de Broglie and Schrödinger) of the creators of that theory shared a commitment to a continuous wave as the basic entity subject to a causal description.

2.3 *The Matrix-Mechanics Route*

The small number of central players (Bohr, Heisenberg, Pauli, Jordan and Born) involved in the program that led to matrix mechanics suggests that we ask whether this, too, was a closed group. This becomes all the more plausible when we realize that Pauli and Heisenberg were both Ph.D. students with Sommerfeld at Munich, each in succession were then Born's assistants at Göttingen and later worked with Bohr at Copenhagen. Each of these young students was greatly impressed by Bohr's 1922 Göttingen lecture. Jordan was also a student at Göttingen at this time. We have already indicated (philosophical) factors in Bohr's own background that inclined him toward, or at the very least made him receptive to, a discontinuous structure in nature at the most fundamental level and, eventually, to a doctrine of complementarity between opposites. This element of discontinuous transitions is a central feature in his 1913 "semiclassical" model for the hydrogen atom. This was certainly the current language for discussing atomic phenomena in Sommerfeld's school. So, *prima facie*, there is a case that Pauli and Heisenberg as young students were impressionable and naturally accepted this central tenet of atomic theory. However, this *inclination* toward credence is scarcely sufficient to account for the prevailing strength of this conviction on the discontinuity vs. continuity issue.

Largely due to the *failure* of certain classical approaches, the main players took up various *philosophical* positions on what was and was not *possible* in principle. That is, these were not logical or in-principle refutations, but strong, practical *beliefs* that became dogma. Thus, Bohr's own Ph.D. dissertation showed that the failure of the classical electron theory of metals was attributable to a fundamental insufficiency of the classical principles themselves. Pauli, in his work on general relativity and related field-theory generalizations, convinced himself (again, because of a *failure*) that a continuum field theory, with the particles as singularities, was not possible.[8] In his famous 1921 *Theory of Relativity*, Pauli already was of the opinion that "... there is no point in discussing ... quantities [that] cannot, in principle, be observed experimentally". This certainly has a strong operationalist air about it. By 1923, in a letter to Eddington, Pauli required operational definitions of anything used in physics and the replacement of continuous concepts by discrete ones. After all, observations are essentially localized and instantaneous, with measurements being discrete and, he felt, this structure should be carried over into the foundation of any theory that accounts for such observations and measurements. He believed this would require major conceptual revisions. Both Pauli and Heisenberg had been involved in Born's program of attempting to apply the old quantum theory, with its orbitals, to molecular systems and the utter failure of this approach convinced them that electron orbitals were meaningless. The Stern-Gerlach experiments (1921/22) (notice the name - *space quantization* - originally used for this effect!) supported quantum predictions, rather than a *continuous* (classical) magnetic moment. Pauli's success with the Zeman effect (1924) in terms of a classically nondescribable two-valuedness in the quantum-theoretical properties of the electron further strengthened his belief in nonvisualizability. He was convinced as well that the exclusion principle could not follow from classical mechanics or from the old quantum rules. Heisenberg began to be converted to Bohr's and Pauli's views on the failure of mechanics. The failure of the Bohr-Kramers-Slater theory (in which energy conservation had been given up) in 1925 indicated to Bohr that "... a complete renunciation of the usual space-time methods of visualization of the physical phenomena..." would be necessary for further progress. Heisenberg's matrix mechanics provided this.

It was the "collision" between these two forms of quantum mechanics that provided the impetus for the formulation of a consistent interpretation of quantum mechanics.[9, 10] It is not uncommon for scientists to believe that there is just one, unique law or theory. Such a belief would justify one in looking for, or attempting to formulate, *the* correct version of quantum mechanics. Heisenberg's faith in the finality of quantum mechanics was essential for his struggle to fashion the "Copenhagen" interpretation via his uncertainty relations.

Born believed that microscopic coordinates were unmeasurable and, therefore, irrelevant. Through his analysis of scattering processes with Schrödinger's formalism, Born came to the opinion that even *perfect* initial information still led to uncertainty in the result and this implied, for him, a lack of causality. Apropos of these issues, Heilbron[11] asks what was the warrant for the complementarity principle as being complete and the final word in forbidding even the *possibility* of a description of microphenomena that is both causal and pictured in a continuous space-time. His response is that (*thus far*) experience has shown the validity of complementary pairs of descriptions and that belief in the ultimate *necessity* of complementarity rests on the *subjective* epistemological criterion of the need for classical concepts and on the indivisibility of atomic phenomena (i.e., Bohr's act of faith). In Heilbron's representation, the Copenhagen interpretation defined itself as true and strengthened its hold on physics, rewriting history so that

Einstein, de Broglie and Schrödinger largely fade from view, thus leaving "Copenhagen" as the only intelligible version of quantum mechanics.

3. Forging the "Copenhagen" Interpretation

With this as background for the historical development that led to two opposing interpretations of the formalism of quantum mechanics, we can now summarize rather briefly how the Copenhagen interpretation came to be formulated under the challenge that Schrödinger's wave mechanics presented to the Göttingen-Copenhagen matrix-mechanics program and how this latter interpretation established its hegemony. Since we have argued that matters of logic and of empirical adequacy alone are insufficient for rejecting out of hand a causal interpretation, it can be little surprise that we must turn to other factors to provide an adequate explanation.

In 1923 Louis de Broglie put forward his concept of wave-particle duality as extended to electrons and he further developed these duality arguments in his doctoral dissertation. Although, as we have seen, Einstein was impressed with de Broglie's thesis, and Schrödinger's own work on wave mechanics was influenced by de Broglie's insight, de Broglie nevertheless had the reputation of being an unorthodox theoretician and that did not condition the scientific community at large to consider his subsequent speculations seriously. Also, the French scientific community had isolated the Germans for reasons of nationalism after World War I and this was not conducive to a flow of ideas between the Germans and the French. A crucial factor in the relevant historical developments was a split in philosophical outlook largely along generational lines: the "older" essentially classical, worldview of people like Einstein, Schrödinger and de Broglie versus a radically different, eventually indeterministic conception of physical processes engendered by the younger generation including Heisenberg, Pauli, Jordan and Dirac.

We have cited some relevant personal, or even psychological, views of prominent members of the Copenhagen school. The prevalence of an empiricist-operationalist philosophical tendency among Heisenberg, Pauli and Bohr can be traced in part (somewhat ironically, given Einstein's later views) back to Einstein's 1905 relativity papers. This operationalist approach seems to have made a great impression and to have exerted a profound influence upon young German physicists. Several such factors help us in understanding the vehemence of the Copenhagen school's reaction against Schrödinger's wave mechanics. Matrix mechanics had been formulated by Heisenberg, and developed by other members of the Copenhagen school, as an essentially abstract mathematical formalism with no physical interpretation.

Heisenberg's views on the nature of a successful mathematical formalism of a physical theory and of its relation to an interpretation were the following.[12] He believed that a successful formalism, such as classical mechanics, was of a piece or whole and that it could not be modified in any essential way without destroying the entire structure. Thus, when such a formalism encounters difficulties (as classical mechanics did with quantum phenomena), it is not possible to modify that formalism successfully. Rather, a radically new formâlism must be found to accommodate these new features of the physical world. (In his later years Heisenberg came to see "Kuhnian" revolutions in science as a natural outcome of this sharp break between formalisms.) Not only did Heisenberg see a successful formalism as unique and of a whole, but he also held a remarkable view of the relation between a formalism and its interpretation. He believed that a careful examination of a successful formalism would provide (uniquely) its own interpretation. This view made the appearance of Schrödinger's (apparently) very different formalism (theory) quite disturbing for Heisenberg. This possibility, coupled with Born's "defection" to the enemy camp when he wrote his paper on the statistical interpreation of (Schrödinger's) wavefunction, produced in Heisenberg the fear that the physics community might take a wrong fork in the road and be led into error.

While these considerations of Heisenberg's remained on the lofty plane of objective truth and concern with physics not being misled into a wrong turn, there were other, much more practical and mundane, factors that lent urgency to the dispute/choice between matrix mechanics (Copenhagen) and wave mechanics. As Beller[9,10] and Cassidy[13,14] have made clear, the formalism of matrix mechanics had *not* had many successful applications (and, in fact, appeared to be bogged down in a mathematical morass) before Schrödinger's wave mechanics allowed theorists to make a stunningly wide variety of well-supported calculations. Wave mechanics, not matrix mechanics, was the formalism employed by most theorists! This danger of losing the war on the calculational front threatened further consequences. Not only did Heisenberg have personal ambitions for advancement, but several chairs in theoretical physics were opening up in Germany. There was a conscious realization by members of the Copenhagen school that control of the future direction of theoretical physics was at stake. This group had the talent, organization and drive to carry the day in establishing the hegemony of the Copenhagen view. Heisenberg's uncetainty relation paper was a major step in accomplishing this. They worked in concert, while their opponents (Einstein, de Broglie, Schrödinger) pulled each in his own direction. The influence of the Bohr Institute in Copenhagen was enormous on an entire generation of leading theoretical physcists who passed through it (e.g., most of those who played dominant roles in establishing theoretical physics in the United States).

A crucial encounter occurred at the 1927 Solvay Congress. At the Fifth Solvay Congress, de Broglie presented his ideas in a form he termed the pilot-wave theory. Here a physical particle is pictured as being guided by the pilot wave. In discussion at that Congress, Wolfgang Pauli criticized de Broglie's theory on the basis of the example of the inelastic scattering of a plane wave by a rigid rotor. Although de Broglie felt he understood the general outlines of a suitable response to Pauli's objection, he in fact did a poor job in attempting to rebut Pauli at the 1927 Congress so that his response appeared *ad hoc* and was not convincing then.[15] In addition to Pauli's negative reaction to de Broglie's paper, neither Einstein nor Schrödinger gave positive support to de Broglie's ideas: Einstein because he did not like the nonlocal (or nonseparable) nature of the theory and Schrödinger because he wanted a theory based only on waves (not on waves *and* particles). And, de Broglie's reputation of being an unorthodox theoretician did not help the situation. Bohr was for a long time against the concept of the photon, so that de Broglie's ideas had never spread rapidly in the Copenhagen school. The Institute at Copenhagen was a very closed community and those invited there were identified as the "respectable" theorists. De Broglie was never a member of this group. By 1930 when he wrote a *very* standard quantum-mechanics book, de Broglie had himself changed his mind about the pilot-wave theory and rehearsed other arguments against it, both general conceptual ones and specific thought experiments. Von Neumann's 1932 impossibility "proof" for hidden variables theories further confirmed de Broglie's position against his own previous theory.

4. An Alternative Scenario?

There matters essentially stood until 1952 when David Bohm published two papers on a causal interpretation of quantum mechanics. Initially, de Broglie was against Bohm's ideas (which were similar to his own pilot-wave theory of 1927) and he raised the same objections against Bohm's theory that had been raised against his own. Interestingly enough, when Bohm sent Pauli a copy of the paper in which Bohm showed Pauli's objections to the causal interpretation to be specious, Pauli never responded. Bohm had produced a causal version of quantum mechanics – one capable of a *realistic* interpretation with a largely classical (micro) ontology. Bohm's work of the early 1950's reconverted de Broglie to his former ideas. For de Broglie the issue at stake was not (classical) determinism, but rather the possibility of a precise space-time representation for a clear picture of microprocesses. In this, his expectations were similar to Einstein's.

Subsequent to Bohm's papers, Edward Nelson showed that a single particle subject to Brownian motion, with a diffusion coefficient ($\hbar/2m$) and no friction, and responding to imposed forces in accord with Newton's second law, $\mathbf{F} = m\mathbf{a}$, obeys (*exactly*) the Schrödinger equation. Although there is randomness, a radical departure from classical physics is unnecessary so that the resulting theory is probabilistic in a *classical* way. William Lehr and James Park generalized Nelson's work to the relativistic case. This alternative program has thus shown a great deal of fertility for generalization within its own resources, not just as *ad hoc* moves. That is, *if*, say in 1927, the fate of the causal interpretation had taken a very different turn and been accepted (over the "Copenhagen" one), it would have had the resources to cope with spin and other generalizations essential for a broad-based empirical adequacy. We could today have arrived at a *very different* world view of micro-phenomena. If someone were then to present the (merely) empirically equally as adequate Copenhagen version, with all of its own counterintuitive and mind-boggling aspects, who would listen!

That is, a highly "reconstructed" but entirely plausible bit of history could run as follows (all around 1925-1927). Study of a classical particle subject to Brownian motion (about which Einstein surely knew something!) leads to a "classical" understanding of the already discovered "Schrödinger" equation, which is then given a "de Broglie-Bohm" realistic interpretation. (Perhaps even a "Bell's" theorem is discovered to convince Einstein of the existence of a harmless type of nonlocality in nature.) These models and theories are generalized to include relativity and spin. The program is off and running! It is essential to appreciate that this "story" is neither *ad hoc* (in the sense of these causal models having as their sole justification an origin in successful results of a rival program) nor mere fancy, since all of these developments exist in the physics literature. However, "Copenhagen" got to the top of the hill first and, to most practicing scientists, there seems to be no point in dislodging it.

It has not been our intention here to argue in favor of a Bohm type of interpretation of quantum mechanics over the standard Copenhagen one. Rather, our interest has been to show that a very different choice (from the actual historical one) *could* reasonably have been made. And, if so, then one can ask what the warrant is for believing that a successful, accepted physical theory is a reliable guide to giving us a true picture of our world.

References

Because of space limitations here, only general references are given to several of the works most relevant to this paper. More complete references can be found in my contribution to the forthcoming *The Wave-Particle Duality* (F. Selleri, ed.).

1. Cushing, J.T. (1990), *Theory Construction and Selection in Modern Physics: The S-Matrix*, Cambridge, Cambridge University Press.
2. Bohm, D. (1952), "A Suggested Interpretation of the Quantum Theory in Terms of 'Hidden' Variables, I and II," *Physical Review* **85**, 166-193.
3. Selleri, F. (1990), *Quantum Paradoxes and Physical Reality*, Dordrecht, Kluwer Academic Publishers.
4. Klein, M.J. (1964), "Einstein and the Wave-Particle Duality," in D.E. Gershenson and D.A. Greenberg (eds.), *The Natural Philosopher* Vol. 3, New York, Blaisdell Publishing Co., pp. 1-49.
5. Jammer, M. (1989), *The Conceptual Development of Quantum Mechanics*, 2nd ed., New York, Tomash Publishing Co.
6. Brush, S.G. (1980), "The Chimerical Cat: Philosophy of Quantum Mechanics in Historical Perspective," *Social Studies of Science* **10**, 393-447.
7. Degen, P.A. (1989), "Einstein's Scientific Orientation as a Search for the God of Spinoza, "*Abstracts of the XVIII Internatinal Congress of History of Science*, Hamburg-München, ICHS.
8. Hendry, J. (1984), *The Creation of Quantum Mechanics and the Bohr-Pauli Dialogue*, Dordrecht, D. Reidel Publishing Co.
9. Beller, M. (1983), *The Genesis of Interpretations of Quantum Physics, 1925-1927*, unpublished Ph.D. dissertation, University of Maryland.
10. Beller, M. (1983), "Matrix Theory Before Schrödinger," *Isis* **74**, 469-491.
11. Heilbron, J.L. (1988), "The Earliest Missionaries of the Copenhagen Spirit," in E. Ullmann-Margalit (ed.), *Science in Reflection*, Dordrecht, Kluwer Academic Publishers.
12. AHQP (1926-1964), *Archive for the History of Quantum Physics*, transcript of interview (number 7 of 12 of Heisenberg by Thomas Kuhn on February 22, 1963, in Munich) on deposit in the Department of History of Science, University of California, Berkeley.
13. Cassidy, D.C. (1976), *Werner Heisenberg and the Crisis in Quantum Theory, 1920-1925*, unpublished Ph.D. dissertation, Purdue University.
14. Cassidy, D.C. (1989), "Science and Biography: Werner Heisenberg and the Route to Uncertainty," unpublished manuscript.
15. Jammer, M. (1974), *The Philosophy of Quantum Mechanics*, New York, John Wiley & Sons.

---

* This work has been supported in part by the National Science Foundation under Grant No. DIR-89-08497.

Some Remarks on *Separability* and *Reality* in Classical and Quantum Physics

Peter Frodl
Institut für Physik
Johannes Gutenberg-Universität
D-6500 Mainz, FRG.

Abstract

The problems which emerge when we investigate the interrelations between classical physics and quantum physics are essentially physical problems. But a physical theory covering the realms of both presupposes a common ontology for their basic referents. The possibility of a concept of reality which is compatible with classical *and* quantum physics is evaluated by scrutinizing what can be inferred from Bell's inequality and various experiments on entangled systems. It is shown that we are neither forced to assume instantaneous actions-at-a-distance, which violate the principle of causality, nor that we must establish a new kind of holism hitherto unknown in physics. But we have to change the classical concept of reality in order to avoid drastic modifications of our space-time picture. We introduce the concept of a *relative reality* which is consistent with both classical and quantum physics and which allows a uniform ontology of the referents of our well confirmed physical theories.

Philosophy or Physics ?

It is often stressed that the problems we have when we try to understand quantum mechanics are no philosophical problems but physical ones, and that we have to solve a lot of physical and mathematical problems *before* the philosophical, epistemological discussion is to be opened. The measurement problem which is only a special case of the interrelations between quantum physics and classical physics, is of course a *physical* problem. If quantum mechanics is a universally valid theory, *physics* should tell us why macroscopically distinct objects exist at all whilst the superposition principle governs the quantum world.

Linking such problems to 'brain states' or the consciousness of possible observers means to explain things we don't understand in terms of things which are much less understood. But the demand for a physical theory which is able to cover both classical and quantum physics presupposes a certain philosophical attitude, namely that the basic referents of both theories *exist* in the very same way, that they are *real* in the same sense, i.e., that they have the same ontological status. Here we can start the philosophical investigation and scrutinize concepts of reality which can be compatible in principle with both classical and quantum physics, even at a time when the physical problems of their interrelations are not yet solved.

Language and Concepts

The words which we use to describe our concepts, physical and philosophical ones as well, are not unbiased. Their meaning is not fixed by definition once and for all. It is rather the usage of the words in a lot of specific contexts which constitutes their meaning. As the contexts represent particular historical and sociological situations, or different intellectual discourses, the meaning of, e.g., *reality*, which is of course an idea of western philosophical thought, has been and is still changing. We cannot hope that a concept of reality which emerges from philosophical discourses connected to classical physics is still able to cover the whole lot of experiences brought about by quantum physics. We shall see that the classical concept of physical reality indeed is not able to fulfill this requirement without running into severe conflicts with well established physical principles.

Classical Metaphysics

Since the advent of quantum mechanics there have been attempts to interprete it in terms of a statistical theory concerning individual 'classical' systems. The very conditions necessary to consider hidden variable theories describing these individual systems as 'classical' had been pointed out by Einstein, Podolsky and Rosen in 1935 [1]: 1. Physical systems are in principle separable. 2. If it is possible to predict with certainty the value of a physical quantity without disturbing the system under consideration, then there exists an element of physical reality corresponding to this physical quantity.

By means of their well known *Gedankenexperiment* EPR were convinced of having shown the incompleteness of quantum theory, which had to be completed by additional, hitherto *hidden*, parameters. In spite of von Neumann's proof [2] that no hidden parameter theory could reproduce all the predictions of quantum theory, such a theory was successfully introduced by Bohm [3] and refined by Vigier and others [4] [1]. Although providing a familiar picture of the microworld

[1] The irrelevance of von Neumann's proof was demonstrated by J.S. Bell [5].

which here consists of classical particles having definite position and momentum simultaneously, thus fulfilling the EPR criterion of physical reality, Bohm's theory and their refined versions are not able to negociate the real obstacles. For, the quantum potential theory is *nonseparable* in the same way as standard quantum theory. Moreover, the nonseparability of the quantum potential is explained in terms of a superluminal *action-at-a-distance*. Hence the very problem is not the suspected incompleteness of quantum theory but their *nonseparability*, as was stressed by Einstein again and again [6].

Classical Physics vs. Quantum Physics

For a couple of decades the struggle for a satisfactory interpretation of quantum theory seemed to be purely metaphysical. But in 1964 Bell came up with his famous inequality [7] which, based on Bohm's version of the EPR experiment [8], for the first time gave a formalization of the EPR conditions for a classical theory of hidden variables in an experimentally testable fashion.

Since then a lot of experiments have been carried out measuring correlation functions of decay products from singlet states [2], the most outstanding ones being those of Aspect and co-workers [10] [3]. More recently another type of experiments, now using superposition states of correlated parametric down converted photons [12], has attracted attention. Almost all of these experiments corroborate quantum mechanics and they seem to force us to drop at least one of the assumptions which had been stated by EPR as necessary conditions for a 'classical' view of the world. This means that either the physical world is not separable or (maybe and) the classical criterion of physical reality is no more tenable [4].

Separability has been taken for self-evident in classical physics. It served as a fundamental principle which made possible some of the greatest advances in physics, from Galilei and Newton up to Bohr whose first atomic theory was to meet the required stability of the *isolated* hydrogen atom. Einstein took separability as one of the necessary conditions of the possibility to do physics at all. On the other hand his reality condition seems to be very natural, virtually matching the notion of reality in everyday life. This familiarity, however, misled some physicists to identify the EPR reality condition with realism as such, being enthusiastic about having to hand Bell's inequalitity as a means to prove or disprove the existence of the external world. But even if we do not overstrain the philosophical categories, a serious change of our view of the physical world seems to be indispensable.

---

[2]For a review up to 1978 see [9].

[3]For a survey see [11].

[4]I refrain from discussing apparent *ad hoc*-explanations like the 'enhancement'-hypothesis (see [13]).

Principles and Concepts: Separability

According to Einstein's view [14], a physical theory describing the physical reality has to obey the following *separability* condition: Two physical systems $S1$ and $S2$ are separated, if the state of $S1$ which is assumed to be at a spacelike distance from $S2$, is not affected by any manipulation at $S2$ whatsoever (and vice versa). This means that the decision which observable is to be measured at $S2$ must not change the state of $S1$. The *locality condition* of special relativity is implicitely contained in 'at a spacelike distance' which excludes only superluminal signal travelling.

Consider the EPRB experiment: A singlet state system $S12$ decays into two subsystems $S1$ and $S2$ obeying spin conservation. It has been argued that quantum mechanics is at variance with special relativity because a measurement on $S2$ instantaneously would change the state of $S1$. Or, more precisely, it would create the system $S2$ and hence $S1$ and their states from the entangled singlet state $S12$. There are no separate states of the subsystems before the measurement, but a measurement on $S2$ would instantaneously create the separate state of $S1$, too.

We cannot properly define *separability* without having clarified the meaning of *state*, which is linked to the *physical reality* of the systems. We have to investigate in which way we are able *in principle* to gain knowledge about the state of a system, i.e., which kind of acces to the physical reality we have.

Principles and Concepts: Reality

Roughly speaking, the state of a system is related to the set of potential outcomes of possible experiments on the system. This description may look purely instrumentalistic, but it is not. It is operationalistic, i.e., the reality of our physical world which is described in terms of states, depends in a sense on what one can or cannot do *in principle* [5]. Reality does not depend only on our abilities, but also on 'something' independent of our endeavour to gain a structured picture of 'the' world. The existence of a physical reality is neither denied (as pure idealists would do) nor considered to be irrelevant (as by pure instrumentalists). *Reality* is not a mere philosophical notion which might be defined and explained by pure reasoning. The way things are real for us depends on our possibilities to get information about their properties. If it is impossible, within a given theoretical or conceptual framework, to distinguish between a world with certain features or without them, it is meaningless to stick to the reality of these features of the world. If we develop a physical theory, it is a well-tried precept not to introduce ontic interpretations of entities which are unobservable in principle. We should take Occam's razor equally serious when we try to infer features of the physical reality from physical theories.

Einstein himself was guided by this principle and employed an operationalistic

[5]This is to the contrary of a verdict by B. d'Espagnat [15].

access to clarify the reality of *time* and *space* by means of possible operations with clocks and rulers [16]. Within the frame of the theory of special relativity some 'elements of physical reality' are not independent of an observer. They depend on the reference frame of the observer's measurement apparatus. The æther once representing an absolute reference system became an obsolete entity which is undetectable within the new theoretical framework. The notion of an *absolute space* has no more sense since all inertial frames are equivalent due to the new principle of relativity. Hence no measurement in any frame can determine the *absolute real* value of a, e.g., kinematical observable, all possible values being equally real *relative* to an inertial frame. We have no access to the absolute space and we may totally drop this concept. "Whoever speaks about an absolute space makes usage of a word which has no sense."[17] Since the *absolute real* values of kinematical observables describe the state of motion relative to the absolute space, it is meaningless to assume the existence of these *absolute real* values any more if an absolute space does not exist.

Due to this view, the value of a spatial (or in some way space dependent) observable is generated by measuring it, since it depends on the state of motion of the inertial frame relative to the system which is investigated. Consider the illustrative example of two observers moving with considerable velocity relative to one another. Each of them is *equally* right claiming that the other one is contracted. In this sense some elements of reality, e.g., velocities, lenghts and time intervals, etc. depend *objectively* on the observer [18]. There are no 'brain states', no conciousness, or other psychological entities envolved in this kind of observer dependence. It is a completely physical affair. Nevertheless the *state* of a system depends on what the observer can or cannot do, namely in which way he is able *in principle* to gain knowledge about the physical reality.

Local Description vs. Global Description

It is convenient to distinguish a *local* experimenter from a *global* one. The local experimenter gains knowledge about the state of the system only from measuring observables on $S1$ or $S2$, respectively, whereas the results from both subsystems are available only to a global experimenter. The world of a global observer is different from the world of a local one. It has been shown by rigorous proofs [19] that the locally registrated outcomes (i.e., the eigenvalues of observables, statistical predictions of their frequencies, expectation values and their variations) of experiments on one subsystem do neither depend on whether a measurement is made on the other subsystem nor on which observable has been measured, if a measurement has been performed on the other subsystem. We may even assume the same sequence $a_1, a_2, ...a_n$ of measured values of an observable A to occur on one side irrespective of whether an observable B or C is measured on the other side, without running into conflicts with the statistical predictions of quantum mechanics [20]. Hence the *local state* of a system is not affected, superluminal signal transmission is ruled out and therefore the EPRB scenario does not violate the locality condition of special relativity.

Thus quantum mechanics fulfills Einstein's requirement of separability, his 'conditio sine qua non' of doing physics, in an essential manner. The separability condition turns out to concern the *states* only via *measurements* which can be performed *in principle*. A locally operating experimenter may divide the world into a part under investigation and the remainder of which he doesn't need to take care. The quantum world is *locally separable* like the classical one.

Now let us turn to the *global* experimenter. It has been claimed that *holism* is a particular feature of quantum mechanics [22]. Holism means, that in a non-trivial way the whole is more than the combination of its parts. Typical holistic effects emerge from *entangled* systems. The state of an entangled system $S12$ is by definition a superposition of product states of the subsystems $S1$ and $S2$, like the singlet state in the EPRB experiment. Any local measurements even on both subsystems cannot yield but informations about the local state of them, i.e., expectation values, variances, relative frequencies which can be inferred from a sequence $a_1, a_2, ..., a_n$ of n measured values of an observable A of any of the subsystems. It is only the global experimenter who is in the position to compare the sequences $a_1, a_2, ...a_n$ and $b_1, b_2, ...b_n$ of values of observables A and B measured on $S1$ and $S2$, respectively, being able to derive correlations between them. In case of quantum mechanics these correlations are due to the non-diagonal elements of the density operator of the entangled system and are called *EPR-correlations*.

We consider now as a classical analogon an ensemble of compound systems which are decaying into their parts, obeying conservation of angular momentum. Local measurements of locally defined projections of the angular momenta of both subsystems yield sequences $a_1, a_2, ..., a_n$ and $b_1, b_2, ..., b_n$, respectively. We have to know the state of motion and orientation of the subsystems relative to one another in order to derive the correlation between the two sequences via a transformation into a common inertial frame. The transformed values then may show for, e.g., a measurement of the same component of the angular momenta in both subsystems the correlation $a_1^t + b_1^t = a_2^t + b_2^t = ... = a_n^t + b_n^t$ in the common reference frame, providing more information than can be inferred from local experiments. Hence even in classical physics the whole is more than the trivial combination of its parts. The correlations cannot be explained in a purely local way, one has to fall back upon nonlocal properties of space-time to account for the conservation laws. However, all these classical correlations obey Bell's inequality; they can be explained as correlations due to a common cause in contrast to the quantum correlations [23].

Counterfactual Definiteness and Local Transformations

So far we have shown that it is not possible to violate the locality condition of special relativity by means of an EPRB experiment. Quantum correlations do not permit signal transmission at any speed whatsoever. This is a feature of classical correlations as well. We have to distinguish *correlations* and *actions*. Actions cause locally detectable physical effects whereas correlations represent global properties or symmetries of space-time. But why are quantum correlations so strange that they gave reason to explanations in terms of nonlocality or nonseparability? We propose that we may explain the suspected nonseparability of quantum mechanics in the very same way as an analogous nonseparability in classical physics, namely by means of global symmetries of space which guarantee for the validity of conservation laws (a necessary condition to have correlated measurements). The very difference between quantum correlations and classical correlations is due to the problem of the *reality* of the measured values of observables.

Bell's inequalities can only be derived if we go on the assumption that the observables are *jointly measurable* [24] or at least that the outcome of every possible measurement is *predetermined* by the state of the system. Every observable must have a definite value irrespective of being measured or not. Therefore it is sufficient for a violation of Bell's inequalities *not* to fulfill this requirement of *counterfactual definiteness* [25] of the observables. This is a purely local affair [26].

It has been shown by M. Hoffmann [18] that the astonishing correlations between spacelike separated 'entangled' systems in the case of the EPRB scenario can be traced back entirely to the behavior of one subsystem under subsequent measurements which are connected through a particular unilateral, and therefore local transformation scheme.

For this purpose one can split the problem into a dynamical and a kinematical part [30]. Dynamics account for perfect anticorrelation in the case of a measurement in the same direction on both sides, just like in classical physics. To obtain the correlated values of measurements in different directions one may employ a *kinematical* transformation to transform the locally obtained values of a measurement in direction A into the values one *would* obtain if one *would* measure in direction B on the same side, provided that perfect anticorrelation is guaranteed for measurements in the same direction on both sides. Clearly, the transformation is deterministic in the classical case: One can adopt the view *as if* an absolutely real value of an observable exist, determining the values in all possible reference frames. In the case of discrete quantum mechanical values like spin, however, the transformation is shown to be necessarily indeterministic, thus avoiding counterfactual definiteness of observables. For, it is not possible to transform a discrete set of fixed angular dependent values into the *same* set of discrete values by means of a rotation in an isotropic space, provided that the transformation law is generally valid for any rotation angles [27].

Consider the simple case of the spin observable in the EPRB experiment measured in the xy-plane with only two possible values $i = \pm 1$ irrespective of the preferred direction of the Stern-Gerlach apparatus in this plane. Assume $\mathbf{T}(\theta)$ to be a deterministic transformation between two reference frames in the xy-plane with an orientation θ relative to one another. Let $\mathbf{a}, \mathbf{b}, \mathbf{c}$ be different directions in the xy-plane with θ being the angle between $\mathbf{a}$ and $\mathbf{c}$, and $\mathbf{b}$ bisecting this angle. Hence $\theta/2$ is the angle between $\mathbf{a}$ and $\mathbf{b}$, and between $\mathbf{b}$ and $\mathbf{c}$, respectively. *Isotropy of space* requires that a measurement first in direction $\mathbf{a}$ and then in direction $\mathbf{b}$ is equivalent to a measurement in directions first $\mathbf{b}$ and then $\mathbf{c}$. Hence $\mathbf{T}(\mathbf{a} \to \mathbf{b}) = \mathbf{T}(\mathbf{b} \to \mathbf{c}) = \mathbf{T}(\theta/2)$, and $\mathbf{T}(\theta) = \mathbf{T}(\theta/2)\mathbf{T}(\theta/2), \forall\theta$. Since we have only a discrete set of measurement results $i \epsilon [+1, -1]$ in any direction within the xy-plane, we have either $\mathbf{T}(\theta/2)i = i$ or $\mathbf{T}(\theta/2)i = -i$, which in both cases results in $\mathbf{T}(\theta)i = \mathbf{T}(\theta/2)\mathbf{T}(\theta/2)i = i, \forall\theta$ $(*)$ for two subsequent transformations $\mathbf{T}(\theta/2)$. We know from the experiments that $\mathbf{T}(0)i = i$ and $\mathbf{T}(\pi)i = -i$ hold, because two subsequent measurements in the same (opposite) direction yield the same (opposite) result. Therefore there exists at least one θ with $\mathbf{T}(\theta)i = -i$, which is in contradiction to $(*)$. Hence either space is not isotropic or the transformation is not deterministic.

Since we need isotropy of space to guarantee for perfect anticorrelation in the case of measurements in the same direction on both subsystems, the transformation has to be indeterministic. A indeterministic transformation, however, cannot enable us to assume absolute values predetermining the outcomes of all possible measurements, which means that counterfactual definiteness is not fulfilled. The reason for the strange correlations therefore need not be an action-at-a-distance. The apparent nonlocality is due to symmetries and conservation laws like in classical physics, whereas purely local, and in the case of the EPRB experiment indeterministic, transformations between distinct reference frames of the measurement devices account for the violation of Bell's inequality.

Relative Reality

Thus we are able to avoid drastic modifications of our space-time picture if we accept to drop the classical concept of pre-existing values for those measurable quantities which are represented by non-commuting operators. The EPR reality criterion does not follow from classical physics nor from special relativity, although being compatible with both. However, the assumption, that the value of an observable is not passively measured but *created* by a measurement is not at variance with classical physics. Moreover, this concept of a *relative reality* is in the very spirit of special relativity theory, as it has been discussed on the occasion of introducing the notion of 'state' above. It enables us to assume a uniform ontology of the referents of our well confirmed physical theories, in contrast to, e.g., the orthodox Copenhagen interpretation according to which classical systems exist independently of an observer whereas quantum systems do not.

Accepting *relative reality* has not the consequence that we have to give up realism in the sense that the external world does not exist, or to assume that 'the' world is created by being observed. Neither physical theories nor empirical evidences can contribute anything to this kind of epistemological problems. They concern classical physics and quantum mechanics in the very same way. But if we have once established the assumption of the existence of an external world as a regulative principle, physical theories which are well corroborated by experiments have to be taken seriously not only as instruments to predict experimental results but also to describe some features of this reality. If we regard the history of physics, we can learn that new kinds of knowledge sometimes require not only a new answer to "what is real?" but also to "what is the meaning of *real*?".

In this sense problems of the interpretation of theoretical entities which describe some aspects of the physical reality arose even in classical physics. E.g., Newton's concept of gravitation as an instantaneous action-at-a-distance led to a lot of objections among his contemporaries. In the 19th century one could hardly accept the reality of the electromagnetical field in empty space. A couple of mechanical models of the æther were developped in order to understand the mechanism of signal transport via oscillations of 'something' [28]. But the term 'interpretation' has been used only since physicists and philosophers are trying to understand quantum mechanics [29], misleading the majority to take the problems only as philosophical ones.

To *understand* quantum mechanics means on the one hand to solve the *physical* problems connected with the measurement process, or, in a broader context, to clarify the interrelations between quantum physics and classical physics by means of a physical theory which should deliver rigorous explanations why classical physics works so well in a lot of contexts. On the other hand there are of course epistemological problems, like 'What can we learn about the reality from measurements?' or 'What can physical theories tell us about the world?'. The concept of a *relative reality* enables us to treat this kind of problems in quantum and classical physics on an equal footing because now the referents of both theories have the same ontological status.

Insofar as the concepts which we use are not only descriptions but define a frame of possible 'speakable' experiences, *our* world is created by being described in terms of our physical concepts. A change of the meaning of these concepts might lead to new ways to develop intelligible pictures of the physical world.

Acknowledgements

I am much indebted to M. Hoffmann for innumerable fruitful and enjoyable discussions. It's a pleasure to thank A. Loidl for providing a travel grant to the Joensuu symposium and G. Schreider for a critical reading of the manuscript.

References

[1] A. EINSTEIN, B. PODOLSKY, N. ROSEN, *Phys. Rev.* **47** (1935) 777.

[2] J. v. NEUMANN, *Mathematische Grundlagen der Quantenmechanik*. (Springer) Berlin 1932.

[3] D. BOHM, *Phys. Rev.* **85** (1952) 166, 180.

[4] J.P. VIGIER, *Astron. Nachr.* **303** (1982) 55.

[5] J.S. BELL, *Rev. Mod. Phys.* **38** (1966) 447.

[6] A. EINSTEIN, *Dialectica* **2** (1948) 320; see also D. HOWARD, *Stud. Hist. Phil. Sci.* **16** (1985) 171; and R. DELTETE, R. GUY, *Am. J. Phys.* **58** (1990) 673.

[7] J.S. BELL, *Physics* **1** (1964) 195.

[8] D. BOHM, *Quantum Theory*. (Prentice Hall) Englewood-Cliffs, N.J. 1951.

[9] J.F. CLAUSER, A. SHIMONY, *Rep. Prog. Phys.* **41** (1978) 1881.

[10] A. ASPECT, J. DALIBARD, G. ROGER, *Phys. Rev. Lett.* **49** (1982) 1804.

[11] P. FRODL, *Annalen der Physik* **46** (1989) 513.

[12] Y.H. SHIH, C.O. ALLEY, *Phys. Rev. Lett.* **61** (1988) 2921; Z.Y. OU, L. MANDEL, *Phys. Rev. Lett.* **61** (1988) 50.

[13] M. FERRERO, T.W. MARSHALL, E. SANTOS, *Am. J. Phys.* **58** (1990) 683.

[14] A. EINSTEIN, *Autobiographisches*. In: P.A. SCHILPP (ed.), *Albert Einstein: Philosopher – Scientist*. (Open Court Publ. Comp.) La Salle, Ill. 1949, 2.

[15] B. D'ESPAGNAT, *J. Stat. Phys.* **56** (1989) 747.

[16] A. EINSTEIN, *Annalen der Physik* **17** (1905) 891.

[17] H. POINCARÉ, *Science et Méthode*. Cit. transl. from the German Edition: *Wissenschaft und Methode*. (Teubner) Leipzig 1914, 80.

[18] M. HOFFMANN, *Found. Phys.* **20** (1990) 991.

[19] G.C. GHIRARDI, A. RIMINI, T. WEBER, *Lett. Nuovo Cim.* **27** (1980) 293; L.E. BALLENTINE, J.P. JARRETT, *Am. J. Phys.* **55** (1987) 696; K. KRAUS, *Found. Phys. Lett.* **2** (1989) 1.

[20] K. KRAUS, in [21], 461.

[21] P. LAHTI, P. MITTELSTAEDT, (eds.), *Symposium on the Foundations of Modern Physics*. (World Scientific) Singapore 1985.

[22] H. PRIMAS, *Chemistry, Quantum Mechanics and Reductionism*. (Springer) Berlin, Heidelberg, New York, Tokyo 1983^2.

[23] B.C. VAN FRAASSEN, in [21], 113.

[24] W.J. LEHR, *Nuovo Cim.* **90B** (1985) 185.

[25] H.P. STAPP, in [21], 673.

[26] W.M. DE MUYNCK, J.T. VAN STEKELENBORG, *Annalen der Physik* **45** (1988) 222.

[27] M. HOFFMANN, *Correlations at a Distance in Quantum Theory – A New Interpretation*. Ph. D. Thesis (in German), University of Tübingen, 1988.

[28] E. WHITTAKER, *History of the Theories of Æther and Electricity*. Vol. I,II. (T. Nelson & Sons) London 1951, 1953.

[29] J.M. LÉVY-LEBLOND, private interchange.

[30] M. HOFFMANN, *Correlations at a Distance in Quantum Theory: A Kinematic Interpretation*. Poster presented at the Symp. Found. Mod. Phys. 1990. Joensuu, Finland. Unpublished.

RELATIVISTIC DYNAMICAL REDUCTION MODELS AND NONLOCALITY

GIANCARLO GHIRARDI
Dipartimento di Fisica Teorica, Università di Trieste and ICTP, Trieste, Italy

RENATA GRASSI
Dipartimento di Fisica Teorica, Università di Trieste, Italy

PHILIP PEARLE
Hamilton College, Clinton, New York 13323

ABSTRACT

We discuss some features of continuous dynamical models yielding state vector reduction and we briefly sketch some recent attempts to get a relativistic generalization of them. Within the relativistic context we analyze in detail the local and nonlocal features of the reduction mechanism and we investigate critically the possibility of attributing objective properties to individual systems in the micro and macroscopic cases. At the nonrelativistic level, two physically equivalent versions of continuous reduction mechanisms have been presented. However, only one of them can be taken as a starting point for the above considered relativistic generalization. By resorting to counterfactual arguments we show that the reason for this lies in the fact that the stochasticity involved in the two approaches has different conceptual implications.

1. Introduction

Quantum mechanics is a theory which makes only conditional predictions about the results of prospective measurements of (in general) incompatible observables. Any attempt to apply quantum mechanics to the measurement process itself can only lead to further conditional predictions about what will be found if ... and so on. Here the problem of the quantum theory of measurement finds its origin.

There is now a widespread agreement about the fact that the standard solution to this problem, i.e. the adoption of the postulate of wave packet reduction (WPR), is inadequate. In particular WPR breaks the linear nature of the theory and it requires the acceptance of a splitting between system and apparatus, (or micro and macro, quantum and classical) which turns out to be basically shifty. Finally WPR gives rise to specific difficulties when one considers changes of reference frame.

An attempt to overcome the first two of the above mentioned difficulties has led to the consideration of Dynamical Reduction Models (DRM). The underlying philosophy is quite simple: one introduces a "small" stochastic modification of the standard dynamics that has a negligible impact for microsystems but which nevertheless leads to a dynamical suppression of macroscopic superpositions. That such a line can be consistently followed (in the nonrelativistic case) has been proved in a series of recent papers [1,2]. An attempt at a relativistic generalization of a particular DRM which is known[2] as CSL (Continuous Spontaneous Localizations) has also been presented[3]. Here we will be mainly concerned with a critical discussion of the implications of such relativistic models for nonlocality and with the possibility of a macro-objective interpretation of the formalism.

There are two physically equivalent versions of nonrelativistic CSL; only one of them, however, can be taken as the starting point for the relativistic generalization. This is due, as we will prove here by resorting to counterfactual arguments, to the fact that the two considered versions of CSL exhibit stochastic features which are conceptually different.

2. Nonrelativistic DRM: Some Remarks

The development of the DRM program necessarily requires a dynamical evolution which yields, under specific circumstances, a suppression of the off-diagonal elements of the statistical operator in an appropriate "preferred basis". It is important to stress, however, that this condition on the statistical operator is not sufficient to guarantee that any individual system must be associated to a state vector belonging to only one of the "preferred manifolds". Typically, there are stochastic dynamical evolution mechanisms which leave individual systems in superpositions of preferred states but which induce a randomization of the relative phases of such superpositions, thus leading to the above mentioned suppression. Following Stapp [4] we say that such mechanisms induce von Neumann reductions. We are not interested in such models since we want the reductions to occur at the individual level, or, in Stapp's language, to have Heisenberg reductions.

CSL models actually give such kinds of reductions. It is useful to give a sketchy description of the two distinct, physically equivalent, versions of CSL which have been mentioned above. The first is based on the consideration of the linear evolution equation with a skew hermitian stochastic term[2],

$$\frac{d\,|\,\psi_V, t>}{dt} = \left[-iH - \lambda\int dx\, N^2(x) + \int dx\, N(x)\, V(x,t)\right] |\,\psi_V, t> \qquad (1)$$

where N(x) is a number density operator averaged on an appropriate volume around **x**, and V(**x**,t) is a white noise in all variables with covariance

$$<< V(x, t)\, V(x', t') >> = \lambda\, \delta(x - x')\, \delta(t - t'). \qquad (2)$$

The second term in the equation is a counterterm guaranteeing the conservation of the stochastic average of the square norm. The physics is obtained by considering the normalized states $|\psi_v(t)>/\,\|\,|\,\psi_v(t)>\|$ and by assuming that the probability density of occurrence of a specific potential V(x,t) is not the (Raw) one *P* [V] associated with the white noise distribution, but the cooked one

$$P_C[V] = P\,[V]\,\|\,|\,\psi_V, t>\|^2 . \qquad (3)$$

We will refer to this description as the *Raw + Cooking* scheme.

Alternatively one can consider the nonlinear, stochastic, norm conserving, evolution equation[2]

$$\frac{d\,|\,\psi_V, t>}{dt} = \left\{-iH - \lambda\int dx\, [N(x) - <N(x)>]^2 + \right. \qquad (4)$$

$$\left. +\lambda\int dx\, [<N^2(x)> - <N(x)>^2] + \int dx\, [N(x) - <N(x)>]\, V(x,t)\right\} |\,\psi_V, t>$$

and assume that $V(\mathbf{x},t)$ is distributed according to the raw distribution $P[V]$. We will refer to this description as the *Nonlinear-Equation* scheme. For a detailed mathematical study of the relations between the two schemes see ref.(5).

The two schemes are equivalent in the following precise sense: for any given initial state $|\psi,0\rangle$ and, e.g., for a given $V^{\sim}(\mathbf{x},t)$ to be used in the *Raw + Cooking* scheme, there exists a $V(\mathbf{x},t)$ to be used in the *Nonlinear- Equation* scheme such that the evolved states at any time t coincide. Moreover $V^{\sim}(\mathbf{x},t)$ and $V(\mathbf{x},t)$ occur with the same probability.

3. Relativistic DRM

Within a stochastic framework one has to formulate in an appropriate way the invariance requirement. Actually what is needed is only stochastic invariance, i.e., while individual processes can look different to different observers, the ensemble of possible individual processes must turn out to be the same. We remark that CSL, as presented in Section 2, exhibits Galileian invariance.

Trying to get a relativistic theory yielding Heisenberg reductions we consider the Tomonaga-Schwinger (T-S) picture in a quantum field theory framework. We assume that the fields evolve according to Heisenberg equations deriving from a hermitian lagrangian density L_0 while the state vector obeys the T-S equation

$$\frac{\delta\,|\,\psi_V(\sigma)\rangle}{\delta\sigma(x)} = \left[L_I(x)\,V(x) - \lambda L_I^2(x)\right]|\,\psi_V(\sigma)\rangle \tag{5}$$

where $L_I(x)$ is a hermitian function only of the fields and $V(x)$ is a c-number white noise with covariance given by eq.(2).

We remark that we are working in the *Raw+Cooking* scheme, and that the equation corresponding to eq.(3) is

$$P_C[V] = P[V]\,\|\,|\psi_V(\sigma)\rangle\|^2 \tag{6}$$

The lagrangian density L_0 can contain interaction terms. We point out that hermitian terms in L_0 which do not depend on the field derivatives can be shifted from the Heisenberg to the T-S equation, while the reducing dynamics can only be described in terms of the T-S equation, due to the skew hermitian coupling of the noise to $L_I(x)$.

One would like the dynamics to induce localization of massive fermions. This cannot be obtained directly by specifying the above formalism only to fermion fields[3]. One can then try to get localization for fermions by coupling their field to a meson field and then introducing a reducing dynamics for this auxiliary field. In refs.(3) the following choice has been proposed

$$L_0 = \text{Free fermions} + \text{Free mesons} + \eta\,\bar{\Psi}(x)\,\Psi(x)\,\Phi(x)$$

$$L_I(x) = \Phi(x). \tag{7}$$

As a quantum field theory the model presents additional divergences with respect to the standard ones, arising from the white noise character of the stochastic potential

V(x). However, one can study the nonrelativistic limit for fermions in a sector with a fixed number of fermions and one gets, with some approximations, a theory which is very similar to CSL. We refer the reader to refs.(3) for a detailed discussion of this model.

4. Objectivism and Nonlocality

Without going into specific details about the proposed model we want to investigate here the consequences of the adoption of the previously outlined relativistic framework for describing dynamical reductions. We will be particularly concerned to discuss the nonlocal features of the scheme and to investigate whether it allows a macro-objective description of natural phenomena. In this discussion we shall make use of concepts like *elements of physical reality*, *chance* and *determinism* which we examine now. It has to be stressed that this investigation requires, to be meaningful, the consideration of the individual level of description of physical processes.

In standard quantum mechanics, when consideration is given to an observable A, one states that a physical system posesses* an element of physical reality a associated to the observable A iff the system is in an eigenstate of A corresponding to the eigenvalue a. Calling P_a the projection operator on the eigenmanifold associated to a one has, in such a case,

$$<\psi,t \mid P_a \mid \psi,t> = 1 \,. \tag{8}$$

We remark that the condition that $<\psi,t \mid P_a \mid \psi,t>$ be extremely close to one has to be considered sufficient for the attribution of the element of physical reality. In fact since the recording of the result of any measurement involves the reading of the position of some pointer, and since the wave functions associated to different pointer positions unavoidably overlap, the probability that a pointer be in a certain interval is never exactly equal to 1.

Again within the standard quantum formalism let us consider a system which is in a linear superposition of two eigenvectors of the observable A. In such a case no element of physical reality corresponding to A can be attributed to the system. Suppose that a measurement of A is performed, and that we use the WPR postulate. One can then say that the definite response of the macroscopic measuring apparatus *is due to chance*. On the contrary, if the system is in an eigenstate of A, one can say that the definite response of the macroscopic measuring apparatus *is deterministically implied* by the pre-existing element of physical reality of the system.

4.1 WPR, Nonlocality and Changes of Reference Frame.

Consider an EPR-Bohm like set up for the singlet state involving a measurement of a spin component of particle 1 by a measuring apparatus M_1 (see Fig.1). In the figure 1 and 2 represent the "world lines" of the two particles. C is the space-time point (region) in which the measurement occurs. Note that if one assumes that WPR takes places instantaneously in the reference frame in which M_1 is at rest, according to the previous definition, the event in C implies the instantaneous emergence of an element of physical reality associated to the same spin component for particle 2 and, in particular, that this

*For better appreciating what follows it might be useful to recall the obvious fact that in the standard quantum framework the entanglement of wave functions for composite systems forbids the attribution of elements of physical reality to the constituents and that the persistence of linear superpositions of far away states forbids the attribution of objective local properties even to a single particle.

element of physical reality can be attributed to the particle at the space-time point B which is space-like separated from C.

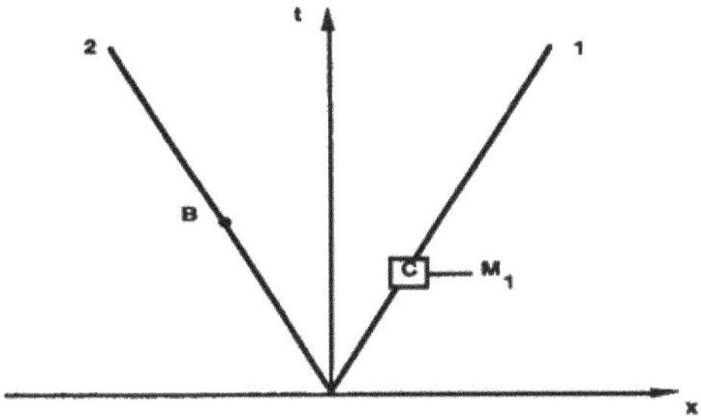

FIG. 1

We must consider the possibility of changes of reference frame. Since the separation $(B\text{-}C)^2$ can be arbitrarly large and negative, *even changes to an extremely slowly moving frame* can reverse the time order of B and C. Let us call O' such a reference frame. If one assumes that WPR takes place instantaneously also for O', the fact that for him B occurs earlier than C implies that (for him) the mean value of the relevant projection operator for the spin component of the particle at B is 1/2. This ambiguity about the mean values of the considered projection operator reflects an ambiguity in the possibility of attributing to particle 2 the corresponding element of physical reality at the given (objective) space-time point B. Note that if consideration is given to a space-time point in the future of C, such an ambiguity does not occur*.

Let us consider the analogous situation in the case in which, beside the apparatus at C, there is an apparatus M_2 at B devised to measure the same spin component. The ambiguity in the attribution of the element of physical reality to the particle at B, in accordance with the considerations of the previous subsection, has as a consequence that while according to O the result of the measurement at B is deterministic, according to O' it is due to chance.

Non relativistic DRM models exhibit analogous ambiguites. It has however to be remarked that the fact that the statement "*the outcome at B is due to chance*" is not covariant does not constitute a dead end for the theory and, in particular, it does not forbid its relativistic generalization. In fact the ambiguity does not involve the response of the macroscopic apparatus but only the nontestable (for space-like separation B-C) modalities yielding this unambiguous response.

4.2 Macro-Objectivism and Nonlocality in Relativistic Reduction Models

We will perform now a critical investigation of relativistic CSL, which parallels the one of the previous Subsection. Such a theory has precise implications, differing from those of standard quantum mechanics, about the behaviour of macroobjects. It is then important to analyze whether these implications are compatible with the adoption of a macroobjective position.

*A detailed discussion of the difficulties met by WPR in connection with changes of reference frame has been presented in refs.(6). B. d'Espagnat[7] has stressed that in trying to generalize DRM one should face analogous difficulties.

To this purpose we start by considering local observables

$$A_I(\sigma) = \int dx\, f_\alpha(x)\, F[\Phi_I(x), \partial_\mu \Phi_I(x)] \tag{9}$$

where $f_\alpha(x)$ is a function having as its support α a compact subset of the space-like surface σ and is of class C^∞ on σ. In eq.(9) the index I is added to recall that we are working in the (T-S) Interaction Picture .

Let us consider now two arbitrary space-like surfaces σ_1 and σ_2 containing α (see Fig.2). In standard quantum field theory the mean value of A_I does not depend on

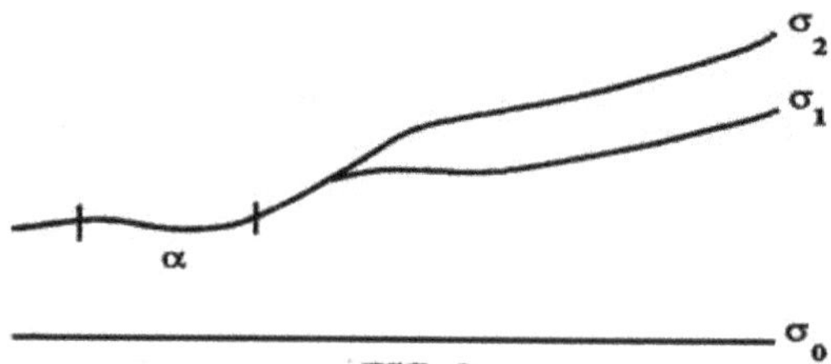

FIG. 2

which one of the two states $|\psi(\sigma_i)\rangle$, i = 1,2, associated to the considered surfaces, is chosen for its evaluation. In relativistic CSL this is no more true. In fact, in such a case, the state vector describing an individual physical process for a specific realization of the stochastic potential V(x) is $|\psi_V(\sigma)\rangle / \| |\psi_V(\sigma)\rangle \|$. If we denote by $S_V(\sigma_2, \sigma_1)$ the evolution operator from σ_1 to σ_2, we have

$$\frac{<\psi(\sigma_1)|S^+_V(\sigma_2,\sigma_1)\, A_I\, S_V(\sigma_2,\sigma_1)|\psi(\sigma_1)>}{\| S_V(\sigma_2,\sigma_1)|\psi(\sigma_1)> \|^2} \neq \frac{<\psi(\sigma_1)|A_I|\psi(\sigma_1)>}{\| |\psi(\sigma_1)> \|^2} \tag{10}$$

even though

$$[A_I, S_V(\sigma_2, \sigma_1)] = 0 \tag{11}$$

since $S^+_V(\sigma_2, \sigma_1)\, S_V(\sigma_2, \sigma_1) \neq 1$. It follows from this that if one relates, as before, the attribution of objective local properties to individuals to the mean value of a projection operator referring to a local observable, one meets difficulties.

We stress, however[3], that no such surface dependence exists for the ensemble average of the mean value of local observables. This lack of ambiguity at the ensemble level is necessary for the consistency of the theoretical scheme.

In analyzing the dependence, at the individual level, of the mean value of a local observable A_I upon the space-like surface over which it is evaluated (among those coinciding on its support), it is useful to discuss separately the ambiguities for microscopic and macroscopic systems.

We start by considering a microscopic system S and two local observables A_1 and A_2 of S having space-like separated supports α_1 and α_2 . Suppose that there is a macroscopic measuring apparatus M_1 devised to measure A_1. The T-S equation describing the unfolding of the physical process is then

$$\frac{\delta | \psi_V(\sigma)>}{\delta \sigma (x)} = \left[i L_{1\text{-}S} (x) + L_I (x) V (x) - \lambda L_I^2 (x)\right] | \psi_V (\sigma) > \tag{12}$$

where $L_{1\text{-}S}$ is the local hermitian system-apparatus interaction which describes the triggering of the apparatus by the microsystem and L_I is the skew hermitian term inducing reductions. Both $L_{1\text{-}S}$ and L_I are assumed to be different from zero only in the region C indicated in Fig.3.

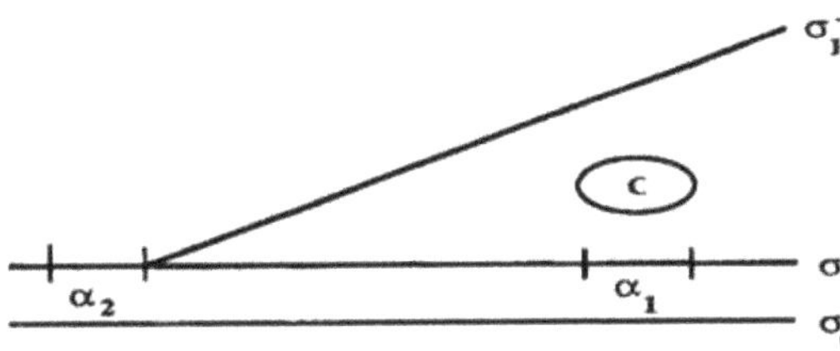

FIG.3

The initial state $|\psi(\sigma_0)>$ is supposed to be

$$| \psi (\sigma_0) > = \frac{1}{\sqrt{2}} [| \psi_1 > + | \psi_2 >] | \chi_1 > \tag{13}$$

with

$$A_i | \psi_j > = \delta_{ij} | \psi_j > ; \qquad i, j = 1, 2 \tag{14}$$

and $|\chi_1>$ describing the untriggered apparatus state. Suppose that for the individual case under consideration the specific realization of V(x) is one of those "yielding the result 1" for the measurement of A_1. Let us consider the two space-like surfaces σ_1 and $\sigma^{\sim}_1$ which contain the support α_2 of A_2 but such that, in going from σ_1 to $\sigma^{\sim}_1$, one crosses the space-time region C (see Fig. 3). According to eqs.(12) and (13) one has

$$<\psi_V(\sigma_1)| \hat{A}_2 | \psi_V(\sigma_1)> = 1/2 \tag{15.a}$$

while

$$<\psi_V(\sigma^\sim{}_1)|\, A_2|\, \psi_V(\sigma^\sim{}_1)> \cong 0. \tag{15.b}$$

Eqs.(15.a) and (15.b) show that there is an ambiguity in the mean value of the local observable A_2 associated to the microsystem S. This is not surprising, it constitutes simply the analogy within relativistic DRM of the situation discussed in Subsection 4.1 for standard quantum mechanics with the WPR postulate. In fact the surface $\sigma^\sim{}_1$ can be approximately identified with a t' = const. hyperplane for a boosted observer O' which moves with a very slow velocity with respect to O but for whom, however, the interaction (and the consequent reduction) at C has taken place at a time earlier than t'.

We pass now to discuss the analogous problem, i.e. the possible occurrence of ambiguities in the mean values of local observables, in the case of a macroobject. To this purpose, we consider a situation strictly analogous to the previous one, in which however a further macroscopic measuring apparatus M_2 devised to measure A_2 is supposed to be present. With obvious meaning of the symbols we have as the analogy of eq.(12) the following T-S evolution equation

$$\frac{\delta\,|\,\psi_V(\sigma)>}{\delta\,\sigma(x)} = \Big[i\,\mathcal{L}_{1\text{-}S}(x) + i\,\mathcal{L}_{2\text{-}S}(x) + \mathcal{L}_{I1}(x)\,V(x) + \mathcal{L}_{I2}(x)\,V(x) -$$

$$-\lambda\,\mathcal{L}^2_{I1}(x) - \lambda\,\mathcal{L}^2_{I2}(x)\Big]\,|\,\psi_V(\sigma)> \tag{16}$$

With reference to Fig.4 we specify that the system-apparatus interactions $\mathcal{L}_{1\text{-}S}$ and $\mathcal{L}_{2\text{-}S}$ are supposed to be different from 0 in C_1 and B_1, respectively, and the reducing terms $\mathcal{L}_{I1}$ and $\mathcal{L}_{I2}$ in C_2 and B_2, respectively. The initial state is now

$$|\,\psi(\sigma_0)> = \frac{1}{\sqrt{2}}\,\big[|\,\psi_1> + |\,\psi_2>\big]\,|\,\chi_1>|\,\chi_2> \tag{17}$$

where $|\chi_2>$ is the untriggered state of the apparatus M_2. As in the previous case we assume that V(x) is such as to lead to the result A_1=1.

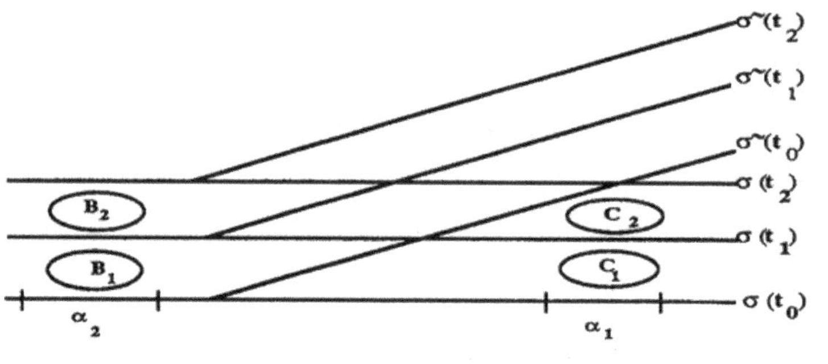

FIG.4

We remark that according to eq.(16), for the considered initial state, we have the following situation.

i) For $t<t_0$, the states associated to $\sigma(t)$ and $\sigma^{\sim}(t)$ are factorized, one of the factors being $|\chi_2>$.

ii) For $t=t_1$

$$|\psi_V(\sigma(t_1))> = \frac{1}{\sqrt{2}}[|\psi_1>|\chi_1^1>|\chi_2^0>+|\psi_2>|\chi_1^0>|\chi_2^1>] \tag{18.a}$$

while

$$|\psi_V(\sigma^{\sim}(t_1))> \cong |\psi_1>|\chi_1^1>|\chi_2^0> . \tag{18.b}$$

At the r.h.s. of eqs.(18a-b) the superscripts in the apparatuses states refer to the eigenvalues of the observables A_1 and A_2.

iii) For $t>t_2$ the states associated to $\sigma(t)$ and $\sigma^{\sim}(t)$ are factorized, one of the factors referring to the apparatus M_2 and corresponding to having registered the definite outcome 0 in the measurement of A_2.

The results (18.a) and (18.b) show that, at t_1, an ambiguity is present even in the mean value of the local observable corresponding to a definite macroscopic outcome of M_2. It has to be stressed, however, that according to iii), this ambiguity lasts only for the typical reduction time of the macroscopic measuring device M_2.

Let us summarize the situation. The consideration of the individual level of description of physical processes within standard quantum mechanics with the WPR postulate gives rise to difficulties when one takes into account changes of reference frame, in particular for what concerns the possibility of attributing to physical systems elements of physical reality corresponding to local observables. Such difficulties find their formal counterpart within relativistic DRM in the fact that, for a given realization of the stochastic potential, the mean values of local observables are ambiguous in the above specified sense. As we have shown, there is however a fundamental difference in the microscopic and the macroscopic cases. For microsystems such ambiguities may last, under appropriate circumstances, for arbitrarly large time intervals, while in the macroscopic case they last only for extremely short intervals. We can then take advantage of this nice feature to overcome the puzzling situation we have discussed in this section.

The analysis of the formal aspects of relativistic CSL leads naturally to the introduction of a more appropriate criterion for the attribution to physical systems of elements of physical reality corresponding to local observables: we say that an individual system has an element of physical reality associated to the local observable A_I with support α iff the mean value of the projection operator on one of the eigenmanifolds of A_I is extremely close to one for all space-like surfaces containing α.

With reference to the previous discussion we stress that, according to this criterion:

i) Macroscopic objects have " almost always" definite local macroproperties. This means that relativistic CSL is compatible with macroobjectivism.

ii) For what concerns micro-objects, contrary to what happens when the standard attitude is taken, no objective local property can emerge as a consequence of a "measurement process" occurring in a space-like separated region. Such property emerges only in the future light cone of the region in which the measurement is performed. This has to be so since the observer performing the measurement could send a message by a light signal, containing the information about the definite outcome of the appropriate prospective measurement.

From the above discussion it also turns out that in a situation like the one sketched in Fig.4, it is impossible to establish objectively whether the outcome of the measurement of A_2 is due to chance or is deterministic.

5. Counterfactual Arguments and the Stochastic Features of the Two Versions of Nonrelativistic CSL.

As discussed in Section 2 there are two distinct, physically equivalent versions of nonrelativistic CSL which are based on the use of the linear equation (1) plus the cooking prescription (3) for the distribution of the stochastic potential and on the use of the nonlinear eq.(4), respectively. The two schemes are physically equivalent in the precise sense specified in the last sentence of Section 2. However, in the remainder of this paper we will show that, by resorting to counterfactual arguments, one can prove that the two schemes exhibit conceptually different stochastic natures, so that the first scheme is suitable for relativistic generalization, while the second scheme is not.

The use of counterfactual arguments requires the consideration of possible worlds. One then makes this notion more precise by imposing appropriate restrictions on these worlds. We do not aim for great generality and thus we feel free to make the following precise assumptions:

a) The "laws of nature" embodied in the specific CSL scheme under consideration (Raw + Cooking or Nonlinear Equation) are assumed to hold for all possible worlds.

b) With reference to the actual and the alternative worlds (among the possible ones) we will consider, we remark that we will always deal with situations involving physical processes in which there are microsystems interacting with macroscopic measuring devices. We assume that the switching on and off of the interactions between the microsystems and the macro apparatuses are free variables whose different choices characterize the actual and the alternative worlds.

5.1 Nonrelativistic Context

Within such a context let us consider the following process taking place within the time interval (t_i,t_f) : we have a system S consisting of one free spin 1/2 particle which at time t_i is in the eigenstate of σ_x belonging to the eigenvalue +1, and two macroscopic measuring apparatuses M_1 and M_2 devised to measure the observable σ_z. As usual, in a certain time interval around t_1 ($t_i< t_1 < t_f$), an appropriate interaction between S and M_1 governed by a coupling constant g_1 yields the triggering of M_1, and, subsequently, the CSL dynamics leads very quickly to a definite outcome, e.g. $\sigma_z = +1$. Analogously around t_2 ($t_1 < t_2 < t_f$), due to an S- M_2 coupling governed by g_2 the measurement of σ_z is repeated . The specific individual process we have described constitutes the actual world.

We raise now the following counterfactual question: *In the considered individual case, if we would have not performed the measurement at t_1, would the result of the measurement at t_2 have still been +1?* The question requires the consideration, besides the actual world, of an alternative world characterized by the choice $g_1 = 0$.

5.1.1 Nonlinear Equation Scheme

Let us take the position that the "laws of nature" are given by the Nonlinear Equation version of CSL. The considered individual physical process which has taken place in the actual world is governed by the evolution equation

$$\frac{d\,|\psi_V, t>}{dt} = \Big\{ -i\, g_1 H_{S\text{-}M_1} - i g_2 H_{S\text{-}M_2} - \lambda \int dx\, \big[N(\mathbf{x}) - <N(\mathbf{x})>\big]^2 \qquad (19)$$

$$+\lambda \int dx\, \big[<N^2(\mathbf{x})> - <N(\mathbf{x})>^2\big] + \int dx\, \big[N(\mathbf{x}) - <N(\mathbf{x})>\big]\, V(\mathbf{x},t) \Big\}\, |\psi_V, t>$$

in which a specific $V(\mathbf{x},t)$ occurs among those yielding the result +1 for both measurements of σ_z . Obviously we do not know which $V(\mathbf{x},t)$ has actually occured, but this does not forbid the analysis which follows. Since, in this scheme, the distribution of $V(\mathbf{x},t)$ is independent of the physical situation under consideration and, in particular, is independent of whether g_1 is zero or not , in the alternative world the same realization of the potential occurs. This being the situation one can think of solving eq.(19) explicitly with the given $V(\mathbf{x},t)$ and with $g_1 = 0$ (alternative world) and consequently, from the conceptual point of view, one can give a precise answer to the counterfactual question. Let us examine what the answer can be. One can easily convince oneself (see Appendix A) that there are specific realizations of the stochastic potential which, when substituted in eq.(19) with g_1 and g_2 different from zero, lead to the results +1 in both measurements, while, when substituted in the same equation with $g_1 = 0$, $g_2 \neq 0$, lead to the result -1 for the measurement by M_2. Moreover such potentials have an appreciable probability of occurrence. If in the considered individual case a potential of this type has occurred the answer to the counterfactual question raised above is : *no* !

Let us confine our attention to the set of all those individual cases, among all the possible ones, for which the above situation occurs. Obviously the identification of such a set is only conceptually and not practically feasible. For all these individual cases one would be allowed to state that there is a cause-effect relation between the switching on of the apparatus M_1 and the result of the apparatus M_2. In fact, if we denote by a and b the two following events

a = switching on of g_1
b = registering the result +1 by M_2

we have, within the above set

i) a can be made to happen at will

ii) If a is made to happen then b happens

iii) If a is not made to happen b does not happen either.

The above analysis can be repeated with reference to an EPR-Bohm like set -up, leading to the conclusion that, for an appropriate subset of all possible realizations of the stochastic potentials, one is allowed to assert that a measurement at a space time point "causes" the particular result obtained in another measurement taking place in a space-like separated region.

5.1.2 Raw + Cooking Scheme

We take now the position that the "laws of nature" are embodied in the Raw+Cooking version of CSL. It is important to remark that , since in this scheme the probability of occurrence of the stochastic potential depends on the specific physical situation under consideration and in particular, for the considered case, on whether g_1 is zero or not, there is no way of relating the realization of the potential in the alternative world to the one occuring in the actual world. As a consequence there is no possibility of identifying, even conceptually, occurrences for which the considered counterfactual question admits an answer, and specifically a negative answer.

5.2 Relativistic Context

Let us consider an EPR-Bohm like set-up involving, with reference to Fig.1, the measurement by two macro-apparatuses M_1 and M_2 of the same spin component at the space time points C and B, respectively. We consider now the problem in the context of the Nonlinear Equation scheme. Suppose that the actual world is characterized by a realization of the stochastic potential which would lead to the result +1 for the measurement at C and which would also lead to the result +1 for the measurement at B when (alternative worlds) only the measurement at C or only the one at B is performed, respectively. As in the case discussed in Subsection 5.1.1 such potentials have a finite probability (1/4) of occurring. In the case in which both measurements are performed (actual world) in the reference frame O of Fig. 1 , the final result would be +1 at C and -1 at B. Let us analyze now the situation from a very slowly moving reference frame O' for which, however, B occurs earlier than C. We can then argue as follows

i) Consistency requirements about the definite outcomes of the macroscopic measuring apparatuses for the considered individual case imply that, when only one measurement (at C or at B respectively) is performed, the result +1 for the measurement at C and for the one at B should occur for O', just as it occurs for O. Analogously one must require that when both measurements are performed the results +1 at C and -1 at B, respectively, are found by O'. Then, since also for O' it is the result at B which changes when one chooses to perform both measurements, for him an event occurring later can be considered "to cause" an earlier event.

ii) If the theory is stochastically Lorentz invariant the recognition of the above fact implies that also for O such a situation should occur in some specific cases . However, according to the Nonlinear Equation, in no situation can O assert that an event occurring later could be considered "cause" of an event occurring earlier.

Concluding, requirements i) and ii) are incompatible. This shows that one could not take as a starting point for a relativistic generalization the non relastivistic CSL in the Nonlinear Equation version.

No similar argument can be developed, as discussed in Subsection 5.1.2, for the Raw + Cooking scheme. Actually the relativistic DRM presented in Section 3 has been based on such a scheme. In the next Subsection we discuss the specific formal difficulties that one meets in trying to get a relativistic generalization of the Nonlinear Equation CSL scheme. They represent the mathematical counterpart of the conceptual difficulties we have just discussed.

5.3 Difficulties of a Nonlinear Equation Scheme

We remark that there is a standard way, at the non relativistic level, to go from the Raw+Cooking scheme to the Nonlinear Equation scheme. If one follows this line, starting from the model presented in Section 3, within the relativistic T-S context, one is led to consider the following nonlinear equation

$$\frac{\delta\,|\,\psi_V(\sigma)>}{\delta\,\sigma\,(x)} = \Big\{[\Phi\,(x) - <\Phi\,(x)>]\;V\,(x) - \lambda\,[\Phi\,(x) - <\Phi\,(x)>]^2 +$$

$$+\,\lambda\,[<\Phi^2\,(x)> - <\Phi\,(x)>^2]\Big\}\;|\,\psi_V(\sigma)> \tag{20}$$

where

$$<F\,(x)> = <\psi_V\,(\sigma)\,|\,F\,(x)\,|\,\psi_V\,(\sigma)> \;. \tag{21}$$

It turns out that this equation is non integrable. This corresponds to the fact that, since the theory is built in such a way as to be stochastically relativistically invariant , it must necessarily violate requirement i) of Subsection 5.2.

Let us sketch here the proof that eq.(20) is non integrable. Let us consider a one parameter family of space-like surfaces $\sigma(\eta)$, labelled by a real continuous parameter η, $0\leq\eta\leq1$ such that $\sigma(0) = \sigma_0$ and $\sigma(1) = \sigma$. We choose an initial state $|\psi(\sigma_0)>$ and we associate through eq.(20) with a given realization V(x) of the stochastic potential a state vector $|\psi_V(\eta)>$ to any surface of the family. Following the lines of ref.(2) it is easy to prove that, for this V(x), there exists a corresponding $V^{\sim}(x)$ to be put in the raw equation, given by

$$V^{\sim}(x) = V(x) \;+ 2\lambda <\psi_V(\eta)|\,\phi(x)\,|\,\psi_V(\eta)> \tag{22}$$

(η being the value of the parameter characterizing the surface passing through x) , such that the states associated to the surface $\sigma(\eta)$, for any η, in the two schemes coincide. If we now consider another one parameter family of space-like surfaces $\sigma'(\tau)$ we can play the same trick. However, since eq.(22) shows that the $V^{\sim}(x)$ to be put in the raw equation depends, besides x, on the particular family of surfaces which is considered, the potential $V^{\sim}(x)$ associated to the same space time point when the family $\sigma'(\tau)$ is considered, turns out to be different from the one associated to the family $\sigma(\eta)$. Since, however, raw equations with different $V^{\sim}(x)$ lead to different final states from the same initial state, this shows that the non linear equation is not integrable or equivalently that the final state depends on the specific way which one follows to go from σ_0 to σ.

6. Conclusions

We have presented a relativistic CSL formalism which describes a stochastically Lorentz invariant reducing dynamics. We have discussed specific problems like nonlocality and macro-objectivism within such a relativistic scheme and we have pointed out the necessity of introducing an appropriate criterion for the attribution of elements of physical reality to physical systems. The considered theoretical scheme makes use of a T-S linear equation with a skew-hermitian coupling to a stochastic potential and of an appropriate rule to evaluate the probability of occurrence of the potential. By making use of counterfactual arguments we have shown that, contrary to the nonrelativistic case in which one has two equivalent versions of CSL, the relativistic generalization must be based on one of them. If one tries to get it from the alternative scheme one meets insurmountable difficulties.

Appendix A

With reference to the situation considered in Subsection 5.1.1 we remark that the values of the potential which are relevant for the reductions are those in very narrow neighbourhoods of t_1 and t_2. Let us consider the probability of occurence of a potential $V_1(\mathbf{x},t)$ for $t \cong t_1$ and of a potential $V_2(\mathbf{x},t)$ for $t \cong t_2$. Due to the white noise character of the distribution of V, the joint probability of occurrence of the two considered potentials is the product of their probabilities

$$P\,[V_1\&V_2] = P\,[V_1]\,P\,[V_2]. \tag{A.1}$$

It has to be stressed that, for *a given potential* in the time interval in which the reduction of the macro-apparatus takes place, the outcome of the measurement depends in a crucial way on the state of the system which triggers it. According to CSL, the probability of occurrence of a $V_1(\mathbf{x},t)$ leading, when M_1 is triggered by the eigenstate of σ_x corresponding to the value +1, to the outcome " M_1 registers that σ_z =+1", is 1/2. Similarly, the probability of occurrence of a $V_2(\mathbf{x},t)$ leading, when M_2 too is triggered by the eigenstate of σ_x corresponding to the value +1, to the outcome " M_2 registers that σ_z= -1", is 1/2. [Note that the same $V_2(\mathbf{x},t)$, when M_2 is triggered by the eigenstate of σ_z corresponding to the value +1 leads to the outcome " M_2 registers that σ_z =+1"]. It follows then from eq.(A.1) that the probability of occurrence of a potential leading when $g_1\neq 0$ and $g_2\neq 0$ to the result +1 and when $g_1=0$ and $g_2\neq 0$ to the result -1 for M_2 is 1/4.

References

1. G.C. Ghirardi, A. Rimini, T. Weber, Phys. Rev. **D 34**, 470 (1986); ibid. **D 36**, 3287 (1987); J.S. Bell, in *Schrodinger: Centenary Celebration of a Polymath*, C.W. Kilmister, ed. (Cambridge University Press, Cambridge, 1987), p. 41; G.C. Ghirardi, A. Rimini, to appear in: *Sixty Two Years of Uncertainty: Hystorical Philosophical and Physics Inquiries into the Foundations of Quantum Physics*, A.I. Miller ed. Plenum Press, NY.

2. P. Pearle, Phys. Rev. **A 39**, 2277 (1989); N. Gisin, Helv. Phys. Acta **62**, 363 (1989); G. C. Ghirardi, P. Pearle, A. Rimini, Phys. Rev. **A 42**, 78 (1990)

3. P. Pearle, to appear in: *Sixty Two Years of Uncertainty: Hystorical Philosophical and Physics Inquiries into the Foundations of Quantum Physics*, A.I. Miller ed. Plenum Press, NY; G.C. Ghirardi, R. Grassi, P. Pearle, IC/90/87 Trieste preprint, to appear in Found. Phys..

4. H.P. Stapp, preprint LBL-26968; see also N. Gisin, Phys. Rev. Lett. **52**, 1657 (1984).

5. D. Gatarek, N. Gisin, Continuous Quantum Jumps and Infinite-Dimensional Stochastic Equations, preprint June 1990.

6. I. Bloch, Phys. Rev. **156**, 1377 (1967); Y. Aharonov, D.Z. Albert, Phys. Rev. **D 21**, 3316 (1980); ibid. **D 24**, 359 (1981).

7. B. d'Espagnat: Conceptual Foundations of Quantum Mechanics, second edition, Addison-Wesley 1989.

QUANTUM MEASUREMENT SCHEME AND NEW EXAMPLES OF GENERALIZED OBSERVABLES

Marian Grabowski
Institute of Physics, Nicholas Copernicus University, Toruń, Poland

ABSTRACT

We provide new examples of generalized observables. Our formulas are closely related to the recent developments of theoretical description and experimental technics in quantum optics. All of them can be interpreted as the quantum measurements.

Introduction

Observables in quantum mechanics are equivalently represented either as a self-adjoint operator A or as a spectral measure E^A. The properties of E^A guarantee that the function p_ϱ^A defined as

$$p_\varrho^A(\Delta) = \mathrm{Tr}\,\varrho\, E^A(\Delta), \qquad /1/$$

where $\Delta \in \mathcal{B}(\mathbb{R})$, ϱ is a density operator describing the state of the system, is a probability measure. A simple observation shows that the property of E^A being a projection is redundant. Any positive operator valued measure (POV) preserves the sense of /1/ as the probability measure and in consequence the probabilistic Born interpretation of quantum mechanics. Such POV-measures we call generalized observables or semispectral measures. This relatively old approach, of which sense, history and aims have recently been described [1] is poor in examples. Only few examples were built and used although the theoretical background of this approach is very well developed. A theory which can only be applied in very few cases is not a good one from methodological point of view. Such a theory may be useless because it is then sufficient to treat and investigate the examples only.

The notion of generalized observable is closely connected to the quantum measurement theory (QMT). Within the scheme of QMT we can easily indicate quite generally the whole class of generalized observables. In the QMT we couple an object system in the initial state ϱ to a measuring apparatus in the initial state σ. The joint evolution is described by a unitary group U(t). Let F be a pointer observable (F describes a quantity in a measuring device which corresponds in any way to the measured quantity of the object system) and let E be the observable in the object system. The natural basic requirement with respect to the Born interpretation is that the pointer observable and the final state of the measuring apparatus reproduce the probabilities of measured quantity in the initial state ϱ of the object system. The "probability reproducibility condition" can be expressed as

$$\mathrm{Tr}\,(I\otimes F(\Delta))U(t)(\varrho\otimes\sigma)U^*(t) = \mathrm{Tr}\,\varrho E(\Delta), \qquad /2/$$

where E is a generalized observable. Even if the $F(\Delta)$ are projection operators, E need not be projection operator. The /2/ can serve as a definition of E.

In concrete experiment we prepare σ, F and U(t) and the measured observable E is accessible by means of /2/. The equality /2/ provides the scheme to obtain the whole class of generalized observables. As we will see from applications F may in generally be a spectral or semispectral measure. We also remark that the role of the measuring apparatus and the object system is not rigorously determined. It can sometimes happen that F and E operate in the same space.

Below we list some examples which can more or less be traced from /2/ and give the explicite expressions for E.

Quantum nondemolition measurement of the photon number

The signal and probe optical waves interact possibly giving rise to the optical Kerr effect. The Hamilton operator is the following

$$H = \omega_1(a_1^*a_1\otimes I) + \omega_2(I\otimes a_2^*a_2) + \alpha(a_1^*a_1\otimes a_2^*a_2), \qquad /3/$$

where the last term describes the interation in the Kerr medium. The idea of this experiment comes from Imoto, Haus, Yamamoto [2]. The experiment of this type has also been reported [3]. The signal we treat is the investigated object system in which we want to measure the photon number N_1

$= a_1^* a_1$. The probe wave is our measuring apparatus, where the optical phase is the pointer observable. The number of quanta $N_2 = a_2^* a_2$ as the read-out observable cannot be accepted. The operator N_2 is a constant of motion and it does not reflect any correlation established by the interaction. To quantum - mechanically describe the phase we accomodate here the standard description as in [4]. It is based on the polar decomposition of $a = P\sqrt{N}$, where the partial isometry P generates in a unique way the semispectral measure F, which satisfyies the property

$$e^{-iN\theta} F(\Delta) e^{iN\theta} = F(\Delta - \theta) \qquad /4/$$

modulo 2π, because $\Delta \in \mathcal{B}(0,2\pi)$ σ-algebra of Borel subsets on $(0,2\pi]$. Thus from /2/ after some calculations with $U(t) = \exp(-itH)$ where H is equal to /3/ we have the following generalized observable for the number of photons in the signal mode

$$E(\Delta) = \sum_{n=0}^{\infty} P_n \mathrm{Tr}\, e^{-iN_2 t\omega_2} \sigma e^{iN_2 t\omega_2} F(\Delta + n\alpha t), \qquad /5/$$

where $P_n = |n\rangle\langle n|$, $N_1 |n\rangle = n|n\rangle$. Let us remark that here we choose the semispectral measure F as the pointer observable. From /5/ we can really construct the spectral measure $\{P_n\}$ with the desired accuracy (for details see [5]). The only question is what kind of measured procedure corresponds to the rather formal expression F.

In 1983 Milburn and Walls proposed an experiment of a

similar type but with the different pointer observable [6]. They took the real part of the complex amplitude of the probe mode

$$Y_2 = \frac{1}{\sqrt{2\omega_2}}(a_2^{*} + a_2) \qquad /6/$$

as the pointer observable. In quantum optics it is widely known that we can measure the observable /6/ using the homodyne detection scheme [7]. Let us assume the initial state of the detector as a coherent state $\sigma = |y\rangle\langle y|$ such that $y^{*} = y$. The operator Y_2 is a symmetric one and we can write the spectral decomposition for Y_2 with respect to the POV measure

$$F(\Delta) = \int_{\Delta} \left(\int_{-\infty}^{+\infty} |z\rangle\langle z| \frac{dp}{\pi} \right) dq, \qquad /7/$$

where $|z\rangle$ is a coherent state. At this stage we assume that the observable /7/ describes adequately the process of homodyne detection. Now from /2/ after calculations we obtain

$$E(\Delta) = \frac{1}{\sqrt{\pi}} \sum_{n=0}^{\infty} P_n \int_{\Delta} \exp\,(q - y\cos nt\alpha)^2 dq. \qquad /8/$$

The semispectral measure /8/ allows to reproduce the result from [6] by means of the integral

$$\int q dE(q) = y\cos a_1^{*}a_1\alpha t. \qquad /9/$$

As in [6] we can infer a value of $\cos \alpha t a_1 a_1$ - a nonlinear function of the number operator N_1 by making measurements on Y_2.

We would like to stress that the formulas /5/, /7/ for generalized observables in quantum nondemolition measurement are completely new. The considerations in [2] as well as in [6] are done at level of mean values and variances and not in terms of spectral and semispectral measures connected with the whole measurement scheme.

Quantum mechanical lossless beam splitter and photon statistics

A quantum mechanical treatment of a beam splitter provides an interesting class of generalized observables. Part of them can serve as the models for the photon counting.

To demonstrate quantum mechanically the mixing of two optical signals by a partially transmitted beam splitter the following description is assumed. We have the beam splitter

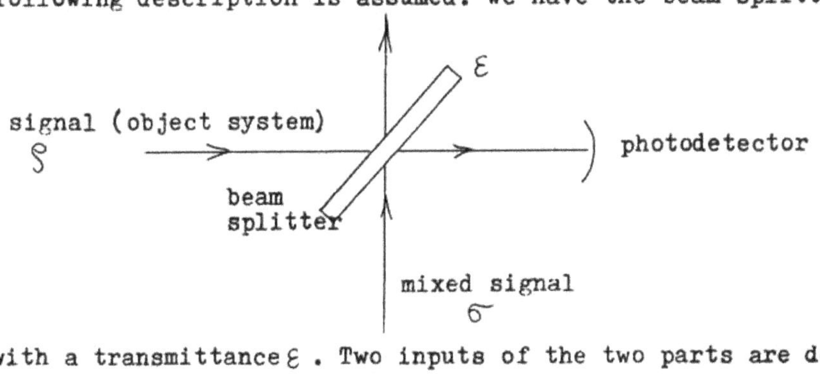

with a transmittance ε . Two inputs of the two parts are described by the two Hilbert spaces $\mathcal{H}_1$ and $\mathcal{H}_2$ which correspond to two different modes of quantum electromagnetic field with the same frequency ω. Let a^* and a act in $\mathcal{H}_1$ as the creation and anihilation operators and similarly let b^* and b act in $\mathcal{H}_2$. The beam splitter transformation acts on

$\mathcal{H}_1 \otimes \mathcal{H}_2$ and is the following

$$U(\alpha) = \exp(\alpha^* a \otimes b^* - \alpha a^* \otimes b), \qquad /10/$$

where $\alpha = re^{i\varphi}$, $\cos r = \sqrt{\varepsilon}$. The expression /10/ is a unitary operator acting on $\mathcal{H}_1 \otimes \mathcal{H}_2$. Let us assume that the signal is in the state ϱ and the mixed signal in the state σ. In the output further identifyied with the space $\mathcal{H}_1$ we have the ideal photodetector. Such detectors are usually described by the spectral measure $\{|n\rangle\langle n|\}_{n=0}^{\infty}$ connected with the number of photons $N = a^* a = \sum_{n=0}^{\infty} n|n\rangle\langle n|$.

Now we can look what kind of semispectral measures we obtain for different but known states σ. In quantum optics it is typical to treat the signal state as unknown and to manipulate with the known signal described by σ. All further calculations we make according to the formula

$$\mathrm{Tr}_{\mathcal{H}_1 \otimes \mathcal{H}_2} (\varrho \otimes \sigma) U(\alpha)^* (|n\rangle\langle n| \otimes I) U(\alpha) = \mathrm{Tr}\, \varrho E_n \qquad /11/$$

we simply eliminate the degree of freedom connected with the mixed signal. As the result we obtain a generalized observable E_n connected with the measurement device: beam splitter + photodetector.

A. Photomixing with the vacuum

We take $\sigma = |0\rangle\langle 0|$, where $|0\rangle$ is the state of vacuum. The calculations based on the formula [8]

$$U(\alpha) = \exp(-x^* a \otimes b^*) \exp\left(-\frac{y}{2}(a^* a \otimes I - I \otimes b^* b)\right) \exp(x a^* \otimes b) \qquad /12/$$

for $x = -(\mathrm{tg}\, r)e^{i\varphi}$, $y = -2\ln\cos r$ give

$$E_n^A = \sum_{m=n}^{\infty} \binom{m}{n} \varepsilon^n (1-\varepsilon)^{m-n} |m\rangle\langle m| . \qquad /13/$$

The spectral decomposition with respect to /13/ is as follows

$$\sum_{n=0}^{\infty} n E_n^A = \varepsilon N . \qquad /14/$$

We see that the part ε of photons from the signal mode is described in terms of /13/. In this case it is simultaneously the total number of all photons in the output arm (that with the detector). In the language of probabilities $p_n = \mathrm{Tr}\, \varrho E_n^A$ corresponds to the binomial distribution. Beam splitter experiments with the vacuum have indeed demonstrated binomial counting statistics [9].

The considered example can also be used as the model for the photodetector with effective quantum efficiency equal to ε. In the limit $\varepsilon \to 1$ we have $E_n^A \to P_n = |n\rangle\langle n|$.

In turn, in the language of semispectral measures /11/ provides the concrete formula for the Naimark dilation of /13/ to the spectral measure in the larger Hilbert space.

B. Photomixing with the coherent signal

We now assume that $\sigma = |v\rangle\langle v|$, where $|v\rangle = D(v)|0\rangle$ is a coherent state, $D(v) = \exp(va^* - v^* a)$ is a displacement operator. Again the calculations according to /11/ by means of /12/ give a compact formula for E_n

$$E_n^B(v) = D^*(xv)E_n^A D(xv), \qquad /15/$$

where $|x|^2 = \frac{1-\varepsilon}{\varepsilon}$ and E^A is equal to /13/. The spectral decomposition with /15/ is equal to

$$\sum_{n=0}^{\infty} nE_n^B(v) = \varepsilon N + (1-\varepsilon)|v|^2 + \sqrt{\varepsilon(1-\varepsilon)}(v^* a + va^*). \qquad /16/$$

Again we have here part of photons counted in the output arm. The last term describes the interference of two signals which takes place owing to mixing in the beam splitter.

Sometimes the following procedure is described [10]. We take an almost transparent beam splitter $\varepsilon \approx 1$ and the coherent signal with a great intensity $|v|^2$. The following limiting procedures $|v| \to \infty$, $|x| \to 0$ $(\varepsilon \to 1)$ and simultaneously $|xv| \to |z|$ precisely express the above mentioned physical situation. These limits on the level of generalized observables /15/ induce the transition

$$E_n^B(v) \to D^*(z)\,|n\rangle\langle n|\,D(z).$$

We have now a new parameter z. The contributions of $|v\rangle\langle v|$ to the output state in the detector arm is merely a classical driving force. This force is described by the displacement operator D(z) and z is a fluctuation parameter.

$$\sum_{n=0}^{\infty} nD^*(z)\,|n\rangle\langle n|\,D(z) = N + |z|^2 + za^* + z^* a \qquad /17/$$

and our limiting procedure preserves the known phase in $|v\rangle$. From /17/ we can extract the information about the number of photons and about the phase. This can also be done at the level of generalized observables [11].

C. Photomixing with a Gaussian noise

Let σ be the Gibbs state $\sigma = z^{-1}\exp(-\beta H)$, where $\beta = (kT)^{-1}$ and $H = b^*b$. The mixed state here describes the thermal noise. We use the Glauber representation

$$\sigma = \int P(v)\,|v\rangle\langle v|\,d^2v/\pi, \tag{18}$$

where

$$P(v) = \frac{1}{\langle N\rangle_G}\exp\left(-|v|^2/\langle N\rangle_G\right). \tag{19}$$

According to /11/ we obtain

$$E_n^C = \int P(v)D^*(xv)E_n^A D(xv)\frac{d^2v}{\pi}. \tag{20}$$

The last semispectral measure yields the following expression for the number of photons

$$N^C = \varepsilon N + (1-\varepsilon)\langle N\rangle_G. \tag{21}$$

The statistic for the photon counting was experimentally investigated. On the photocatode the signal is mixed with the Gaussian optical field with almost the same frequency [12]. The semispectral measure which reproduces that statistics is the following

$$A_n = \int P(y)D^*(y)\,|n\rangle\langle n|\,D(y)\,d^2y/\pi. \tag{22}$$

For A_n we have $N^A = N + \langle N\rangle_G$. N^A is always different from N because of quantum Gaussian noise. E_n^C can be used as the model for /22/ when $\varepsilon \approx 1$.

Complementarity in a quantum nondemolition measurement

Sanders and Milburn have proposed a model experiment which reveals a continuous transition from wavelike to corpuscular behaviour of the photon [13].

Quantum nondemolition measurement is follows

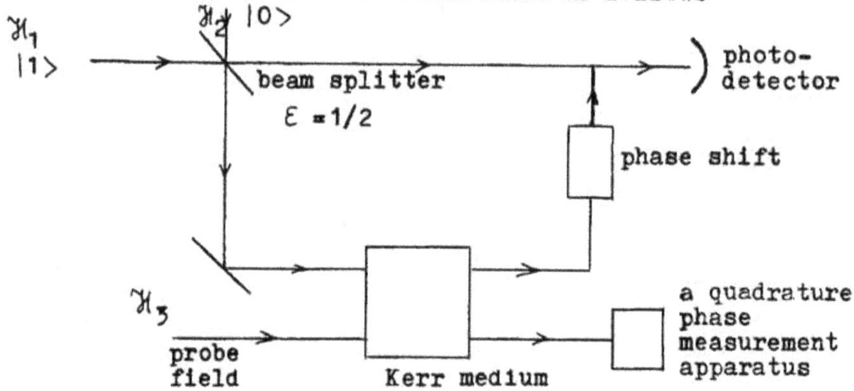

We use a Mach - Zehnder interferometer. One arm of the interferometer contains the Kerr medium. The probe field interacts with the field in this arm via the Kerr medium. The Hilbert spaces $\mathcal{H}_1$ and $\mathcal{H}_2$ correspond to the mode of field in two arms of the interferometer and $\mathcal{H}_3$ corresponds to the mode of the field probe. The Hamiltonian for this system is given by the formula

$$H = \hbar\omega_a(I\otimes N_2\otimes I) + \hbar\omega_b(I\otimes I\otimes N_3) + 2\chi(I\otimes N_2\otimes N_3) + \\ + \frac{1}{2}\hbar\chi_a(I\otimes N_2^2\otimes I) + \frac{1}{2}\hbar\chi_b(I\otimes I\otimes N_3^2), \qquad /23/$$

where $N_i = a_i^* a_i$; ω_a, ω_b are angular frequencies for the modes and χ, χ_a, χ_b are parameters of nonlinearity, which are frequency dependend. The interaction term $(I \otimes N_2 \otimes N_3)$ describing the Kerr medium does not change the number of photons but alters the phase. A presence of a photon in the arm of the interferometer coupled with the probe field produces a phase shift in the probe. In turn, an absence of a photon does not change the phase. The quantum nondemolition measurements of the probe field (measurement of the phase) determine the presence of a photon without destroying the photon. The whole proposition shows how the certainty of measuring the photon along the arm coupled to the probe field affects the observation of the interference pattern at the interferometer output (photodetector).

We are here interested only in the description of the measurement of the probe field. We show how this procedure is connected with the generalized observables. The probe field undergoes a quadrature phase measurement, which measures the phase shift. The operator associated with it is the following

$$Y(\varphi) = \frac{1}{2}(e^{-i\varphi}a_3 + e^{i\varphi}a_3^*), \qquad /24/$$

where φ is the reference phase of the apparatus. It is a similar choice of the pointer observable as in /7/. Let us remark that $Y(\varphi)$ is proportional to $\cos\varphi Q + \sin\varphi P$, where P,Q have the properties of momentum and position operators. $Y(\varphi)$ is a selfadjoint operator and has a unique spectral measure E on the real line. The initial state for the whole set is equal to $|\Psi\rangle\langle\Psi|\otimes\varrho_3$, where ϱ_3 is a state of the probe field. We assume the situation when only one photon is inside the interferometer. Thus $|\Psi\rangle$ is equal to

$$|\Psi\rangle = \frac{1}{2}(|1\rangle_1 \otimes |0\rangle_2 - i\,|0\rangle_1 \otimes |1\rangle_2), \qquad /25/$$

because the transmittance $\varepsilon = 1/2$. We can also adjust the measuring device so that only the interaction $2\chi(I\otimes N_2 \otimes N_3)$ plays the role. Now the probability of finding the outcome in the set Δ is equal to

$$p(\Delta) = \mathrm{Tr}\,(I\otimes I\otimes E(\Delta))U(t)(|\Psi\rangle\langle\Psi|\otimes\varrho_3)U(t)^* = \mathrm{Tr}\,A(\Delta)\varrho_3, \qquad /26/$$

where

$$A(\Delta) = \frac{1}{2}(E(\Delta) + e^{-2i\chi N_3 t}E(\Delta)e^{2i\chi N_3 t}). \qquad /27/$$

The semispectral measure /27/ carries information whether a photon is present or not in the second arm of the interferometer.

References

[1]. P.Busch, M.Grabowski, P.J.Lahti, Found. of Phys. Lett. 2 (1989) 331.

[2]. N.Imoto, A.H.Haus, Y.Yamamoto, Phys. Rev. A32 (1985) 2287.

[3]. N.Imoto, S.Watkins, Y.Sasaki, Optics Commun. 61 (1987) 159.

[4]. A.S.Holevo, Probabilistic and Statistical Aspects of

Quantum Theory (North Holland, Amsterdam 1982).

[5]. M.Grabowski, Quantum Measurements and Dynamics, Annalen der Physik (to appear).

[6]. G.J.Milburn, D.F.Walls, Phys. Rev. A28 (1983) 2065.

[7]. H.P.Yuen, J.H.Shapiro, in Coherence and Quantum Optics IV, ed. by L.Mandel, E.Wolf (Plenum, New York 1980).

[8]. A.M.Perelomov, Generalized Coherent States and Their Applications (Springer Verlag 1986).

[9]. J.Brendel, S.Schütrumpf, R.Lange, W.Martienssen, M.O. Scully, Europhys. Lett. 5 (1988) 223.

[10]. M.Kitagawa, Y.Yamamoto, Phys. Rev. A34 (1986) 3974.

[11]. M.Grabowski, The Phase in Quantum Mechanics (in preparation).

[12]. F.T.Arecchi, V.Degiorgio, in Laser Handbuch Vol. I ed. by F.T.Arecchi, E.D.Schultz-Dubois, (North Holland, Amsterdam 1972).

[13]. B.C.Sanders, G.J.Milburn, Phys. Rev. A39 (1989) 694.

COLLAPSE IN RELATIVE-STATES QUANTUM MEASUREMENT THEORY WITHOUT THE MANY-WORLDS ASSUMPTION

FEDOR HERBUT

Faculty of Physics, University of Belgrade, P.O.Box 550

11001 Beograd, Yugoslavia

ABSTRACT

A resolution of the problem of measurement (or of objectification) is suggested along the lines of considering quantum mechanics correct and valid for the measuring instrument, but incomplete as far as individual-system properties are concerned. The pointer observable is taken to be a beable (an observable with hidden, i.e., definite premeasurement values). The suggested resolution is shown to be a natural further development of the almost forgotten Jauch objectification theory.

1. Introduction

I chose the title of this talk under the influence of the long history of the quantum measurement problem: of its pre-Everett stage,[1,2] of the renowned paper of Everett,[3] and of the post-Everett studies,[4,5,6,7] especially of Whitaker.[8] Actually, in view of the newest terminology,[9] the title should be "Objectification in discrete unitary premeasurement".

The quantum measurement problem, as well known, consists in the baffling theorem that if quantum mechanics (QM) is correct and complete, then there is no measurement in nature (though QM does not make sense, i.e., it cannot be interpreted, without measurement). The proof of this important conclusion can be found in references 10 and 11, or in the clear presentation of prof. Mittelstaedt at this symposium.[12]

I will now shortly set up the basic equations of quantum measurement theory (equivalently as in references 9 or 12). The indices 1 and 2 refer to the measured object and the measuring instrument (MI) respectively. The (discrete) measured observable in spectral form is

$$A_1 = \Sigma_k a_k P_1^{(k)} \tag{1}$$

$(k \neq k' \Rightarrow a_k \neq a_{k'})$. The pointer observable, in partial spectral form, is given by

$$B_2 = \Sigma_k b_k Q_2^{(k)} + b_0 Q_2^{(0)} + B_2' , \tag{2}$$

where again $k \neq k' \Rightarrow b_k \neq b_{k'}$, and $Q_2^{(l)} B_2' = 0$ for all $l=0,k$ (b_k are the so-called pointer positions). Let an arbitrary initial state of the object be expressed by the state vector $|\psi_0\rangle_1$, and let the initial state of the MI be $|\chi_0\rangle_2$, statisfying

$$B_2 |\chi_0\rangle_2 = b_0 |\chi_0\rangle_2 . \tag{3}$$

If QM is correct, then there exists a unitary evolution operator U_{12} that contains the object-MI interaction and that maps the initial state of the composite system into its *final state*:

$$U_{12}(|\psi_0\rangle_1 \, |\chi_0\rangle_2) \stackrel{d}{=} |\phi\rangle_{12}. \tag{4}$$

(This aspect of correctness of QM is often called its "universal validity".)

The decomposition of $|\psi_0\rangle_1$ with respect to Eq. (1) is

$$|\psi_0\rangle_1 = \Sigma_k w_k^{1/2} \, (w_k^{-1/2} P_1^{(k)} |\psi_0\rangle_1), \tag{5}$$

where

$$\forall k : \qquad w_k \stackrel{d}{=} \langle\psi_0| P_1^{(k)} |\psi_0\rangle_1 \tag{6}$$

are the probabilities of the results a_k. Linearity of U_{12} implies (upon replacement of Eq. (5) in Eq. (4)):

$$|\phi\rangle_{12} = \Sigma_k w_k^{1/2} |\phi^{(k)}\rangle_{12} , \tag{7}$$

where the definite-k components $|\phi^{(k)}\rangle_{12}$ of the final state $|\phi\rangle_{12}$ are by definition given by

$$\forall k : \quad |\phi^{(k)}\rangle_{12} \overset{d}{=} U_{12}[(w_k^{-1/2} P_1^{(k)} |\psi_0\rangle_1)|\chi_0\rangle_2] \tag{8}$$

(unless $w_k = 0$), and obviously

$$\langle\phi^{(k)}|\phi^{(k')}\rangle_{12} = \delta_{k,k'} \quad . \tag{9}$$

For U_{12} to be the evolution operator of measurement it is evidently necessary that $|\phi^{(k)}\rangle_{12}$ is a definite-pointer-position state:

$$\forall k : \quad (1 \otimes B_2)|\phi^{(k)}\rangle_{12} = b_k |\phi^{(k)}\rangle_{12} \quad . \tag{10}$$

A glance at Equations (5) and (7) (that contain the same $w_k^{1/2}$) reveals that Eq. (10) is also sufficient for the general property of U_{12} that consists in defining unitary premeasurement: that the probability distribution implied by $|\psi_0\rangle_1$ with respect to A_1 is the *same* as that entailed by $|\phi\rangle_{12}$ regarding $(1 \otimes B_2)$.[9]

2. Objectification Through a Beable Pointer Observable

Taking up the problem of quantum measurement theory again, one cannot help noticing that the final state $|\phi\rangle_{12}$ is, in general, symmetrical in all k values. Hence, if QM is complete, i.e., if nothing more can be said about the *individual* systems than what is implied by the state vector, then there are no definite measurement results. Using the new term,, this is the problem of objectification.

In this talk I want to give a short progress report about a search for a solution of the objectification problem along the lines of the following assumptions:

A) The state vector $|\phi\rangle_{12}$ is treated from the point of view of the *realist*, i.e., it is taken to represent knowledge on individual (1+2)-systems that are assumed to possess some objective (premeasurement) properties. This knowledge actually applies to the quantum ensemble that is described by $|\phi\rangle_{12}$. (This ensemble is prepared in two stages: the first is the preparation of the initial state $|\psi_0\rangle_1|\chi_0\rangle_2$, and the second is the evolution, see Eq. (4).)

B) It is assumed that QM is correct but *incomplete*, i.e., when it comes to individual systems, there may exist objective properties not comprised by the information that is contained in the quantum-mechanical state, in particular in $|\phi\rangle_{12}$.

The essential idea is that in this approach the quantum measurement problem is understood as a symptom of the wrongness of the completeness assumption.

I am advocating a *minimal* extension of QM: the latter is considered as entirely correct and as little as possible should be added to it. Evidently, this is achieved if one assumes that at the end of measurement each (1+2)-system is described by

$$|\phi\rangle_{12} \text{ plus a } b_{k_0} \text{ value.} \tag{11}$$

Let us clarify that (11) does not make B_2 a superselection observable, i.e., it does not require $(1\otimes B_2)$ to commute with all (1+2)-observables of interest (making B_2 thus an essentially classical variable). If this were the case, we would still be on purely quantum-mechanical ground, and $|\phi\rangle_{12}$ would have to be replaced by the mixed state $\rho_{12} \stackrel{d}{=} \Sigma_k w_k |\phi^{(k)}\rangle_{12}\langle\phi^{(k)}|_{12}$ (as it is done by explicit ad hoc assumption in conventional QM).

Stipulation (11) requires that only "reading the pointer position" (measurement of $(1\otimes B_2)$) is to be understood as learning about a *pre-existing property* of the (1+2)-system. It allows the measurement of all other (1+2)-observables to be possibly understood as creation (not discovery) of results. We are actually assuming that $(1\otimes B_2)$ is a *beable*[13] (i.e., that the b_k values are "able to be" or exist prior to measurement).

3. Nonquantum-Mechanical Subensembles

Though I am discussing the project of a minimal extension of QM, we are bound to be plagued by a host of new ideas and problems. The first that occurs to me is the following.

Let me picture a laboratory ensemble of N (1+2)-systems that is described by $|\phi\rangle_{12}$. Owing to (11), I can think of the subensemble of all those (1+2)-systems in the ensemble that have the pointer position b_{k_0} (they must objectively exist if (11) does make sense). Let there be N_{k_0} of them. (Then $N=\Sigma_k N_k$, and $\lim N_k/N=w_k$, when $N\ \infty$.) I am talking about nonquantum-mechanical subensembles because the quantum-mechanically homogeneous ensemble $|\phi\rangle_{12}\langle\phi|_{12}$ does not contain quantum-mechanical subensembles inequivalent to the entire ensemble itself. How to describe these new subensembles?

Ensembles and subensembles are described by statistical states. Hence (11) requires the extension of the set of all statistical operators ρ_{12} at least by the possibly nonquantum-mechanical definite-b_k subensembles (states) $\bar{\rho}_{12}^{(k)}$ of quantum-mechanical ensembles (states) ρ_{12}. Incidentally, if and only if $[\rho_{12}, (1\otimes Q_2^{(k)})]=0$ then $\bar{\rho}_{12}^{(k)} = (1\otimes Q_2^{(k)})\rho_{12}/Tr_{12}(1\otimes Q_2^{(k)})\rho_{12}$ is a statistical operator.

One should have

$$|\phi\rangle_{12}\langle\phi|_{12} = \Sigma_k w_k \bar{\rho}_{12}^{(k)} \tag{12}$$

(cf. (4), (6), (7) and (10)) in the *extended set* of states. Decomposition (12) is ill defined because at present we lack the nonquantum-mechanical states $\bar{\rho}_{12}^{(k)}$.

4. On the Missing Nonquantum-Mechanical States

If C_{12} is a general (1+2)-observable, and $\bar{\rho}_{12}^{(k)}$ is a nonquantum-mechanical definite-b_k state, then we do not know if the average $\langle C_{12}\rangle_{\bar{\rho}_{12}^{(k)}}$ exists, and if it does, how to evaluate it.

We now make a simple and natural nonquantum-mechanical assumption on the beable B_2. We assume that the premeasurement value b_{k_0} that any individual (1+2)-system in the state $\bar{\rho}_{12}^{(k_0)}$ possesses and that is revealed when measuring B_2 is *not influenced* by either the choice of the first-subsystem observable C_1 that is measured in coincidence with B_2, or by the value c_1 of C_1 that is thus obtained.

Without this assumption the beable B_2 would hardly be of any use to us at all. In spite of its simplicity, the assumption is actually very nontrivaial. This is so due to the so-called contextuality,[14] a nonquantum-mechanical new concept that appears indispensably when one extends QM into a completely causal hidden-variables (HV) theory. Acceptance of contextuality is an especially high price to pay for composite systems, when it takes on the special form of *nonlocality*: if C_1 and C_2 are measured on one and the same individual (1+2)-system (i.e., are measured in coincidence), then, in general, the result c_1 of C_1 depends on the choice of C_2 and on its result c_2, and symmetrically (interchanging C_1 and C_2). Hence, the above assumption means partial locality: it is confined to the pointer observable B_2 and it is one-way, i.e., the symmetrical noninfluence assumption of (B_2, b_k) on the result c_1 of C_1 is not made.

Since we are dealing with only one beable, there is no clash with Bell's theorem.[15] Bell's inequality applies to two pairs of mutually incompatible observables, one for each subsystem.

It will be shown elsewhere that the above assumption implies

$$\langle C_1 \otimes 1 \rangle_{\bar{\rho}_{12}^{(k)}} = \langle C_1 \otimes Q_2^{(k)} \rangle_{\bar{\rho}_{12}^{(k)}} \tag{13}$$

$$= w_k^{-1} \langle C_1 \otimes Q_2^{(k)} \rangle_{\rho_{12}} \tag{14}$$

$$= \mathrm{Tr}_1 C_1 \rho_1^{(k)}, \tag{15}$$

where

$$w_k \overset{d}{=} \mathrm{Tr}_{12} \rho_{12} (1 \otimes Q_2^{(k)}), \tag{16}$$

$$\rho_1^{(k)} \overset{d}{=} w_k^{-1} \mathrm{Tr}_2 \rho_{12} (1 \otimes Q_2^{(k)}), \tag{17}$$

and $\bar{\rho}_{12}^{(k)}$ is a nonquantum-mechanical subensemble of the quantum-mechanical ensemble ρ_{12}.

Thus, as far as first-subsystem prediction is concerned, the evaluation in the nonquantum-mechanical state $\bar{\rho}_{12}^{(k)}$ reduces to a calculation with the help of the quantum-mechanical state ρ_{12}. The statistical operator $\rho_1^{(k)}$ (see (15) and (17)) is the conditional state of subsystem 1 under the condition that the quantum event $Q_2^{(k)}$ takes place for subsystem 2 when the (1+2)-system is in the state ρ_{12} (see subsection 2.1. in reference 16). In the context of measurement theory, i.e., when $\rho_{12} \overset{d}{=} |\phi\rangle_{12}\langle\phi|_{12}$ (cf. (4) and (7)), $\rho_1^{(k)}$ is the *collapsed state* of the measured object (after measurement) corresponding to the pointer position b_k.

5. Connection with Jauch's Statistical States

Jauch's elegant attempt at a resolution of the objectification problem is nowadays almost forgotten. His theory was originally restricted to ideal measurement,[17,18] but it could be extended to any discrete unitary premeasurement.[19] The trouble with his theory is that it is really convincing only if the classicalness of the measuring instrument is taken quite seriously (but then it is not a quantum-mechanical theory of measurement).

Putting it roughly, we may assume with Jauch that only B_2 (and functions of B_2) can actually be measured on the MI (this is the assumption of pure classicalness of the MI). The most general actually measurable observable on the composite system is coincidence of C_1(arbitrary) and B_2 (the pointer observable). This gives rise to a redundancy in the quantum-mechanical description.

To get ridof the redundancy, Jauch makes use of the following equivalence relation in the set of all (1+2)-statistical-operators:

$$\rho_{12} \sim \rho'_{12} \text{ if } \mathrm{Tr}_{12}\rho_{12}(C_1 \otimes Q_2^{(k)}) = \mathrm{Tr}_{12}\rho'_{12}(C_1 \otimes Q_2^{(k)}), \tag{18}$$

where C_1 is an arbitrary first-subsystem observable, and k takes on all of its values.

The quotient set that consists of all equivalence classes (Jauch called them macrostates) is ipso facto a set of statistical states because it is provided with a natural convex structure via arbitrary class representative. (Empirically, convex combinations correspond to mixing of subensembles.) Jauch was able to show that the macrostate $M_{|\phi\rangle_{12}\langle\phi|_{12}}$ corresponding to $|\phi\rangle_{12}$, the final state of premeasurement, has a unique and disjoint decomposition

$$M_{|\phi\rangle_{12}\langle\phi|_{12}} = \Sigma_k w_k M^{(k)} \tag{19}$$

into definite-b_k macrostates (w_k is given by (6), and $M^{(k)}$ is the equivalence class of $|\phi^{(k)}\rangle_{12}\langle\phi^{(k)}|_{12}$, cf. (7), (8) and (10)).

Now, how do Jauch's macrostates relate to the missing nonquantum-mechanical states $\bar{\rho}_{12}^{(k)}$? Owing to (13) and (14), one can introduce Jauch's equivalence relation (18) in our extended set of states (that encompasses the $\bar{\rho}_{12}^{(k)}$). It turns out that this has the interesting consequence that some of Jauch's equivalence classes become larger (containing also a nonquantum-mechanical state $\bar{\rho}_{12}^{(k)}$) but the quotient structure, i.e., the set of Jauch's statistical states, *does not change at all.*

Returning to the more restricted set of quantum-mechanical states, we notice that decomposition (19) has a purely quantum-mechanical basis (cf.(18)). Then it has a formal appearance because it is not parallelled by a decomposition of $|\phi\rangle_{12}\langle\phi|_{12}$ (precisely this is the objectification problem). If one views the set of macrostates as a quotient structure on the extended set of states, then (19) *is parallelled* by the (hypothetical) decomposition (12) of $|\phi\rangle_{12}\langle\phi|_{12}$ into actual (though nonquantum-mechanical) states. So, (19) is no longer formal, but gives a restricted view of (12). The restriction applies to the observables C_1 in coincidence with B_2 and not to any more general observable C_{12}.

6. Concluding Remarks

A) The extended set of states (that we want to achieve) can, in principle, be obtained in two ways: By constructing in some way the definite-b_k nonquantum-mechanical states $\bar{\rho}_{12}^{(k)}$, or by restricting a full HV theory to the desired one-beable case.

Within the second alternative one can probably work out a rather simple model of two spin-1/2 particles (with the spatial degrees of freedom abstracted away), because there exists the simple and noncontextual Kochen-Specker HV theory of a spin-1/2 particle.[20] The spin-up and the spin-down values of the z-projection of the spin of the second particle would be the pointer positions of the beable $B_2 \stackrel{d}{=} s_{z,2}$. This is expected to give a mathematically rigorous model. Though it will have the physically unsatisfactory feature that the MI is not a classical system, and B_2 is not a macroscopic variable.

It is my conjecture (at the present state of insight into the research project at issue) that the beable property of B_2 has to do with the *macroscopic environment*[21,22] of B_2 (or more precisely, of a more restricted subsystem of the MI on which B_2 is actually defined).

B) A very important feature of the outlined approach is the preservation of *all the interference* in $|\phi\rangle_{12}$ (cf. (7)) in spite of the beable property of B_2. It would be very desirable to confirm *experimentally* the existence of this interference at the end of measurement. This can be done[23,6] by measuring in coincidence two suitable subsystem observables: C_1 and C_2 such that $[C_1,A_1] \neq 0$ (A_1 being the measured observable), and $[C_2,B_2] \neq 0$. It is, of course, hard to find an observable C_2 that is incompatible with the pointer observable. Finding it would also give a clear demonstration of the quantum-mechanical nature of the MI.

7. References

1. N.F. Mott, *Proc. Roy. Soc.* **A 126** (1929) 79.
2. W. Heisenberg, *Die Physikalischen Prinzipien der Quantentheorie*, 3. Auflage (S.Hirzel, Leipzig, 1942), pp.50-53.
3. H. Everett III, *Rev. Mod. Phys.* **29** (1957) 454.
4. Ph. Pearle, *Am. J. Phys.* **35** (1967) 742.
5. L.N. Cooper and D.Van Vechten, *Am. J. Phys.* **37** (1969) 219.
6. P.A. Moldauer, *Phys. Rev.* **D 5** (1972) 1028.
7. J.R. Fox. *Am. J. Phys.* **51** (1983) 49.
8. M.A.B. Whitaker, *J. Phys. A: Math. Gen.* **18** (1985) 253.
9. E.G. Beltrametti, G. Cassinelli and P.J. Lahti, *J. Math. Phys.* **31** (1990) 91.
10. A. Fine, *Phys. Rev.* **D 2** (1970) 2783.

11. B. D'Espagnat, *Conceptual Foundations of Quantum Mechanics,* 2nd edition (Benjamin,Reading, 1976).
12. P. Mittelstaedt, in *Symposium on the Foundations of Modern Physics* (Wordl Scientific, Singapore, 1990).
13. J.S. Bell, *Phys. Rep.* **137** (1986) 49.
14. A. Shimony, *Brit. J. Phil. Sci.* **35** (1984) 25.
15. J.S. Bell, *Physics* **1** (1964) 196.
16. F. Herbut, *Int. J. Theor. Phys.* **25** (1986) 1215.
17. J.M. Jauch, *Helv. Phys. Acta* **37** (1964) 293.
18. J.M. Jauch, *Fuondations of Quantum Mechanics,* (Addison-Wesley, London, 1968).
19. F. Herbut, *Int. J. Theor. Phys.* **25** (1968) 863.
20. S. Kochen and E.P. Specker, *J. Math. and. Mech.* **17** (1967) 59.
21. W.H. Zurek, *Phys. Rev.* **D 26** (1982) 1862.
22. E. Joos and H.D. Zeh, *Z. Phys.* **59** (1985) 223.
23. W.H. Furry, *Phys. Rev.* **49** (1936) 393.

THEORY OF NON-IDEAL QUANTUM NONDEMOLITION MEASUREMENT AND CONTINUOUS STATE REDUCTION OF PHOTONIC STATE IN VIEW OF UNSHARP MEASUREMENT

N. IMOTO, M. UEDA, and T. OGAWA
NTT Basic Research Laboratories,
3-9-11 Midori-cho, Musashino-shi, Tokyo 180, Japan

ABSTRACT

There seemingly is a difficulty in describing unsharp measurement and continuous measurement within the present framework of the quantum mechanics because the so called projection postulate assumes an instantaneous, sharp measurement. We show, however, that there is no such difficulty. In this paper, we formulate a coherent theory for unsharp and continuous measurements. Unsharp measurement can be described as a combination of (1) unitary interaction in which the system and apparatus are weakly correlated and (2) a projection process for the apparatus by reading out its state. Continuous measurement can be described as a sequence of infinitesimally weak unsharp measurements. Examples are given for measurement of photonic field with some numerical examples.

1. Introduction

The main argument in quantum mechanics is the conceptual mystery of quantum measurement process. One approach is to find limitations of the applicability of von Neumann's projection postulate to examine the validity of the present quantum mechanics. A limitation of the projection postulate have often been pointed out: in particular, its inability to describe photon counting, which is a continuing measurement process and is not a first kind measurement.[1,2] The projection postulate also assumes a sharp measurement. Recent exeriments have also investigated unsharp measurement.[3]

We should, however, also ask whether the limitation are real or not because if the present quantum mechanics can be applied to describing unsharp measurement and continuous measurements, it means that these are not real problems. It is also worthwhile to develop a theory for unsharp and continuous measurement processes within the present framework of quantum mechanics.

This paper gives a theoretical description of unsharp and continuous measurements as a combination of unitary coupling with the measuring apparatus and readout process of the apparatus. Examples are given in the area of photon number measurement such as familiar photon counting and some quantum nondemolition (QND) measurement schemes.

There are preceding works related to the present work. The concept of unsharp measurement is indirectly included in "general quantum measurement" by Yuen.[4] Theory of photon counting as a continuous measurement has been initiated by Srinivas and Davies[1] and developed by the authors.[5,6] A general theory of continuous measurement has been developed by Barchielli et al.[7] and specific models are examined by Caves and Milburn for the position of a free particle.[8] Theory of QND measurement of photon number has been developed by Milburn and Walls[9] and Imoto et al.[10,11] These apparently unrelated subsects are unified in this paper.

2. Unsharp Mesurement

2.1 *General description*

The basic idea is that unsharp measurement is decomposed into a combination of two steps; step 1: unitary interaction between the system to be measured and the measuring appartus and step 2: readout process of the apparatus. In step 1, a weak (imperfect) correlation is introduced between the system and the apparatus and in step 2, the system state changes in a nonunitary way associated with the readout process of the apparatus via the correlation. Let us consider that the system and the apparatus are coupled via an interaction Hamiltonian $\hat{H}_{\text{int}}$ in step 1. The coupled density operator after the interaction time t then becomes

$$\hat{\rho}_{s-a}(t)=\hat{U}(t)\,\hat{\rho}_s(0)\otimes\hat{\rho}_a(0)\,\hat{U}^{\dagger}(t), \tag{1}$$

where, $\hat{\rho}_{s-a}(t)$ is the coupled density operator after interaction time t, $\hat{U}(t)$ is a unitary operator defined as $\hat{U}(t)\equiv\exp(-iH_{\text{int}}\,t/\hbar)$, and $\hat{\rho}_s(0)$ and $\hat{\rho}_a(0)$ are the initial density operator for the system and apparatus, respectively.

In step 2, the apparatus state is read out by measuring an apparatus observable. When we only know that the measurement of the apparatus has been done but do not know the readout itself, the signal density operator for us is obtained simply by taking the trace of the coupled density operator over the apparatus as

$$\hat{\rho}_s(t)=\frac{\mathrm{Tr}_a[\hat{\rho}_{s\text{-}a}(t)]}{\mathrm{Tr}_{s\text{-}a}[\hat{\rho}_{s\text{-}a}(t)]}, \tag{2}$$

where $\hat{\rho}_s(t)$ is the system density operator just after the measurement, symbols Tr_a and $\mathrm{Tr}_{s\text{-}a}$ respectively denote traces over the apparatus, and total coupled system. Let us refer to this process as non-referring measurement process (NMP) as we do not refer to the measurement result in this process.

When the readout is recorded, the change of the system state is different from the NMP process. The state of the system, after the measurement of the apparatus, depends on the readout value X. The density operator of the system is obtained, using the operator-valued measures for the coupled system,[12] as

$$\hat{\rho}_s(t)=\frac{{}_a\langle X|\hat{\rho}_{s\text{-}a}(t)|X\rangle_a}{\mathrm{Tr}_s[{}_a\langle X|\hat{\rho}_{s\text{-}a}(t)|X\rangle_a]}=\frac{\mathrm{Tr}_a[\hat{\rho}_{s\text{-}a}(t)\,\hat{\rho}_a^{(\mathrm{read})}]}{\mathrm{Tr}_{s\text{-}a}[\hat{\rho}_{s\text{-}a}(t)\,\hat{\rho}_a^{(\mathrm{read})}]}, \tag{3}$$

where $|X\rangle_a$ is an eigenvector of apparatus observable $\hat{X}$ with an eigenvalue of X, and $\hat{\rho}_a^{(\mathrm{read})}=|X\rangle_a\,{}_a\langle X|$. Eq.(3) means that the system state is obtained by partial projection of the total density operator in the apparatus Hilbert space. If we write the above measurement process by an X-dependent "unsharp measurement" operator $M(X)$, the total process of the unsharp measurement is written as

$$\hat{\rho}_s(0)\rightarrow\hat{\rho}_s(t)=M(X)\left[\hat{U}(t)\,\hat{\rho}_s(0)\otimes\hat{\rho}_a(0)\,\hat{U}^{\dagger}(t)\right], \tag{4}$$

We call this process a referring measurement process (RMP) as we refer the measurement result X.

2.2 *Quantum nondemolition (QND) measurement*

The concept of QND meas-urement was proposed as a measurement which does not disturb the free motion of an observable to be measured.[13] The recent proposal of a QND measurement scheme[10] gives a good example of unsharp QND measurement. These schemes also have been investigated experimentally.[14] Here, we illustrate the scheme in the context of unsharp measurement.

As is shown in Fig.1, the signal beam, whose photon number is going to be measured, passes through a transparent optical Kerr medium without being absorbed. Since an optical Kerr medium is such a medium whose refractive index increases in proportion to the field intensity, (that is, photon nmber) the phase of the probe beam]is also increases inproportion to the signal photon number. Thus, if the probe phase is detected by a phase sensitive detector such as homodyne detection, the signal photon number can be measured without being disturbed. The unitary coupling is derived from the interaction Hamiltonian of the optical Kerr effect. This interaction keeps the signal or probe photon number constant, but correlates the probe phase with the signal photon number and the signal phase with the probe photon number.

If there is no phase noise in the probe beam, that is, if the probe beam is in a phase state, the correlation between the probe phase and the signal photon number is perfect. In this case, there is no measurement error, which is the feature of a sharp measurement. Also, since a phase state has an infinite uncertainty in photon number, the phase of the signal beam becomes totally uncertain due to the cross-phase modulation. This means that a sharp QND measurement of photon number is only possible at the cost of an infinite increase in phase uncertainty.

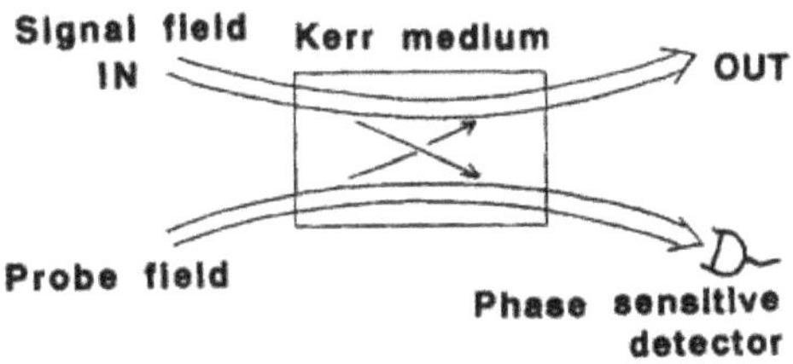

Fig. 1. Principle of QND measurement of photon number via the optical Kerr effect.

In practice, however, a phase state cannot be prepared but a coherent state, for example, is used as the probe beam. In this case, a finite Kerr effect and interaction time cause an imperfect correlation between the probe phase and the signal photon number. Thus, there is an intrinsic measurement uncertainty, which is the feature of an unsharp measurement. In this case, uncertainty increase int the signal phase is also finite. In general, there is an uncertainty relation between the unsharpness in photon number measurement and the disturbance in the optical phase. This Kerr-QND scheme may be called an ideal unsharp measurement because the uncertainty product is minimum and thus there is no excess unsharpness or disturbance. More detailed analyses can be found in Ref.10 in Heisenberg picture and Ref.11 in Schrödinger picture.

3. Continuous measurement as a sequence of unsharp measurements

3.1 *General Description*

When the coupling between the system and apparatus is infinitesimally weak, the measurement will be quite unsharp and also the disturbance will be very small. If such a very unsharp measurement is repeated many times we will gain information about the original state of the system more and more as the number of measurement steps increases. In this case, we can obtain the system state evolution for each step provided that the initial state is known. Even though the readouts are not predictable, once given the series of readouts, the state evolution of the sytem under the progressing measurement steps is obtained.

If the time interval of each step as well as the magnitude of the system-apparatus coupling is made infinitesimally small with an appropriate limit operation, the above sequence of the unsharp measurements will yield continuous readouts, $X(t)$, instead of a X_n, and the real time evolution of the system state can be obtained as a functional of readout function $X(t)$. This is the concept of the continuous measurement in this paper.

As we did before, let us consider nonreferring continuous measurement process (NMP) and reffering continuous measurement process (RMP). Let us assume that N is the number of the successive steps and Δt is the time interval of each step so that the total time T is expressed as $T = N\Delta t$. We also assume that the apparatus state is reset to an initial state $\hat{\rho}_s(0)$ each time. Then, the system state evolution in NMP is easily obtained from Eq.(2), as

$$\hat{\rho}_s(N\Delta t) = \mathrm{Tr}_a\left[\cdots\mathrm{Tr}_a\left[\hat{U}\mathrm{Tr}_a\left[\hat{U}\hat{\rho}_s(0)\otimes\hat{\rho}_a(0)\,\hat{U}^\dagger\right]\otimes\hat{\rho}_a(0)\,\hat{U}^\dagger\right]\otimes\cdots\right], \quad (5)$$

where, the density operator is not yet normalized (and so, one needs to normalize it for practical calculations). By making the step Δt infinitesimally small, the whole process becomes a non-referring continuous measurement process. In this case, as will be described later, Eq.(5) may lead to a diffrential equation, which determines the real time nonunitary evolution of the system.

In RMP, the evolution depends on the readouts. Let us assume that the readouts are $X_1, X_2, \cdots, X_N$. The evolution of the system density operator is written as

$$\hat{\rho}_s(N\Delta t\,; X_1, X_2, \cdots, X_N) = M_N\left[\cdots M_2\left[\hat{U}M_1\left[\hat{U}\hat{\rho}_s(0)\otimes\hat{\rho}_a(0)\,\hat{U}^\dagger\right]\otimes\hat{\rho}_a(0)\,\hat{U}^\dagger\right]\otimes\cdots\right], \tag{6}$$

where, "unsharp measurement operator" M_N is the abbreviation of $M(N)$. If we make $\Delta t \to 0$ (and $N \to \infty$ with $N\Delta t = T$), the readouts form a function of time as $X(t)$. The nonunitary state evolution of the system depende on all readouts $X(t)\,(0 \le t \le T)$ and the initial state $\hat{\rho}_s(0)$. Therefore, evolution $\hat{\rho}_s(T)$ is expressed in a functional form as $\hat{\rho}_s(T) = f\left[X(t)\,(0 \le t \le T),\, \hat{\rho}_s(0)\right]$.

In order to obtain an explicit expressions for Eqs.(5) and (6), a more specific model for the system and apparatus is needed. In case of photon counting or QND photon counting described later, the readouts forms a binary function because the photon counter only tells whether it count a photon or not. When the detector counts a photon, the process is called one-count process, which is an instantaneous event. When the detector does not count a photon, the process is called no-count process, which continues for a certain period in time. So, all we need to know is a set of times of one-count processes. This makes it possible to obtain analytic formulae of state evolution for continuous photon counting.

3.2 *Photon Counting*

The quantum theory of continuous photon counting was initiated by Srinivas and Davies,[1] in which they postulated a model for one-count and no-count processes, namely, the Srinivas-Davies model (S-D model, hereafter). In this section, we deliniate the *derivation* of the postulated S-D model as an example of the general continuous measurement. Since the detailed calculation can be found in Ref.6, we here give an explanation without much equations.

The photon counting to be considered is shown in Fig.2(a). The photodetector is so weakly coupled to the optical field in the closed cavity that the detetor does not absorb more than one photon during an inifinitesimal time interval. The output of the detector is a time sequence of photoelectric current pulses, each of which represents the detection of one photon.

Fig. 2(b) shows the physical model that we propose for the model of 2(a). Here, two level atoms, which are pre-pared initially in the ground state, are driven into the cavity one by one. Each atom passes through the cavity within the same time duration Δt, during which it interacts with the cavity field via the electric-dipole interaction (Step 1). This interaction is described by the Jaynes-Cumings Hamiltonian as

$$\hat{H}_{int} = \hbar g (\hat{a}\hat{\sigma}^{+} + \hat{a}^{+}\hat{\sigma}) \qquad (7)$$

where g is the coupling constant between the field and the atom. Using the second order perturbation theory, the mixing of the upper and lower states through the interaction can be calculated.

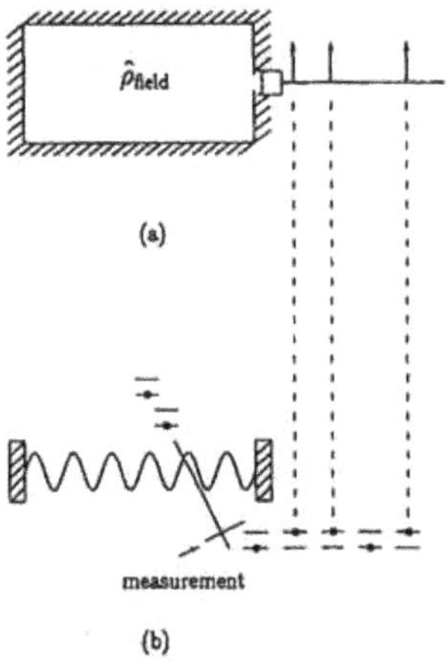

Fig. 2. Photon counting as continuous measurement. (a) Photon counting for a closed cavity field. (b) A model with two-level atom stream as the probe for continuous measurement.

After passing through the field, the level of each atom is measured. This causes a state change in the optical field (step 2). In nonreferring process (NMP), Eq.(5) is reduced to a difference equation as

$$\frac{\hat{\rho}(t+\Delta t)-\hat{\rho}(t)}{\Delta t} = -\frac{g^{2}\Delta t}{2}\left\{\left[\hat{\rho}_{f}(t)\hat{n} + \hat{n}\hat{\rho}_{f}(t)\right] - 2\hat{a}\hat{\rho}_{f}(t)\hat{a}^{+}\right\}, \qquad (8)$$

If we require that the evolution by Eq.(8) does not depend on the choice of step Δt in the limit of $\Delta t \to 0$, then $g^{2}\Delta t$ should be a constant. This means that when the time interval is made smaller, the coupling constant should be made larger in order to keep the photon counting rate constant, which is similar situation to the quantum Zeno effect. Setting this constant as λ, Eq.(8) becomes to a differential equation as

$$\frac{d\hat{\rho}}{dt} = -\frac{\lambda}{2}\left[\hat{\rho}_{f}\hat{n} + \hat{n}\hat{\rho}_{f}\right] + \lambda\hat{a}\hat{\rho}_{f}\hat{a}^{+} . \qquad (9)$$

This is a well known master equation for the photonic state under linear dissipation.

In the reffering process, we should consider one-count process and no-count process. In one-count process, the readout state of the apparatus in Eq.(3) is the excited state of the atom, and in no-count process, it is the ground state. Substituting these states into Eq.(3), the photonic state change in one-count process is described as $\hat{\rho}_f(t) \rightarrow \hat{\rho}_f(t+0) = \lambda \Delta t \cdot \left[\hat{a}\hat{\rho}_f(t)\hat{a}^+\right]$, which is nothing but the Srinivas-Davies postulate for one-count process, and the photonic state change in no-count process is described as

$$\hat{\rho}_f(t) \rightarrow \hat{\rho}_f(t+T) = \exp\left(-\frac{\lambda}{2}\hat{n}\Delta t\right)\hat{\rho}_f(t)\exp\left(-\frac{\lambda}{2}\hat{n}\Delta t\right), \tag{10}$$

which is nothing but S-D postulate for no-count process.

The state evolution for an arbitrary sequence of one-count and no-count processes has been extensively studied in Ref.5. Fig. 3 shows the evolution of the average photon number $\langle\hat{n}(t)\rangle$ for several initial states: number, squeezed, coherent and thermal states. The initial average photon number and X are chosen to be the same. One-count processes are assumed to occur at τ_1, τ_2, τ_3, τ_4. Fig. 3 tells that, for number state, the photon number does not change during the no-count process and decreases by one when a one-count process occurs. For coherent state, however, the average photon number decreases even during the no-count process and does not change when a one-count process occurs. When the initial state is a thermal state, the average photon number even increases for one-count processes.

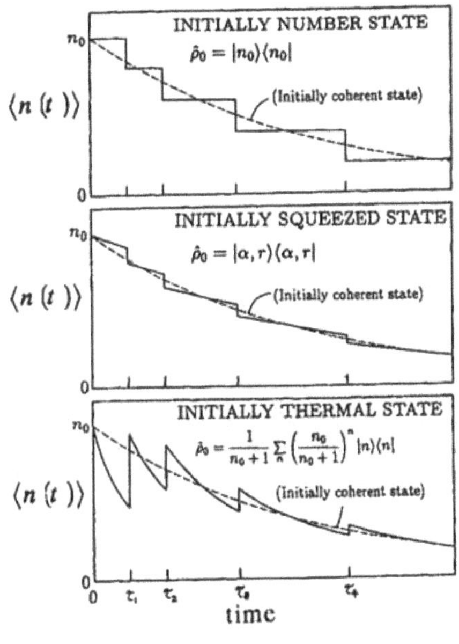

Fig. 3. Evolution of average photon number under photon counting. One-count process is assumed to occur at τ_1, τ_2, τ_3, τ_4.

This apparent mystery dissolves if we consider a simple example of initial state, $\frac{1}{2}(|0\rangle\langle 0|+|100\rangle\langle 100|)$. It is obvious that the initial average photon number is 50 for this state. When a photon is detected, we suddenly know that the possibility of $|0\rangle\langle 0|$ is zero and conclude that the initial state was $|100\rangle\langle 100|$, and now, the state is $|99\rangle\langle 99|$. If, on the other hand, no photon is counted for a long time, we will conclude that $|0\rangle\langle 0|$ is more probable than $|100\rangle\langle 100|$, and so, we modify the weight of $|0\rangle\langle 0|$ and $|100\rangle\langle 100|$, which makes the average photon number smaller. This tells that a density operator expresses our knowledge about the system, and we always modify it according to the (continuous) measurement results.

Figure 3 also shows that the field state evolves toward the vacuum state for any case, which is a reflection of the fact that photon counting is a destructive measurement. Thus, photon counting may be called as "second kind" continuous measurement.

3.3 *QND Photon Counting*

The main object of this section is to construct a "first kind" continuous measurement, in which the change from the initial to the final ($T \rightarrow \infty$) state can be expressed by the projection postulate. Let us assume an interaction Hamiltonian of the form

$$\hat{H}_{int} = \hbar g \hat{a}^{+} \hat{a} (\hat{\sigma}^{+} + \hat{\sigma}) \quad . \tag{11}$$

This Hamiltonian satisfies the QND condition, that is, it commutes with the photon number, so that photons are not absorbed by the detector in one-count processes. This Hamiltonian, of course, is not for two-level atoms. An example of microsystem having this Hamiltonian may be a bistable molecule having two potential valleys. The transition from the left valley to right valley is caused by a photon, but without absorbing it.

Assuming Eq.(11), we can construct a new model, which is similar to but different from S-D model. Here, we only present the results. With $\lambda = g^{2}\Delta t$, the state evolution under non-referring process (NMP) is obtained from Eq.(6) as

$$\frac{d\hat{\rho}}{dt} = -\frac{\lambda}{2}\left[\hat{\rho}_t \hat{n}^{2} + \hat{n}^{2}\hat{\rho}_t\right] + \lambda \hat{n}\hat{\rho}_t \hat{n} \tag{12}$$

This is the same master equation obtaind in Ref.17. The solution is obtained in a density-matrix-component formula as

$$\rho_{mn}(t) = e^{-i\omega(m-n)t} \exp\left[-\frac{\lambda}{2}(m-n)^2 t\right] \rho_{mn}(0) . \tag{13}$$

This solution explicitely shows the con real time diagonalization of the density matrix as a function of time.

In the referring process, the one-count process and no-count processes are expressed as

$$\hat{\rho}_f(t) \rightarrow \hat{\rho}_f(t+0) = \lambda \Delta t \cdot \left[\hat{n} \hat{\rho}_f(t) \hat{n}\right] . \tag{14}$$

and

$$\hat{\rho}_f(t) \rightarrow \hat{\rho}_f(t+T) = \exp\left(-\frac{\lambda}{2}\hat{n}^2 \Delta t\right) \hat{\rho}_f(t) \exp\left(-\frac{\lambda}{2}\hat{n}^2 \Delta t\right) . \tag{15}$$

Fig. 4 shows the evolution of the average photon number for initially coherent, squeezed and thermal states. One-count process is assumed to occur at τ_1, τ_2, τ_3, τ_4, τ_5. The evolution for a nujber state is shown as a broken line in each figure. Different from the case of usual photon counting (Fig.3), the average photon number does not decreases but converges to a certain number in the long range. This is due to the fact that QND photon counting does not absorb photons, and thus the state approaches a number state. Note, however, that the final photon number is not predictable beforehand be-cause the times when one-count process occurs are not predictable.

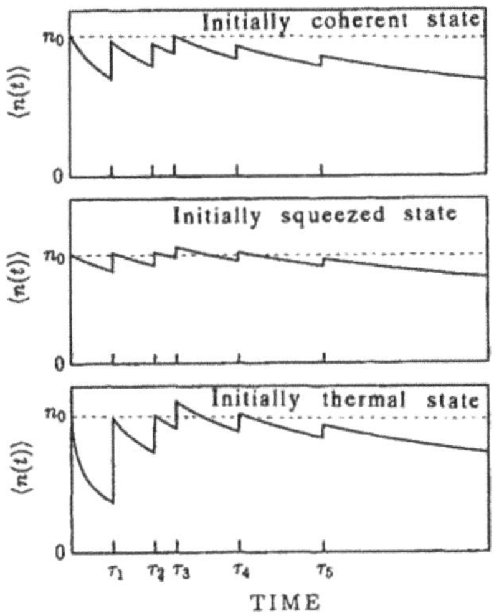

Fig. 4. Evolution of average photon number under QND photon counting. One-count process is assumed to occur at τ_1, τ_2, τ_3, τ_4, τ_5.

4. Conclusion

In conclusion, we have presented a theoretical framework for unsharp measurement and continuous measurement. Unsharp measurement is described as the result of two succesive processes:

step 1: a unitary interaction between the system and apparatus and step 2: reading process for the apparatus. Continuous measurement is expressed as a limit of sequence of unsharp measurements with an appropreate limit operation for the step and the strength of the coupling. This means that the present framework of quantum mechanics is capable of treating unsharp and continuous measurements, and thus, these are not real difficulties.

Photon counting is an example of continuous measurement of photon number, in which the Jaynes-Cummings Hamiltonian can be used as a model. Since the photodetector absorbs photons in photon counting, the final state is always the vacuum state. The state evolution for non-referring process is simply expressed by a master equation for the linear loss. A QND photon counting process is also proposed, in which the Jaynes-Cummings Hamiltonian is modified to one that satisfies the QND condition. In this case, photons are not absorbed by the detector, and thus, the field state approaches a number state, rather than a vacuum state. Computer simulation results also show the above feature of the QND photon counting. The field evolution for non-referring process is expressed as real-time diagonalization of the density matrix, which is analytically obtained.

References

1. M. D. Srinivas and E. B. Davies, *Optica Acta* **28** (1981) 981; *Optica Acta* **29** (1982) 235.
2. C. A. Holmes, G. J. Milburn, and D. F. Walls, *Phys. Rev.* **A39** (1989) 2493.
3. H. Rauch, in *Proc. 3rd Int. Symp. Foundations of Quantum Mechanics* (ISQM-Tokyo'89) (1990) p.3.
4. H. P. Yuen, in *Proc. 2nd Int. Symp. Foundations of Quantum Mechanics* (ISQM-Tokyo'86) (1987) p.360.
5. M. Ueda, *Quantum Optics* **1** (1989) 131.; M. Ueda, *Phys. Rev.* **A41** (1990) 3875; M. Ueda, N. Imoto, and T. Ogawa, *Phys. Rev.* **A41** (1990) 3891; M. Ueda, N. Imoto, and T. Ogawa, *Phys. Rev.* **A41** (1990) 6331.
6. N. Imoto, M. Ueda, and T. Ogawa, *Phys. Rev.* **A41** (1990) 4131.
7. A. Barchielli, L. Lanz, and G. M. Prosperi, *Nuovo Cimento* **B7** (1982) 79.
8. C. M. Caves and G. J. Milburn, *Phys. Rev.* **A36** (1987) 5543.
9. G. J. Milburn and D. F. Walls, *Phys. Rev.* **A28** (1983) 2065.
10. N. Imoto, H. A. Haus and Y. Yamamoto, *Phys. Rev.* **A32** (1985)

2287; N. Imoto and S. Saito, *Phys. Rev.* **A39** (1989) 675; Y. Yamamoto, N. Imoto, and M. Machida, *Phys. Rev.* **A33** (1986) 3243.

11. M. Kitagawa, N. Imoto, and Y. Yamamoto, *Phys. Rev.* **A35** (1987) 5270.
12. C. W. Helstrom, *Quantum Detection and Estimation Theory*, (Academic Press, New York,1976)pp.83-85.
13. V. B. Braginsky and Y. I. Vorontsov, *Sov. Phys.-Usp.* **17** (1975) 644; V. B. Braginsky, Y. I. Vorontsov, and K. S. Thorne, *Science* **209** (1980) 547; C. M. Caves K. S. Thorne, R. W. P. Drever, V. D. Sandberg, and M. Zimmermann, *Rev. Mod. Phys.* **52**(1980)341.
14. M. D. Levenson, R. M. Shelby, M. Reid, and D. F. Walls, *Phys. Rev. Lett.* **57** (1986)2473; N. Imoto, S. Watkins, and Y, Sasaki, *Optics Communications* **61** (1987)159.
15. N. Imoto, in *Proc. 3rd Int. Symp. Foundations of Quantum Mechanics* (ISQM-Tokyo'89) (1990) p.306.
16. M. Sergent III, M. O. Scully, and W. E. Lamb, Jr., *Laser Physics* (Addison-Wesley, Reading, 1974).
17. D. F. Walls, M. J. Collet, and G. J. Milburn, *Phys. Rev.* **D3** (1985) 3208.

TWO REMARKS ON NONRELATIVISTIC QUANTUM MECHANICS

SUSUMU KAMEFUCHI

Institute of Physics, University of Tsukuba
Ibaraki 305, Japan

ABSTRACT

Remarks are made on (1) quantization and (2) Galilei covariance in nonrelativistic quantum mechanics. As for (1) a general theoretical framework is presented for quantizing the system of a simple harmonic oscillator, such that all known, as well as some new, methods of quantization are accommodated in a unified form. As for (2) it is pointed out that when a 5-dimensional form is adopted of Galilei transformations, quantum mechanics is cast in a manifestly Galilei-covariant form.

0. Introduction

In this report I would like to make two remarks on nonrelativistic quantum mechanics (NQM). The remarks are not mutually related, but have something in common in that both are concerned with 'reliance' of NQM on classical mechanics (CM) by which I mean the following.

As a physical theory, NQM or at least the one in the usual presentation is unsatisfactory in that its theoretical structure is not closed within itself: that is to say, NQM has to rely on CM at very essential places. This is seen firstly in its physical interpretation. As advocated by Niels Bohr, physical measurements are to be made by macroscopic apparatus and hence to be interpreted in terms of classical physics. Obviously, CM is more satisfactory in this respect. We may say, however, that such a situation is, in fact, inevitable when viewing from the standpoint of the structural strata of nature, where each stratum has its own language. Since we, human beings, and the atoms belong to the strata that are far apart, an extra procedure becomes necessary, in order to understand the atoms, of translating their language, i.e., NQM into our own, i.e., classical physics.

Another unsatisfactoriness lies in its mathematical formulation. Here also, we have to rely on the formulation of CM: when considering a quantal system we usually start with the corresponding classical system, and then go over to the quantum theory through the procedure called quantization. This is another manifestation of the reliance.

As for quantization we shall remark in section 1 that the procedure is by no means unique, and that there are in fact infinite many possibilities thereof. Further, in section 2 we shall show that the reliance is greatly lessend when NQM is reformulated, on the basis of the symmetry principle, in a manifestly Galilei-covariant form.

1. Quantization

1.1. The question we here want to ask ourselves is: to what extent, or under what conditions the usual, canonical method of quantization can be generalized in NQM. This is a question similar to the one posed by Wigner in 1950[1]. For this purpose let us consider one of the simplest quantal systems, that is, a one-dimensional simple harmonic oscillator.

The canonical quantization fulfills a number of physical conditions, and when some of them are weakened or lifted, there arise various possibilities of generalization. Those conditions are to maintain i) the classical form of the equation of motion $\ddot{q}+q=0$ (in units such as $\hbar=m=\omega=1$); ii) the classical form of the Hamiltonian $H=(p^2+q^2)/2$; iii) the energy spectrum of the form $E_n=E_0+n$ with $n=0,1,2,\cdots$; and iv) the right correspondence-theoretic limit for matrix elements of q and p. Needless to say, condition iii) is to make possible the energy-quanta interpretation. All the conditions are fulfilled for the canonical or Bose quantization as well as for Wigner quantization[1], which includes Bose and para-Bose quantization[2] as special cases. For Fermi and para-Fermi quantization[2], on the other hand, conditions ii) and iv) are not fulfilled.

1.2. Our problem is formulated in terms of an algebraic system consisting of three operators a, $a^\dagger$ and $N=N^\dagger$. The position coordinate q and its conjugate momentum p are defined by

$$q \equiv (a + a^\dagger)/\sqrt{2}, \quad q \equiv (a - a^\dagger)/\sqrt{2}i. \tag{1.1}$$

For this algebraic system we now require (I)

$$[a, N] = a, \quad [a^\dagger, N] = -a^\dagger; \tag{1.2}$$

(II) N is given as

$$N = F([a^\dagger, a]_\alpha) \tag{1.3}$$

or

$$[a^\dagger, a]_\alpha = F^{-1}(N) \equiv G(N), \tag{1.3'}$$

where G is a real function and $[A,B]_\alpha \equiv AB+\alpha BA$ with real $\alpha\neq 0$; and (III) the existence of at least one representation in which the spectrum of N is bounded below.

Requirements (I) and (III) already guarantee that N is of the spectrum $N_n=N_0+n$ with $n=0,1,2,\cdots$, and N_n being the n-th eigenvalue. Thus, a, $a^\dagger$ and N play, as usual, the roles of annihilation, creation and number operators, respectively. Let us denote by $|n\rangle$ the n-th eigenstate of N with eigenvalue N_n, and assume that the ground state $|0\rangle$, normalized to unity, is non-degenerate. From requirement (II) it then follows that the state-vector space of the system is spanned by $|n\rangle$'s and that each $|n\rangle$ is non-degenerate as well: thus $|n\rangle \propto (a^\dagger)^n|0\rangle$.

When the spectrum of N has the (no) upper bound, the case is said to be Fermi-like (Bose-like). In order for condition iii) to be fulfilled we have to put $H=\hbar\omega N=N$: thus $E_0=\hbar\omega N_0$ represents the 0-point energy. Then condition ii) is no longer fulfilled, except for the case $\alpha=1$, $G(N)=2N$ which corresponds to Wigner quantization (Bose-like spectrum with $N_0>0$ being arbitrary).

1.3. By taking the expectation value of Eq. (1.3') with respect to $|n\rangle$ we obtain the matrix elements of a and $a^\dagger$ as follows:

$$\langle n|a|n+1\rangle = \langle n+1|a^\dagger|n\rangle = \sqrt{I(n)} \tag{1.4}$$

and all other elements = 0, where

$$\begin{aligned} I(n) &= \| a^\dagger|n\rangle\|^2 \\ &= \sum_{k=0}^{n} (-1)^k \alpha^{-(k+1)} G(N_{n-k}) \geq 0. \end{aligned} \tag{1.5}$$

If $I(n)>0$ for all n, then the quantization is Bose-like. On the other hand, if $I(n)>0$ up to $n=n'-1$ and $I(n')\leq 0$, we then adjust the parameter N_0 so that $I(n')=0$ is realized. In this case the spectrum of N terminates at $n=n'$ or the maximal eigenvalue of N is given as $N_{max}=N_0+n'$, hence the quantization becomes Fermi-like. And, if neither of the above is the case, such sets of α and G must be rejected.

1.4. As seen below, all known cases of quantization fall within our framework.

a) Wigner quantization (Bose-like, arbitrary $N_0>0$)[1]: $\alpha=1$, $G(N)=2N$;

$$I(n) = \begin{cases} 2N_0+n & \text{for } n=\text{even} \\ n+1 & \text{for } n=\text{odd}. \end{cases} \tag{1.6}$$

Here, the special case of $N_0=p/2$ with $p=1,2,\cdots$ corresponds to the canonical or Bose quantization[2] if $p=1$, and to para-Bose quantization if $p\geq 2$.

b) Fermi or para-Fermi quantization[2] (Fermi-like, $N_0=-p/2=-N_{max}$ with $p=1,2,\cdots$): $\alpha=-1$, $G(N)=2N$;

$$I(n) = -(N+1)(N_0+\frac{n}{2}). \tag{1.7}$$

The special case of p=1 (≥2) corresponds to Fermi (para-Fermi) quantization.

c) q-deformed quantization by Biedenharn[3] (Bose-like, $q>0$, N_0=real): $\alpha=-q^{1/2}$, $G(N)=-q^{-(N+1)/2}$;

$$I(n) = (n+1)q^{-(N_0+n+2)/2}. \tag{1.8}$$

d) q-deformed quantization by Macfarlane[4] (Bose-like, $q>0$, N_0=real): $\alpha=-1$, $G(N)=[N]-[N+1]$, where $[x]\equiv(q^x-q^{-x})/(q-q^{-1})$;

$$I(n) = \frac{1}{q-q^{-1}}[1-q^{-(n+1)}][q^{N_0+n+1}+q^{-N_0}]. \tag{1.9}$$

Further, it is not a difficult matter to invent new kinds of quantization along this line. For example,

e) q-deformed Wigner quantization (Bose-like, $q>0$, $N_0>0$): $\alpha=1$, $G(N)=[2N]$;

$$I(n) = \frac{q^{2N_0}}{(q-q^{-1})(1+q^2)}[q^{2(n+1)}+(-1)^n][1+(-1)^{n+1}q^{-2(2N_0+n)}]. \tag{1.10}$$

f) q-deformed Fermi or para-Fermi quantization (Fermi-like, $N_0=-p/2=-N_{max}$ with $p=1,2,\cdots$): $\alpha=-1$, $G(N)=[2N]$;

$$I(n) = \frac{q^{-2(N_0-1)}}{(q-q^{-1})(1-q^2)}[q^{-2(n+1)}-1][1-q^{2(2N_0+n)}]. \tag{1.11}$$

In all the cases, excepting a), the conditions ii) and iv) are no longer fulfilled. Further, any of the q-deformed cases in the above reduces, as $q\to1$, to the corresponding q-nondeformed case.

1.5. The next question we must ask is: how to generalize the above discussions to the case of f oscillators. Perhaps, the most naive way is to assume that for each oscillator the "commutation relations" remain the same as in the case of f=1, whereas between different oscillators the corresponding operators simply commute or anticommute. Unfortunately, however, such a set of commutation relations, though consistent mathematically, are of very limited applicability physically, thus necessitating a more sophisticated way.

In this respect it is instructive to recall the situation in para-quantization. For the case f=1 the operator N is defined by $[a^\dagger,a]_\mp=2N$ where the upper (lower) sign corresponds to the para-Fermi (para-Bose) case. We now note that the above relation and Eqs.(1.2) are precisely those to be satisffied by the Lie algebra so(3) (Lie super-algebra $osp(1/2)$)[2]. Thus, in order to obtain the required commutation relations for the para-Fermi (para-Bose) case with $f>1$ we have only to enlarge the above algebra to $so(2f+1)(osp(1/2f))$. We expect that such an approach will also be possible for q-deformation of general commutation relations.

Lastly, it is indeed an interesting historical fact that in spite of such infinite multitude of possibilities in quantization Heisenberg somehow found the simplest and hence the right one.

2. Galilei Covariance

2.1. The paradigm of the present-day theoretical physics is that theories be founded on the symmetry principle. That is, starting with a group of symmetry transformations we formulate a theory in such a way that the basic entities correspond to (irreducible) representations of the group, and all equations be covariant under those symmetry transformations. In the realm of relativistic quantum mechanics (RQM) we are very faithful to the paradigm, but not at all so in NQM.

The symmetry that underlies NQM is, of course, Galilean symmetry. Although a number of authors[5] have so far discussed Galilean aspects of NQM, their discussions are unsatisfactory in that Galilei covariance is not manifestly seen. This is basically due to the fact that in the usual, 4-dimensional formulation G_4 of Galilei transformations the Schrödinger wavefunctions ψ are given as projective representations of the Galilei group: the phase factor f that arises when ψ is subject to G_4, where $\psi \to \psi' = \exp(if/\hbar)\psi$, masks covariance properties of the equations concerned.

Thus we shall show below that such inconvenience can be removed when a 5-dimensional form G_5 is employed for Galilei transformations. In this case ψ's correspond to vector representations of the Galilei group, and NQM can be formulated in a manifestly Galilei-convariant manner, just as RQM is done so for Lorentz transformations. Moreover, many of the equations of NQM here take the forms similar to the corresponding ones of RQM, so that the formulation of the former can proceed in a way quite parallel to that of the latter. As is well known, manifest covariance becomes very important when we go over to quantum field theory.

2.2. The usual, 4-dimensional form G_4 of Galilei transformations for $\vec{x}$ and t is:

$$\vec{x} \to \vec{x}' = R\vec{x} - \vec{v}t, \quad t \to t' = t, \tag{2.1}$$

where R is a 3x3 matrix such that $R^tR=I_3$ and $\vec{v}$ is a constant vector. In reformulating the above transformations let us start by considering the system of a particle with mass m with the Lagrangian $L=m\dot{\vec{x}}^2/2$. Obviously, the equations of motion $\ddot{\vec{x}}=0$ remain invariant under the transformations (2.1), but the Lagrangion L does not: in fact the latter is transformed as

$$L \to L' = L + \dot{f} \tag{2.2}$$

with

$$f \equiv -m(R\vec{x})\cdot\vec{v} + \frac{m}{2}\vec{v}^2 t. \tag{2.3}$$

Schwinger's variation principle[6] then tells us that in such a case the invariance of the theory is recovered if ψ is subject to the phase change as mentioned above, or ψ should be taken to be a projective representation of G_4. In view of this we may expect that the problem will be greatly simplified if we have an invariant Lagrangian.

Thus, instead of L let us adopt[7] $\tilde{L}\equiv L-m\dot{s}$. In this case $\delta\tilde{L}\equiv\tilde{L}'-\tilde{L}=0$ provided that s is transformed, together with Eq. (2.1), as

$$s \to s' = s + \frac{1}{m}f = s - (R\vec{x})\cdot\vec{v} + \frac{1}{2}\vec{v}^2 t. \tag{2.4}$$

Our system is thus described by the variables $\vec{x}$, s and their time derivatives.

2.3. In view of the above we now consider a 5-dimensional space with contravariant coordinates $x^\mu(\mu=1,2,\cdots,5)$ such that $(x^1,x^2,x^3)\equiv\vec{x}$, $x^4\equiv t$ and $x^5\equiv s$, and define the 5-dimensional form G_5 of Galilei transformations as $x^\mu \to x'^\mu$:

$$\begin{aligned} x'^i &= R^i{}_j x^j - v^i x^4, \qquad x'^4 = x^4, \\ x'^5 &= x^5 - v_i(R^i{}_j x^j) + \frac{1}{2}\vec{v}^2 x^4; \end{aligned} \tag{2.5}$$

or in a unified form as

$$x'^\mu = \Lambda^\mu{}_\nu x^\nu. \tag{2.5'}$$

Here, $i,j=1,2,3$, $(v^1,v^2,v^3)=(v_1,v_2,v_3)\equiv\vec{v}$, and the matrix elements $\Lambda^\mu{}_\nu$ are obtainable from (2.5). Group-theoretically, the transition $G_4 \to G_5$ corresponds to a central extension of the group concerned.

Under the transformations (2.5) the bilinear expression $\eta_{\mu\nu}x^{\mu}x^{\nu}$ remains invariant, with the metric tensor $\eta_{\mu\nu}=\eta^{\mu\nu}$ being such that $\eta_{ii}=-\eta_{45}=-\eta_{54}=1$ and all other elements=0. Then $\eta_{\mu\nu}x^{\mu}x^{\nu}=\vec{x}^2-2ts$. In this connection it is of interest to note that $\eta_{\mu\nu}x^{\mu}x^{\nu}=\vec{x}^2+x_-^2-x_+^2$ for $x_\pm\equiv(x^4\pm x^5)/2$: thus the group G_5 is a subgroup of the 5-dimensional Lorentz group in which x^4 is kept invariant. Incidentally, the momentum vector $p^{\mu}\equiv(\vec{p},m,E)$ with $E=\vec{p}^2/2$ provides another contravariant vector. On the other hand, the derivative operator $\partial_\mu\equiv\partial/\partial x^{\mu}$ behaves as a covariant vector, which is defined in the usual way: thus we have

$$\partial_i' = \partial_i + v_i\partial_5, \qquad \partial_5' = \partial_5,$$
$$\partial_4' = \partial_4 + v^i\partial_i + \frac{1}{2}\vec{v}^2\partial_5. \tag{2.6}$$

Scalars $\chi(x)$ under G_5 are defined as those satisfying $\chi'(x')=\chi(x)$.

2.4. We are now ready for introducing the Schrödinger wavefunctions. Let us begin by considering a scalar function $\chi(x)$ and assume for this an equation of the Klein-Gordon type:

$$\eta^{\mu\nu}\partial_\mu\partial_\nu\chi(x) = 0. \tag{2.7}$$

Further, to eliminate the extra degree of freedom s we impose on χ a subsidiary condition:

$$(\partial_5 + i\frac{m}{\hbar})\,\chi(x) = 0, \tag{2.8}$$

which is also an invariant equation under G_5. We now introduce the wavefunction $\psi(\vec{x},x^4)$ through

$$\chi(x) = \exp(-i\frac{mx^5}{\hbar})\,\psi(\vec{x},x^4), \tag{2.9}$$

thereby finding, from Eqs.(2.7) and (2.9), the required Schrödinger equation:

$$i\hbar\,\frac{\partial\psi(\vec{x},x^4)}{\partial x^4} \doteq -\frac{\hbar^2}{2m}\,\Delta\psi(\vec{x},x^4). \tag{2.10}$$

On the other hand, the relation $\chi'(x')=\chi(x)$ and Eq.(2.9) lead us to

$$\psi'(\vec{x}',x'^4) = \exp\,[\frac{im}{\hbar}(x'^5-x^5)]\psi(\vec{x},x^4) = \exp(\frac{if}{\hbar})\psi(\vec{x},x^4). \qquad (2.11)$$

In this way we now reconfirm that while $\chi(x)$ is a vector representation with respect to G_5, the Schrödinger wavefunction $\psi(\vec{x},x^4)$ is a projective representation with respect to G_4.

In terms of $\chi(x)$ various relations are given in Galilei-covariant forms. For example, the current 5-vector is defined by $j^\mu(x)\equiv(\hbar/2im)(\chi^*(x)\overleftrightarrow{\partial}^\mu\chi(x))$ and its conservation law is given as $\partial_\mu j^\mu(x)=0$ (summed over $\mu=1,2,\cdots,5$).

Since the electromagnetic field is of essentially relativistic nature, we have to regard the usual electromagnetic potentials $\vec{a}(\vec{x},x^4)$ and $\phi(\vec{x},x^4)$ as given external potentials and assume that the electromagnetic 5-vector $A^\mu(x)$ is independent of x^5 (in accordance with Eq.(2.8) with m=0) and is given by $A^\mu(x)\equiv(\vec{a}(\vec{x},x^4)/c,0,\phi(\vec{x},x^4))$. Note that $A^4(x)=-A_5(x)=0$ holds true in every coordinate system. Introduction of the electromagnetic interaction can then be made by replacing ∂_μ with $D_\mu\equiv\partial_\mu-(ie/\hbar)A_\mu(x)$ in Eqs.(2.7) and (2.8). Since, however, $D_5=\partial_5$, Eq.(2.8) and hence Eq.(2.9) need not be changed. The resulting equation for the $\psi(\vec{x},x^4)$ in interaction turns out to be of the right form. In this way we are assured that $\chi(x)$ rightly describes a spinless particle with mass m.

2.5. For the case of a particle with mass m and spin 1/2 we take a spinor function $\psi(x)$ which is to satisfy an equation of the Dirac type

$$\gamma^\mu\partial_\mu\psi(x) = 0 \qquad (2.12)$$

as well as a subsidiary condition of the same form as Eq.(2.8). Further, it is assumed that under a G_5 transformation as given by (2.5') $\psi(x)$ transforms as

$$\psi(x) \to \psi'(x') = S\psi(x) \qquad (2.13)$$

with S satisfying

$$S^{-1}\gamma^\mu S = \Lambda^\mu{}_\nu\gamma^\nu . \qquad (2,13')$$

In the above γ^μ's are numerical matrices to satisfy

$$\gamma^{\mu}\gamma^{\nu} + \gamma^{\nu}\gamma^{\mu} = 2\eta^{\mu\nu}. \tag{2.14}$$

The γ^{μ}'s thus form a Clifford algebra, and without loss of generality we may employ the representation

$$\gamma^{i} = \begin{bmatrix} \sigma_i & 0 \\ 0 & -\sigma_i \end{bmatrix}, \quad \gamma^{4} = \begin{bmatrix} 0 & 0 \\ -\sqrt{2} & 0 \end{bmatrix}, \quad \gamma^{5} = \begin{bmatrix} 0 & \sqrt{2} \\ 0 & 0 \end{bmatrix}, \tag{2.14'}$$

with σ_i being Pauli matrices.

Introducing 2-component wavefunctions $\hat{\varphi}_j(\vec{x},x^4)(j=1,2)$ by

$$\psi(x) = \exp\left(-\frac{imx^5}{\hbar}\right) \begin{bmatrix} \hat{\varphi}_1(\vec{x},x^4) \\ \hat{\varphi}_2(\vec{x},x^4) \end{bmatrix}, \tag{2.15}$$

we find that $\hat{\varphi}_1$ satisfies the usual Schrödinger equation, whereas $\hat{\varphi}_2$ is a redundant component, being expressible in terms of $\hat{\varphi}_1$. From the tranformation property of $\hat{\varphi}_1$ under space rotation we can confirm that $\hat{\varphi}_1$ does describe a particle with spin 1/2.

The formal similarity to the Dirac electron theory enables us to proceed in the same way as in the Dirac case. For example, defining the adjoint $\bar{\psi}(x)$ of $\psi(x)$ by $\bar{\psi}(x)\equiv\psi^{*}(x)\zeta$ with the matrix ζ satisfying $\zeta\gamma^{\mu\dagger}\zeta=-\gamma^{\mu}$, we can write Galilei contravariant tensors of rank n as $\bar{\psi}(x)\gamma^{\lambda_1}\gamma^{\lambda_2}\cdots\gamma^{\lambda_n}\psi(x)$. In particular, the current 5-vector is given as $j^{\mu}(x)\equiv\bar{\psi}(x)\gamma^{\mu}\psi(x)/\sqrt{2}i$, and its conservation law as $\partial_{\mu}j^{\mu}(x)=0$.

The electromagnetic intereaction can be introduced in the same way as in the previous case, with the result that the gyromagnetic ratio g equals 2, as first noted by Lévy-Leblond[1]. Thus, g=2 is a consequence of Galilean (not Einsteinian) symmetry.

The case of higher spins can be dealt with easily by adapting the Bargmann-Wigner method[8] developed for RQM to our formalism of NQM. It is found in this way that the basic equations of NQM can be derived within its own framework, that is, without recourse to CM.

2.6. We may regard $\chi(x)$ or $\psi(x)$ discussed above as a nonrelativistic field quantity and can construct its quantum field theory again in a manifestly Galilei-covariant form. In this case, for dealing with the subsidiary condition, we have to resort to Dirac's general method. In many respects the resulting nonrelativistic quantum field theory (NFT) formally resembles the usual relativistic quantum field theory

(RFT). In spite of this, however, the two have some essential differences, the most notable of them being the spin-statistics connection. As is well known, NFT has no such connection, and in this sense its theoretical framework is wider than that of RFT. The reason for this is as follows.

In RFT the basic symmetry is Poincaré symmetry, and what is to be called "local covariance" requires a local field $\psi(x)$ to be a direct sum of two irreducible representations of Poincaré group: $\psi(x)=\psi^{(+)}+\psi^{(-)}(x)$, where $\psi^{(+)}$ and $\psi^{(-)}$ correspond to particle and antiparticle states, respectively: only with $\psi^{(+)}$ or $\psi^{(-)}$ the resulting theory cannot be a local field theory. In this way the field $\psi(x)$ obtains a symmetry with respect to exchange of $\psi^{(+)}$ and $\psi^{(-)}$, that is, C-symmetry (particle-antiparticle conjugation), and this leads eventually to the spin-statistics connection. In NFT, on the contrary, we have seen above that a satisfactory local field theory can be constructed with use only of $\psi^{(+)}$ to describe particle states. Of course, we may define a quantity such as $\psi(x)=\psi^{(+)}+\psi^{(-)}$ by artificially introducing $\psi^{(-)}$ to describe antiparticle states. However, there is no a priori reason for doing this at all. This implies that NFT has no unique connection between spin and statistics.

The remarks given in sections 1 and 2 are based on the works done in collaboration with K. Odaka and T. Kishi[9] and with M. Omote, Y. Takahashi and Y. Ohnuki[10], respectively, to whom the author is greatly indebted.

3. References

1. E.P. Wigner, Phys. Rev. 77 (1950) 711.
2. See for example, Y. Ohnuki and S. Kamefuchi, Quantum Field Theory and Parastatistics (Univ. of Tokyo Press/Springer Verlag, 1982).
3. L.C. Biedenharn, J. Phys. A: Math. Gen. 22 (1989) L873.
4. A.J. Macfarlane, J. Phys. A: Math. Gen. 22 (1989) 4581.
5. E. Inönü and E.P. Wigner, Nuovo Cimento 9 (1952) 705. J.M. Lévy-Leblond, J. Math. Phys. 4(1963) 776; Commun. Math. Phys. 4 (1967) 107. J. Voisin, J. Math. Phys. 6 (1965) 1519. C.R. Hagen, Commun. Math. Phys. 18 (1970) 97. C.R. Hagen and W.J. Hurley, Phys. Rev. Letters 24 (1970) 1381. J.J. Aghassi, P. Roman and R.M. Santilli, Phys. Rev. D1 (1970) 2753.
6. J. Schwinger, Phys. Rev. 82 (1951) 914.
7. G. Marmo, G. Morandi, A. Simoni and E.C.G. Sudarshan, Phys. Rev. D37 (1988) 2196.
8. V. Bargmann and E.P. Wigner, Proc. Nat. Acad. Sci. U.S. 34 (1948) 211.
9. K. Odaka, T. Kishi and S. Kamefuchi, to be published.
10. M. Omote, S. Kamefuchi, Y. Takahashi and Y. Ohnuki, Fortschr. Phys. 37 (1989) 933.

INTERNAL TRACE FOR DETECTOR VARIABLES AND THE QUANTUM THEORY OF MEASUREMENT

TSUNEHIRO KOBAYASHI

Institute of Physics, University of Tsukuba,
Ibaraki-305, Japan

We propose a construction of the quantum theory of measurement which is definable and calculable in the ordinary quantum mechanical treatment. It is pointed out that the change of internal states of detectors represented by the internal trace must be described precisely. The wave-packet collapse is realized in the macroscopic limit where a physical quantity to be measured becomes a macroscopic number.

§1. Introduction

In the quantum theory of measurement so many discussions for the derivation of the wave-packet collapse have been carried out but only a very little has been done for deriving the mechanism to produce the macroscopic measured-quantities. It is clear that the mechanism producing the macroscopic measured-quantities is also one of the necessary conditions in the measurement processes. It is also well-known that in the macroscopic limits for some physical quantities the original structure of Hilbert space is broken and then the unitary time-evolution by the original Hamiltonian is no more imposed there. It is, therefore, one of the interesting ideas that the wave-packet collapse is realized in the macroscopic limit which is determined by the mechanism extracting the macroscopic measured-quantity in each detector.[1,2]

In this paper I shall show the existence of the macroscopic limit where both mechanisms, that is, (A) the extraction of the macroscopic measured-quantity and (B) the realization of the wave-packet collapse, are simultaneously derived.

Throughout this paper I shall follow the following notations: The measurement process is described by the transition from the initial state (Φ_I) to the final state (Φ_F)

$$\Phi_I \equiv \phi \times \Psi_0 = \sum_n c_n u_n \times \Psi_0$$

$$\rightarrow \Phi_F = \sum_n c_n u_n \times \Psi_n, \tag{1}$$

where the state for the total system is written in terms of the direct product of the state for the quantum object (ϕ) and that for the detector (Ψ). In Eq.(1) $\phi = \sum_n c_n u_n$ stands for the spectral decomposition of the object state in terms of the quantum number to be measured and Ψ_n is the detector state corressponding to the eigen state u_n. The wave-packet collapse is understood as the change of the density matrix from the pure state (ρ_F) to the mixed state ($\hat{\rho}_F$)

$$\rho_F \equiv |\Phi_F\rangle\langle\Phi_F| = \sum_n \sum_m c_n c_m^* |u_n\rangle\langle u_m| \times |\Psi_n\rangle\langle\Psi_m|$$

$$\rightarrow \hat{\rho}_F = \sum_n |c_n|^2 |u_n\rangle\langle u_n| \times |\Psi_n\rangle\langle\Psi_n|, \tag{2}$$

where the wave-packet collapse is described by the disappearance of the off-diagonal terms in the density matrix ρ_F.

§2. Change of Detector States and Internal Trace

Let Q be the physical quantity to be measured and $|\Psi_n\rangle$ be the detector state corresponding to the measured-number Q_n for Q. The number Q_n is, of course, a macroscopic number. We can summarize impotant points in the measurement processes as follow:[2]

(i) In order to describe $|\Psi_n\rangle$ presisely, $|\Psi_n\rangle$ must be written in terms of the superposition of "huge number" of different internal states as

$$|\Psi_n\rangle = \sum_{\{q\}} c_n (\{q\}) |Q_n : \{q\}\rangle, \tag{3}$$

where $\{q\} \equiv \{q_1, q_2, \cdots\}$ stands for all eigenvalues required to describe the detector state completely and the sum should be carried out over all combinations constructing the complete set.

(ii) Not only Q_n but also the difference, $\Delta Q_{nm} \equiv Q_n - Q_m (n \neq m)$, must be macroscopic numbers. This means that the difference between $|\Psi_n\rangle$ and $|\Psi_m\rangle$ must represent "the macroscopic difference" which induces the macroscopic number ΔQ_{nm}.

(iii) The possibilities for producing the macroscopic numbers are categorized into the following two cases:

(I) A certain enhancement mechanism by the external field works in the detector.

(II) A macroscopic accumulation of microscopic differences takes place in the detector.

(I shall show two models corresponding to (I) and (II) in §3.) From these considerations I insist that we must "trace" the change of detector states "precisely". Now the problem is how to introduce such effects in the density matrix.

In an experiment where the same measurement processes must be repeated by using a lot of samples to derive the right number of the expectation value of the measured-quantity $\langle Q\rangle$, the same quantity is always extracted and other quantities required to descibe the detector state are "deliberately ignored". It is clearly seen in the evaluation of $\langle Q\rangle$ as

$$\langle Q\rangle = \mathrm{Tr}\{\rho_F Q\} = \widetilde{\mathrm{Tr}}\{\tilde{\rho}_F Q\}. \tag{4}$$

were $\tilde{\rho}_F$ is defined by the density matrix partially traced over all internal quantum numbers describing the detector state, $\{q\}$, except that corresponding to Q. We may write $\tilde{\rho}_F$ as

$$\begin{aligned}\tilde{\rho}_F &\equiv \prod |\Phi_F\rangle\langle\Phi_F| \\ &= \sum_n \sum_m c_n c_m^* |u_n\rangle\langle u_m| \times \mathrm{Tr}_{\{\tilde{q}\}}[|\Psi_n\rangle\langle\Psi_m|].\end{aligned} \tag{5}$$

The partial-trace operation is called as "internal trace". That is to say, the internal trace means "the trace operation over internal quantum numbers of the detector which are deliberately ignored in the measurement process". We may say that we can use $\tilde{\rho}_F$ instead of ρ_F in every experiments. The important point is that the internal trace includes a lot of the trace operations, of which number increases with the number of the constituent particels of the detector (N). The accumulation of the differences produced by such trace operations may become a macroscopic quantity in the macroscopic limit of $N\to\infty$.

§3. Calculable Examples and Mechanism of the Wave-Packet Collapse

Let us present two simple but typical measurement processes corresponding to the two possibities noted in (iii) of the last section. (As for the details of the evaluation, see Ref. 1.) In this section I shall discuss the models in the measurement processes for the typical on-off experiments with the object described by two different eigen states. In this case the object is written by

$$\phi(r,t)=\phi^{(1)}(r,t) + \phi^{(2)}(r,t), \tag{6}$$

where the normalization constants are included in $\phi^{(1)}$ and $\phi^{(2)}$. The detector state is given by

$$\Psi(R,t)\equiv\psi^{I}(r_1\cdots r_N;t) \times \psi^{II}(s_1\cdots s_N;t), \tag{7}$$

where ψ^{I} detects only $\phi^{(1)}$ and ψ^{II} does only $\phi^{(2)}$.

(1) Model I -Detectors with an enhancement mechanism by an external field-

In this model detectors have stable internal states. Then, in order to extract macroscopic quantities to be measured, some enhancement mechanisms by external fields are needed.

A simple example. The detector constructed from the same neutral atoms is put in the constant electric external field (E). In the initial state, therefore, the detector does not couple to the electric field. By the interaction of the detector with the object some atoms are ionized in the detector. Then after the interaction the detector couples to the electric field. In this case the movement of the center-of-mass of the detector can be the macroscopic quantity to be measured in the limit of $|E|\to\infty$. It is stressed that in model I the macroscopic limit is always defined "by the limit of the external field ($|E|\to\infty$) not by that of the constituent-particle number ($N\to\infty$)". In general we may write the ionization process as the change of the detector state

$$\psi_I(r_1\cdots r_N:t)\to\psi_F=\sum_{n=0}^{N} c_n\psi_n(r_1\cdots r_N:t). \tag{8}$$

where n in ψ_F denotes the number of the ionized atoms and c_n is the coefficient. Taking account of that the final total system is described by $\Phi_F=\phi^{(1)}\times\Psi_F^{(1)}+\phi^{(2)}\times\Psi_F^{(2)}$ and ψ^{I} and ψ^{II}, respectively, detect only $\phi^{(1)}$ and $\phi^{(2)}$, we see

$$\Psi_F^{(1)}=[\sum_{n=0}^{N} c_n\varphi_n^{I}(r_1\cdots;t)] \times \varphi_0^{II}(s_1\cdots;t)$$

$$\Psi_F^{(2)}= \varphi_0^{I}(r_1\cdots;t) \times [\sum_{n=0}^{N} c_n\varphi_n^{II}(s_1\cdots;t)]. \tag{9}$$

The Hamiltonian for the detector with the n-ionized atoms is given by

$$H_D^{n}= H_D^{0} + e\sum_{i=1}^{n} r_i\cdot E, \tag{10}$$

where the Hamiltonian for the neutral detector is generally written by $H_D^0= \frac{1}{2m}\sum_{i=1}^{N}P_i^2 + V(r_1\cdots r_N)$ with $\hbar$=1. Taking into account that the measured quantity is the movement of the center-of-mass of the detector, the internal trace is defined by the integration over all internal relative-coordinate variables of the detector as

$$\tilde{\rho}_F=\int\prod_{i=1}^{N} dr_i\delta^{(3)}(G_I-\frac{1}{N}\sum_{i=1}^{N} r_i)\int\prod_{j=1}^{N} ds_j\delta^{(3)}(G_{II}-\frac{1}{N}\sum_{j=1}^{N} s_j)$$

$$\times [\varphi^{(1)}\varphi^{(1)*}\times \Psi_F^{(1)}\Psi_F^{(1)*}+ \varphi^{(2)}\varphi^{(2)*}\times \Psi_F^{(2)}\Psi_F^{(2)*}+(\varphi^{(1)}\varphi^{(2)*}\times$$

$$\Psi_F^{(1)}\Psi_F^{(2)*}+c.c.)]. \tag{11}$$

Provided that all atoms are bounded by the following harmonic oscillator potential

$$H_D^0= \frac{1}{2m}\sum_{i=1}^{N} P_i^2 + \frac{k}{2N}\sum_{j>\ell=1}^{N}\sum |r_j-r_\ell|^2,$$

the integration given by Eq.(11) is exactly calculable. (See Ref. 1.) Since we are interested in the on-off experiment, the final density matrix for the object derived by the integration over two center-of-mass coordinates, G_I and G_{II}, is obtained in the macroscopic limit of $|E|\to\infty$ as

$$\rho_F= (1-|c_0|^2)(\varphi^{(1)}\varphi^{(1)*}+ \varphi^{(2)}\varphi^{(2)*}) + |c_0|^2\varphi\varphi^*. \tag{12}$$

In Eq.(12) the pure state proportial to $|c_0|^2$ still remains. It is sure that $|c_0|^2$ has a meaning of the inefficiency of the detector which is interpreted as the probability of the elastic forward scattering between the object and the detector[3]. In this type of the dectors this inefficiency is unavoidable. Though it is very interesting to measure this inefficienty in realistic experiments, it will be discussed elsewhere.

(2) Model II -Detectors with quasistable internal states-

This type of detectors is characterized by quasistable internal states in the initial state like the bubble chamber or the Wilson cloud chamber. A simple example is represented as follows (for the details, see Ref. 1.): The potential binding the atoms in the detector has a real minimum and a local minimum. In the initial state most of the atoms stay in the local minimum and then a certain number of the atoms (say n) transites from the local minimum to the real minimum by the interaction of the detector with the object. In this process the energy emitted by the transition of the atoms may be estimated by

$$E_e = nV_0, \tag{13}$$

where V_0 is the energy difference between the real and the local minimums. The final state of the detector is in general written down by the linear combination of the states characterized by the number of the transition atoms (φ_n) as $\varphi_F = \sum_{n=0}^{N} c_n \varphi_n(r_1 \cdots r_N; t)$, where c_n is the coefficient. The expectation value of the emitted energy is given by $\langle E_e\rangle \sim \langle n\rangle V_0$, where $\langle n\rangle$ is the expectation value of the transition number. Provided that the detector is set up so as to have a property

$$\langle n\rangle \sim N \text{ (n is proportinal to N)}, \tag{14}$$

$\langle E_e\rangle$ can be a macroscopic quantity to be measured in the macroscopic limit defined by "the number of the constituent atoms of the detector, $N\to\infty$." That is, the apparatus which do not have the property $\langle n\rangle \sim N$ do not work as detectors. "The relation of Eq.(14) is the criterion for the detector in this model."

The internal trace for the density matrix $\tilde{\rho}_F$ is given by the integration over all detector variables because we are studying the on-off experiment where the magnitude of the emitted energy is not required as the information of the measurement. When the potential is represented by two harmonic oscillator potentials and $c_n = e^{i\theta_n}[B(t)/(\sqrt{2\pi}\,\sigma)^{1/2}]\ \exp\{-(n-\langle n\rangle)^2/4\sigma^2\}$ with $\langle n\rangle = qN$ $(o<q\leq 1,\ q\sim O(1))$ and $\sigma=\sigma_0 N$ is taken (for details, see Ref. 1), the all quantities in the internal trace are calculable. In the macroscopic limit of $N\to\infty$ we obtain

$$\lim_{N\to\infty} \tilde{\rho}_F = \phi^{(1)}\phi^{(1)*} + \phi^{(2)}\phi^{(2)*}. \tag{15}$$

The complete wave-packet collapse is realized.

The mechanism of the wave-packet collapse is understood as follows: In the internal trace the transition of one atom induces a number being smaller than 1, which is derived from the overlapping integral between φ_0 and φ_1 as $|I_{0,1}| = |\int \prod_{i=1}^{N} dr_i \varphi_1^*(r_1\cdots r_N)\varphi_0(r_1\cdots r_N)| \equiv e^{-A/2} < 1$ with $A>0$. Of course, it is non-zero even in the macroscopic limit, $\lim_{N\to\infty}|I_{0,1}| \neq 0$. Since the overlapping integral for the transition of $\langle n\rangle$-atoms is evaluated as $|I_{0,\langle n\rangle}| \simeq |I_{0,1}|^{\langle n\rangle}$ in the internal trace, then it vanishes in the macroscopic limit, i.e. $\lim_{N\to\infty}|I_{0,\langle n\rangle}| = \lim_{N\to\infty} e^{-\langle n\rangle A/2} = 0$ if $\langle n\rangle$ is proportional to N, $\langle n\rangle \sim N$.

From the above two processes we may summarize the mechanism of the wave-packet collapse as follows:

(1) The off-diagonal terms of a density matrix partially traced over the internal variables of the detector become zero operators in the macroscopic limit determined by the mechanism inducing a macroscopic physical quantity to be measured in the detector.

(2) The state vector of a quantum system coupled to a detector having measurable macroscopic degrees of freedom is collapsed by taking the internal trace and then going to the macroscopic limit.

§4. Another Possibility for the Wave-Packet Collapse -Statistical Collapse-

In general the initial states of detectors should be written in terms of the superposition of different internal states as $|\Psi_0\rangle = \sum_{\{q\}_I} c_I(\{q\}_I)|\{q\}_I\rangle$. In realistic experiments, especially at finite temperatures, it is impossible to determine all of $c_I(\{q\}_I)$ uniquely in the repetition of the same measurement process. (Note that the expectation value in experiments is given by the average $\langle Q\rangle = \lim_{L\to\infty} \frac{1}{L} \sum_{\ell=1}^{L} Q_\ell$, where Q_ℓ is the observed number in the ℓ-th measurement process.) For the ℓ-th measurement, therefore, the density matrix for the detetor must be labeled by ℓ as

$$\tilde{\rho}_F^{(\ell)} \equiv |\Psi_F^{(\ell)}\rangle\langle\Psi_F^{(\ell)}|$$

$$= \sum_{\{q\}_F} \sum_{\{q\}'_F} c_F^{(\ell)}(\{q\}_F) c_F^{(\ell)*}(\{q\}'_F) \quad |\{q\}_F\rangle\langle\{q\}'_F|, \tag{16}$$

where the coefficients of the final state are related to the initial ones via the S-matrix as

$$c_F^{(\ell)}(\{q\}_F) = \sum_{\{q\}_I} \langle\{q\}_F|S|\{q\}_I\rangle c_I^{(\ell)}(\{q\}_I). \tag{17}$$

The average of $\tilde{\rho}_F^{(\ell)}$ over the L-times of the repetition is evaluated as

$$\frac{1}{L}\sum_{\ell=1}^{L} \hat{\rho}_F^{(\ell)} = \sum_{\{q\}_F} \sum_{\{q\}'_F} \langle c_F^{(\ell)}(\{q\}_F) c_F^{(\ell)*}(\{q\}'_F)\rangle_L \quad |\{q\}_F\rangle\langle\{q\}'_F|, \tag{18}$$

where

$$\langle c_F^{(\ell)}(\{q\}_F) c_F^{(\ell)*}(\{q\}'_F)\rangle_L \equiv \frac{1}{L}\sum_{\ell=1}^{L} c_F^{(\ell)}(\{q\}_F) c_F^{(\ell)*}(\{q\}'_F). \tag{19}$$

It is noted that the internal trace does not change in this average at all. It is, however, very interesting that, if the coefficients, $c_I^{(\ell)}$, are completely random with respect to their phases and possibly their magnitudes also, the average has the property

$$\lim_{L\to\infty} \langle c_F^{(\ell)}(\{q\}_F) c^{(\ell)*}(\{q\}_F')\rangle_L \to \delta_{\{q\}_F,\{q\}_F'} \tag{20}$$

in the limit of $L\to\infty$. (For the details, see Ref. 2.) The off-diagonal terms vanish in the limit. This means that the wave-packet collapse is realized. This collapse, however, does not describe the measurement process, because this mechanism is not accompanied by any macroscopic measurable-quantities even in the limit of $L\to\infty$.

Actually we may write the average as a statistical average

$$\lim_{L\to\infty} \langle \rho_F^{(\ell)}\rangle_L = \int_{\{q\}} \Pi\, dc(\{q\}) e^{-F(c(\{q\});T)} \prod |\Psi_F\rangle\langle\Psi_F|/A, \tag{21}$$

where $c(\{q\})=c_I(\{q\}_I)$, $A=\int_{\{q\}}\Pi dc(\{q\})e^{-F(c:T)}$ and $F(c:T)$ stands for the statistical destribution function for the coefficients $c(\{q\})$ at the temperature T. Then we may call this type of the wave-packet collapse "statistical collapse". It is stressed that " the statistical collapse does not describe the real quantum collapse in the measurement processes." Many Hilbert space theory of quantum measurement proposed by Machida and Namiki[4] may be understood to be one of the statistical collapses.

It should, however, be noted that the effect of this average appears not only in the measurement processes but in all processes including many-particle system as well. The investigation of this average is important to analize experimental results precisely[5].

§5. Conclusion

The density matrix in quantum measurements should be defined by the density matrix including the internal trace for detector variables which is defined by the trace operations over all detector variables deliberately ignored in the measurement process

$$\tilde{\rho}_F = \Box\, \rho_F .$$

(1) All quantities in $\tilde{\rho}_F$ are definable and basically calculable in the ordinary quantum mechanics.

(2) The wave-packet collapse is realized in the macroscopic limit determined by the choice of the quantity to be measured. Then the density matrix for the objects is derived as

$$\rho_{ob}=\text{Macroscopic limit of } \tilde{\rho}_F .$$

"The quantum theory of measurements is well-defined in quantum mechanics as the macroscopic limit of the measured-quantity."

Finally we would like to remark that the complete wave-packet collapse is realized only in the macroscopic limit defined by that of the external field strength ($|E|\to\infty$) or the number of the constituent particles ($N\to\infty$). It means that the collapse is always incomplete in realistic experiments with $|E|$ finite or N finite. Since the density matrix with the internal trace is basically calculable even at $|E|$ or N finite, examining these incompleteness is very interesting and important problems in this scheme. The gradual change of the external field strength might be possible in realistic processes, while that of N may be difficult. To construct models possibly realized in experiments is one of the next tasks to be carried out. It should, however, be noted that the effect of the average discussed in §4 must precisely be taken into account in the analyses of such processes.

References

1. T. Kobayashi and K. Ohmomo, Phys. Rev. A 41 (1990) 5798.
2. T. Kobayashi, preprint of University of Tsukuba, Ibaraki, Japan (1990).
3. R. Fukuda, preprint of Keio University, Kanagawa, Japan (1989).
4. S. Machida and M. Namiki, Prog. Theor. Phys. 63 (1980) 1457 and 1833. M. Namiki, Found. Phys. 18 (1988) 29.
5. M. Namiki and S. Pascazio, preprint of Waseda University, Tokyo, Japan (1990).

REPRESENTATIONS OF QUANTUM MECHANICS BY MEANS OF TRUE PROBABILITY DISTRIBUTIONS IN PHASE SPACE

JAN KRÜGER
Institute for Theoretical Physics,
State University of Ghent, Proeftuinstraat 86,
B 9000, Ghent, Belgium

ABSTRACT

We present a complete analysis of the central question to consider phase space representations of quantum mechanics as true probability distributions. Their existence depends critically on Wigner's theorem, which is examined in the light of different interpretations of joint probability distributions. Although Wigner's theorem bans a large class of hidden variable theories, it allows the construction of joint probabilities in each of the possible interpretations. An review of existing and new theorems and constructions is given and the relation with the usual and the operational version of Heisenberg's uncertainty principle in Schrödinger form is pointed out.

1. Introductory remarks

Perhaps the most convenient way towards a realistic underpinning of non-relativistic quantum phenomena is the representation of quantum mechanics bij means of joint probability distributions (j.p.d.) in phase space (for a review see e.g. Hillery et al. [1]). In this paper we present a complete analysis of the central question to consider these distributions as true (positive) probability distributions in the phase space of position and momentum.

Let us make some introductory remarks concerning the validity of the concept of j.p.d..

a) Already in the early discussions on the uncertainty principle Heisenberg states, in a letter to Pauli (October 28, 1926) [2] : " There is no point in talking about the position of a particle of a definite velocity. However, it makes sense if one does not consider the velocity and the position too accurately."

b) The statistical interpretation [3] of quantum theory is formulated for the purpose of providing a sound interpretation using a minimum of assumptions. The quantum state description applies only to an ensemble of similarly prepared systems, rather than supposing that it exhaustively represents an individual system. Ballentine [3] comes to the conclusion that with regard to the uncertainty principle, quantum theory is not inconsistent with the supposition that a particle has at any instant both a definite position and a definite momentum, although there is a widespread folklore to the contrary.

c) It is now realized that noncommuting quantum observables [4,5], if considered as unsharp observables, are simultaneously measurable, in measurement theory as well as in concrete experimental realizations [6].

2. Interpretation of joint probability distributions

Three ways seems to be open to a theoretical approach of j.p.d. as functionals of the state function [7].

a) The first one considers the j.p.d. in the sense of classical probability theory of stochastic variables. The j.p.d. of the first kind f(q,p) is then interpreted as the probability that the variables q and p have certain values, the variable considered as a property possessed by the object system.

b) Looked upon as a j.p.d. of the second kind, f(q,p) is the probability that on simultaneous measurement of the observables Q and P the values q and p are found.

c) For j.p.d. of the third kind, f(q,p) is not function of the object system alone but may also depend on the measurement arrangement. The distribution of the third kind is then considered as the probability that, in a disturbance theory of joint measurements of two incompatible observables Q and P,the values q and p are obtained.

3. Realization of phase space distributions

We mention three ways in which these phase space distributions can be realised :

a) They can be introduced as Galilei invariant functionals of the density matrix (generalized Wigner distributions) [8,9] giving a definite joint probability distribution associated with each correspondence rule. Requiring Galilei invariance, the most general form of the j.p.d. f(q,p) is given by

$$f(q,p) = \frac{1}{\hbar^n} tr(\hat{U}_{q,p}\, \hat{\rho}\,'\, \hat{U}_{q,p}\, \hat{\rho}) \quad ,$$

$$\hat{U}_{q,p} = e^{i(q\hat{p} + p\hat{q})/\hbar}$$

is an element of the unitary representation of the Galilei group, $\hat{\rho}$ is the density matrix representing the object system, $\hat{\rho}\,'$ stands for an arbitrary operator normalized so that tr $\hat{\rho}\,' = 1$ and which is completely determined by giving a correspondence rule:

$$e^{i(\mu . q + \nu . p)} \to \Omega(\mu,\nu) e^{i(\mu . \hat{q} + \nu . \hat{p})} \quad ,$$

where $\Omega(\mu,\nu)$ is an arbitrary function of the vectors μ and ν with $\Omega(0,0) = 1$. The correct quantum mechanical marginal distributions are guaranteed by requiring $\Omega(0,\nu) = \Omega(\mu,0) = 1$. E.g. for the Weyl correpondence, which leads to the Wigner distribution, $\Omega(\mu,\nu) = 1$ and $\hat{\rho}\,'$ is nothing but the parity operator, which by the way is the reason why the Wigner distribution is not always non-negative.

b) Stochastic phase space distributions are introduced by adopting the attitude that an exhaustive description of the measurement outcome can be achieved by providing in addition to the point (q,p) two confidence functions (fuzzy sample point) [10]. This approach can be identified with a subclass of the generalized Wigner distributions.

c) It is possible to introduce an autonomous quantization in phase space by deformation of the structure of the Poisson Lie algebra and of the deformations of the associative algebra defined by the usual product of functions [11]. (star product). The choice of the star product is equivalent to the choice of a correspondence rule.

4. Phase space in classical and quantum mechanics

Already in classical dynamics, the consideration of single trajectories in connection with problems in the large ceases to be meaningfull. This functional form of dynamics is statistical mechanics. The evolution of mechanical systems, classical or quantal, can be described by a Liouville equation in master Liouville space. Differential cross sections can be obtained from the time dependence of phase space packets. This procedure is valid for classical and quantum mechanics.

What is the difference between a quantum state and a classical state in phase space e.g. for the Weyl correspondence rule ?

THEOREM *(Narcowich and O'Connell* [12]*). The necessary and sufficient conditions for a normalized phase space function f(p,q) to be a Wigner function are*

$$\int f(q,p)\, f_\psi(q,p) d^n q d^n p \geq 0 \qquad \text{for all } f(q,p)\,,$$

where $f_\psi(q,p)$ denotes the Wigner distribution corresponding to the pure state ψ .

We introduce now the

DEFINITION. $\tilde{f}(a)$ *is of* η *-positive type if for every choice of* $a_1, a_2, \ldots a_m$, *the mxm matrix*

$$e^{i\eta\sigma(a_k,a_j)/2}\, \tilde{f}(a_j - a_k) \qquad \textit{is non-negative} \quad .$$

Here $\tilde{f}(a)$ is the symplectic Fourier transform

$$\tilde{f}(a) = \int f(z) e^{i\sigma(a,z)}\, dz \quad ,$$

where $\sigma(z_1,z_2)$ denotes the symplectic form with the shorthand notation : $z_i = (q_i,p_i)$.With the help of this definition we can formulate the equivalent

THEOREM *(K.L.M. conditions* [13] *). The necessary and sufficient conditions for a normalized phase space function to be a Wigner distribution are that* $\tilde{f}(a)$ *is continuous and of* $\hbar$-positive type.

5. Wigner's theorem

The existence of true probabilities in phase space depends critically on Wigner's theorem[14], which considers the following five plausible requirements :

a) The j.p.d. is the mean value of an hermitian operator:

$$f(q,p) = tr(\hat{K}\hat{\rho}) \quad .$$

b) The j.p.d. is a linear functional of the density matrix. (For a pure state the j.p.d. is sesquilinear in the wavefunction.) This means that the hermitian operator $\hat{K}$ is independent of $\hat{\rho}$.

c) It is a true probability function :

$$f(q,p) \geq 0 \quad .$$

d) When integrating over momentum space, the marginal distributions coincide with the proper quantummechanical probabilities in q:

$$\int f(q,p)\, d^n p = \langle q|\hat{\rho}|q\rangle \quad .$$

e) When integrating over position, the marginal distributions coincide with the proper quantummechanical probabilities in p:

$$\int f(q,p) d^n q = \langle p|\hat{\rho}|p\rangle \quad .$$

THEOREM *(Wigner). The five requirements a),b),c),d), and e) are incompatible.*

The requirement b) is not explicitely present in the original version of Wigner's theorem. The necessity of this requirement was observed by Mugur-Schächter [15]:

PROPOSITION *(Mugur-Schächter). In the absence of the arbitrary restriction b) Wigner's theorem cannot be realized.*

In the stronger version of Kruszynski and de Muynck [16] the requirement on one marginal distribution suffices:

COROLLARY *(Kruszynski and de Muynck). There is no phase space representation of quantum mechanics which satisfies the requirements a),b),c), and d) or e).*

J.p.d. of the first and second kind require that the proper quantummechanical marginal distributions are obtained. This leads to a severe restriction on account of Wigner's theorem, but three possibilities are left open for the construction of true j.p.d.:

a) j.p.d. of the first and second kind which are linear functionals of the density matrix and have the correct quantum mechanical marginals exist only for a restricted class of functions;

b) j.p.d. of the first and second kind which are nonlinear functionals of the density matrix and have the correct quantum mechanical marginals are not restricted by Wigner's theorem;

c) for j.p.d. of the third kind it is not desirable that the marginals coincide with the correct quantum mechanical distributions because the j.p.d. of the third kind are considered as the result of a joint measurement procedure. Hence there is also no restriction on account of Wigner's theorem.

We shall now consider these three possibilities in more detail.

6a. Construction of positive distributions: case a

(j.p.d. of the first and second kind which are linear functionals of the density matrix and which have the correct quantum mechanical marginals)

For possibility a) we restrict ourselves to the Weyl correspondence rule. In this case we have the

THEOREM *(Hudson, Soto and Claverie [17]). For systems with an arbitrary number of degrees of freedom a necessary and sufficient condition for the Wigner distribution function of a pure state to be non-negative is that the corresponding wave function is the exponential of a quadratic form.*

This theorem was proven by Hudson for one-dimentional systems and generalized by Soto and Claverie for systems with an arbitrary number of degrees of freedom. As a consequence of this theorem the wave-function is Gaussian and represents a coherent or a squeezed state, hence the j.p.d. is a multivariate normal (Gaussian) distribution in phase space. The characteristic function, which is defined as the symplectic Fourier transform of the j.p.d., is therefore also a Gaussian function. In the context of C^*-algebra this means that the state is quasi-free [18].

A complete characterization of Gaussian Wigner distributions

$$f(q,p) = \frac{1}{\pi^n} \sqrt{\det G}\; e^{-z^T G z}$$

is given by the

THEOREM *(Simon, Sudarshan ,and Mukanda [19]). The necessary and sufficient conditions for a Gaussian phase space distribution to be a Wigner distribution are that the matrix G is positive definite and that the Sp(2n,R) invariant powers of JG satisfy n inequalities.*

Compare this with the infinite set of K.L.M.conditions! We will not write down these rather complicated inequalities because we will present now a new theorem equivalent to the theorem of Simon et al.,but which entails a considerable simplification and gives a physical insight because it is based on the uncertainty principle. Take for symplicity $\mathrm{tr}(\hat{q}\,\hat{\rho}) = \mathrm{tr}(\hat{p}\,\hat{\rho}) = 0$ then the uncertainty relation can be derived from the inequality

$$tr\,[(a.\hat{q} + b.\hat{p})^+(a.\hat{q} + b.\hat{p})\hat{\rho}] \geq 0 \quad .$$

Introducing the covariance matrix

$$\sigma = \begin{pmatrix} \sigma_{q_i q_j} & \sigma_{q_i p_j} \\ \sigma_{p_i q_j} & \sigma_{p_i p_j} \end{pmatrix} \quad ,$$

it is easily seen that the last inequality is satisfied if and only if the matrix

$$\sigma - \frac{1}{2} i \hbar J \qquad \textit{is non-negative,}$$

where J is the matrix

$$J = \begin{pmatrix} 0 & 1 \\ -1 & 0 \end{pmatrix} \quad .$$

For one-dimensional systems the condition reduces to

$$\sigma_q \, \sigma_q - \sigma^2_{q,p} \geq \frac{\hbar^2}{4} \quad .$$

This relation, as well as the foregoing extention in matrix form, is nothing but the expression of Heisenberg's uncertainty relation in Schrödinger form, which is of course a necessary (but not sufficient) requirement for all j.p.d. in phase space. Note that the uncertainty relation in Schrödinger form is invariant for Sp(2n,R) transformations, in contrast to the usual,less strong formulation. We can derive the

THEOREM. *The necessary and sufficient condition for a Gaussian phase space function to be a Wigner distribution is that the covariance matrix satisfies Heisenberg's uncertainty relation in Schrödinger form :* $\sigma - \frac{1}{2} i \hbar J \geq 0.$

Note that f(p,q) may now be a mixed or a pure state.The difference between a pure and a mixed state is given by the

THEOREM *(Littlejohn [20]). The necessary and sufficient condition for a Gaussian Wigner distribution to be a pure state is that the matrix $\hbar G$ is a symplectic matrix:*

$$(\hbar G)\, J\, (\hbar G) = J$$

For quadratic hamiltonians and for the Weyl correspondence rule, the quantum Liouville operator reduces to the classical one. This allows us to prove in a simple way the following theorem [21]

THEOREM. *Any evolution operator which changes a Gaussian Wigner distribution into a Gaussian Wigner distribution is necessarily a quadratic Hamiltonian.*

This theorem reduces the Gaussian Wigner distributions to free particle and/or essentially harmonic oscillator states. A Gaussian Wigner distribution representing a mixed state is a thermal state of harmonic oscillators and/or free particles.

We know already that pure states necessarily correspond to Gaussian j.p.d.;for mixed states, Gaussian as well as non-Gaussian positive j.p.d. exist. The construction of these positive j.p.d. depends on the following theorems :

THEOREM. *(Gracia-Bondía and Várilly [22]) The convolution of a classical state with a quantum state is a quantum state.*

By "quantum state" we mean a Wigner distribution (positive or not),a "classical state" is any positive j.p.d.(satifying the K.L.M. conditions or not).

THEOREM. The convolution of a quantum state with another quantum state is a classical state.

As a consequence of these two theorems, we derive the

COROLLARY. *The convolution of a Gaussian Wigner distribution with an arbitrary Wigner distribution is a positive Wigner distribution.*

More generally, we have the

COROLLARY. *The convolution of a positive Wigner distribution with an arbitrary Wigner distribution is another positive Wigner distribution.*

These corollaries allow us to construct a true j.p.d. by smoothing any Wigner distribution with a Gaussian or any other positive Wigner distribution.

6b. Positive joint probability distributions: case b

(j.p.d. of the first and second kind which are nonlinear functionals of the density matrix and which have the correct quantum mechanical marginals)

These j.p.d. are not restricted by Wigner's theorem, as can be seen from the trivial example

$$f(q,p) = |\langle q|\hat{\rho}|q\rangle|^2\, |\langle p|\hat{\rho}|p\rangle|^2$$

An example with correlations is given by Cohen and Zaparovanny [23] :

$$f(q,p) = |\langle q|\hat{\rho}|q\rangle|^2\, |\langle p|\hat{\rho}|p\rangle|^2\, [1 + g(q,p)] \quad ,$$

where

$$g(q,p) = h(v,w) - \int_0^1 h\,dv - \int_0^1 h\,dw + \int_0^1\int_0^1 h\,dv\,dw \quad ,$$

with

$$v = \int_{-\infty}^{q} |\psi(q)|^2\,dq \quad , \quad w = \int_{-\infty}^{p} |\phi(p)|^2\,dp \quad .$$

We now introduce a new construction of j.p.d. only based on simple transformation properties.Consider a stationnary state

$$\psi_i(q,E_i) = \sqrt{\rho}\, e^{i S(q,E_i)/\hbar}$$

with energy E_i occurring with probability P_i, ρ is the probability density.In the classical limit $S(q,E_i)$ is a generator of the motion, and the probability density expressed in the canonical variables q and p is obtained by eliminating E_i .Quantum mechanically the corresponding eliminating procedure is replaced by the transformation

$$\psi(q,p) = \sum_i \sqrt{P_i}\,\psi_i(q,E_i)\,\phi_i^*(p,E_i) \quad ,$$

where ϕ_i denotes the wavefunction in momentum representation and which defines a phase space amplitude. The j.p.d. is then defined by the absolute square of the phase space amplitude. It is easy to see that the correct marginals are obtained. Indeed,

$$\int f(q,p)\,dp = \sum_i P_i\,|\psi_i(q)|^2 \quad .$$

A new, more general construction based on a phase space amplitude is given for a density matrix which can always be written as

$$\hat{\rho} = \hat{a}^+\,\hat{a} \quad .$$

Define now the phase space amplitude

$$\psi(q,p) = \langle p|\,\hat{a}\,|q\rangle \quad .$$

The j.p.d. is then defined as

$$f(q,p) = |\langle p|\,\hat{a}\,|q\rangle|^2 \quad .$$

Again we obtain the correct marginal distribution:

$$\int f(q,p)\,dp = \langle q|\,\hat{a}^+\,\hat{a}\,|q\rangle \quad .$$

6c. Positive joint probability distributions depending on the measurement arrangement : case c

(j.p.d. of the third kind)

If the measurement process can be described as a quantum mechanical process, then in a joint measurement of incompatible observables the measurements influence each other in such a way that the singly measured quantum mechanical probability distributions cannot be reproduced from the measurement results. Hence it is no longer desirable that the marginal probability distributions equal the single measured ones. It follows that the construction of these j.p.d. is not restricted bij Wigner's theorem and f(q,p) may be a linear function of the density matrix. Requiring Galilei invariance and positivity for any density matrix $\hat{\rho}$ describing the object system, we have for the most general form of the j.p.d.:

$$f(q,p) = \frac{1}{\hbar^n} tr\,(\hat{U}^{+}_{q,p}\ \hat{\rho}_{meas}\ \hat{U}_{q,p}\ \hat{\rho}_{obj}) \quad ,$$

where $\hat{\rho}_{meas}$ is not longer an arbitrary operator but a density matrix describing exhaustively the measurement apparatus. If both $\hat{\rho}_{meas}$ and $\hat{\rho}_{obj}$ are pure states then f(q,p) reduces to the transition probability

$$f(q,p) = \frac{1}{\hbar^n} |(\psi_{meas},\ \hat{U}_{q,p}\ \phi_{obj})|^2 \quad .$$

The marginal distributions are given by the convolution of two true probability densities:

$$\int f(q,p)\, dp = \langle q|\,\hat{\rho}_{meas}\,|q\rangle * \langle q|\,\hat{\rho}_{obj}\,|q\rangle \quad ,$$

$$\int f(q,p)\, dq = \langle p|\,\hat{\rho}_{meas}\,|p\rangle * \langle p|\,\hat{\rho}_{obj}\,|q\rangle \quad ,$$

which can be interpreted as accuracy calibrations given bij the confidence functions $\langle q|\,\hat{\rho}_{meas}\,|\rangle$ and $\langle p|\,\hat{\rho}_{meas}\,|p\rangle$. The couple q,$\langle q|\,\hat{\rho}_{meas}\,|q\rangle$ together with p,$\langle p|\,\hat{\rho}_{meas}\,|p\rangle$ can also be seen as a fuzzy sample point in phase space. Note also that, for j.p.d. of the third kind linear in $\hat{\rho}$, the ordening of operators is equivalent to a measuring procedure. One can also write the j.p.d. as a convolution of two Wigner distributions:

$$f(q,p) = f_{meas}(q,p) * f_{obj}(q,p) \quad ,$$

the first one representing the measurement procedure and the second one describing the object system.This "smoothing" or "coarse graining" of the Wigner distribution eliminates fast oscillations in $\hbar$ and gives therefore a better representation in the classical limit [24]. Another consequence of the last formula is that the covariance matrix is the sum of the covariance matrix of the object system and the one of the measurement procedure.Hence we obtain the operational uncertainty relation

$$\sigma - i\hbar J \geq 0$$

which reduces in one dimention to

$$\sigma_q \sigma_p - \sigma^2_{q,p} \geq \hbar^2$$

This operational uncertainty relation is in accordance with the experimental uncertainty relation [25]

$$(\Delta q)_{ex}\,(\Delta p)_{ex} \approx \hbar \quad .$$

Comparing this with the uncertainty relations for the j.p.d. of the first and the second kind, we observe that the inequalities are analogeous but with $\hbar$ replaced by $2\hbar$.

Taking for $\hat{\rho}_{meas}$ a coherent state we obtain the special case of stochastic distributions treated by Prugovecki [26]. These Gaussian wave packet or coherent state representations correspond to the Husimi transform and to the correspondence rule of anti-normal ordening.

7. Conclusion

To summarize we state that Wigner's theorem bans a large class of hidden variable theories but allows also, in contradistinction to a widespread belief, the construction of t.p.d. in each of the three possible interpretations : t.p.d. considered as a property of the object system, as a description of measurement results or as a result of joint measurements. A review of all the possibilities for the construction of j.p.d. is given in table 1. A more extensive account of new theorems will be published elsewhere.

Table 1. Construction of non-negative joint probability distributions

distribution	first kind	second kind	third kind
dependence on measurement arrangement	no		yes
correct q.m. marginals	yes		no
f(q,p) linear function of $\hat{\rho}$	yes	no	yes
restriction by Wigner's theorem	yes	no	no
pure states	Gaussian Wigner distributions with symplectic covariance matrix	examples with correlations $\lvert\psi(q,p)\rvert^2$	transition probability $\lvert(\psi_{meas}, \hat{U}_{qp}\,\phi_{obj})\rvert^2$
mixed state	Gaussian or non-Gaussian Wigner distributions	$\lvert\langle p\rvert\,\hat{a}\,\lvert q\rangle\rvert^2$	$tr\,(\hat{U}^{+}_{qp}\,\hat{\rho}_{meas}\,\hat{U}_{qp}\,\hat{\rho}_{obj})$
uncertainty relation	$\geq \hbar/2$		$\geq \hbar$

References

1. M.Hillery, R.F.O'Connell, M.O.Sully and E.W.Wigner, *Phys.Rep.* **106** (1984) 121.
2. W.Pauli, *Scientific Correspondence*, Vol.I (Springer Verlag, Berlin and New York 1979).
3. L.E.Ballentine, *Rev.Mod.Phys.* **42** (1970) 358.
4. P.Busch, *Found.Phys.* **17** (1987) 905.
5. H.Martens and W.M.de Muynck, *Found.Phys.* **20** (1990) 375.
6. P.Mittelstaedt, A.Prieur, and R.Schieder, *Found.Phys.* **17** (1987) 891
7. W.de Muynck, P.Janssen and S.Santman, *Found.Phys.* **9** (1979) 71.
8. L.Cohen, *J.Math.Phys.* **7** (1966) 781.
9. G.S.Agarwal and E.Wolf, *Phys.Rev.* **10** (1970) 2161.
10. E.Prugovecki, *Ann.Phys.* **7** (1978) 102.
11. Bayen, M. Flato, C. Fronsdal, A. Lichnerowicz, D.Sternheimer, *Ann. of Phys.*, **111** (1978) 61.
12. F.J.Narcowich and O'Connell, *Phys.Rev.* **A34** (1986) 34.
    F.J.Narcowich and O'Connell, *Phys.Lett.* **133** (1988) 167.
13. D.Kastler, *Commun.Math.Phys.* **1** (1965) 14;
    G.Loupias and S.Miracle-Sole,*Commun.Math.Phys.* **2** (1966) 31; Ann.Inst.H.Poincaré 6 (1967) 39
14. E.P. Wigner, in *Perspectives in Quantum Theory*,
    W. Yourgrau and A. Van der Merwe, eds., MIT Press (1971).
15. M. Mugur-Schächter, *Found. Phys.* **9** (1979) 389.
16. P.Kruszynski and W.M.de Muynck, *J. Math. Phys.*, **28** (1987) 1761.
17. R.L. Hudson, *Rep. Math. Phys.*, **6** (1974) 249;
    F. Soto and P. Claverie, *J. Math. Phys.*, **24** (1983) 97.
18. H.Scutaru, *Phys.Lett.* **A141** (1989) 223.
19. R.Simon, E.C.G.Sudarshan, and N.Mukanda, *Phys.Lett.* **A124** (1987) 223; *Phys.Rev.* **A36** (1987) 3868.
20. R.G. Littlejohn, *Phys. Rep.* **138** (1986) 193.
21. F. Soto and P. Claverie, *J. Math. Phys.*, **24** (1983) 1104;
    M.Gadella, J.M.Gracia-Bondia, L.M.Nieto, and J.C.Várilly, *J.Phys.A:Math.Gen.* **22**(1989) 2709.
22. J.M.Gracia-Bondía and J.C. Várilly, *Phys.Lett.* **A128** (1988) 20.
23. L.Cohen and Y.L.Zaparovanny, *J.Math.Phys.* **21** (1980) 794.
24. K.Takahashi, *J.Phys.Soc.Japan* **55** (1986) 762.
25. K.Wódkiewicz, *Phys.Lett.* **A124** (1987) 207.
26. E.Prugovecki, *Stochastic Quantum Mechanics and Quantum Spacetime*, D.Reidel Publishing Company (1984)

QUANTUM MEASUREMENT AND GENERALIZED COHERENT STATE

Shoju KUDAKA
Department of Physics, University of the Ryukyus
Okinawa 903-01, Japan

and

Shuichi MATSUMOTO
Department of Mathematics, University of the Ryukyus
Okinawa 903-01, Japan

ABSTRACT

We present and discuss some models for the quantum measuring process. Our consideration indicates that the concept of the generalized coherent state is useful to clarify the various measuring processes. It enables us to calculate explicitly the whole process of interaction between object system and apparatus.

1. Introduction

Many quantum theories possess the number of degrees of freedom as a free parameter. If N is such a parameter, then for a wide class of these theories the quantum theories with large N behave like classical theories in some sense. We can find a classical system (i.e., a classical phase space, Poisson bracket, and classical Hamiltonian) whose dynamics equivalent to the $N \to \infty$ limit of a given quantum theory [1-11]. Following Yaffe [4], we review a method used to explore this subject: (a) We choose an appropriate group of unitary transformations, and (b) generate a set of coherent states by applying elements of this group to a suitable initial state. These coherent states are called generalized coherent states. (c) Different coherent states become orthogonal in the $N \to \infty$ limit. (d) By using group theoretical method, we can define a symplectic structure on the space of all coherent states. This structure allows us to define classical dynamics on this space, and show that the $N \to \infty$ limit of the quantum dynamics is equivalent to this classical dynamics.

In this paper we would like to discuss quantum measuring process whose measuring apparatus has such an $N \to \infty$ limit.

We stand on a point of view that any macroscopic state of apparatus is one of generalized coherent states. Then a measurement is formally described as a change

$$\left[\sum_i |\phi_i\rangle\right] \otimes |u_0\rangle \longrightarrow \sum_i |\phi_i\rangle \otimes |u_i\rangle,$$

where $\sum_i |\phi_i\rangle$ is an initial state of object system and $|u_i\rangle$ are generalized coherent states. Here we may think that the values of $|\langle u_i|u_j\rangle|^2$ $(i \neq j)$ express the degree of "completeness of the corelation" between the states of object and the states of apparatus, because the probability of finding the state $|\phi_i\rangle\otimes|u_j\rangle$ $(i \neq j)$ in the final state $\sum_i |\phi_i\rangle \otimes |u_i\rangle$ is proportional to $|\langle u_i|u_j\rangle|^2$. One of our aims is to show that the detection of the interference effects between $|\phi_i\rangle \otimes |u_i\rangle$ and $|\phi_j\rangle \otimes |u_j\rangle$ $(i \neq j)$ becomes quite difficult as the degree of completeness of the corelation is increased.

In order to sketch the view, we treat a model in §§2 and 3. A little more general formalism is presented in §4. The result of this paper is an extension of one of Ref. 12. We omit the proofs of theorems in this article. They will be given elsewhere [13].

2. A model for measuring apparatus

We consider an interaction between an incoming particle and an apparatus consisting of AgBr molecules at N sites of the emulsion with equal spacing a [14]. After an interaction, some of AgBr molecules dissociate into Ag and Br atoms. These two states may be designated as up (AgBr molecule) and down (Ag and Br atoms), and can be described as two eigenstates of σ_3. The Hilbert space of each molecule in the apparatus is C^2 (C is the space of all complex numbers) and any state of the apparatus corresponds to a vector in their tensor product $C^2 \otimes \cdots \otimes C^2$ (N factors).

Suppose that the total Hamiltonian of the incoming particle and the apparatus is given by

$$H = H_S + H_A + H'$$

with

$$H_S = cp, \qquad H_A = \frac{1}{2}\hbar\omega \sum_{n=1}^{N}(1 - \sigma_3^{(n)}), \qquad H' = \sum_{n=1}^{N} V(x - na)\sigma_1^{(n)},$$

where x and p are the position and momentum operators of the particle, respectively, c the velocity of light, $V(x)$ a real function , ω a constant, and $\sigma_j^{(n)}(j =$

1, 2, 3) the Pauli matrices acting on the n-th site. The Hamiltonian H is invariant under any exchange of AgBr molecules. (This is called S_N invariance, where S_N is the group of permutations of the set $\{1, 2, \cdots, N\}$).Thus we may restrict our attention to the S_N invariant sector of the theory. The Hilbert space $\mathcal{H}_N$ is generated by S_N invarianr states $|k\rangle_N$ ($\langle k|k\rangle_N = 1$), $k = 0, 1, \cdots, N$, where k is the number of dissociated molecules (i.e., the number of down state molecules).

The S matrix is calculable and written as

$$S = \exp\left[-i\frac{\bar{V}a}{\hbar c}\sum_{n=1}^{N}\sigma_1^{(n)}\right],$$

where

$$\bar{V}a \equiv \int_{-\infty}^{+\infty} V(x)dx.$$

Now we consider the group of unitary transformations in the Hilbert space C^2, that is, we consider the special unitary group $G = SU(2)$. The corresponding Lie algebra g is $su(2)$ generated by the Pauli matrices $\sigma_j (j = 1, 2, 3)$. G acts in C^2 in the natural way, and we obtain the tensor product representation in $C^2 \otimes \cdots \otimes C^2$ (N factors);

$$u \cdot (\xi_1 \otimes \cdots \otimes \xi_N) \equiv (u \cdot \xi_1) \otimes \cdots \otimes (u \cdot \xi_N) \qquad u \in G, \quad \xi_n \in C^2.$$

Since the Hilbert space $\mathcal{H}_N$ is an invariant subspace in $C^2 \otimes \cdots \otimes C^2$ under the G-action, we obtain a representation (say, D_N) of G in $\mathcal{H}_N$. D_N is irreducible and this fact is essential for our following discussion.

We parameterize almost all elements in $SU(2)$ as follows:

$$u = u(\alpha, \beta) \equiv \begin{pmatrix} e^{\beta} & -\alpha^* e^{\beta^*} \\ \alpha e^{\beta} & e^{\beta^*} \end{pmatrix}$$

with

$$(1 + |\alpha|^2)e^{\beta+\beta^*} = 1,$$

where α and β are complex numbers.

The "generalized coherent states" we will use are obtained by applying elements of G to a "base state" of the apparatus. We take the state $|0\rangle_N$ as the base state, hence the coherent states are

$$|u\rangle_N \equiv D_N(u)\,|0\rangle_N \quad (u \in G).$$

In paticular, if we denote

$$|u(\alpha,\beta)\rangle_N = D_N(u(\alpha,\beta))\,|0\rangle_N$$

by

$$|\alpha,\beta\rangle_N,$$

then we have

$$|\alpha,\beta\rangle_N = e^{\beta N}\sum_{k=0}^{N}\alpha^k\binom{N}{k}^{1/2}|k\rangle_N.$$

Before turning to the measuring process, we list some properties of the coherent states.

(1) We have

$$|\langle k|\alpha,\beta\rangle_N|^2 = \binom{N}{k} w^k(1-w)^{N-k},$$

where w is the density of dissociated molecules in the state $|\alpha,\beta\rangle_N$. This equation is the binomial distribution.

(2) Define operators Σ_j by

$$\Sigma_j \equiv \sum_{n=1}^{N}\sigma_j^{(n)} \qquad (j = 1,2,3),$$

then they are 1-particle operators. (We note that the operator $\frac{1}{2}(1-\frac{1}{N}\Sigma_3)$ corresponds to the density of dissociated molecules.) Their expectation values and standard deviations are written as

$$\langle\alpha,\beta|\Sigma_j|\alpha,\beta\rangle_N = N\,c_j(\alpha,\beta),$$

and

$$\sigma(\Sigma_j) \equiv \left[\langle\alpha,\beta|\Sigma_j^2|\alpha,\beta\rangle - \langle\alpha,\beta|\Sigma_j|\alpha,\beta\rangle^2\right]^{1/2} = \sqrt{N}\,d_j(\alpha,\beta) \qquad (j = 1,2,3),$$

where $c_j(\alpha,\beta)$ and $d_j(\alpha,\beta)$ are constants which do not depend on N. This shows that each operator Σ_j/N has an almost definite value for each coherent state.

More generally, we consider an l-particles operator

$$A = \sum_{n_1,\cdots,n_l} A_{n_1,\cdots,n_l},$$

where $A_{n_1,\cdots,n_l}$ acts on n_1-th, n_2-th,$\cdots$,n_l-th sites and $\sum$ represents the symmetralization procedure. (It will be natural to assume that each operator $A_{n_1,\cdots,n_l}$ does not depend on N explicitly.) Then we can show that the operator

$$\frac{A}{\binom{N}{l}}$$

has an almost definite value for each coherent state, that is, the expectation value

$$\langle u|\frac{A}{\binom{N}{l}}|u\rangle_N$$

has a finite limit as N tends to ∞ and the standard deviation

$$\sigma\left(\frac{A}{\binom{N}{l}}\right)$$

tends to 0 as N becomes large.

(3) Different coherent states are not orthogonal; their overlap is

$$|\langle\alpha,\beta|\alpha',\beta'\rangle|^2 = \left[1 - \frac{|\alpha-\alpha'|^2}{(1+|\alpha|^2)(1+|\alpha'|^2)}\right]^N.$$

Thus, when $N \gg 1$ and $\alpha \neq \alpha'$, these two different states are approximately orthogoinal. More generally, we can prove the following theorem [13].

Theorem 1. Let $A = \sum_{n_1,\cdots,n_l} A_{n_1,\cdots,n_l}$ be an l-paticles operator, then we can estimate the matrix elements of A;

$$|\langle u|A|u'\rangle_N| = C_{N,u,u'}\,|\langle u|u'\rangle_N|^{1-l/N},$$

where $C_{N,u,u'}$ has the same order as $|\langle u|A|u\rangle_N|$ and $|\langle u'|A|u'\rangle_N|$ have with respect to N.

3. A model for quantum measuring process

We are now ready for discussing the measuring process between the incomming particle and the apparatus. It will be natural to take the the state $|0\rangle_N$ as an initial state of the apparatus, i.e.,

$$|\Psi_i\rangle \equiv |0\rangle_N.$$

As stated above, the S matrix is written as

$$S = \exp\left[-i\frac{\bar{V}a}{\hbar c}\sum_{n=1}^{N}\sigma_1^{(n)}\right] = D_N\left[\exp\left(-i\frac{\bar{V}a}{\hbar c}\sigma_1\right)\right].$$

Here we have

$$\exp\left(-i\frac{\bar{V}a}{\hbar c}\sigma_1\right) = \begin{pmatrix} e^{\beta_f} & -\alpha_f^* e^{\beta_f^*} \\ \alpha_f e^{\beta_f} & e^{\beta_f^*} \end{pmatrix} = u(\alpha_f, \beta_f)$$

with $\alpha_f = -i\tan(\frac{\bar{V}a}{\hbar c})$ and $e^{\beta_f} = \cos(\frac{\bar{V}a}{\hbar c})$. Applying S to the initial state, we find the final state

$$|\Psi_f\rangle \equiv S|\Psi_i\rangle = D_N(u(\alpha_f, \beta_f))\,|0\rangle_N = |\alpha_f, \beta_f\rangle_N.$$

Thus the initial and final states of the apparatus are both given by coherent states.

To clarify our argument, we assume that the incomming particle has spin $\frac{1}{2}$ and its initial state is given by

$$|\psi\rangle = |\phi_+\rangle + |\phi_-\rangle,$$

where $|\phi_+\rangle$ and $|\phi_-\rangle$ are well-localizesd wave packets with eigenvalue $\frac{1}{2}$ and $-\frac{1}{2}$ of the z-component of the particle spin. In order to measure the z-component of the particle spin, we arrange so that only the wave packet $|\phi_+\rangle$ interacts with our apparatus after a spectral decomposition through an applied magnetic field. The interaction between the particle and the apparatus gives a change of the state vector:

$$|\psi\rangle \otimes |\Psi_i\rangle \longrightarrow |\phi_+\rangle \otimes |\Psi_f\rangle + |\phi_-\rangle \otimes |\Psi_i\rangle,$$

where, for simplisity, we assume that the wave packets $|\phi_\pm\rangle$ do not change their forms.

The statistical operator ρ for the final state is then given by

$$\rho = \rho_0 + |\phi_+\rangle\langle\phi_-| \otimes |\Psi_f\rangle\langle\Psi_i| + |\phi_-\rangle\langle\phi_+| \otimes |\Psi_i\rangle\langle\Psi_f|,$$

where

$$\rho_0 \equiv |\phi_+\rangle\langle\phi_+| \otimes |\Psi_f\rangle\langle\Psi_f| + |\phi_-\rangle\langle\phi_-| \otimes |\Psi_i\rangle\langle\Psi_i|$$

is a mixed state statistical operator.

Let us consider the expectation value of an observable O of object system. Then we get

$$\mathrm{Tr}[\rho(O \otimes I)] = \mathrm{Tr}[\rho_0(O \otimes I)] + \mathrm{Re}\langle\phi_-|O|\phi_+\rangle\langle\Psi_i|\Psi_f\rangle,$$

where I is the identity operator of the apparatus, and we arrive at

$$|\mathrm{Tr}[\rho(O \otimes I)] - \mathrm{Tr}[\rho_0(O \otimes I)]| \sim |\langle\Psi_i|\Psi_f\rangle|.$$

Similarly, for any operator A of the apparatus, the expectation value of $O \otimes A$ becomes

$$\mathrm{Tr}[\rho(O \otimes A)] = \mathrm{Tr}[\rho_0(O \otimes A)] + \mathrm{Re}\langle\phi_-|O|\phi_+\rangle\langle\Psi_i|A|\Psi_f\rangle.$$

If the operator A is an l-particles operator:

$$A = \sum_{n_1,\cdots,n_l} A_{n_1\cdots,n_l},$$

then we can conclude from Theorem 1 that

$$|\mathrm{Re}\langle\phi_-|O|\phi_+\rangle\langle\Psi_i|A|\Psi_f\rangle| = C_N \; |\langle\Psi_i|\Psi_f\rangle|^{1-l/N}$$

where C_N has the same order as $|\mathrm{Tr}\rho_0(O \otimes A)|$. Thus

$$|\mathrm{Tr}[\rho(O \otimes A)] - \mathrm{Tr}[\rho_0(O \otimes A)]| = C_N \; |\langle\Psi_i|\Psi_f\rangle|^{1-l/N}.$$

This equation shows that when l is rather small as compared with N then

$$\mathrm{Tr}[\rho(O \otimes A)] \simeq \mathrm{Tr}[\rho_0(O \otimes A)],$$

that is, as long as we measure only an l-particles operator such that

$$l/N \sim 0$$

the quantum interference efects between $|\phi_+\rangle \otimes |\Psi_f\rangle$ and $|\phi_-\rangle \otimes |\Psi_i\rangle$ is almost undetectable.

4. A generalization of our measuring apparatus

In the previous section, we argued about a model for quantum measuring process and compared the non-diagonal terms of the statistical operator with the mixed state operator. The essential point was the estimate of the matrix elements of an l-particles operator A (see, Theorem 1):

$$|\langle u|A|u'\rangle_N| = C_{N,u,u'}\ |\langle u|u'\rangle_N|^{1-l/N}.$$

In this section, we show that a similar result can be obtained even if a little general case is considered.

We consider a measuring apparatus consisting of N identical particles. Let $\mathcal{H}$ be the Hilbert space of each particle in the apparatus. In order to simplify the calculations, we assume

Assumption 1. The Hamiltonian of total system "object + apparatus" under consideration is invariant with respect to any exchange of N particles in the apparatus, and we may restrict our attention to the S_N invariant sector of the system.

Concequently, we consider the space of S_N invariant vectors in $\mathcal{H}\otimes\cdots\otimes\mathcal{H}$ (N factors). It is denoted by $\mathcal{H}_N$. Let G be a group of unitary transformations in $\mathcal{H}$. G acts in the tensor product $\mathcal{H}\otimes\cdots\otimes\mathcal{H}$ (N factors), which is the Hilbert space of the apparatus, in such a way that

$$u\cdot(\xi_1\otimes\cdots\otimes\xi_N) \equiv (u\cdot\xi_1)\otimes\cdots\otimes(u\cdot\xi_N), \qquad u\in G,\ \ \xi_n\in\mathcal{H}.$$

$\mathcal{H}_N$ is invariant under the G-action, so that we have a representation D_N of G in $\mathcal{H}_N$. Next assumption is essential for our discussion.

Assumption 2. The representation D_N is irreducible.

Our final assumption is on the Hamiltonian H_A of the apparatus.

Assumption 3. H_A is given by $H_A = \sum_{n=1}^{N} H^{(n)}$, where $H^{(n)} = H$ is the Hamiltonian of each particle. (Hence the interactions between different particles are assumed to be negligiable.) Moreover, the lowest energy level of H_A is non-degenerate.

These assumptions suffice to show that the quantum theory of the apparatus reduces to classical mechanics as N becomes very large. The definition of generalized coherent states is identical to that presented for the previous model: They are defined by

$$|u\rangle_N \equiv D_N(u)\,|0\rangle_N, \qquad u\in G,$$

where $|0\rangle_N$ is the normalized lowest energy state. We introduce a subclass, K, of operators in $\mathcal{H}_N$; an operator A is a member of K if there is a non-negative integer

m such that

$$\langle u|A|u'\rangle_N / \langle u|u'\rangle_{N-m} \longrightarrow \text{a finite limit} \qquad (N \to \infty)$$

for all $u, u' \in G$. We call each operator in K a classical operator. For any pair of classical operators A and B, we can prove the factorization:

$$\langle u|AB|u\rangle_N \simeq \langle u|A|u\rangle_N \cdot \langle u|B|u\rangle_N \qquad (N \to \infty).$$

We would like to emphasize that if A is an l-particles operator, then we have

$$\frac{A}{\binom{N}{l}} \in K.$$

In order to introduce a "phase space", we consider the Lie algebra g of the group $G = SU(2)$. Let g^* be the dual space of g. Then we can consider two natural representations of G in g and g^* given by

$$Ad(u)\lambda \equiv u\lambda u^{-1}, \qquad u \in G, \qquad \lambda \in g,$$

and

$$\langle Ad^*(u)\zeta, \lambda\rangle \equiv \langle \zeta, Ad(u^{-1})\lambda\rangle, \qquad \zeta \in g^*.$$

They are called the adjoint representation and the coadjoint representation of G, respectively. Let Γ be an orbit generated by the action of $Ad^*(u)$ applied to any element in g^*. For any element ζ in Γ, the tangent space to Γ at ζ is spanned by tangent vectors to the curves $Ad^*(e^{t\lambda})\zeta$ $(-\infty < t < +\infty)$ at ζ, where λ runs over g. Therefore we can define a 2-form ω on Γ by

$$\omega_\zeta([\lambda],\ [\mu]) \equiv \langle \zeta,\ [\lambda, \mu]\rangle,$$

where $[\lambda]$ denotes the tangent vector to the curve $Ad^*(e^{t\lambda})\zeta$ at ζ, and $[\lambda, \mu]$ is the commutator of λ and μ. As is well known [15], this 2-form is called a symplectic form and it enables [4] us to define the Poisson bracket of any pair of functions on Γ.

We can show the following theorem [13].

Theorem 2. For each coherent state $|u\rangle$, define an element ζ_u in g^* by

$$\langle \zeta_u, \lambda \rangle \equiv -\frac{i}{N}\langle u|D_N(\lambda)|u\rangle_N, \qquad \lambda \in \mathrm{g}.$$

Then we have:

(a) $\zeta_u = \zeta_{u'}$ if and only if there is a constant c such that

$$|u\rangle_N = c|u'\rangle_N, \quad |c| = 1.$$

(b) All ζ_u $(u \in G)$ constitute a paticular coadjoint orbit in g^*. (This orbit is denoted by Γ from now on.)
(c) For any classical operator A, define a function $a(\zeta)$ on Γ by

$$a(\zeta_u) \equiv \lim_{N\to\infty} \langle u|A|u\rangle_N.$$

Then the quantum equation of motion

$$\frac{dA}{dt} = \frac{i}{\hbar}[H_A, A]$$

reduces to a classical equation

$$\frac{da(\zeta)}{dt} = \{h_{cl}(\zeta), a(\zeta)\}_{PB},$$

where the classical Hamiltinian $h_{cl}(\zeta)$ is given by

$$h_{cl}(\zeta_u) \equiv \frac{1}{N\hbar}\langle u|H_A|u\rangle_N.$$

Estimating matrix elements of an l-particles operator, we have [13]next theorem. It has the same meaning as Theorem 1.

Theorem 3. Let $|u\rangle_N$ and $|u'\rangle_N$ be two generalized coherent states, and let us define

$$\phi(u, u') \equiv -\frac{1}{N}\log\langle u|u'\rangle_N.$$

Then we have:

(a) $\mathrm{Re}\phi(u,u') > 0$ if and only if $\zeta_u \neq \zeta_{u'}$.
(b) If A is an arbitrary l-particles operator of the apparatus, then

$$|\langle u|A|u'\rangle_N| = C_{N,u,u'} \cdot e^{-(N-l)\mathrm{Re}\phi(u,u')},$$

where the constant $C_{N,u,u'}$ has the same order that $|\langle u|A|u\rangle_N|$ and $|\langle u'|A|u'\rangle_N|$ have.

5. Summary

Focussing our attention to the model in §§2 and 3, we summarize our results.

(a) Suppose that the simultaneous measurement of the two quantities (the z-component of particle spin and the state of the apparatus) is done for the final state

$$|\phi_+\rangle \otimes |\Psi_f\rangle + |\phi_-\rangle \otimes |\Psi_i\rangle.$$

Then the probability of finding the state $|\phi_+\rangle \otimes |\Psi_i\rangle$ (i.e., the state in which the z-component of particle spin is up, nevertheless, the measuring apparatus is still in the initial state $|\Psi_i\rangle$) is proportional to

$$|\langle \phi_+ \otimes \Psi_i|\ \phi_+ \otimes \Psi_f + \phi_- \otimes \Psi_i\rangle|^2 = |\langle \phi_+|\phi_+\rangle\langle \Psi_i|\Psi_f\rangle|^2$$

because $\langle \phi_+|\phi_-\rangle = 0$. That is, the magnitude of $|\langle \Psi_i|\Psi_f\rangle|^2$ expresses the degree of possibility that the correlation between the state of object and the state of apparatus is not complete. In other words, the magnitude of $|\langle \Psi_i|\Psi_f\rangle|^2$ can be a measure of "completeness of the correlation ".

We know that $|\langle \Psi_i|\Psi_f\rangle|^2 = (1-w)^N$, where w is the density of dissociated molecules in the state $|\Psi_f\rangle$, that is, w is the probability for the transition from an AgBr molecule to Ag and Br atoms in each site. Therefore the magnitude of $|\langle \Psi_i|\Psi_f\rangle|^2$ is influenced by the intensity of the interaction between the incomming particle and the apparatus.

As stated in §3, for any operator O of the object system and for any l-particles operator A of the apparatus, we have

$$|\mathrm{Tr}[\rho(O\otimes A)] - \mathrm{Tr}[\rho_0(O\otimes A)]| = C_N\, |\langle \Psi_i|\Psi_f\rangle|^{1-l/N}.$$

This equation shows how the difficulty of the detection of non-diagonal efects of ρ depends on the degree of completeness of the correlation and the magnitude of N. The detection of the interference efects between $|\phi_+\rangle \otimes |\Psi_f\rangle$ and $|\phi_-\rangle \otimes |\Psi_i\rangle$ becomes quite difficult as the completeness of the correlation is increased as long as we measure an l-particles operator such that $l/N \simeq 0$.

(b) As stated in §4, each generalized coherent state corresponds to a point on a "phase space" in which we can consider classical dynamics determined by the Poisson bracket. When N becomes large, quantum equations of motion for classical operators in K approach the corresponding classical equations on the phase space. On the other hand, each coherent state $|u\rangle$ is parametrized by the expectation values $\langle u|A|u\rangle$, where A runs over K. Hence we can conclude that the time evolution of each coherent state is represented by a classical trajectory in the phase space.

(c) For an arbitrary classical operator A, the expectation value

$$\langle u|A|u\rangle_N$$

has a finite limit as N becomes large, and we have

$$\sigma(A) \equiv [\langle u|A^2|u\rangle_N - \langle u|A|u\rangle_N^2]^{1/2} \longrightarrow 0 \qquad (N \rightarrow \infty)$$

This means that when we measure a classical operator for a coherent state, we always obtain a definite value without being aware of fluctuations about this value as long as N is sufficiently large. In this sense, the classical operators can be taken as "macroscopic quantities". For example, let us define a 1-particle operator A by

$$A \equiv \sum_{n=1}^{N}(1 - \sigma_3^{(n)})/2,$$

then A/N is a classical operator and we have

$$\langle \alpha, \beta|\frac{A}{N}|\alpha, \beta\rangle_N = \frac{|\alpha|^2}{1+|\alpha|^2},$$
$$\sigma\left(\frac{A}{N}\right) = \frac{1}{\sqrt{N}}\frac{|\alpha|}{1+|\alpha|^2}.$$

In fact A/N corresponds to the density of the dessociated molequles.

6. Acknowledgemens

It is a pleasure for us to thank Prof. K. Kakazu and Dr. H. Nakazato for many valuable discussions.

7. References

1. J.R.Klauder, J. Math. Phys. 4 (1963) 1055, 1058; 5 (1964) 177.
2. A.M.Perelomov, Commun. Math. Phys. 26 (1972) 222; 44 (1975) 197.
3. F.A.Berezin, Commun. Math. Phys. 63 (1978) 131.
4. L.G.Yaffe, Rev. Mod. Phys. 54 (1982) 407.
5. J.R.Klauder and B.Skagerstam, *Coherent States*, (World Scientific Publishing Co Pte Ltd., 1985).
6. R.Gilmore, Critical Phenomena. 1983 Brasov School Conference (1985) 213.
7. F.R.Brown and L.G.Yaffe, Nucl. Phys. B271 (1986) 267.
8. A.Perelomov, *Generalized Coherent States and Their Applications*, (Springer-Verlag Berlin Heidelberg, 1986).
9. A.Bach, Lett, Math. Phys. 15 (1988) 75.
10. S.D.Howard and S.K.Roy, Found. Phys. 19 (1989) 1285.
11. J.Kurchan, P.Leboeuf and M.Saraceno, Phys. Rev. A40 (1989) 6800.
12. S.Kudaka, S.Matsumoto and K.Kakazu, Prog. Theor. Phys. 82 (1989) 665.
13. S.Matsumoto, to be published.
14. K.Hepp, Helv. Phys. Acta 45 (1972) 237.
15. A.A.Kirillov, *Elements of the Theory of Representations*, (Springer Verlag Berlin, 1976).

Entropy and Information in Quantum Theory of Measurement

PEKKA J. LAHTI
Department of Physical Sciences
University of Turku
SF-20500 Turku, Finland

ABSTRACT

Information theoretical characterizations of measurement induced state and probability transformations are studied.

1. Introduction.

Quantum theory of measurement induces two types of transformations, *transformations of states* and *transformations of probability measures*. The properties of such transformations reflect the properties of the inducing measurements. Characterizations of these transformations lead to characterizations of the measurements. In this paper we shall investigate such characterizations with the hope that they could further clarify the role of various requirements posed on measurements in quantum mechanics.

The paper will apply the usual Hilbert space formulation of quantum mechanics. In its usual formulation the theory of measurement is a part of the theory of compound systems with its own specific questions. In order to formulate the questions to be studied here, we shall recall first the relevant ingredients – terminology, notations, and basic facts – of the Hilbert space quantum mechanics with its measurement theory. The monographs of Beltrametti and Cassinelli[1), Davies[2), and Reed and Simon[3) serve as our basic references.

Let $\mathcal{H}$ be a complex separable Hilbert space, with the inner product $<\cdot|\cdot>$. We let $\mathcal{L}(\mathcal{H})$ denote the set of bounded operators on $\mathcal{H}$, and $\mathcal{L}(\mathcal{H})^+$ the subset of its positive elements. A normalized positive operator valued (POV) measure $E:\mathcal{F}\to\mathcal{L}(\mathcal{H})^+$ on a measurable space $(\Omega,\mathcal{F})$ is defined through the properties: (i) $E(X)\geq E(\emptyset)$ for all $X\in\mathcal{F}$, (ii) if (X_i) is a countable collection of disjoint sets in $\mathcal{F}$ then $E(\cup_i X_i)=\sum_i E(X_i)$, the series converging in the weak operator topology; (*iii*) $E(\Omega) = I$, where I is the identity operator on $\mathcal{H}$. The space of trace class operators on $\mathcal{H}$ will be denoted $\mathcal{T}(\mathcal{H})$, and $\mathcal{T}(\mathcal{H})_1^+$ consists of the positive trace one operators on $\mathcal{H}$. The set $\mathcal{T}(\mathcal{H})_1^+$ is convex, and its extremal points are the one-dimensional projection operators $P[\varphi]$ on $\mathcal{H}$, where $P[\varphi]\psi \doteq < \varphi|\psi > \varphi$ for all $\varphi \in \mathcal{H}$.

In the Hilbert space formulation of quantum mechanics the description of a physical system $\mathcal{S}$ is based on a (complex separable) Hilbert space $\mathcal{H}_\mathcal{S}$. Any *observable* of $\mathcal{S}$ is represented as (and identified with) a POV measure $E : \mathcal{F} \to \mathcal{L}(\mathcal{H}_\mathcal{S})^+$ on a Borel space $(\Omega, \mathcal{F})$ – the value space of the observable. We recall that a POV measure E is a projection operator valued (PV) measure if and only if it is multiplicative, that is, $E(X \cap Y) = E(X)E(Y)$ for all $X, Y \in \mathcal{F}$. According to the spectral theorem, the usual notion of an observable as a self-adjoint operator in $\mathcal{H}_\mathcal{S}$ refers to the special case of E being a PV measure on the real Borel space $(\Re, \mathcal{B}(\Re))$. Clearly, the value spaces $(\Omega, \mathcal{F})$ of observables are, as a rule, real Borel spaces $(\Re, \mathcal{B}(\Re))$, or $(\Re^n, \mathcal{B}(\Re^n))$, $n = 1, 2, \cdots$, or their Borel subspaces. Throughout this paper we shall assume that this is the case. *States* of $\mathcal{S}$ are represented as (and identified with) the elements of $\mathcal{T}(\mathcal{H}_\mathcal{S})_1^+$, and we refer to the unit vectors of $\mathcal{H}_\mathcal{S}$, as well as to the one-dimensional projection operators they generate, as the *vector states* of $\mathcal{S}$. Any pair (E, T) of an observable E and a state T defines a *probability measure* $E_T : \mathcal{F} \to [0,1]$, $X \mapsto E_T(X) \doteq tr[TE(X)]$, where [0,1] is the unit interval of the real line $\Re$. According to the probability interpretation of the theory, the number $E_T(X)$ is the probability that a measurement of E performed on $\mathcal{S}$ in a state T leads to a result in a value set X. It is a basic fact of the Hilbert space formulation of quantum mechanics that the above representations of observables and states are the most general ones which are compatible with the probability structure of quantum mechanics.

Let $\mathcal{H}_\mathcal{A}$ be another complex separable Hilbert space. The Hilbert space tensor product of $\mathcal{H}_\mathcal{S}$ and $\mathcal{H}_\mathcal{A}$ is denoted as $\mathcal{H}_\mathcal{S} \otimes \mathcal{H}_\mathcal{A}$. The sets $\mathcal{L}(\mathcal{H}_\mathcal{S} \otimes \mathcal{H}_\mathcal{A})$, $\mathcal{T}(\mathcal{H}_\mathcal{S} \otimes \mathcal{H}_\mathcal{A})$, and so on, are defined as above. The partial trace over $\mathcal{H}_\mathcal{A}$, say, is the positive linear map $\mathcal{R}_\mathcal{S} : \mathcal{T}(\mathcal{H}_\mathcal{S} \otimes \mathcal{H}_\mathcal{A}) \to \mathcal{T}(\mathcal{H}_\mathcal{S})$ defined as follows: $tr[\mathcal{R}_\mathcal{S}(W)S] = tr[WS \otimes I_\mathcal{A}]$, where $W \in \mathcal{T}(\mathcal{H}_\mathcal{S} \otimes \mathcal{H}_\mathcal{A})$, $S \in \mathcal{L}(\mathcal{H}_\mathcal{S})$, and $I_\mathcal{A}$ is the identity operator on $\mathcal{H}_\mathcal{A}$. Similarly, one defines the partial trace over $\mathcal{H}_\mathcal{S}$, and we denote it as $\mathcal{R}_\mathcal{A}$.

Let $\mathcal{A}$ be another physical system, the description of which is based on $\mathcal{H}_\mathcal{A}$. The description of the compound system $\mathcal{S} + \mathcal{A}$ is then based on $\mathcal{H}_\mathcal{S} \otimes \mathcal{H}_\mathcal{A}$ along the usual ideas of the theory of compound systems.[1)] We recall here only that any state W of $\mathcal{S} + \mathcal{A}$ determines uniquely the states of the subsystems $\mathcal{S}$ and $\mathcal{A}$ as the *reduced states* $\mathcal{R}_\mathcal{S}(W)$ and $\mathcal{R}_\mathcal{A}(W)$, respectively.

In its usual formulation, the theory of measurement of an observable E of the object system $\mathcal{S}$ starts with fixing a *measuring apparatus* $\mathcal{A}$ (with a Hilbert space $\mathcal{H}_\mathcal{A}$), its *initial state* $T_\mathcal{A} \in \mathcal{T}(\mathcal{H}_\mathcal{A})_1^+$, a *pointer observale* $P_\mathcal{A} : \mathcal{F}_\mathcal{A} \to \mathcal{L}(\mathcal{H}_\mathcal{A})^+$ (with its value space $(\Omega_\mathcal{A}, \mathcal{F}_\mathcal{A})$) together with a (measurable) *pointer function* $f : \Omega_\mathcal{A} \to \Omega$ (which correlates the pointer values with the values of the measured observable), and a *measurement coupling* $V : \mathcal{T}(\mathcal{H}_S \otimes \mathcal{H}_A) \to \mathcal{T}(\mathcal{H}_S \otimes \mathcal{H}_A)$ (positive linear trace preserving mapping).[1,4,5,6,7] The 5-tuple $< \mathcal{H}_A, P_A, T_A, V, f >$ is a *premasurement* of E if it satisfies the *probability reproducibility condition*

$$E_T(X) = P_{A,\mathcal{R}_A(V(T\otimes T_a))}(f^{-1}(X)) \tag{1}$$

for all $X \in \mathcal{F}$, $T \in \mathcal{T}(\mathcal{H}_S)_1^+$. The term premeasurement is used in order to emphasize the fact that condition (1) is not enough to specify all the properties of a measurement. Other important aspects of measurements are discussed, for example, in the references 6 and 7. For short, we let $\mathcal{M}$ stand for a premeasurement $< \mathcal{H}_A, P_A, T_A, V, f >$.

Any premeasurement $\mathcal{M}$ of an observable E determines an *instrument* [2], that is, a state transformation valued measure $\mathcal{I}_\mathcal{M} : \mathcal{F} \to \mathcal{L}(\mathcal{L}(\mathcal{H}_S))^+$, $X \mapsto \mathcal{I}_\mathcal{M}(X)$, with

$$\mathcal{I}_\mathcal{M}(X)T \doteq \mathcal{R}_S(V(T \otimes T_A) \cdot I \otimes P_A(f^{-1}(X))) \tag{2}$$

for all $T \in \mathcal{T}(\mathcal{H}_S)_1^+$. This instrument is E-compatible: it defines uniquely the observable E via the relation $tr[\mathcal{I}_\mathcal{M}(X)T] = tr[TE(X)]$, $X \in \mathcal{F}$, $T \in \mathcal{T}(\mathcal{H}_S)_1^+$. If $tr[TE(X)] \neq 0$ we denote $T_\mathcal{M}^X = \mathcal{I}_\mathcal{M}(X)T/tr[TE(X)]$, so that, in particular, $T_\mathcal{M}^\Omega = \mathcal{I}_\mathcal{M}(\Omega)T = \mathcal{R}_S(V(T \otimes T_A)) \doteq T_\mathcal{M}$. Thus, if T is the initial state of $\mathcal{S}$, then $T_\mathcal{M}$ is its final state. Moreover, the state $T_\mathcal{M}^X$ can be interpreted as the final state of $\mathcal{S}$ with the condition that the measurement led to a result in X.[5] Similarly, the reduced state $\mathcal{R}_A(V(T \otimes T_A)) \doteq T_{A,\mathcal{M}}$ is the final state of $\mathcal{A}$.

Any premeasurement $\mathcal{M}$ of E induces, among others, the following state and probability transformations:

$$T \mapsto T_\mathcal{M} = \mathcal{R}_S(V(T \otimes T_A)) \tag{3}$$

$$T_A \mapsto T_{A,\mathcal{M}} = \mathcal{R}_A(V(T \otimes T_A)) \tag{4}$$

$$E_T \mapsto E_{T_\mathcal{M}} \tag{5}$$

$$P_{A,T_A} \mapsto P_{A,T_{A,\mathcal{M}}} \tag{6}$$

Several aspects of measurements have been studied by investigating the properties of these transformations. References 1, 2, 5, 6, 7, 8, and 9 are examples of such a work. In Section 2 and 3 some further, information theoretical properties of these transformations will be studied. We remind ourselves that the basic measurement condition (1) connects the initial object probability measure E_T in (5) with the final apparatus probability measure $P_{A,T_{A,\mathcal{M}}}$ in (6).

We close this introduction with pointing out an important class of premeasurements. Let $A = \sum_{i=1}^N a_i E^A(\{a_i\}) = \sum_{i=1}^N \sum_{j=1}^{n(i)} P[\varphi_{ij}]$ be a discrete ordinary observable. (Here N denotes the – finite or infinite – number of distinct eigenvalues of A, $n(i)$ is the degeneracy – finite or infinite – of the ith eigenvalue a_i, and $\{\varphi_{ij}\} \subset \mathcal{H}_S$ is a complete orthonormal set of eigenvectors of A.) Consider a pointer observable $P_A : a_i \mapsto P[\Phi_i]$, and let Φ be a unit vector of $\mathcal{H}_A$. Then the unitary mappings $U : \mathcal{H}_S \otimes \mathcal{H}_A \to \mathcal{H}_S \otimes \mathcal{H}_A$ which define premeasurements of A are exactly the unitary extensions of the mapping

$$\varphi_{ij} \otimes \Phi \mapsto \psi_{ij} \otimes \Phi_i, \tag{7}$$

where $\{\psi_{ij} : i = 1,\cdots,N,\ j = 1,\cdots,n(i)\} \subset \mathcal{H}_S$ is a set of unit vectors which satisfy the orthogonality conditions $< \psi_{ij}|\psi_{ik} > = 0,\ j \neq k$, for all $i = 1,\cdots,N$.[6] The resulting of premeasurements of A are denoted as $\mathcal{M}_U^m$. Such measurements are distinguished by a *unitary* measurement coupling and a *minimal* pointer observable.[6,7]

2. State-entropy.

The notion of entropy of a state can be used to characterize the state transformation $T \mapsto T_{\mathcal{M}}$ induced by a measurement $\mathcal{M}$ of an observable E. In particular, this concept allows one to single out premeasurements of E which lead to an optimal separation of the final component states (a posteriori states[14]) of the object system.

2.1. Preliminaries. The *entropy of a state* $T \in \mathcal{T}(\mathcal{H})_1^+$ is defined[4] as the (nonnegative) number

$$S(T) \doteq -tr[Tln(T)] \tag{8}$$

Using the spectral decomposition of $T = \sum t_i E^T(\{t_i\})$ one gets the expression

$$\begin{aligned} S(T) &= -\sum t_i ln(t_i) tr[E^T(\{t_i\})] \\ &= -\sum n_i t_i ln(t_i) \end{aligned} \tag{9}$$

where n_i is the degeneracy degree of the eigenvalue t_i. If $Tln(T)$ is trace class, then $S(T)$ is finite. In particular, if the support projection P_T of T is trace class (finite dimensional) then

$$0 \leq S(T) \leq ln(tr[P_T]) \tag{10}$$

where the lower bound is obtained iff T is pure, and the upper bound iff T is totally degenerate (maximally mixed).[10]

The entropy of a state $T \in \mathcal{T}(\mathcal{H})_1^+$ is a measure for the *degree of mixing* in T. Let

$$T = \sum t_i P[\varphi_i] \tag{11}$$

be an orthogonal decomposition of T into some vector states $P[\varphi_i]$. As an orthogonal decomposition (11) is unique iff T is nondegenerate, that is, all the eigenvalues t_i are nondegenerate. Apart from that there is uncountable many decompositions of T into vector states

$$T = \sum \lambda_i P[\psi_i] \tag{12}$$

or, even more generally, arbitrary decompositions

$$T = \sum w_i T_i \tag{13}$$

where the weights t_i / λ_i / w_i are nonnegative and sum up to one. The number $S(T)$ is independent of any decomposition of T, but due to the nonuniqueness of such decompositions this number cannot, as a rule, be interpreted as the *lack of information on the state* $P[\varphi_i]$ / $P[\psi_i]$ / T_i of the system if the system is known to be in the state

$$T = \sum t_i P[\varphi_i] = \sum \lambda_i P[\psi_i] = \sum w_i T_i \tag{14}$$

Such an interpretation seems to presuppose that the ignorance interpretation of T with respect to a particular decomposition can be given.

In the context of the measurement theory a measurement $\mathcal{M}$ of an observable E leads to a natural decomposition of the final state $T_{\mathcal{M}}$ of $\mathcal{S}$ into the component (a posteriori) states corresponding to the possible pointer readings. If the measurement leads, as it perhaps should, to the objectification of the result, then $T_{\mathcal{M}}$ admits an ignorance interpretation with respect to this particular decomposition so that also $S(T)$ admits an epistemic interpretation as the lack of information on the actual state of the system. We assume nowon that this is the case. However, the formal results presented below do not depend on this particular interpretational assumption.

Before closing these preliminaries we remind ourselves on the concavity and subadditivity of the entropy functional:

$$\sum w_i S(T_i) \leq S(\sum w_i T_i) \leq \sum w_i S(T_i) + S(\{w_i\}) \tag{15}$$

with $S(\{w_i\}) \doteq -\sum w_i ln(w_i)$. The first inequality is an equality iff $T_i = T_j$ for all i and j, whereas the second inequality is an equality iff $T_i T_j = O$ for all $i \neq j$.[11]

2.2. Measurement context. Consider a measurement $\mathcal{M}$ of an observable E on the system $\mathcal{S}$ in the state T. Then

$$T_{\mathcal{M}} = \mathcal{R}_{\mathcal{S}}(V(T \otimes T_{\mathcal{A}})) \tag{16}$$

is the final state of $\mathcal{S}$ and

$$T_{\mathcal{M}} = \sum N_i^2 T_i \text{ with } N_i^2 = E_T(X_i) \tag{17}$$

is the natural decomposition of $T_{\mathcal{M}}$ with respect to $\mathcal{M}$ and with respect to the given reading scale $\mathcal{R}$ (a partition of the value space $\Omega = \cup X_i$ of E into disjoint sets $X_i \in \mathcal{F}$). The state change

$$T \mapsto T_{\mathcal{M}} = \sum N_i^2 T_i \tag{18}$$

is associated with the state-entropy change

$$S(T) \mapsto S(T_{\mathcal{M}}) \tag{19}$$

This transformation provides various possibilities to study the measuring process $\mathcal{M}$ by the entropy functional S. For example, one may compute the difference

$$S(T_{\mathcal{M}}) - S(T)$$

which should show the increase in the lack, caused by $\mathcal{M}$, of information on the actual state of the system. Secondly, one may compute the more involved quantity

$$\begin{aligned} S(T) - \sum N_i^2 S(T_i) = \\ & [S(T_{\mathcal{M}}) - \sum N_i^2 S(T_i)] - [S(T_{\mathcal{M}}) - S(T)] \end{aligned}$$

which, under some further assumptions, is the difference of the potential information gain upon reading the result and the increase in the lack of information on the actual state of the system.[7] The results of Groenewold[12], Lindblad[13], and Ozawa[14] characterize measurements for which this difference is always nonnegative. Thirdly, one may study the relative state-entropy

$$S(T|T_{\mathcal{M}}) \doteq tr[T\ln(T) - T\ln(T_{\mathcal{M}})]$$

associated with the measurement $\mathcal{M}$. We recall that this quantity is finite only if $P_T \leq P_{T_{\mathcal{M}}}$.[11]

We remark next that if the initial state of $\mathcal{S}$ is a vector state $T = P[\varphi]$, so that $S(T) = 0$, then

$$\begin{aligned} S(T_{\mathcal{M}}) - S(T) = S(T_{\mathcal{M}}) &= S(\sum N_i^2 T_i) \\ &\leq \sum N_i^2 S(T_i) + S(\{N_i^2\}) \end{aligned}$$

where now also inequality (15) is applied. The Groenewold-Lindblad-Ozawa inequality

$$-\sum N_i^2 S(T_i) \geq 0$$

becomes trivial since the component states T_i must, in that context, be vector states[14], whereas the relative state-entropy obtains a simple formal expression

$$S(P[\varphi]|T_{\mathcal{M}}) = - < \varphi | \ln(T_{\mathcal{M}})\varphi >$$

We note that the assumption that the initial state of the object system is a vector state is neither a formal nor a conceptual restriction on the measuring process. Thus we shall go on with that assumption.

In order to explicate the above options we shall consider now a unitary measurement $\mathcal{M}_U^m$ of a discrete ordinary observable $A = \sum a_i E^A(\{a_i\}) \doteq \sum a_i P_i = \sum_i a_i \sum_j P[\varphi_{ij}]$. In that case the pointer eigenvalues define a natural (finest) reading scale, and we assume that this reading scale in now chosen. Let $P[\varphi]$ be the initial state of $\mathcal{S}$. Then

$$\begin{aligned} T_{\mathcal{M}_U^m} &= \sum |\psi_{il}><\varphi_{il}|\varphi><\varphi|\varphi_{ij}><\psi_{ij}| \\ &\doteq \sum N_i^2 P[\gamma_i] \end{aligned} \tag{20}$$

with

$$N_i^2 = E^A_{P[\varphi]}(\{a_i\}) = <\varphi|P_i\varphi> \tag{21}$$

is the final state of $\mathcal{S}$. Hence

$$S(T_{\mathcal{M}_U^m}) = S(\sum N_i^2 P[\gamma_i]) \leq S(\{N_i^2\}) \tag{22}$$

where the equality holds iff the vectors γ_i are orthogonal. We say that a measurement $\mathcal{M}_U^m$ is a *maximal state-entropy* measurement if $S(T_{\mathcal{M}_U^m}) = S(\{N_i^2\})$ for all initial states φ of $\mathcal{S}$. A measurement $\mathcal{M}_U^m$ of A is a maximal state-entropy measurement iff the component vectors γ_i are always (that is, for each φ) orthogonal. Maximal state-entropy measurements are thus characterized by the fact that such measurements lead to an optimal separation of the (a posteriori) component states. We recall that such measurements are exactly the *strong state-correlation* measurements of A – measurements of A which for each initial state φ of $\mathcal{S}$ lead to strong correlations between all the component states $P[\gamma_i]$ and $P[\Phi_i]$ of $\mathcal{S}$ and $\mathcal{A}$.[6)] We summarize these results as a theorem.

THEOREM 1. *A unitary measurement $\mathcal{M}_U^m$ of a discrete ordinary observable is a maximal state-entropy measurement if and only if it is a strong state-correlation measurement.*

We observe next that, in the present context, the Groenewold-Lindblad-Ozawa inequality

$$S(P[\varphi]) - \sum N_i^2 S(P[\gamma_i]) \geq 0 \tag{23}$$

is, indeed, trivial since all the involved terms are now equal to zero. Secondly, the relative state-entropy

$$S(P[\varphi]|T_{\mathcal{M}_U^m}) = - <\varphi|ln(\sum N_i^2 P[\gamma_i])\varphi> \tag{24}$$

is finite for each φ only if $\mathcal{M}_U^m$ is a Lüders measurement.[7)] In that case $S(P[\varphi]|T_{\mathcal{M}_U^m}) = S(\{N_i^2\})$.

3. Average information.

Consider an E-measurement $\mathcal{M}$. Accompanied with the state transformation $T \mapsto T_{\mathcal{M}}$ there is also the transformation of the probability measures $E_T \mapsto E_{T_{\mathcal{M}}}$. This transfromation leads to various probabilistic characterizations of $\mathcal{M}$. We recall first that $\mathcal{M}$ is of the ***first kind*** if this transformation is identical, that is, $E_T = E_{T_{\mathcal{M}}}$ for all $T \in \mathcal{T}(\mathcal{H})_1^+$. For ordinary (PV) observables E the first kind property of $\mathcal{M}$ is equivalent with its *repeatability*:

$$tr[\mathcal{I}_{\mathcal{M}}(X)T] = tr[\mathcal{I}_{\mathcal{M}}(X)^2T], \quad X \in \mathcal{F},\ T \in \mathcal{T}(\mathcal{H})_1^+ \tag{25}$$

In general, repeatable measurements are first kind measurements, but the converse need not be true.[7)] We recall also that observables which admit repeatable unitary measurements are necessarily discrete.[5)] Finally, we remind ourselves that a unitary measurement $\mathcal{M}_U^m$ of a discrete ordinary observable A is of the first kind, and thus repeatable, iff it is a ***strong value-correlation*** measurement – a measurement which leads to strong correlations between the corresponding values of the pointer observable $A_{\mathcal{A}}$ and the measured observable A.[6)]

The notion of average information of a probability measure can also be used to study the properties of a measurement. Indeed, consider a measurement $\mathcal{M}$ of an observable E with a given reading scale $\mathcal{R}$. As above,

$$T_{\mathcal{M}} = \sum N_i^2 T_i$$

with

$$T_i = N_i^{-2}\mathcal{I}_{\mathcal{M}}(X_i)T$$

and

$$N_i^2 = E_T(X_i)$$

(where, in fact, we put $T_i = O$ if $N_i^2 = 0$). Then

$$H(E,T;\mathcal{R}) \doteq S(\{N_i^2\}) \tag{26}$$

is the *average information* of the probability measure $\{N_i^2\} = \{E_T(X_i) : X_i \in \mathcal{R}\}$.[10)] This number can be interpreted as the average *lack of information for predicting a measurement outcome* X_i in the initial state T of $\mathcal{S}$. Similarly, one has

$$H(E,T_{\mathcal{M}};\mathcal{R}) = S(\{E_{T_{\mathcal{M}}}(X_i)\}) \tag{27}$$

Since $T_{\mathcal{M}} = \sum N_i^2 T_i$, we have the basic inequalities

$$\begin{aligned}\sum N_i^2 H(E,T_i;\mathcal{R}) &\le H(E,T_{\mathcal{M}};\mathcal{R})\\ &\le \sum N_i^2 H(E,T_i;\mathcal{R}) + S(\{N_i^2\})\\ &= \sum N_i^2 H(E,T_i;\mathcal{R}) + H(E,T;\mathcal{R})\end{aligned} \tag{28}$$

This inequality gives, again, a possibility to compare the initial and final average informations $H(E,T;\mathcal{R})$ and $H(E,T_{\mathcal{M}};\mathcal{R})$.

The case

$$H(E,T;\mathcal{R}) = H(E,T_{\mathcal{M}};\mathcal{R}) \tag{29}$$

is important. This condition is required in order the average lack of information for predicting a measurement outcome (coded in T) can be interpreted as the average potential information gain obtained by reading a result (coded in $T_{\mathcal{M}}$). Clearly, the first kindness and repeatability of $\mathcal{M}$ are sufficient for that. By (27), the conditions

$$H(E,T_i;\mathcal{R}) = 0, \quad i = 1,2\cdots \tag{30}$$

are necessary for (29). But $H(E,T_i;\mathcal{R}) = 0$ only if $E(X_j)T_i = T_i$ for some $X_i \in \mathcal{R}$. We observe that $E(X_j)T_i = T_i$ implies that

$$tr[\mathcal{I}_{\mathcal{M}}(X_i)T] = tr[\mathcal{I}_{\mathcal{M}}(X_j)\mathcal{I}_{\mathcal{M}}(X_i)T] \tag{31}$$

which should be compared with the repeatability condition (25). Requiring (30) for each initial state T of $\mathcal{S}$ and for any partition $\mathcal{R}$ of Ω implies that the final componet states T_i are always eigenstates of E. This is a strong condition and it holds, in particular, for repeatable measurements. Repeatable measurements are thus *optimal* in the sense that the average lack of information for predicting a measurement result equals with the average gain of information upon reading the measurement result. Since unitary measurements $\mathcal{M}_U^m$ of discrete ordinary observables A are of special importance we shall summarize the above results for them in a form of a theorem. The reading scale is, again, chosen to be the natural one.

THEOREM 2. *A unitary measurement $\mathcal{M}_U^m$ of an ordinary discrete observable A has the property*

$$H(A,P[\varphi]) = H(A,T_{\mathcal{M}_U^m}) \tag{32}$$

for each initial state φ of $\mathcal{S}$ if and only if the U-generating set $\{\psi_{ij}\}$ is a complete set of eigenvectors of A. In that case

$$H(A,P[\varphi]) = H(A,T_{\mathcal{M}_U^m}) = S(T_{\mathcal{M}_U^m}) = S(\{N_i^2\}) \tag{31}$$

which shows, in particular, that $\mathcal{M}_U^m$ is also a maximal state-entropy measurement.

4. Concluding remarks.

It has been argued elsewhere[7)] that repeatable measurements are necessary for a realistic interpretation of quantum mechanics. In this lecture we have seen some further peculiar features of such measurements. Repeatable measurements lead to an optimal separation of the final component states and for them the lack of information for predicting a measurement outcome is exactly the potential information gain obtained by reading the result of the measurement.

References.

1. BELTRAMETTI, E., AND CASSINELLI, G., *The Logic of Quantum Mechanics*, Addison-Wesley, 1981.

2. DAVIES, E.B., *Quantum Theory of Open Systems*, Academic Press, 1976.

3. REED, M., AND SIMON, B., *Methods of Modern Mathematical Physics. I: Functional Analysis*, Addison-Wesley, 1972.

4. VON NEUMANN, J., *Mathematical Foundations of Quantum Mechanics*, Princeton UP, 1955. Original German text in 1932.

5. OZAWA, M., Quantum measuring process of continuous observables, *J. Math. Phys.* **25** (1984) 79-87.

6. BELTRAMETTI, E., CASSINELLI, G., AND LAHTI,P., Unitary measurements of discrete quantities in quantum mechanics, *J. Math. Phys.* **31** (1990) 91-98

7. BUSCH, P., LAHTI, P., AND MITTELSTAEDT, P., The problem of measurement in quantum mechanics. Forthcoming (1990).

8. BUSCH, P., CASSINELLI, G., AND LAHTI, P., On the quantum theory of sequential measurements, *Foundations of Physics* **20** (1990) 757-775.

9. CASSINELLI, G., AND LAHTI, P., Strong correlation measurements in quantum mechanics, *Il Nuovo Cimento* (1990) in print.

10. FEINSTEIN, A., *Foundations of Information Theory*, McGraw-Hill, 1955.

11. LINDBLAD, G., Entropy, information and quantum measurements, *Commun. math. Phys.* **33** (1973) 305-322.

12. GROENEWOLD, H.J., A problem of information gain by quantal measurements, *International Journal of Theoretical Physics* **4** (1971) 327-338.

13. LINDBLAD, G., An entropy inequality for quantum measurements, *Commun. math. Phys.* **28** (1972) 245-249.

14. OZAWA, M., On information gain by quantum measurements of continuous observables, *J. Math. Phys.* **27** (1986) 759-763.

Quantum Theory and the Problem of Free Will

K.V.Laurikainen

Research Institute for Theoretical Physics, University of Helsinki
Siltavuorenpenger 20 C, SF-00170 Helsinki, Finland

ABSTRACT

The statistical laws of quantum mechanics give rise to a new conception of causality, and, *a fortiori*, to a new ontological view, as Wolfgang Pauli in particular pointed out. What is characteristic of this new conception of reality is a *freedom* to choose between different alternatives in individual phenomena. Thus the "material world" is presupposed to have also "psychic aspects" such as *free will* which influences phenomena in the "material world". If this freedom is considered a basic characteristic of the world, a possibility is opened for reintroducing *teleology* and the idea of *continuous creation* into the scientific view of the world.

Three hundred years ago, when empirical methods had been generally accepted in science, the conception of causality became a philosophical problem. Best known is the dispute between Leibniz and Newton concerning the foundations of the scientific view of the world. The dispute took place by letter between Leibniz and the philosopher and theologian Samuel Clarke, who was perhaps the most distinguished pupil of Newton's.

The correspondence was opened with some remarks by Leibniz, written in November 1715 about the way Newton seemed to describe the concept of causality. Here is an extract from Leibniz's first letter:

- - - - - - - Mr. Newton and his followers have also an extremely odd opinion of the work of God. According to them God has to wind up his watch from time to time. Otherwise it would cease to go. He lacked sufficient foresight to make it a perpetual motion. This machine of God's is even, on their view, so imperfect that he is obliged to clean it from time to time by an extraordinary concourse, and even to mend it, as a clockmaker might his handiwork; who will be the less skilful a workman, the more often is he obliged to mend and set right his work. According to my view, the same force and vigour goes on existing in the world always, and simply passes from one matter to another, according to the laws of nature and the beautiful pre-established order. And I hold that, when God performes miracles, it is not to satisfy the needs of nature, but those of grace. To think otherwise would be to have a very low opinion of the wisdom and power of God.[1]

In the religious language characteristic of this correspondence, here two opinions about causality governing occurrences in the world have been presented. The essential thing is that Newton did not see the laws of nature as something absolute: they are expressions of God's will and prevail to such an extent and for as long a time as God finds it appropriate. Newton thought that God must interfere from time to time with the goings-on in the world and "restore the harmony". Thus, the general opinion that the world view of Newton's was strictly deterministic is not correct. Newton was not a determinist, because he thought that God maintains everything that exists and participates in everything influencing phenomena when it is necessary.[2]

Consistent with this view was that Newton emphasized the role of man's free will. Thus, he said that not even God can foresee the future because it depends on our choices. According to Newton, God still takes care of the key events: He decides what He finds important.

Leibniz, on the other hand, considered the world to be absolutely causal. It was governed by a "beautiful pre-established harmony". God never needs to interfere in what happens in the world because everything takes place according to the predestined plan of God's.[3]

Here one can see two basic views concerning the nature of causality, which have always divided opinions among thinkers. If we continue to use religious language, Leibniz's attitude can be described as a *belief in Fate* and in the *predestination* of everything. Newton, however, believed in the *Providence of God* and in His participation in everything that happens by actively guiding all goings-on. These two basic attitudes, which fundamentally are religious, divide opinions of researchers even today, although one usually does not use religious expressions in this connection.

Statistical causality

In this century physics has encountered problems concerning the nature of human knowledge, which have led to very far-reaching philosophical discussions. In my opinion, the clearest way to describe the new views opened by quantum theory in particular is to state that the laws of nature are, at the atomic level, probabilistic, and even *genuinely probabilistic*. The attribute "genuine" refers to the fact that the laws are not probabilistic merely because our knowledge is in some respect incomplete but because the very nature of the events is such that it is not possible to describe them in a more complete way.

It is said that causality is the "cement" which binds the view of the world to a whole and gives it form. The conception of causality required in atomic physics can be called *statistical causality*. Sufficient statistics concerning similar events are needed before it is possible to compare "expectation values" derived from a statistical law with experimental results. Wolfgang Pauli states that these statistical prognoses must be understood as the "last experimental fact", which cannot be vindicated in any other way than by stating that these prognoses are always found correct when sufficient statistics are available.

It seems to be difficult to truly understand and accept the idea of statistical causality. The greatest difficulty is perhaps that people are no longer accustomed to searching for unifying principles in science. Concepts such as causality which belong to the basis of a coherent world view have become unclear. People do not understand why results concerning atoms could have any significance when discussing the phenomena of the human behaviour. *The belief in a coherent wholeness of the world has disappeared.* This is an extremely severe fact, and we must begin to oppose these implications of the specialization in science, otherwise the most fundamental and most important questions of existence will completely disappear from our mental horizon.

The change in the conception of causality has revolutionary

implications for the view of the world. These implications will occasion criticism against many research programs in all fields of empirical science. This also concerns physics research.

A new philosophy of freedom

Probabilistic laws do not describe the progress of an individual event in an unambiguous way; they only give expectation values for quantities to be measured. Therefore, in each individual event a *choice between different alternatives* must take place: one of the alternatives allowed by the probabilistic law becomes materialized and the other alternatives become excluded in a way which in principle cannot be described. This is a situation characteristic of genuinely probabilistic laws, which for many people seems to be unacceptable.

In quantum theory the term *reduction of state* is used in this connection. This change always takes place in measurement. It can be characterized as *irrational*, because one cannot present any rational explanation for it.

Werner Heisenberg described this situation by saying that the theory always concerns a *potential reality* only, *i.e.* it describes possible results of measurement and their probabilities. In measurement, one of the potential events becomes *actualized*.[4] This mode of expression originates from Aristotelian philosophy. It must be emphasized that the "actualization" of one alternative can be interpreted as an important expression of the *nature of reality*. One can see this as an "ability to fix a choice" between different alternatives and "to accomplish" the alternative which was chosen.

This is the pith of the matter, and it is wrong to try to hide it behind different attempts to find a rational explanation for these "choices" which are in principle irrational. Even Niels Bohr was unclear about this point. Perhaps it was because of the influence of positivistic philosophy that he refused to discuss ontological problems, *i.e.* the nature of reality.[5] He always spoke of the *epistemological lesson* which nature has given us in atomic research, instead of an ontological lesson.

Among the founding fathers of quantum mechanics, only Wolfgang Pauli tried to ponder explicitly the conception of reality implied by this theory. He ended up emphasizing the *irrationality of reality*: there are questions concerning reality which cannot be described within rational theories.[6]

When the younger generations after the World War II came to use quantum mechanics and to lecture about the subject, they encountered conceptual difficulties: the conception of reality presupposed by the theory was not clear. This has resulted in a multitude of new interpretations for quantum mechanics and in a simply overwhelming amount of literature aimed at eliminating "paradoxes burdening the Copenhagen interpretation".

Unfortunately the philosophical remarks by Pauli have remained unnoticed so far. The ontological remarks made in Pauli's articles have not been understood as they are very concise and written in German. I find that Pauli has clearly described the basic features of the reality presupposed by quantum mechanics according to the Copenhagen interpretation in his writings.

Instead of using the term "irrationality" as Pauli does, I would like to speak of the *freedom characteristic of reality,* i.e. *choices not bound by any laws*. Even a probabilistic law limits the course of processes, but it allows different possibilities for individual events, and one of these possible events will become "freely chosen" or it "actualizes". It is in these free choices concerning individual events where the new feature of reality implied by statistical causality finds its expression. I would like to call this feature *freedom.*

However, here we encounter a terminological difficulty. Whatever term is used to describe this new feature of reality it will seem unsatisfactory and easily leads to misunderstandings. This is a fundamentally "irrational" feature which cannot be described satisfactorily by using rational theories and exact concepts. One can only describe it by using certain analogies, and I would prefer the term "freedom": *laws of nature allow free choices between different alternatives.*

The new approaches to quantum theory which have been offered to replace the Copenhagen interpretation try to "explain" in some way or another the "irrationality" or these "free choices". The representatives of the Copenhagen school, however, emphasized that the description given by quantum mechanics is, in this respect, *complete.* Of course, this does not mean, that they considered quantum theory to be a final theory and the end of the development of physics! They just wanted to emphasize that the nature of phenomena seems to be such that it is necessary to change the concept of causality: one has to use probabilistic laws instead of deterministic ones, and a more complete description of phenomena does not seem possible. This implicates a new ontological view: the "irrationality" or "freedom" mentioned above is understood as belonging to the nature of reality. Therefore one speaks of *genuinely* probabilistic laws.

In this respect Einstein had another view. Einstein himself described his opinion as being based on a certain *belief* concerning the nature of reality and the possibilities of the human knowledge. Like Leibniz, Einstein was convinced that determinism governed the world: everything takes place as "according to a pre-established harmony".

Here Einstein represents in a consistent way a very strong Western tradition, which at the beginning of modern times was represented by Descartes, Spinoza, and Leibniz. It can be described by the terms "realism" and "rationalism".

The Copenhagen philosophy, on the other hand, was characterized by an idea of "freedom" or "choices", and this was not compatible with the idea of strict causality. This was the reason why Einstein could not accept quantum mechanics as a complete theory concerning the basic questions of the atomic world.

During the last two decades, several experiments have been carried out in order to ask nature to give its doom in the discrepancy between Bohr and Einstein. In these so-called *EPR experiments* one has put into practice an idea which Einstein,

Podolsky, and Rosen used in 1935 in order to illuminate their deterministic attitudes. The results of these experiments seem clearly support Bohr's view instead of Einstein's. Thus, there is now even more reason than before to consider carefully the possibility that the occurrences in fact contain an aspect of freedom, as implied by the Copenhagen interpretation.[7]

The problem of free will

In the deterministic picture of the world which was generally adopted three hundred years ago, the world was described as a clockwork constructed by the Creator at the beginning of the time. Newton's idea that God influences phenomena by restoring the harmony when necessary was quite forgotten in the 18th century. Leibniz's view that God had created a finished world where no restoring was necessary corresponded better to the view of the world which became prevalent. In the complete world machine there was also no place for free will, either of God or of man, in the sense that it would really change the processes in the world. Thus the problem of free will emerged.

This is why René Descartes, who has been called the father of modern philosophy, was forced to divide the world in two parts: the *material world*, which was a deterministic world machine, and the *spiritual world*, which was quite independent of causal laws. Immanuel Kant developed further this picture of two worlds of basically different nature. Free will was understood as something belonging to the spiritual world. The material world - or phenomenal world as Kant called it - was governed by *strict causality*, *i.e.* determinism, but this did not whatsoever limit the freedom which was characteristic of the spiritual world.

The relation between matter and spirit, however, remained unclear. With good reason Pauli has called the dualism of Descartes and Kant an "intellectual cloud of fog in the Western thought".[8] The psychophysical problem, *i.e.* the question of the relation between matter and spirit, Pauli regarded as the most important question today.

The situation is completely different if we consider causality to be statistical, rather than deterministic. Then the material world (or phenomenal world) remains open to "influences" which do not belong to this world. The *principle of sufficient reason* which Leibniz especially emphasized *is not valid any more*: one cannot find sufficient reasons for individual events, only for expectation values.

This *freedom*, which has not had any place in the modern sciences, can now be met in all individual events. One can see in this freedom an expression of the spiritual element of the world. It appears as free choices which are able to give direction to processes, and therefore Pauli has in this connection used the term *will* in referring to the philosophy of Arthur Schopenhauer.[9]

Insofar as we speak of man and his behaviour, this kind of thought seems natural, and obviously it offers a satisfactory solution to the problem of free will. Man is not a machine, although such ideas have been presented on the basis of positivistic philosophy. Man has free will, and this makes him capable of influencing the phenomena of the material world; this is quite consistent with the picture of the world implied by the idea of statistical causality.

But what about inorganic nature? Must we say that we can see "will" and "free choices" in the phenomena of dead matter as well? Do elementary particles also have the capacity of "willing" something?

Here we run against very profound questions, and it is true that quantum mechanics has created wild speculations indeed concerning the expressions of spirit. Speculations of this sort are often associated with *Eastern philosophies and religions.* I think that this is caused by the fact that the representatives of the Christian theology have not been willing to discuss these questions with natural scientists. Therefore many people who realize the insufficient attitude of Western thought with respect to the relation between matter and spirit are led to the sphere of the Eastern religions. Personally I find this very regrettable since I do not believe that Eastern religions are the right solution to the cultural crisis of the Western world: they can be helpful but they are not the right models for Western people.

Another alternative is *materialism*, which denies the independent existence of spirit. This one-sided view concerning reality has found unbelievably strong support in the Western world during modern times, but now its insufficiency begins to be obvious. *In the foundational research of physics the materialistic view of the world is, however, so far prevailing.*

Quantum theory, instead, seems to lead to some kind of *panspychism, i.e.* to the view that in all phenomena there are both material and spiritual aspects. Pauli said, indeed, that "matter" and "spirit" are complementary elements of reality in the same way as "particle" and "wave" are complementary pictures in quantum mechanics. In fact, he was unable to develop this view into a consistent theory, but he repeatedly referred to this *complementary conception of reality*, which he expected to replace Cartesian dualism in the future.

According to this conception of reality, "matter" and "spirit" are essentially different basic elements of reality which, however, can exist only together. They are like two different projections of reality comprehensible for human consciousness. This view is a kind of intermediary form between monism and dualism, where reality is considered as totally abstract, incomprehensible for us. We cannot form any illustrious picture of reality and of its "objects", in the same way as it is not possible to form any visual picture of the "particles" of quantum mechanics which "simultaneously are waves". The Yin-Yang symbol of Taoism, which Bohr adopted for his coat-of-arms in the Danish Elephant order, symbolizes very well this kind of abstract whole, which is created by the combined effect of two complementary basic elements.[10] One is forced to state that quantum mechanics lead to this kind of view about the basic nature of reality, but there is still much work left - even common efforts of the representatives of science and religion - before this conception of reality is assimilated into the Western thought.

Science and teleology

The deterministic view of the world has resulted in the exclusion of *teleology* from scientific research. If the world is a machine, there cannot be any struggle towards a goal, towards the realization of a purpose, in the phenomena, in any other sense than the Leibnizian one. Leibniz believed in a "beautiful pre-established harmony", which gave meaning to everything that takes place, but such thoughts have been excluded from science when materialism and atheism have become prevailing beliefs among scientists.

When interest in the ontological implications of quantum theory now is obviously increasing, this has connection with the increasing interest in the role of spirit in the human existence. In atomic physics, psychological problems concerning observations have been important. However, physicists today tend to exclude such questions from their science. Therefore, physics has become a firm supporting pillar for the belief that all problems concerning reality can be solved by using scientific methods. This is the prevailing belief in all natural sciences.

This fact can be seen very clearly in the application of the *principle of evolution* in natural sciences. Both in biology and in the exact sciences, theories of evolution now receive great attention. Thoughts of teleology and of creation are in both cases generally excluded as unscientific. The only acceptable attitude is that one projects religious ideas into a spiritual world which is - as in Cartesian dualism - independent of the material world of science. Pauli describes this (very general) attitude by saying that "one speaks about one thing on Sundays and about something else on weekdays".

In this connection one does not notice the *freedom* which I have emphasized above. The solution presupposed in the evolutionary theories of today can be called a *religion of chance*. It is so cleverly hidden under the carpet of scientific argumentation that one does not see the basic decision which determines the direction of the investigation. In fact, the argumentation is based on one definite belief: that in reality there is nothing like free choices which could steer the progress of phenomena. If we really understand the ontological perspective opened by quantum theory, *we cannot accept this attitude as scientifically motivated.* Only a deterministic conception of causality would give the right to claim that belief in teleology is not allowed in science.

The "religion of chance" means that one certainly accepts the use of statistical laws in the description of the evolution but the choices, which these laws presuppose, are understood as being made by "pure chance". The idea of *creation*, so essential in the Western religious thought, is rejected as unscientific. The conception of causality presented above does not motivate this attitude. It is quite permissible to include the idea of creation in the scientific picture of the world as a *continuous creation* (*creatio continua*), which attains its expression in those free choices presupposed by the idea of statistical causality. This means that evolution is not steered by a "blind clockmaker" but by a *creative will* tenacious of its purpose.

The world is not a deterministic machine. One can salvage the idea of a mechanical machinery by interpreting free choices as being made by a "blind chance". Then one cuts off from the world its spiritual element which cannot be governed by rational methods. It is not possible to measure spirit, but in spite of this it can be a real element of the world. Phenomena can have an aim and there can be meaningfulness in the evolution. *One should not imagine that the mystery of creation could be described with the aid of a rational theory.*

However, in *cosmological theories* one imagines this to be so, which is why one cannot take them seriously as scientific theories. If one correctly takes into account the epistemological/ontological lesson of quantum theory, evolution must be understood as a series of "stochastic steps", each of these being an unanalyzable whole. Each step includes a choice, and there is no scientific motivation for excluding teleology from these choices. Occam's razor which is here used as a motivation, is a very dangerous weapon because it cuts off the spirit from the world. In the world described in present cosmological theories beings which have consciousness and spiritual life will never develop and, *a fortiori*, there will scarcely be any living beings at all. To claim something else would be purely unscientific speculation based on a materialistic belief.

Occam's razor is a very good tool for making rational theories clean and clear. I greatly appreciate it, but the world cannot be described by purely rational theories. Therefore it is not right that logical requirements are used for atheistic propaganda and for strengthening materialistic beliefs. Science has not any right to deprive people of the faith on the meaningfulness of existence and the basis of their individual moral. It is not unscientific to believe, like Newton, in God as a transcendental basis of existence and in His Providence which guides phenomena in the world.

References

1. G.H.R. Parkinson (ed.): *Leibniz Philosophical Writings*, p. 205, (J.M. Dent & Sons Ltd, London, 1973; first edition 1934).

2. Markus Fierz: "Über den Ursprung und die Bedeutung der Lehre Isaac Newtons vom absoluten Raum", *Gesnerus* **11** (1954), 62-120, (Sauerländer, Aarau, Schweiz).

3. Gottfried Wilhelm Leibniz: "Monadology (1714)", Ref. 1, p. 179.

4. Werner Heisenberg: *Physik und Philosophie*, "III Die Kopenhagener Deutung der Quantentheorie", (S. Hirzel Verlag, Stuttgart, 1959).

5. Henry J. Folse: *The Philosophy of Niels Bohr. The Framework of Complementarity*, Ch. I,3, Ch. 5, and Ch. 8, (North-Holland, Amsterdam, 1985).

6. K.V.Laurikainen: *What Did Wolfgang Pauli Mean by the Irrationality of Reality*. Lecture given at the Pauli Colloquium on 26.4.1990 arranged by the Finnish Society for Natural Philosophy. Private report distributed to the participants of the Symposium on the Foundations of Modern Physics 1990.

7. P. Lahti & P. Mittelstaedt (eds.): *Symposium on the Foundations*

of Modern Physics: 50 Years of the Einstein-Podolsky-Rosen Gedankenexperiment, Joensuu, Finland, 16-20 June 1985, (World Scientific, Singapore, 1985).

8. Wolfgang Pauli, letter to M. Fierz on June 3, 1952 (PLC 0092.092), in K.V.Laurikainen: *Beyond the Atom. The Philosophical Thought of Wolfgang Pauli*, (Springer-Verlag, Heidelberg,1988), p.141 (translation p. 224).

9. Arthur Schopenhauer: *Die Welt als Wille und Vorstellung I-II*, (Brockhaus, Altenburg, 1818).

10. K.V.Laurikainen, *ibid.*, p. 87.

ENIGMAS OF THE SP$\hbar$IN$_x$

Some *other* problems of quantum theory

JEAN-MARC LEVY-LEBLOND
Physique Théorique, Université de Nice
Parc Valrose, F. 06034 Nice-Cedex, France

ABSTRACT

Quantum theory offers, besides its over-emphasized "interpretation" problems, construction problems too often neglected. Several enigmas of conventional quantum mechanics, most of them concerned with the notion of spin and its implications, are discussed here.

I must start with a double apology. First, I fear this talk to be ill-placed at the beginning of our Symposium. Indeed, it does not deal with the issues of the (so-called) interpretation problem of quantum theory, measurement processes, reduction of the state vector, and so on. If I remember it correctly, however, the first announcement of this conference defined its goal as the discussion of the foundations of quantum theory, *mainly* the problem of measurement, etc. ; the adverb leaves room for some other problems and allows me to step outside of the main stream. Still, at the opening of the Symposium, my talk may appear as a delaying diversion, while later on, in two or three days from now, we might all be so confused and fed up with measurements and collapses, that the present contribution could have been welcome as a relief. My second cautionary satement is that I have no new result to claim here, no new idea to offer - only questions to ask, some of them already rather ancient but still unsolved. I do hope at least to convince you that these are indeed relevant and challenging problems, so that some of you might come with the answers which have eluded me for many years now.

What I will try to show is that quantum theory, in its standard form, offers several enigmas. Most of them are concerned with the notion of *spin*. As they are of a specifically quantal nature and thus related to Planck's constant $\hbar$, and often linked to the position variable x, we see that, indeed, the spin should be viewed as a sp$\hbar$in$x$, and we must not be surprised, but still appalled, by its enigmas.

1. The architecture of quantum theory

In order to explain the nature of the problems I am concerned with, I find it convenient to distinguish two different aspects in the architecture of quantum theory, namely, the *Universal Framework* and the *Specific Structures*[1]. a) *the Universal Framework* is the set of general and abstract statements common to any quantum theory, whatever its scope and setting (whether it be galilean quantum mechanics for a single particle or quantum field theory for strong interactions, etc.). Of course there is no unique way of expressing this universal framework, as it depends upon the choice of a suitable mathematical "codification" (in the words of H. Primas), with a more or less encompassing generality. Let me be content here to rely on the standard hilbertian formalism. The uiversal framework, then, consists of the general representation rules according to which : - the states of quantum systems are represented by rays in a Hilbert space,

- scalar products of state vectors yield probability amplitudes,
- physical properties of the system are represented by hermitian operators,
- numerical values of these physical properties (magnitudes) correspond to the eigenvalues of the operators, etc.

It is this set of statements which gives rise to the *"interpretation" problems*, due to the necessity of understanding the probabilistic nature of the theory, the consistency of the collapse rule[2], the nonclassical correlations (quantum nonseparability, etc.) which will be discussed at length in this Symposium

b) *the Specific Structure* is the set of particular and concrete staments which allow the explicit formulation of a specific quantum theory. These are the rules giving :

- a definite realization of the state space (as, for instance, a space of square integrable of wave functions),
- the selection of operators adequate for representing concrete physical properties of interest (momentum, position, spin, etc.), with the relevant commutation rules,
- the dynamical law of evolution through the expression of the hamiltonian, etc.

Here we are concerned no longer with interpretation problems, but with *construction problems* - some of which I want to investigate here.

2. The Correspondence Principle

The first attempts, historically speaking, to solve these construction problems, have been through the Correspondence Principle. Heuristically introduced and brillantly used by Bohr, and later on mathematically formalized by Dirac, the idea was to use the (specific) structure of classical mechanics in order to build that of a corresponding quantum mechanics. The usual formulation is to set up a correspondence rule between classical dynamical variables $A_c, B_c, \ldots$ and quantum ones (hermitian operators) $A_q, B_q, \ldots$ such that it maps the Poisson bracket of the former onto the commutator of the latter :

$$\text{if } A_c, B_c \to A_q, B_q \quad \text{then} \quad \{A_c, B_c\} \to (i\hbar)^{-1}[A_q, B_q]. \tag{1}$$

But it has been known for a long time, although not by most textbooks, that the Correspondence Principle has several redhibitory drawbacks. Indeed, it is both inconsistent and insufficient.

a) *epistemological inconsistency* : It was heuristically quite justified (and even inescapable) to aim at constructing the new quantum theory starting from the knowledge of the old classical one. However, the Correspondence Principle, if it provides the necessary scaffolding for this building, should not be confused with its foundations. Indeed, we trust nowadays quantum theory to be of a more general scope than classical theory - the latter being but an approximation of the former (the conditions of validity of this approximation, by the way, being poorly understood). We are thus led to a very vicious circle in which classical theory is an approximation to quantum theory, while serving as a foundation for it...

b) *formal inconsistency* : It may be shown in a rather elementary way, by using only the momentum and position dynamical variables P and X, that no correspondence rule exist which fulfills the requirement (1) for all polynomials in P and X[3]. Namely, it is easy to find two pairs of such dynamical variables (A, B) and (A', B'), such that, for any correspondence rule obeying (1) for P and X, one has nevertheless :

$$\{A_c, B_c\} = \{A'_c, B'_c\} \quad \text{and} \quad [A_q, B_q] \neq [A'_q, B'_q]. \tag{2}$$

c) *insufficiency* : Even if one were to limit the use of the correspondence Principle to heuristic purposes (so as to avoid a)) and to simple dynamical variables (so as to avoid b)), it is quite clear that one could recover but the expression of those quantum dynamical variables which do have a classical analog ! Insofar as classical physics indeed is an approximation to quantum theory, it is most plausible - and indeed true - that it has lost some specifically quantum features. This is clearly the case for properties such as spin (which clearly vanishes already in the formal limit $\hbar \rightarrow 0$) or parity (which, as a discrete dichotomic property, has no classical continuous limit). Those properties certainly cannot appear in a quantum theory built out of the Correspondence Principle - as shown by the confused history of the discovery of spin. Furthermore, the Correspondence Principle cannot apply in a consistent way when the classical theory is not satisfying (for instance, electodynamics where the divergence problems are better understood in the quantum domain than in the classical one) or simply does not preexist (such as chromodynamics).

3. Wave-equations

In fact, the Correspondence Principle, despite its still prominent place in many textbooks, fell quite short of allowing the development of quantum theory, which, instead, took a different turn. It is the writing of explicit wave-equations ruling the behaviour of specific quantum objects of increasing complexity which paved the way for this development. In the galilean (so-called "nonrelativistic") domain first, Schrödinger's equation for spinless quantons was soon followed by the Pauli equation for spin 1/2. Extending this procedure to the einsteinian ("relativistic") domain, the Schrödinger-Klein-Gordon equation for zero spin, was followed by the breakthrough of the Dirac equation for spin 1/2, and then by higher spin formalisms (such as the Bargmann-Wigner equations).

To the present day, this empirical approach is riddled with many difficulties :

a) the procedures used are variously *ad hoc* ; think, for instance, of the rather artificial Dirac's linearization.

b) the construction is not fully consistent as the appearance, at least in the einsteinian case, of antiparticles ("negative energy solutions") leads from quantum mechanics to quantum field theory, leaving the initial wave-equations in "first quantized" form with a rather uncertain status.

c) for spins higher than 1/2 (starting with $s = 1$), there are many possible wave-equations, which are not equivalent and furthermore exhibit troubles as superluminal propagation.

4. Invariance Principles (Wigner's way)

There seems to be a way out of the contingencies, inconsistencies and difficulties of the Correspondence Principle or the wave-equations program. It consists in fully exploiting the Invariance Principles of the theory, as forcefully advocated and exemplified by Wigner. The task of building the specific structure of quantum theory then may proceed through the following steps :

a) identify the relevant space-time symmetries, that is, the *relativity group* of the theory (for instance, Galilei or Poincaré group)

b) relying upon Wigner's theorem, construct the *unitary representations* (possibly projective ones) of the group ; more specifically, classify the *irreducible* unitary representations, which correspond to the (kinematically) "elementary" objects of the theory

c) by using the Stone-Naïmark-Ambrose-Godement theorem, derive the infinitesimal generators of the one-parameter subgroups, which are to be identified with the standard extensive physical properties : time translations → energy (hamiltonian), space translation → momentum, rotations → angular momentum. The Lie algebra of the group (or its extensiòn) automatically yields the commutators

d) by requiring the adequate transformation under the action of the group, obtain the other physical properties and their commutation rules - for instance, the position operator.

As an example, the standard "nonrelativistic" quantum theory is straight forwardly obtained by exploiting galilean invariance [(4)]. The Galilei group Lie algebra has a natural basis consisting of the generators $(H, \mathbf{P}, \mathbf{K}, \mathbf{J})$ for time translations, space translations, galilean boosts and rotations respectively. Considering its projective irreducible representations introduces a scalar parameter m, which clearly is the mass, and leads to the following form of the hamiltonian :

$$H = \frac{1}{2m}\mathbf{P}^2 + U_0 \tag{3}$$

(U_0, constant internal energy). The position operator, defined through its behaviour under translations, is

$$\mathbf{R} = \frac{1}{m}\mathbf{K}, \tag{4}$$

the canonical commutation rule automatically following. The angular momentum finally reads

$$\mathbf{J} = \mathbf{R} \times \mathbf{P} + \mathbf{S}, \tag{5}$$

leading to the inescapable and natural introduction of the intrinsic spin (at the *galilean* level, it is to be emphasized [(4)]).

More generally, as shown by Wigner in his seminal work on the Poincaré group [5], this procedure leads to a complete characterization and an extensive description of *free* quantons.

5. Interactions

From this perspective, the usual wave-equations are but a (very) particular way of specifying a representation of the relativity group ; namely, the solutions of wave-equations invariant under the relativity group define a realization of the Hilbert space for the representation. However, no new information is contained in the wave-equation compared to the representation itself - as long as one deals with *free* quantons...

But free particles are of little interest, and we need to describe and discuss interactions. We will, in the following, restrict our considerations to the simplest of all cases, that is, interactions of a single quanton with an external field. In such a situation of course, the symmetries are broken down, and the invariance principle leading to a description of the free quanton is no longer sufficient.

There comes the revenge of the wave-equations ! In effect, no other systematic way is known for introducing the interaction with an external field than to start from the wave-equation and effect the following replacement of the momentum operator :

$$P \to P - A(x) \tag{6}$$

where A is the 4-potential of the field. This recipe of course can be derived by requiring the *gauge-invariance* of the theory, which expresses the possibility of arbitrary local changes in the phase of the wave-function. As such, this is a kind of invariance principle, but of a nongeometric nature, obviously very different from the spacetime invariance principles.

This procedure is most successful, since it leads to the Schrödinger equation and to the Dirac equation, with the potentials, the many applications of which are well-known. It is here, however, that our problems begin, since this very success, after all, is somewhat of a mystery, as we will presently see.

6. Enigma 1 : Wigner vs. Dirac, or the riddle of the position

Let us indulge in a piece of retroactive science-fiction and suppose that Wigner's work on the representations of the Poincaré group [5] had been accomplished before Dirac proposed his wave-equation... It would then have been learnt that an elementary quanton is to be characterized by its mass and spin, both parameters appearing in a straightforward way from purely kinematical arguments. In

Wigner's construction, the state space is built most simply as a space of momentum wavefunction $\hat{\varphi}(\mathbf{p}) =< \mathbf{p}|u>$, where u is the (abstract) state vector. In order to discuss the localization properties of the quanton under investigation, we would then ask for configuration space wavefunctions $\varphi(\mathbf{r})$. This requires the knowledge of the position operator $\mathbf{R}$; its eigenstates $|\mathbf{r}>$ define the wave-function $\varphi(\mathbf{r}) =< \mathbf{r}|u>$ and allow its computation by a Fourier-type transformation from the momentum wave-function :

$$\varphi(\mathbf{r}) = \int d^3p < \mathbf{r}|\mathbf{p}> \hat{\varphi}(\mathbf{p}) \, . \tag{7}$$

Now, it is natural to construct the position operator through group-theoretical arguments, asking first that it has the expected transformation properties under space translations and rotations and adding some equally innocuous, or so its seems, requirements. This indeed leads in a unique way to a position operator $\mathbf{R}_{NW}$ first derived by Newton and Wigner[(6)].

And here start the troubles. First, the Newton-Wigner operator $\mathbf{R}_{NW}$ has awkward transformation properties under Lorentz transformations ; namely, an eigenstate of $\mathbf{R}_{NW}$ is not transformed into another one, so that a localized state in some reference frame is no longer localized in another frame obtained by a Lorentz transformations. As a consequence, the Newton-Wigner wavefunctions $\varphi(\mathbf{r}_{NW})$ do not transform locally : in the space of these wavefunctions, a Lorentz transformation takes a most complicated integral expression. It is clear, then, that these wavefunctions cannot obey an explicitly covariant differential wave-equation and that the introduction of interactions *via* the recipe of local gauge-invariance is not possible...

It is seen how lucky we have been that Dirac came before Wigner ! The situation may be clarified, precisely, by starting from the Dirac equation. It is well-known that the Dirac position operator $\mathbf{R}_D$ (that is, the operator corresponding to the spatial argument of the Dirac wavefunctions) while leading, indeed, to local and covariant wavefunctions, has the strange feature of mixing the particle and antiparticle components of the wavefunction ; in other words, it is *not* diagonal in the one-particle (or antiparticle) state spaces (by the way, this accounts for the famous *zitterbewegung*). Another position operator may be found which is diagonal, through the Foldy-Wouthuysen transformation : it is precisely the Newton-Wigner position operator, with all its difficulties.

There is a second major trouble with the Newton-Wigner operator : it does not exist for massless particles with spin higher than 1/2. That is to say, *the photon is not localizable* - according to that theory. There is no way, then, to go naturally from the representation of the Poincaré group with mass $m = 0$ and spin $s = 1$, in Wigner's form, to the local and covariant Maxwell equations for the electromagnetic field... In more empirical but no less disquieting terms, we do not know how to describe the physical processes involving spatial localization of the photons - for instance, photography.

It may be concluded that we do not really understand the basis for the sucess of the Dirac and Maxwell theories. Their elegance (locality and covariance) is in clear conflict with some apparently more natural requirements based on symmetry arguments. In such a situation, it is tempting to think that we are missing some deep conceptual element which would give to the formal conditions of local covariance a physical basis. I would venture the hypothesis that it is gauge invariance which we need to understand better and its relationship with spacetime symmetries. After all, there is, indeed, a close link between gauge-invariance and galilean invariance at the "nonrelativistic" level[(7)] - even though its extension to the einsteinian domain is not that natural[(8)].

Another, more radical, possibility exist, namely, to reconsider the localization problem. Bacry has recently discussed in depth several arguments in favour of a new position operator, which presents some new and uncommon features (such as commuting components), but has the virtue of existing for the photon[(9)]. Of course, it remains, in that perspective, to explain the success of the Dirac equation.

7. Enigma 2 : the puzzle of the magnetic moment

The simplest problem to be solved by using wave-equations (and by no other known means !) is the computation of the intrinsic magnetic moment of quantum particles. It suffices, by using the gauge-invariance recipe, to introduce in the wave-equation the potential for a static and uniform magnetic field. Deriving the reduced wave-equation ruling but the $(2s + 1)$ physical spin components of the wave function, one then identifies the magnetic moment. It is well-known that, applied to the Dirac equation, this procedure leads, for a spin 1/2 particle with charge q and mass m, to an intrinsic magnetic moment $\mu_{1/2} = q\hbar/2m$, that is, to the famous "anomalous" gyromagnetic factor $g_{1/2} = 2$. This has been held, erroneously as we will shortly see, as a specific success of the Dirac equation as a relativistic (einsteinian) theory. Now, it is interesting to investigate the corresponding situation for higher spins. It turns out that for $s = 1$, one obtains $\mu_1 = q\hbar/2m$, that is the "normal" value $g_1 = 1$. But for $s \geq 1$, where the computations rapidly increase in complexity, so that no absolutely general derivation exists, for all known cases, one obtains

$$\mu_s = q\hbar/2m, \qquad \text{i.e. } g_s = 1/s \tag{8}$$

In other words, it *seems* that the intrinsic value of the magnetic moment for a quanton has the universal value[8], irrespective of the value of the spin s (provided $s \neq 0$)[10] [We are, of course, concerned here with charged particles with purely electromagnetic interactions, and not with the much more complicated effects due to other interactions, such as the strong forces which underlie the magnetic moments of the nucleons]. How not to think that some deep reason exist for such a simple result, behind the present complication of its derivation ?

The problem is somewhat clarified (but, by far, not solved) when one realizes that it is *not* a specifically einsteinian relativistic problem. Indeed, as shown in (8) by the use of the adequate units, and contrarily to most textbooks formulas, the limit velocity c does not enter the expression for the universal magnetic moment. It should not come as a surprise then, although it was recognized rather late, that similar results may be obtained in galilean relativistic theories. This was first shown for spin 1/2 where the (*ad hoc*) Dirac linearization procedure performed on the Schrödinger equation leads to a 4-component galilean wave-equation equivalent to the Pauli equation *with the correct value* (8) *of the magnetic moment* [11]. In this case, the generalization to higher spins may be accomplished in full generality [12]; the galilean wave-equations for any spin, indeed, are simpler than the einsteinan ones. Nevertheless, the result (8) which obtains yet comes out at the end of a rather lengthy and contrived calculation, still pointing to the need for a deeper understanding.

This situation, once more, is indicative (to my mind, at least) of some missing connexion between gauge-invariance and geometrical (i.e. spacetime) considerations. It may be hoped that the arguments linking gauge-invariance and galilean (or einsteinian) invariance [7-8], may be deepened to the point where they will reveal the full dynamical status of the spin, and the value of the intrinsic magnetic moment (8) will result from simple and general considerations of a geometrical nature.

8. Enigma 3 : the Spin-Statistics connexion

The connexion between the value of the spin of a quanton (integer of half-integer) and the "statistics" - one should rather speak of the "permutability" - of states with several such identical quantons (respectively, symmetrical or anti-symmetrical), certainly is one of the deepest properties of the quantum world. It is a theorem in our present theories, resulting from very general and compelling hypotheses, which satisfactorily explain the universality of this connexion. Our satisfaction, however, verges on frustration as we go from the statement to the proof of the theorem. Indeed, here again, we face a situation where a very simple and general result emerges from complicated, indirect and highly intricate consideration. While the initial proofs (Stückelberg, Pauli) within old-fashioned quantum field theory were somewhat restricted and particular, the modern proofs in the framework of axiomatic quantum field theory are of a great generality and power [(13)]; they are, nevertheless, very sophisticated, and certainly not appropriate to elemantary explanations, for students or laymen. The situation is to be contrasted with that of the other general result of quantum field theory, namely the TCP theorem, or, in more common parlance, the existence of antiparticles. In the latter case, there exist simple arguments, non rigorous but quite convincing, showing how the requirements of einsteinian relativity in a quantum theory force the appearance of antiparticles [(14–15)].

The need for a deeper understanding of the spin-statistics connexion, would it be in the form of a heuristic derivation, is not only a pedagogical necessity, but probably could bring to light some new theoretical ideas. There have been several attempts towards such an approach, relying simply on the geometrical properties of half-integer spin [(15–16)]. The strategy is to show how the (-1) factor appearing in the state vector of a half-integer spin quanton after a full turn (2π rotation) is but the same phase factor that has to be introduced in the superposition of the amplitudes for direct and exchange processes involving two such particles. However, these purely geometrical arguments cannot be valid as they do not imply Lorentz invariance ! For the spin-statistics connexion does indeed require einsteinian relativity and does not hold in galilean relativistic theories : explicit consideration of galilean quantum field theories prove that spin 1/2 bosons or spinless fermions may be described with full consistency [(17)]. It is clear, furthermore, either in the rigorous proofs or in the more concrete approach by Feynman [(15)], that relativistic invariance (and, more specifically, the non-absolute nature of simultaneity) plays a crucial role, and has to be brought in the heuristic derivation which I am calling for.

9. Enigma 4 : the Spin-Forces connexion

In our present theories, the forces between two pieces of matter, that is, at the most elementary level, two fermions, are mediated by the "exchange" of bosons - the quanta of the interaction field. The characteristics of the mediating boson (mass, charges, spin...) are reflected by those of the force it transmits (strength, range, etc.). A general feature of the force, namely its attractive and/or repulsive character, seems to be strongly correlated with the value of its spin, or rather, its even or odd character. This is more a folk result than a theorem, and although I remember having seen it in print, I could not find an explicit reference [18]. The simplest way of stating it is probably to restrict oneself to the long-range forces ($\sim r^{-2}$) mediated by massless bosons. Our folk lemma then states that :

- for a boson with even spin ($s = 0, 2, ...$), the transmitted force either has a universally attractive character or a universally repulsive one (but this is a case of no physical interest, since no bound states, that is, no new physical objects would result). A foremost example is the theory of gravitation, the universally attractive character of which requires an even-spin boson. It is known that theories with spin 0 cannot explain all the features of einsteinian relativistic gravitation (in particular the perihelion precession), so that one has to conclude that the graviton has spin 2 (hence the tensor character of the gravitation field). A spin 1 boson could not describe universal *attraction*
- for a boson with odd spin ($s = 1, ...$), the transmitted force is attractive or repulsive, depending on the relative sign of the two types of charges allowed for in such theories. The standard instance is electromagnetism with its spin 1 photon mediating Coulomb attraction *or* repulsion.

The question here is whether it is possible to produce (or to find) a more explicit statement of this connexion and to give a simple, even if not rigourous, explanation. No doubt, it would have to rely on some deep feature of the spin notion - perhaps a hitherto unnoticed one.

10. Enigma 5 : the Spin-Charges connexion

The laws of physics as we know them are subjected to some conservation "superlaws". Indeed, there are charges subject to universal conservation laws, which may also be expressed as *superselection rules*. This seems to be the case for the electric charge Q, the leptonic charge L and the baryonic number, or charge, B (we disregard here the possibility of baryon nonconservation ; also, our statements may easily be generalized to the "quark number" $G = B/3$). These conservation laws derive from invariance under associated gauge groups (of the first kind, that is, with global phase changes). Now, the full symmetry of the theory has to take into account the spacetime symmetries as well as the internal symmetries. The relationship between the two associated groups may not be trivial. In effect, the full invariance group *a priori* is some, possibly nontrivial, extension of the Poincaré group by the gauge group of the charges. It has been shown by Lurçat and Michel, in a too little known paper [19], that in this framework, one is led to *predict* the existence of a relationship between a spacetime property of any system, namely the integer or half-integer character of its spin (again !), and the values of its charges. More precisely, one may define a sort of "combined charge" Γ :

$$\Gamma = \epsilon_q Q + \epsilon_b B + \epsilon_\ell L \quad (\epsilon = 0, 1) \tag{9}$$

numerical combination, with coefficients 0 or 1, of the charges Q, B, L, such that Γ is even or odd as the angular momentum is integer or half-integer. In other words :

$$\Gamma \equiv 2j \pmod{2} \tag{10}$$

Indeed, such a relationship *does* hold in reality, with the following expression for the combined charge

$$\Gamma = B + L \tag{11}$$

What the Lurçat-Michel relationship now tells us is that mesons ($B = 0, L = 0$) necessarily have integer spins and are bosons, while baryons ($B = 1, L = 0$) and leptons ($B = 0, L = 1$) have half-integer spins and are fermions [observe that a leptobaryon ($B = 1, L = 1$), were it to exist, would necessarily be a boson].

This telling statement clearly has a most profound meaning and, as such, is rather enigmatic - like the preceding ones. We would like to know why it is the particular (11) combination which is singled out by Nature, but, more deeply, to understand in more direct physical terms that the abstract and elaborate group-theoretical arguments by Lurçat and Michel, what is the connexion between the notions of spin and charges.

11. Enigma 6 : Spin renormalization ?

That there could be in the notion of spin more dynamical contents than has hitherto been conceived, may be hinted at from the following argument. Consider a charged particle with spin. It then has an intrinsic magnetic moment (see Section 7). Its charge gives rise to a (static) electric field, while its magnetic moment generates a (static) magnetic field. The geometry of the radial electrical field and the dipolar magnetic field gives rise to a nonzero Poynting vector with rotational symmetry. This vector field, physically a momentum density, has a nonzero moment, that is an angular momentum density. In other words, in the same way as the electromagnetic field contributes to the total mass of the particle, it has an angular momentum of its own which contributes to the total angular momentum of the particle.

The argument is easily made more precise in classical electromagnetic theory (20). It is not difficult to compute, for a Coulomb field $\mathbf{E}$ and a dipolar magnetic field $\mathbf{B}$ the total angular momentum of the electromagnetic field :

$$\mathbf{S}_{\text{field}} \simeq \int d^3r \; \mathbf{r} \times (\mathbf{E} \times \mathbf{B}) \neq 0 \tag{12}$$

Of course, for a point particle, the integral diverges, as does the field energy, so that one has to introduce some physically meaningful cut-off procedure. The total spin of the particle then must be evaluated by adding the field spin to the "bare" spin. In classical theory, this program is a perfectly reasonable one, and the discussion of the spin renormalization, along with the more usual mass renormalization, lead as to some new considerations upon the "classical radius" of charged spinning particles [20]. Namely, it may be argued that it is quite reasonable to re-define such a radius, instead of the conventional expression based on the exclusive mass of the field, to wit the "mass radius" :

$$R_M \simeq \frac{Q^2}{Mc^2} \quad (?) \tag{13}$$

(Q, M : charge, mass of the particle), by the new "spin radius" :

$$R_S \simeq \frac{S}{Mc} \tag{14}$$

(S : spin of the particle). Such an expression is much closer to later quantum arguments, since with $S \;\; \hbar$, the "spin radius" (14) is of the order of the Compton length of the particle - its natural quantum length scale (contrarily to the "mass radius" (13)).

This makes it even more puzzling that spin renormalization is not taken into consideration in quantum field theory. The problem, of course, is not only that it does not seem necessary, since quantum electrodynamics, after all, is a most successful theory as it stands. It is, more to the point, that the discrete nature of spin in quantum theory does not seem to allow for a renormalization process which depends, and continuously so, on the strength of the coupling. Could it be that the consistency of such a spin renormalization precisely is the contraint which fixes the specific value of the electromagnetic coupling constant ? There have been, after all, wilder attempts at explaining the value of the fire-structure constant... In any case, even though the question might turn out to be an ill-posed one, I find it challenging to ask for the status of spin renormalization within conventional quantum electrodynamics, or, more generally, in quantum field theory [21].

12. Conclusion : cornestones and stumbling blocks

To these six enigmas of the spin, I could add some more ; but they are already more numerous than the enigmas of the sphinx, and no Oedipus has shown up yet.

In any case, I hope to have shown that there are, at the basis of quantum theory, interesting *open* questions within our most trustworthy theories, from the Schroödinger quantum mechanics to quantum electrodynamic). Indeed, it might be said that some familiar cornerstones of quantum theory, also are stumbling blocks for our understanding. According to the present considerations, there are at least three themes which require deeper investigations :

- spatial localization and manifest covariance
- description of interactions through gauge invariance
- role of spin (kinematical and dynamical).

Along with the more frequently discussed interpretation problems, these questions certainly deserve our attention if we are to understand the "unreasonable effectiveness" ofquantum theory.

The present work was made possible by the generous help of the organizers of the Symposium and the editors of *Tiedepolitikka*.

REFERENCES

(1) see J.-M. Lévy-Leblond, in *Quantum Mechanics, Half a Century Later*, J. Leite Lopes and M. Paty eds. (Reidel, Dordrecht 1977), p.171.

(2) for recent contributions to the discussion of the status of the "collapse", see *Quantum theory without reduction*, M. Cini and J.-M. Lévy-Leblond eds., (Adam Hilger, Bristol 1990).

(3) see L. Cohen, in *Contemporary Research in the Foundations and Philosophy of Quantum Theory*, C.A. Hooker ed. (Reidel, Dordrecht 1973), p.71 and original references therein.

(4) J.-M. Lévy-Leblond, *Riv. Nuovo Cimento* 4, 99 (1974).

(5) E.P. Wigner, *Ann. of Math.* 40, 149 (1939).

(6) T. D. Newton and E.P. Wigner, *Rev. Mod. Phys.***21**, 400 (1949) ; A.S. Wightman, *Rev. Mod. Phys.* **34**, 845 (1962).

(7) J.-M. Jauch, *Helv. Phys. Acta* **37**, 284 (1964) ; *Foundations of Quantum Mechanics* (Reading 1968), Sect. 12.5 and 13.4.

(8) J.-M. Lévy-Leblond, *Ann. Phys. (N.Y.)* **57**, 481 (1970).

(9) H. Bacry, *Localizability and Space in Quantum Physics* (Springer-Verlag, 1988) ; Ann.Inst.Henri Poincaré **49**, 245 (1988) ; contribution to *Proceedings of the International Symposium on Spacetime Symmetries*, Y.S. Kim and W.W. Zachary eds.(North Holland, 1989) ; preprints CPT-90/P. 2356 and 2405 (Centre de Physique Théorique, Marseille-Luminy 1990).

(10) This result was proved for $s = 1$ already by A. Proca, *Compt. Rend.* **202**, 1420 (1936) ; *Journ. Phys. Radium* **9**, 61 (1938). The conjecture that it is generally true for any s is due to F.J. Belinfante, *Phys. Rev.* **92**, 997 (1953) who proved it for $s = \frac{3}{2}$; it was checked for any half-integer spin by P.A. Moldauer and K.M. Case, *Phys. Rev.* **102**, 279 (1956) and for $s = 2$ by V.S. Tumanov, *Sov. Phys. JETP* **19**, 1182 (1964)

(11) J.-M. Lévy-Leblond, *Comm. Math. Phys.* **6**, 286 (1967). See also ref. (4)

(12) C.R. Hagen, *CommMathPhys.* **18**, 97 (1970) ; C.R. Hagen and W.J. Hurley, *Phys. Rev. Lett.* **24**, 1381 (1970) ; W.J. Hurley, *Phys. Rev.* D**3**, 2339 (1971)

(13) R.F. Streater and A.S. Wightman, *PCT, Spin and Statistics and All That* (Benjamin, New-York 1964)

(14) S. Weinberg, *Gravitation and Cosmology* (Wiley, New-York 1972), pp 61-63 ; see also, for a perhaps more careful discussion, J.-M. Lévy-Leblond and F. Balibar, *Quantics (Rudiments)* (North Holland, Amsterdam 1990), Chapter 3, pp.123-129.

(15) R.P. Feynman, "Why antiparticles ?" in R.P. Feynman, S. Weinberg, *Elementary Particles and the Laws of Physics* (Cambridge University Press, Cambridge 1987).

(16) R.W. Hartung, *Am. J. Phys.* **47**, 900 (1979)

(17) J.-M. Lévy-Leblond, *Commun. Math. Phys.* **4** 145 (1967).

(18) The closest argument I could find is by S. Weinberg, *Phys. Rev.* **138**, 988 (1965).

(19) F. Lurçat and L. Michel, *Nuovo Cim.* **21**, 574 (1961)

(20) J.-M. Lévy-Leblond, *Eur. J. Phys.* **10**, 265 (1989)

(21) It may be recalled that an attempt was made, years ago, by F. Lurçat, to generalize quantum field theory so as to treat mass and spin on the same dynamical footing : F. Lurçat, *Physics* **1**, 95 (1964).

Possible Worlds, Quantum Mechanics, and Middle Way Buddhism

Victor Mansfield
Department of Physics and Astronomy
Colgate University
Hamilton, NY 13346, USA

Abstract

Within possible world semantics I review counterfactuals and use them to explore some crucial philosophic implications of quantum mechanics. I focus on the lack of trans-world identity of quantum mechanical systems and show how this is an important distinction between classical and quantum physics. After a brief review of the philosophic heart of Middle Way Buddhism I discuss the problems of trans-world identity from within that system. Employing possible worlds and counterfactuals as a bridge in this way reveals deep underlying similarities between some central philosophical implications of quantum mechanics and Middle Way Buddhism. It also shows that this branch of Buddhism offers a congenial philosophic framework within which to understand these implications of quantum mechanics. This approach offers a different view of the relationship between the conceptual foundations of modern physics and the philosophical structure of Middle Way Buddhism than my earlier papers.

I: Possible Worlds and Counterfactuals

I was strolling down the sidewalk in a fashionable part of a foreign capital lazily enjoying the sunshine and surge of this colorful city. Completely without warning, a two foot square section of the sidewalk gave way beneath my foot and fell into a pool of foul, black water about six feet below the sidewalk. Fortunately, because my weight never fully transferred to that piece of sidewalk, I could pull back and avoid falling into the pool of fetid water with its steel reinforcing rods sticking straight up at me. I stared incredulously into the hole and in my mind's eye saw myself impaled on the reinforcing rods splashing helpless in the evil water. I could only giggle stupidly and give thanks that "what might have been" did not occur. Possible worlds in which I had fallen onto those rusty rods in the malodorous pool were terrifyingly real and close at hand.

> The trouble with possible worlds, we want to say, is that they represent an exotic piece of metaphysical machinery, the armchair invention of a speculative ontologist lacking what Bertrand Russell called "a robust sense of reality." But the fact is that reflection on some of our most deep-seated intuitions suggest that the appeal to possible worlds is nothing more than a formalization of generally held prephilosophical views about matters modal. All (or at least most) of us think that things might have gone otherwise, that there are different ways things might have been, alternative ways things might have gone; and we tend to think that these different ways things might have been constitute the touchstones for determining the truth value of sentences prefaced by modal operators[1].

If I had fallen through the sidewalk then I would certainly have been injured. The previous sentence is a counterfactual conditional. It is a conditional (an if-then expression) and contrary to fact--for I did not actually fall through (the antecedent of the conditional is false). A more famous example is the opening line of the foremost book on counterfactuals[2]: "*If kangaroos had no tails, they would topple over.*" Counterfactuals involve certainty or strict necessity. It must happen that kangaroos topple when tailless or pedestrians get injured when falling into sewers. In other words, in all possible worlds in which the antecedent of the conditional holds, the consequent must follow.

Possible worlds models have become sufficiently fashionable in a variety of philosophical analyses that they have even become the focus for a well known introductory text in philosophy[3]. Although possible worlds are especially useful in the semantic analysis of counterfactuals, there is still much lively debate about the ontic status of such worlds[4]. I touch on some of that debate later. Fortunately, counterfactuals can be formulated largely independent of the detailed nature of the possible worlds within which they are embedded. In this section, I follow David Lewis[5], the most prominent spokesman for *possibilia*, in my brief presentation of counterfactuals.

The figures below represent the logical operators of necessity, $\Box$, and possibility, $\Diamond$. The circle S_i represents a set of possible worlds in possibility space. A single point in possibility space represents a complete world or universe. At the center of S_i is, i, representing our actual world. The range of possible worlds in S_i is restricted by *accessibility* relations. Possible worlds are said to be accessible from i. For example, if we study physical necessity and physical possibility, S_i includes worlds physically accessible from i, those constrained by the same physical laws governing our world. S_i then embraces all worlds in which I fall through the sidewalk or earn a million dollars yearly salary (unlikely, but not violating any physical laws). S_i does not, however, include worlds in which gravity is suddenly inoperative or has a repulsive rather than attractive nature.

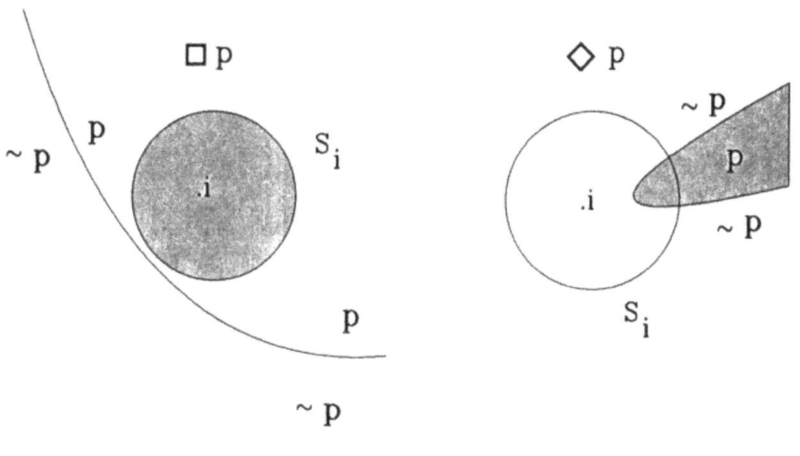

Figure 1

If instead of physical necessity and possibility we consider logical necessity and possibility, then less stringent accessibility requirements constrain S_i. It then includes all worlds in which there is no gravity or in which all living beings are Buddhas. Although these worlds differ greatly from our own, they are logically permissible. Similarly, we could delimit S_i by considering only morally perfect worlds, in which philosopher kings rule benevolently with utmost wisdom and justice. The accessibility requirement for S_i determines whether $\Box$ and $\Diamond$ represent physical necessity and possibility respectively or logical necessity and possibility, and so on. For now we may neglect which accessibility requirement is under consideration.

In the left figure of Figure 1 above the line sloping from the upper left to the lower right represents the division between worlds in which the proposition p is true and false. Since p is always true in all possible worlds within S_i, p is necessarily true at i, symbolized by $\Box$ p. Using physical necessity as an example, p could represent the proposition that gravity is an attractive force, which is true throughout S_i and therefore necessarily true at i.

In the right figure of Figure 1 above the shaded region represents worlds were the proposition p is true. Since S_i is only partly covered by the shaded region, proposition p is possibly true at i. Symbolize this by $\Diamond$p. Again, using physical possibility as an example, p could represent the proposition that there are ten planets in the solar system. Although this could be true in some worlds, it need not be, and thus p is possibly true.

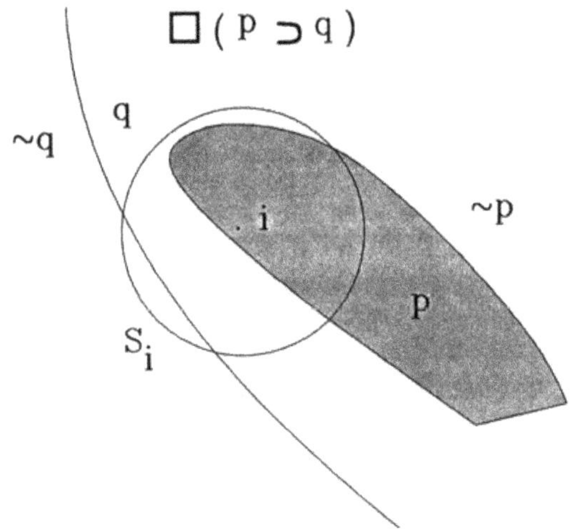

Figure 2

The necessity operator, $\Box$, acts like a universal quantifier over all possible worlds within S_i. $\Box$p is necessarily true at our world, i, when p is true for all worlds within S_i. Similarly, $\Diamond$p acts like an existential quantifier, since there exists at least one world in S_i in which p is true. The two operators are interdefinable. For example, saying that it is not possible that the negation of p holds means that p is necessarily true with the appropriate quantification over S_i. Using the standard negation operator, ~, and the logically defines symbol, $=^{df}$, these equivalences become: $\Box p =^{df} \sim\Diamond\sim p$ or $\Diamond p =^{df} \sim\Box\sim p$.

Counterfactuals are closely related to strict conditionals illustrated in the Figures 2 and 3. Begin with the idea of a material conditional, the standard p implies q, or if p then q, usually symbolized by $p \supset q$. The material conditional or implication is only false with p true and q false, all other combinations of p and q make it true. A strict conditional at i is a material conditional true throughout S_i. Figure 2 above shows a material conditional true throughout S_i, since whenever p is true so is q. Here, in our world, i, both p and q are true and thus $p \supset q$ is true, while in the worlds in the lower left of S_i we have ~p and ~q, which still yields true for $p \supset q$. The material conditional true throughout S_i is the strict conditional at i, symbolized by $\Box (p \supset q)$, also called a necessary implication or entailment. Figure 3 below shows another example of a strict conditional at our world, i, one in which both p and q are false at i, but $p \supset q$ is nevertheless true throughout S_i.

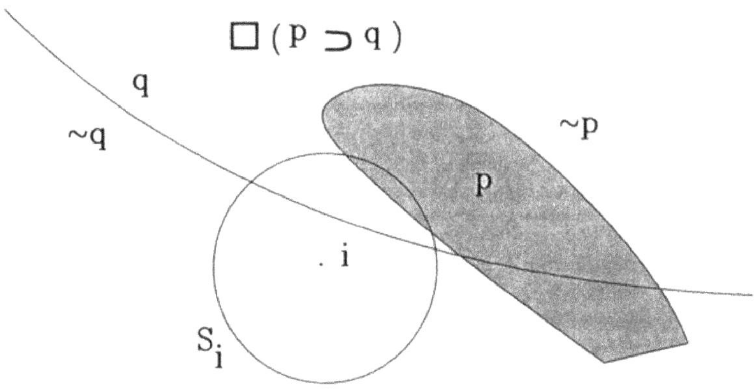

Figure 3

Counterfactuals are strict conditionals with an accessibility requirement of overall comparative similarity of possible worlds. For counterfactuals to hold, all worlds in S_i must be closely similar to our world, i. For example we understand, "If I had fallen through the sidewalk then I would certainly have been injured" in the context where things are pretty much as they actually are. We do not consider possible worlds in which I am too large to fit through two foot square holes or have the high molecular density of superman, which repels both bullets and rusty reinforcing rods. An appropriate world in S_i would be one in which all else is the same as this world except that a greater fraction of my weight transfers to the weak piece of sidewalk. Counterfactual conditionals, symbolized by $p \Box\!\!\rightarrow q$, are strict conditionals where S_i

has the accessibility requirement of overall similarity to our world, i. Or, in terms of kangaroos:

> '*If kangaroos had no tails, they would topple over*' is true (or false, as the case may be) at our world, quite without regard to those possible worlds where kangaroos walk around on crutches, and stay upright that way. Those worlds are too far away from ours. What is meant by the counterfactual is that things being pretty much as they are--scarcity of crutches for kangaroos being pretty much as it actually is, the kangaroos' inability to use crutches being pretty much as it actually is, and so on--if kangaroos had not tails they would topple over.[6]

Unfortunately, it is not easy to specify a metric telling us how similar one world is to the next. I return latter to this problem of similarity of worlds.

II: Trans-World Identity and Quantum Mechanics

An important application of counterfactuals occurs in the definition of dispositional properties such as fluidity, electrical conductivity, or fragility. Dispositional properties inherently bridge worlds. They are securely in the actual world, since we empirically investigate them. But dispositional properties are also firmly planted in other possible worlds, because our assertions about them envisage other conditions or other worlds than those holding for their original empirical investigation. A dispositional property is a feature that something has in our actual world, but it also refers to how the thing would respond in other environments, in other possible worlds.

> Precisely because their bearing is actuality-transcending, dispositions are never determinable in terms of observations alone (observation is, of course an inherently reality-restricted process--one cannot observe the nonactual). Rather, dispositions are *imputed* to things on the basis of *theoretical* considerations. When dispositions are at issue, a reference to actualities alone is never sufficient, because dispositional claims transcend the boundaries of the actual.[7]

For a simple example, consider electrical conductivity. It can be operationally defined using the strict implication: necessarily if current flows in a wire, then it is conductive. Symbolically, let p = current flows in a wire and q = the wire is conductive, then $\Box(p \supset q)$. Counterfactually, if current were to flow in a wire, it would be conductive. The counterfactual definition is weaker (it assumes less) because S_i is constrained by overall similarity to our world, i. The counterfactual definition is therefore preferable to the strict implication. For another simple example, consider the definition for electrical charge: if a particle were to experience a force in an electrical field, then it would be charged. (I intentionally neglect other possible sources of force in this overly simple example.)

The reference to the actual world is clear, since we measure conductivity or electrical charge by standard empirical procedures. But we also wish to assert that the dispositional property of conductivity exists without a current flowing or that charge exists even without an electric field. These dispositional assertions are well-framed using counterfactuals where the contrary-to-fact condition (no current or no field) is clear.

As I discuss in more detail later, counterfactually defining dispositional properties is an attempt to define them within a realistic philosophic framework. At least in classical physics we usually assume there is an object that is the bearer of properties. For example, the classical point particle is the underlying object that bears the properties of mass and spacetime location. More complex objects obviously have a richer set of properties, but we assume, both in empirical life and in classical physics, that the object and its properties have an existence independent of our measurement or knowing. Relativity teaches us that in another possible world in which the observation is made from a different reference frame many crucial properties are radically different. However, because we have a covariant formulation of classical physics and can express the crucial properties as appropriate rank tensors, the independence of the object and its covariant properties is not compromised. As a simple example, rather than consider space and time intervals as defining properties of physical systems we use the invariant spacetime interval.

It is important to recall that counterfactuals employ the necessity operator, a universal quantifier over the possible worlds in S_i, those worlds that are closely similar to our own. This implies that the entity in the counterfactual statement, whether a pedestrian, kangaroo, or classical particle, must have some identity across the possible worlds in questions. Otherwise, what sense does it make to say that we quantify over possible worlds? The identity across possible worlds of classical particles is less complicated than that of more complex entities such as kangaroos or persons. But even for the more complex some identity must be understood. I become giddy with fear in narrowly escaping serious injury because I directly intuit that the fellow impaled on the rusty rods in other possible worlds is me. No matter how empathetic I might be, if it did not seem that there is some form of identity between me in this world and the fellow seriously injured in other possible worlds, it would be difficult to explain the intensity of my fear.

Stated negatively in terms of kangaroos, what sense would it make to say that kangaroos topple when tailless if when tailless they cease to be kangaroos? The precise form of identity implied by universal quantification across the worlds in S_i is the subject of much learned and occasionally acrimonious debate.[8] This literature is too extensive and technical to do justice to here. Rather, I will take a point of view motivated by my dual concerns of quantum mechanics and Middle Way Buddhism and connect with this literature on some of its essential points.

Several authors have recently studied counterfactuals in quantum mechanics. They take two approaches: First, are those such as Michael Redhead[9] who, building on the work of Stapp[10] and Eberhard[11] frames a physical assumption underlying Bell's Inequality in terms of counterfactuals. Since laboratory measurements violate Bell's Inequality and confirm the predictions of quantum mechanics, this casts doubt on the applicability of counterfactuals to quantum mechanics. Bernard d'Espagnat appreciates this point, but emphasizes a second approach[12]. He interprets Bohr's demand that the whole experimental setup be included in the analysis of quantum phenomena as implying the impossibility of counterfactually defining dispositional properties in quantum mechanics.

Although d'Espagnat does not distinguish between strict implications and counterfactuals, I agree with his analysis. Here I expand on his work and show, by appealing directly to the nature of counterfactuals and their possible world setting, why they and their underlying commitment to trans-world objects are troublesome in quantum mechanics. This approach also permits comparisons with Middle Way Buddhism.

Let me assume, at least for the purposes of this analysis, that in standard quantum mechanics, just as in classical physics and daily life, objects are the bearers of dispositional properties. The quantum state is the maximal description we can give of the object and its properties. Wave functions are various mathematical formulations of the underlying quantum state, the most fundamental structure "closest" to the object.

Because of the notorious complementarity of quantum mechanics we must exert care in counterfactually defining dispositional properties. For example, consider particles prepared with spin s_z. Spin could be counterfactually defined with, if particles with spin s_z were put through a Stern-Gerlach apparatus then they would be deflected in the positive z-direction. We also could generalize this into a probability statement. This counterfactual definition implicitly asserts that the z-spin, s_z, transcends the actual z-direction deflection measurement. In other words, the spin in a given direction should exist even when not being measured in that direction. But spin in one direction is complementary to that in another, so we must be very careful in defining the similar worlds in S_i. By restricting the analysis to physically accessible worlds we automatically rule out worlds that attempt simultaneously to measure complementary properties, since such measurements are physically impossible. However, there must be no worlds in S_i that make a complementary spin measurement before measuring s_z. Also, as we well know from the Bell type experiments, we must be very careful in counterfactually defining the spin of correlated particles. For example, in S_i with the x-direction being the line along which the correlated particles separate we must never attempt to measure complementary properties (such as s_y) of the particle correlated to the one on which we are counterfactually defining s_z. With this type of care in limiting S_i we could counterfactually define dispositional properties in quantum mechanics. Equivalently we could say that our definition of closest worlds, S_i, explicitly rules out the troublesome kind of measurements outlined above. But with all these limitations on S_i or our definition of closest worlds, we wonder whether we have truly captured the realistic intuition underlying dispositional properties.

Relativity within quantum mechanics presents us with a fundamentally different situation from that in classical physics. In classical physics we could maintain independent objects with trans-world identity by covariantly formulating their properties. However, this cannot be done in quantum mechanics. Aharonov and Albert[13] have elegantly shown that it is impossible to provide a covariant history of the evolution of quantum states. Because the collapse of quantum states always occurs on a time-like hyperspace, quantum states and their properties cannot be covariantly formulated in relativistic quantum mechanics. The notion of a quantum state only makes sense from within one Lorentz frame, from one possible world. The covariance rests strictly in the probabilities for actual measurements. For example, returning to the Bell type experiments, consider two possible observers (or in possible worlds language two possible worlds). The first observer is at rest with respect to the source of correlated particles and the measuring apparatus and another observer has a large velocity along the line joining the spin measuring devices. There is no quantum state that covariantly gives the history of both observers, yet both agree on the precise correlations found in the experiment. The noncovariance of quantum states compromises their independence and we cannot assert that quantum states have a trans-world identity. If we accept standard quantum mechanics as fundamentally correct and especially its claim that the quantum state is the fullest description of the underlying object we do not have independence or trans-world identity of the underlying object. Since counterfactual statements always assume some level of trans-world identity, they either are not usable in quantum mechanics or they must be used with severe limitations.

Let me try to explicate this fundamental result by slightly modifying a frequently given example by John Wheeler[14] to illustrate the nature of quantum objects. Consider redefining the game of twenty questions. In this version several people query an individual, call him the holder of the word. "Is it alive?" "Is it edible?" and so forth. But in this version there never is a well-defined word in the holder's mind. Although we believe that the holder knows the word from the beginning of the game he really only gives answers which are always consistent with all his previous answers. We as questioners suppose that a well-defined word exists independent of our questions (what reference frames, what set of measurements we make, or more generally, in S_i) but in truth there is only consistent answers to our questions (the dynamical equations and their predictions for the probabilities are covariant). We elicit answers and upon them guess the word into being (impute it) and grant it a false independence which it never had. Nevertheless, despite the creative power of our questioning (what world we are in) there is a consistency underlying all responses which is easily confused with independence or trans-world identity.

Philosophically considering individuals rather than entities studied in physics, Chisholm[15] gives an oft quoted argument against trans-world identity. His argument shows the contradiction between trans-world identity and the transitivity of identity. (If the first individual has a transitive relation to the second, the second has it to the third, then the first has it to the third.) Consider Albert Einstein, A_1, and Bertrand Russell, B_1, to exist in our world, W_1. A_1 and B_1 obviously have very different characteristics. Now consider a different world, W_2, in which Albert and Bertrand exchange some small attribute, let's say that Albert gets another hair on his head and Bertrand gets one fewer. In W_2 they become A_2 and B_2. World W_2 differs from W_1 in other ways, too, because Mrs. Einstein is then married to a man with one more head hair, Albert's children have a father with one more head hair, and so on. Surely, in W_2 Albert and Bertrand maintain their identities through the shift of a single hair. Next consider another world, W_3, in which Albert and Bertrand shift another hair to become A_3 and B_3. Continue in this identity preserving way until you get to a world, W_h, in which A_h has the same number of hairs on his head as B_h. Keep up these small identity preserving exchanges where A in any world is identical with A in the previous world and by repeated application of transitivity is always identical with the original A_1, similarly for B.

Eventually you get to a world W_x where A_x has all the attributes of B_1 and B_x has all the attributes of A_1. This conflicts with the transitive nature of identity, which demands A_x be identical with A_1, not the very different B_1. An analogous conflict occurs for B. Therefore individuals inhabit only one world, despite our prephilosophical notions.

Other conclusions also might be drawn from this paradox but, before doing so, let me briefly discuss an even simpler argument by Michael Loux[16] against trans-world identity. Take Albert Einstein in our world and now move to another world in which Einstein is supposed to exist. How are we to recognize just which individual in this new world is Einstein? Loux claims that we cannot rely simply on any of his properties we normally use in picking him out: his wild hair, mismatched socks, gait, German accent, and so on. In this new world we might have someone else who called himself Einstein, had the same background, appearance, talents, and so on as our Einstein, but there is no "empirically manifest property" allowing us unambiguously to identify him as our Einstein. We have no clear criteria for settling questions of identity and difference. "The claim, then that individuals exist in different possible worlds only appears to be intelligible; it takes our concepts of identity and difference and employs them in contexts where they cannot intelligibly be applied."[17]

Some philosophers conclude from these arguments and others that individuals are world-bound, that they cannot maintain their identity across worlds. If individuals cannot maintain identity through changing conditions in other possible worlds, then there may be problems with an individual persisting through changes in time. The arguments are technical and disputatious[18]. On the other hand, some writers believe that these problems can be handled. David Lewis[19] and others satisfy their prephilosophic intuitions and the needs of modal operators for quantifying over possible worlds by subscribing to world-bound individuals with counterpart relations. It is my counterpart in another possible world that falls through the sidewalk and gets injured. But if this be so, why tremble in fear about what might have been or happens to my counterpart, if that counterpart in another world is not really me, bound to this world as I am? Much technical controversy surrounds these ideas, a little of which I discuss later, but now I wend my way through this rugged terrain by taking cues from quantum mechanics and latter from Middle Way Buddhism.

Since David Lewis and others treat counterfactuals without requiring trans-world identity, perhaps we can rescue the counterfactual definition of dispositional properties and their underlying quantum states using the counterpart relation. Exactly what are counterparts?

> In general: something has for *counterparts* at a given world those things existing there that resemble it closely enough in important respects of intrinsic quality and extrinsic relations, and that resemble it no less closely than do other things existing there. Ordinarily something will have one counterpart or none at a world, but ties in similarity may give it multiple counterparts. Two special cases: (1) anything is its own unique counterpart at its own world, and (2) the abstract entities that exist alike from the standpoint of all world, but inhabit none, are their own unique counterparts at all worlds.
>
> . . . the counterpart relation ought to be used as a substitute for trans-world identity in explaining *de re* modality. The realm of essence and accident is the realm of the vicarious. What

something *might* have done (or might have been) is what it does (or is) vicariously; and that is what its counterparts do (or are). What is essential to something is what it has in common with all its counterparts; what it nowhere vicariously lacks.[20]

Can quantum states have counterparts? At another world, in which all else is the same except the observer has a different Lorentz frame, can we find a state function that "resemble it closely enough in important respects of intrinsic quality and extrinsic relations, and that resemble it no less closely than do other things existing there?" The lack of a covariant history of the state function rules out counterparts for quantum states.

We can alternately frame the question by asking whether there is an essence for the quantum state satisfying the definition: "What is essential to something is what it has in common with all its counterparts; what it nowhere vicariously lacks." Quantum states will of course resist any such formulation because other possible worlds involve observers with different Lorentz frames and it is only within one frame that the state function makes any sense. Rather then attempt to explore the wider implications for possible worlds theory of this quantum mechanical essencelessness, let me next turn to,a full blown theory of essencelessness. The Middle Way Buddhist view of emptiness exalts and celebrates essencelessness rather than hides it as we do in quantum mechanics.

III: A Brief Sketch of Middle Way Buddhism

The central philosophic concern of Middle Way Buddhism (the English translation for Mādhyamika Buddhism) is the principle of emptiness, the lack of independent existence or essence in all phenomena. As in all Buddhism, the ultimate concern is not with philosophy per se, but only as the chief aid in freeing individuals from the beginningless round of the suffering of birth and death, *samsāra*. In this all too brief presentation of Middle Way Buddhism I emphasize the Consequentialist interpretation of Middle Way Buddhism (Prāsaṅgika-Mādhyamika), considered by many Tibetan Buddhists to be the Everest of Buddhism. (Hereafter Middle Way Buddhism will be my translation for Prāsaṅgika-Mādhyamika Buddhism.) For Middle Way Buddhism, emptiness is a denial of a particular type of existence, inherent existence, something that never could be. Middle Way analysis first requires a careful identification of inherent existence. If inherent existence is too broadly defined, nihilism follows, while if too narrowly defined, eternalism results, two heinous philosophic crimes for a view claiming to be the "Middle Way." The Middle Way is not a blending of the two extremes but a thoroughgoing refutation of both.

What then is this special existence which Middle Way Buddhism denies? The major translators and commentators upon Middle Way Buddhism such as Hopkins[21] and Thurman[22] and Tibetan scholar-monks such as Tenzin Gyatso, the fourteenth Dalai Lama,[23] or Geshe Kelsang Gyatso[24] who teach extensively in the West use a variety of words to discuss the "negatee" as they call it. They use inherent existence (*svabhāvasiddhi*) interchangeably with such terms as independent existence, intrinsic existence, substantial existence, or intrinsic self-nature or essence to mean our most innate, unreflective, and pragmatic belief about how subjective and objective phenomena exist. This innate and unreflective belief in inherent existence may be divided into two related convictions: first, that phenomena exist independent of mind or knowing, that "underneath" or "behind" the names, linguistic conventions, or mental designations we apply to all objects there is something objective and substantial, which exists fully and independently from its own side--the essence of the thing. Granted that sense functioning with all its limitations and peculiarities contributes to our experience, nevertheless the instinctive belief is in objects existing independently of this functioning with intrinsic self-nature or essence. Inherent existence seems to provide the objective basis for our shared world. Second, these objects are also believed to be independent of each other. Each object being fundamentally nonrelational, it exists in its own right without essential dependence upon other objects or phenomena. In other words, the essential nature of these objects is their nonrelational unity and completeness in themselves.

As stressed below, the denial of inherent existence does not mean that objects do not exist (the extreme of nihilism). They certainly have a conventional or nominal existence, but they totally lack independent or inherent existence. This denial of independent existence in phenomena is the principle of emptiness. Middle Way Buddhists contend that all our suffering and limitation ultimately flow from our erroneous belief in inherent existence. Belief in independently existent phenomena, whether our own ego or a material possession, is the basis of our craving and aversion, which can only be overcome by deeply understanding emptiness. This personal transformation requires an immense and revolutionary conceptual reeducation and renewal at every level of the personality.

Middle Way Buddhists employ three types of arguments of increasing power to deny inherent existence. First, they show that all objects result from a complicated web of interrelated causes and conditions and thus lack independent existence. Second, they argue for the unfindability of the object either among its parts or as the whole of its parts and thereby deny its independent existence. (This argument is similar to Michael Loux's point that there is no "empirically manifest property" that would allow us unambiguously to identify Einstein in another world.) Third, they show that every object is mentally designated, named, or imputed and thus lacks the inherent existence we believe is the hallmark of reality. Without recapitulating any of these arguments, but nonetheless capturing the spirit of the Middle Way Buddhism dialectic, we may simply analyze inherent existence to see that in fact it could never exist.[25]

Take any object considered to exist inherently or independently, whether it be a tree or a photon. Inherent existence is an essential property and must always apply to the entire object without qualifications.[26] Thus if an object inherently exists during a given time interval, it must have equally been inherently existent in any portion of that interval. Such an object is complete and fully itself in any subinterval, no matter how short. Assume with the Buddhists, that instants or moments of time are indivisible, without internal transitions or change, and that temporal transition or the flow of time is the succession of instants. Since an inherently existent object is fully contained and self-sufficient in any one moment, by definition, it has at that instant no intrinsic or essential reference or relationship to any other objects at any other time. The intrinsically existent object lacking these essential relations to other times and yet enduring for more than one instant, must then be intrinsically changeless, unable to produce or be acted upon. Such impervious and impotent objects are locked in their immutable self-nature--frozen in eternity.[27]

Parallel arguments can be developed for objects with no intrinsic spatial relations. Such objects must exist complete and self-contained in a limited region of space, independent of interaction with other regions. Just as in the argument based on time, this inevitably leads to unacceptable isolation and immutability.

The innate and inveterate belief in inherent existence entails a commitment to intrinsically nonrelational entities unaffected by the dynamism of nature, unable to change, interact, or be objects of knowledge. Such entities cannot exist in our world of flux. Although Middle Way Buddhists firmly deny inherently existent objects, conventional objects do most emphatically exist as dependent and relational. Emptiness is not nihilism. There is valid knowledge of conventional realities and right action. The "I" does conventionally exist as a mental designation or imputation upon the mind and body. Nor does emptiness fall to the extreme of permanence or eternalism because objects lack inherent existence. Objects abide in the "Middle Way," as neither nonexistent nor as inherently existent, but as dependently arisen, interdependent, and relational.

When we distill normal experience do we find persons committed to a belief in independent existence which ensnares them in *samsāra*? Any doubt we may have is not shared by the Middle Way Buddhists. As Hopkins puts it, "The pivot of the practice of emptiness and of the generation of the wisdom that realizes emptiness is the identification that objects appear as if they exist in and of themselves."[28] This independent, nonrelational nature that we falsely believe makes objects seem so substantial and real, is just what makes inherent existence contradictory and self-condemned. If phenomena lack independence, what then do they depend upon?

Phenomena are defined by and depend upon their interrelations, their dynamic connections with other phenomena and minds. The essence of phenomena is their connectedness and dependency. Intrinsic nonrelatedness is a contradictory property. Although we incorrectly assume inherent existence is essential for appearance, the Middle Way adherents claim an object's lack of inherent existence--its emptiness, its radical dependency--is what makes phenomena both possible and functional and simultaneously non eternal.

If phenomena are void of independent existence, why then do they appear to exist so objectively, so fully invested with an independent self-nature or essence? The Middle Way followers assert that the mind, along with building the details of sense perception, reifies and concretizes experience so that we are presented with objects and subjects that falsely appear to exist inherently. While naming or designating, the mind invests each phenomena with the false attribute of inherent existence. But inherent existence is merely a conceptual designation, imputation, or mental superimposition--it never has existed nor ever will exist. Yet we are deeply committed to nonexistents, to conceptual designations or fabrications, attributions in thought that we instinctively and compulsively make. From the diaphanous elements of experience the mind falsely fabricates inherently existent objects and then superimposes attraction and repulsion, the twin shackles of *samsāra*, upon them. This tendency to reify experience into independently existing objects and subjects is the core of our congenital ignorance and the root of suffering. Trying to sustain these false expectations about experience leads to endless suffering.

While imputation is the source of the false appearance of independently existent objects, the mind in its normal activity of naming or designating is also implicated in the conventional (as opposed to inherent) existence of objects. In other words, along with the mind's illegitimate imputation of inherent existence it legitimately functions in designating or naming conventionally existent objects. The mutually dependent conventional existence of the mind and its objects assures their mutual emptiness. Thus Middle Way Buddhists say that dependent arising, the king of reasonings, in its most powerful form is: "All phenomena do not inherently exist because of being dependently imputed."[29] All other reasonings proving lack of inherent existence derive from this king of reasons that can, without residues, overcome the extremes of eternalism and nihilism.[30]

Despite the importance of mind, Middle Way Buddhism scrupulously avoids idealism. In the Middle Way, mind also lacks independent existence and there must be conventionally existent objects external to consciousness so that the interdependent minds and objects can conventionally exist[31].

Middle Way Buddhists make their distinction between inherent existence and conventional existence within the framework of the two truths, which states that there are two complementary views of phenomena, one ultimate, wherein all is empty, and the other conventional, where objects have an existence, functionality, and efficiency in the realm of everyday life. All phenomena or objects of knowledge are included in the division of the two truths. Our normal life and discourse is in the realm of conventional truth but we unfortunately contaminate conventional existence by imputing inherent existence to it.

Emptiness is the Ultimate Truth of phenomena. No higher or more transcendent principle is promulgated in place of the false belief in inherent existence. Emptiness is a nonaffirming negation--the mere denial of independent existence in phenomena. Emptiness itself is empty, nevertheless it is the foundation for the cultivation of limitless compassion for all sentient beings.

Tsongkhapa, an important fourteenth century Middle Way commentator and founder of the Gelukba order of Tibetan Buddhism gave an oft quoted example to illustrate emptiness as a nonaffirming negation. He defines space, not with metric coefficients, affine connections, or another mathematical structure as physicists would, but as the pure lack of obstructive contact. The highest aspect of space for him is that it does not obstruct or block any object that might be placed in it. Space is defined by what it does not do--obstruct. Similarly, emptiness is a lack, an absence of inherent existence. Neither space nor emptiness imply any superior principles beyond themselves. However, phenomena cannot be just characterized negatively by solely emphasizing their emptiness. This could easily lead to nihilism, therefore the complementary description of phenomena requires an appreciation of their radical dependency and interrelatedness--their dependent arising.

If Middle Way Buddhists deny self-standing or inherent existent to individuals in one world, they would never countenance trans-world individuals nor quantum systems with a well-defined identity through worlds with different experimental arrangements. Relations are more fundamental, more "real" than the relata. Contrary to our common belief, relations are not built upon independently existing entities. Instead, we falsely impute or mentally designate independent existence to phenomena that have no existence beyond interdependence and relations. Going from one world to the next the relations change and thus the entity, which is no thing beyond these relations, fails to have identity across worlds.

Some philosophers who believe there are no trans-world individuals fear the boiling cauldron of essentialism with its bath of insoluble problems. (But the water is *aqua vita* for essentialists like Alvin Plantinga[32] or Nicholas Rescher[33]!) With world-bound individuals every relation is essential to the individual, whether it be a human or a quark. In other words, all relations become internal to the definition of the entity. Middle Way Buddhism agrees that the entity is only relational, but it firmly denies that there are essences or qualities that independently exist. In Middle Way Buddhist language, essences are never findable upon analysis, or in the closely equivalent more modern expression, there is no "empirically manifest property" that allows us to identify unambiguously an essence, whether it be the alleged essence of a person or a proton.

According to the proponents of the Middle Way, profoundly appreciating essencelessness and the interdependence of all humans that this implies, inevitably leads to compassion. Emptiness dialectic must issue in moral injunctions. If my neighbor and I both lack independent existence, if our relationship is more primary, more real, then our isolated or independent existence, then how can I possibly profit at his expense? How can I attain ultimate liberation while he suffers? Compassion in turn empowers and deepens the understanding of emptiness. When attention turns toward my neighbor's welfare and away from myself then I make progress toward appreciating the emptiness of the very fiend himself--my ego, my self-cherishing. In other words, emptiness provides the rational support for both the cultivation of compassion and the tantric practices that divinize all. Destroying the false belief in inherent existence is the prerequisite for recognizing the world as sacred.

IV: The Middle Way, Trans-World Identity, and Quantum Mechanics

In this section I primarily focus on the problem of trans-world identity from the Middle Way Buddhist point of view and use this theme to connect the previous sections together. Of particular concern is how, faced with world-bound individuals, or the lack of inherent existence, we can understand the continuity seen both in the inner and outer world.

Let me start with the lack of trans-world identity of quantum states. From the Buddhist perspective, the problem lies in the implicit assumption of an independent existence to quantum states and their alleged underlying object both in this world and throughout others. If the ultimate nature of phenomena is their relational dependency and interrelatedness, then properties cannot be understood as independent of the exact experimental configuration. (I have made this point in detail in previous papers[34] As the Rescher quote above says, "Rather, dispositions are *imputed* to things on the basis of *theoretical* considerations." From the Buddhist perspective, the "*theoretical*" consideration that motivates the imputation of trans-world identity is the belief in inherently existing properties, which transcend the conditions of the exact experimental configuration within which they are measured.

d'Espagnat makes a similar point. He defines *physical realism* as the belief in a mind independent world that can be scientifically known. Several of his works note how physical realism is close to our "common-sense realism," or what the Middle Way proponents call the belief in inherent existence, and that it is naturally connected to counterfactual thinking.

> The kind of [physical] realism we are discussing is essentially, albeit somewhat implicitly, linked to the notion of counterfactual implication. More precisely, we could say that it has close connections with the method of defining the properties of systems based on counterfactual implication. As we have seen, this realism is a kind of sublimation of common-sense realism that does not radically change its nature. When in everyday life we say that an object, or an attribute of an object, is real (and, *a fortiori*, when we say it is "material"), what we are implying is not only that we or others will actually observe it but, more basically that we *could* observe it, or observe its existence, *if* certain conditions are fulfilled, conditions that may actually not be so. Even if it is not actually seen, a mountain is said to be real as soon as it is granted that if we went to the country where it is supposed to be we would see it. The same holds good for properties.[35]

This view is echoed by Paul Teller[36] who calls this pervasive belief in the isolated and independent existence of properties particularism, which he convincingly claims is at the root of so many conceptual difficulties in interpreting quantum mechanics. He is especially effective at pointing out the essential relatedness of quantum objects and how we must avoid particularism to gain a proper understanding of quantum mechanics. Whether you call this reifying tendency of the mind an innate, unreflective imputation of inherent existence, or physical realism, or particularism, it illegitimately invests objects (or responses to our twenty questions!) with a trans-world identity not sustained by quantum mechanical experiment and analysis.

The connections among noncovariant quantum states, world-bound individuals, and emptiness deepens (along with our discomfort) if we turn to impermanence, a principle stressed by all Buddhists. For Middle Way Buddhism emptiness entails extreme impermanence. If there are no inherently existing entities than all must perforce be process and change. The coarse impermanence seen in the continuous transformation of nature is easy to appreciate. Subtle impermanence, that the manifestation of every phenomena is co-temporal with its dissolution or disintegration, is more difficult to accept--especially if applied to ourselves. There seems to be a slow, continuous evolution in our personalities. Emptiness demands, however, that the person who started reading this sentence is different from the person who finishes it!

Is the extreme impermanence entailed by Middle Way emptiness too much at odds with our normal sense of a continuous self-identical person to be true? Could we be both limited to one world and changing from moment to moment in this one? We certainly feel continuous and all Buddhists stress the notion of *karma*, that actions reap effects of an appropriate kind, that my development depends upon my effort. For *karma* to be meaningful there must be some kind of continuity. But can we comprehend continuity without self-standing identity?

A typical move in Middle Way Buddhism is to turn the argument around 180 degrees and thereby find a solution. For example, they are often asked, "How can objects totally lacking in inherent existence perform their function?" For example, if a piece of wood is totally lacking in independent existence, how can we make tables or toothpicks from it? Middle Way Buddhists claim that the emptiness of the object is precisely what allows the object to function and have efficiency at all. If it were not lacking in self-standing being or independent essence then it could not relate to other objects or serve as wood. An inherently existing entity would be a nonfunctional monolith, neither the subject nor object of action. In a similar way, if an individual inherently existed then there would be no possibility for development, no functioning of karma, and no transforming ignorance into enlightenment. In short, emptiness makes continuity and change possible.

A quantum mechanical example shows how continuity can be maintained without self-standing being. A laboratory exercise in our elementary physics course has students find particle properties using bubble chamber photographs and the conservation laws of momentum and energy. The photographs clearly show well-defined trajectories from which particle properties can be deduced. In fact, the notion of particle hinges upon the theoretical idea of trajectory, that some self-identical object has a continuously changing position, or equivalently, that we can simultaneously define position and momentum for an entity. But in fact, particleness is a dispositional property, a mix of actuality and imputation based upon the theoretical notion of trajectory. In other words, simultaneously having well-defined position and velocity is required for us to have a concrete or objective notion of particle. Of course, we are just happy to have students understand and apply the conservation laws to the bubble chamber photographs and we rarely tell them that the continuous trajectory is only an appearance and particlehood is a mental designation.

The physics teacher knows that the sharply defined tracks are really the eye's stringing together of many small bubbles generated in the supersaturated liquid because of individual ionizations. Between each ionization there is no particle trajectory. The best we can do is develop wave mechanical calculations for the probability of ionization of neighboring atoms in the liquid. We might persist in holding incorrect mental images of particle trajectories, when all we have is probability distributions for ionization events, but the Uncertainty Relations will always thwart our particle picture, our image of a self-existent entity with a continuous trajectory.

Faced with this lack of a self-identical particle trajectory, we nonetheless *impute* a trajectory to the series of bubbles. Upon a series of discrete events we mentally designate a continuous trajectory. Nature seems to exhibit continuity which physics so beautifully embodies in its differential equations, but we provide the notion of a substantial, self-identical particle. We are usually conscious when we do this and it is often helpful in making calculations or giving explanations. However, this powerful tendency of the mind to reify a process into a self-standing or inherently existent thing, the disease of "particularism" that Paul Teller warns us against, is often at the root of our difficulties in understanding quantum mechanics.

John Wheeler claims, "What we call reality consists of a few iron posts of observations between which we fill in by an elaborate papier-maché construction of imagination and theory.[37]" Middle Way Buddhists ask us to analyze the iron posts and find them, along with the whole series, just as diaphanous, just as empty of inherent existence as the intervening mental designations, imputations, or "elaborate papier-maché construction of imagination and theory." As a Middle Way Buddhist would say, " . . . the phenomena are imputed to a series of moments and are unfindable among the various moments or as the series itself."[38]

Not only do we mentally designate continuity and the false notion of an inherently existent object undergoing continuous change in location, but according to the Buddhists we do something very similar with our sense of self, our personal identity. Our lives are punctuated with a few vivid emotional experiences that we weave together into a semblance of personal identity, to which we tenaciously cling. What is this object of our self-cherishing? Can we find it apart from our mental designation upon mind and body? Does it exist separate from the myriad relations of attraction and repulsion? Is there any self that is findable upon analysis or empirically manifest? What happens to it during deep sleep or when we die? Surely such a construction, so thoroughly empty of independent existence, cannot have trans-world identity no matter how much we crave such substantiality. Since the self or person is empty, lacking in independent existence, it is necessarily changing from moment to moment. Special meditational exercises have been developed to give the practitioner powerful experiential confirmation of selflessness, to ground the emptiness dialectic in living experiences.

Rapidly changing, momentary gleams of perception and awareness are unconsciously and inveterately welded into a personality with a seeming self-identity, essence, or inherent existent. This "I" conventionally exists, although like all phenomena it is empty. This nominal I is the subject of modal statements and all other references to self. A hypnotic imputation upon the mind and body, this conventional I, is illegitimately invested with inherent existence that becomes the basis for our self-cherishing, our personal fears and cravings. It is this falsehood that we believe has trans-world identity. Upon this construction we lavish all our attention and fear in this world and in all others with rusty rods and other demons of destruction. Can such a construction support all the intuitions we have about modal operators being quantifications over possible worlds, with or without counterpart theory? This is the subject for much more research.

According to this school of Buddhism, the challenge is to take the middle way between the two truths, ultimate and conventional; to appreciate the thoroughgoing emptiness of self and other and yet honor the functionality and efficacy of conventional phenomena. Despite (or more properly because) phenomena lack inherent existence they function, are objects of help and harm, and exhibit continuity. There may be neither trans-world individuals nor quantum systems, but suffering is still authentic and bitter cries ring out for our enlightened action.

Acknowledgements

It is a pleasure to thank my friends at Wisdom's Goldenrod Center for Philosophic Studies in Valois, NY. They have helped me to understand Buddhism and other great philosophic systems. I wish especially to thank professors Jon Jarrett and Mark Crimmins and graduate student Michael Wakoff of the Cornell University Philosophy Department for several extended discussions of an earlier draft of this paper. I also wish to thank professor Shimon Malin of the Department of Physics and Astronomy at Colgate University for comments upon an earlier draft. Of course, any foolishness in this work is all mine. I offer my deepest thanks to the late Anthony Damiani of Goldenrod whose uncompromising devotion to philosophy taught us to draw inspiration from many great traditions without limiting ourselves to any one. Finally, I wish to acknowledge the generosity of the Colgate Research Council which has partly funded this research.

--- References ---

1 Michael J. Loux, "Introduction: Modality and Metaphysics," in *The Possible and the Actual* (Cornell University Press, Ithaca, 1979) p. 30.

2 David K. Lewis, *Counterfactuals* (Harvard University Press, Cambridge, 1973).

3 Raymond Bradley and Norman Swartz, *Possible Worlds* (Hackett Publishing Co. Indianapolis, 1979).

4 For a recent example of the intensity and fundamental nature of the debate about possible worlds see the review of David K. Lewis's book *On the Plurality of Possible Worlds* by Nathan Salmon, *Philosophical Review XCVII, no. 2* 237, 1989.

5 David K. Lewis, *Counterfactuals* and *On the Plurality of Possible Worlds* (Basil Blackwell Ltd., Oxford, 1986).

6 Lewis, *Counterfactuals*, pp. 8-9.

7 Nicholas Rescher, *A Theory of Possibility* (University of Pittsburgh Press, Pittsburgh, 1975) pp. 132-3.

8 See for example, the introductory essay and the references therein written by Michael J. Loux in the book he edited *The Possible and the Actual* (Cornell University Press, Ithaca, 1979) or the final chapter in David K. Lewis, *On the Plurality of Possible Worlds.*

9 Michael Redhead, *Incompleteness, Nonlocality, and Realism* (Clarendon Press, Oxford, 1987), pp. 90-96.

10 Henry Stapp, "S-Matrix Interpretation of Quantum Theory," *Physical Review, D3* 1303, 1971.

11 P. H. Eberhard, "Bell's Theorem without Hidden Variables," *Il Nuovo Cemento, 38B* 75, 1977.

12 Bernard d'Espagnat, *In Search of Reality* (Springer-Verlag, Berlin, 1983) chapter 12; "Nonseparability and the Tentative Descriptions of Reality," *Physics Reports, 110, No. 4* 201, 1984, especially pp. 232-244; and *Reality and the Physicist (Cambridge University Press, Cambridge, 1989) chapter 4.*

13 Y. Aharonov and D. Z. Albert, "Can we make sense out of the measurement process in relativistic quantum mechanics?", *Phys. Rev. D, 24,* 359, 1981.

14 J. A. Wheeler, "Law Without Law," *Quantum Theory and Measurement* (Princeton University Press, Princeton, 1983) pp. 201-202.

15 Roderick M. Chisholm, "Identity through Possible Worlds: Some Questions" *Noûs, 1* 1, 1967.

16 Michael Loux, "Introduction: Modality and Metaphysics," in *The Possible and the Actual*, p. 39.

17 *ibid* p. 39.

18 See the introduction by Michael Loux in *The Possible and the Actual.*

19 David K. Lewis, *Counterfactuals*, pp. 39-43; *On the Plurality of Possible Worlds*, chapter 4.

20 David Lewis, *Counterfactuals*, pp. 39-40.

21 Jeffrey Hopkins, *Meditation on Emptiness* (Wisdom Publications, London, 1983).

22 Robert Thurman, *Tsong Khapa's Speech of Gold in the Essence of True Eloquence* (Princeton University Press, Princeton, 1984).

23 Tenzin Gyatso, *The Buddhism of Tibet* Translated by J. Hopkins and L. Rimpoche (George Allen and Unwin, London, 1975) and *Transcendent Wisdom* Translated by B. A. Wallace (Snow Lion Publications, Ithaca, NY 1988).

24 Kelsang Gyatso, *Heart of Wisdom* (Tharpa Publications, London, 1986) and *Meaningful to Behold* (Tharpa Publications, London, 1986).

25 The next few paragraphs are edited from my paper "Mādhyamika Buddhism and Quantum Mechanics Beginning a Dialogue."

26 Richard Robinson, "Did Nāgārjuna really refute all philosophical views?," *Philosophy East and West, 22* 325, 1972.

27 Thurman, pp. 157-8.

28 Hopkins, p. 10.

29 ibid. p. 168.

30 ibid. p. 167; Thurman, pp. 146-7.

31 Hopkins, Ibid., pp. 367-368. Thurman, Ibid., p. 316.

32 Alvin Plantinga, *The Nature of Necessity* (Oxford University Press, Oxford, 1989).

33 Nicholas Rescher, *A Theory of Possibility*.

34 Victor Mansfield, "Mādhyamika Buddhism and Quantum Mechanics: Beginning a Dialogue," *International Philosophical Quarterly, XXIX, no. 4*, 371, 1989; V. Mansfield, "Relativity in Mādhyamika Buddhism and Modern Physics," *Philosophy East and West, XL, no. 1*, 59, 1990.

35 d'Espagnat, *Reality and the Physicists*, p. 75.

36 Paul Teller, "Relational Holism and Quantum Mechanics,"*British Journal for the Philosophy of Science 37* 71, 1985 and "Relativity, Relational Holism, and the Bell Inequalities," in *Philosophical Consequences of Quantum Theory* edited by James T. Cushing and Ernan McMullin (University of Notre Dame Press, Notre Dame, 1989.

37 John A. Wheeler and W.H. Zurek, *Quantum Theory and Measurement* (Princeton University Press, Princeton, 1983) p. 194.

38 Hopkins, p. 219.

THE OBJECTIFICATION IN THE MEASURING PROCESS

AND THE MANY WORLDS INTERPRETATION

Peter Mittelstaedt
Institut für Theoretische Physik
der Universität zu Köln
D-5000 Köln 41, Federal Republic of Germany

Abstract

The quantum mechanical measuring process should fulfill two requirements: (i) It should objectify a property P which is nonobjective in the preparation and (ii) it should reproduce the probability propositions about P in the statistics of measuring results. It is shown that whithin the framework of Hilbert space quantum mechanics condition (ii) can be justified only for those classes of measurements for which condition (i) cannot be fulfilled. Consequently there is no obvious interpretation for the probability statements. In this situation the only way out seems to be to abandon the objectification postulate and to accept the simultaneous existence of all possible measuring results. It is briefly discussed whether this "many worlds interpretation" is an inescapable consequence of the measurement theory first conceived by J. von Neumann.

§1. INTRODUCTION

Within the framework of the well known quantum mechanics in Hilbert space one can make several predictions about the properties of a physical system. If a system S is given with the preparation $W = P[\varphi]$, where $P[\varphi]$ is the projection operator onto the pure state φ, then the properties a_1, a_2, corresponding to the projection operators $P(a_1), P(a_2), \ldots$ which commute with $P[\varphi]$, $[P[\varphi], P(a_i)]_- = 0$, are *objective*, i. e. either the property a_i or its contrary $\neg a_i$ pertains to the system. All other properties corresponding to projection operators which do not commute with the preparation W, are *nonobjective* in the system $S = S(W)$, i. e. it is not possible to assume that the system S possesses a property of this kind or its contrary.

However, even for these nonobjective properties quantum mechanics provides the following somewhat weaker prediction: If on the system S with preparation W the nonobjective property a is measured, then the probability to obtain this property as a result is given by $p_W(a) = \mathrm{tr}\{W \cdot P(a)\}$. This means that if the observable $P(a)$ would be measured on a large number N of equally prepared systems $S(W)$ one would obtain results $a^{(i)}$ $(i = 1, \ldots, N)$ with $a^{(i)} \in \{a, \neg a\}$ such that the relative frequency $f_W^N(a)$ of outcomes a would approach – in some weak sense – the probability $p_W(a)$ for $N \to \infty$. The justification of this relative frequency interpretation of quantum probabilities will be discussed in §4.

According to J. von Neumann the quantum mechanical measuring process should be considered as a quantum mechanical process which can be treated completely within the framework of Hilbert space quantum mechanics. The system S and the measuring apparatus $\mathcal{M}$ are first brought into contact and are then again separated such that the measuring result can be read off from the apparatus $\mathcal{M}$. This program, first formulated by J. von Neumann in 1932 [12], turned out to be very complicated, and in spite of many important contributions, there are still some unsolved open problems. (§2). In particular a quantum mechanical treatment of the measuring process should make clear that first the measuring process actually leads to a well defined objective result and secondly that this result would appear in a series of measurements with the predicted relative frequency. Both problems – the objectification and the relative frequency interpretation – will be discussed in detail in §3 and §4.

The attempt to understand the objectification and the statistical behaviour of possible outcomes on the basis of the quantum mechanical description of the measuring process does not lead to the desired result. The unitary dynamics of the compound system which consists of the object system and the measuring instrument does not explain the objectification of an observable which was nonobjective in the initial state of the object system. As a consequence the interpretation of probabilities $p_W(a)$ as relative frequencies of measuring outcomes becomes highly problematical. In this hopeless situation the only way out seems to consist in assuming that the objectification does not take place at all and that the complete variety of possible results persists even after the measuring process. Each observer would then register precisely one definite result – but there are many other observers reading different results in many alternate worlds.

This rather bizarre "many worlds interpretation" seems to be an inescapable consequence of the von Neumann program to describe the measuring process as a quantum mechanical process in the Hilbert space of the compound system $\mathcal{S}+\mathcal{M}$. This interpretation opens the possibility to interpret quantum probabilities even in the case of an individual system like the universe, but at the price of a very strange ontology. For this reason several proposals for alternative solutions of the measuring problem have been made. In §6 some of these proposals will be briefly mentioned. For an extensive discussion of a very promising attempt in this direction we refer to the lecture of P. Busch in this volume.

§2 THE MEASURING PROCESS

Quantum theory provides predictions about the properties of an object system $\mathcal{S}$ and about the results of a measuring process. On the other hand there are measuring instruments $\mathcal{M}$ which allow to confirm or to falsify the theoretical predictions. The theory of the quantum mechanical measuring process tries to describe and to explain this process in terms of physics. In this way the relation between quantum theory and experimental outcomes is partly reduced to the dynamical behaviour of the systems $\mathcal{S}$ and $\mathcal{M}$.

According to J. von Neumann the quantum mechanical measuring process should be described exclusively within the framework of quantum mechanics, i. e. the measuring process is nothing else but a particular quantum mechanical phenomenon. Consequently the object system $\mathcal{S}$ and the apparatus $\mathcal{M}$ are considered as quantum mechanical systems with Hilbert spaces $\mathcal{H}_{\mathcal{S}}$ and $\mathcal{H}_{\mathcal{M}}$ respectively . Let $\varphi \in \mathcal{H}_{\mathcal{S}}$ be the initial preparation of $\mathcal{S}$, and Φ the initial state of $\mathcal{M}$ before the measurement. Here we consider the measuring process of an ordinary discrete observable $A = \sum a_i P(a_i)$ with eigenvalues a_i and a pointer observable $Z = \sum Z_i P[\Phi_i]$ with a nondegenerate spectrum Z_i. If $W = P[\varphi]$ is the preparation of the system $\mathcal{S} = \mathcal{S}(W)$ and $A = \sum a_i P(a_i)$ an observable of $\mathcal{S}$, then quantum theory provides the following predictions: If $[W, A]_- = 0$, then the observable A is *objective* with respect to W and the eigenvalue $\mathrm{tr}\,(P[\varphi]A)$ pertains to the system $\mathcal{S} = \mathcal{S}(W)$. If $[W, A]_- \neq 0$ then A is *nonobjective* with respect to $\mathcal{S}(W)$ and the eigenvalue a_i pertains to the system with the probability $p_W(a_i) = \mathrm{tr}\,(W \cdot P(a_i))$. It is a minimal requirement on the measuring process that these assertions are reproduced by the probabilities of the pointer values Z_i. This *probability reproducibility condition* can be justified for a large class of measuring processes.

For the quantum mechanical description of the measuring process we distinguish three subsequent steps: In step (I), the preparation, the two systems $\mathcal{S}$ and $\mathcal{M}$ with preparations φ and Φ respectively are conceptually comprehended to a compound system $\mathcal{S}+\mathcal{M}$ with the state $\psi = \varphi \otimes \Phi$ in the tensor product Hilbert space $\mathcal{H} = \mathcal{H}_{\mathcal{S}} \otimes \mathcal{H}_{\mathcal{M}}$. In step (II), the interaction, the two systems are brought into contact at time $t = t_0$ and in addition to the free Hamiltonian $H_{\mathcal{S}} + H_{\mathcal{M}}$ some interaction Hamiltonian H_{int} is turned on. After a short time interval Δt the interaction is turned off at $t = t_1$ and the state $\psi(t_0)$ of $\mathcal{S}+\mathcal{M}$ has been changed into $\psi'(t_1)$. This state transformation will be described here by a unitary operator $U(t) = \exp\left(\frac{i}{\hbar}Ht\right)$ with $H = H_{\mathcal{S}} + H_{\mathcal{M}} + H_{int}$ such that $\psi'(t_1) = U(t_1 - t_0)\psi(t_0)$ holds. The state $\psi'(\mathcal{S}+\mathcal{M})$ cannot be written generally as a product state of two states $\varphi' \in \mathcal{H}_{\mathcal{S}}$ and $\Phi' \in \mathcal{H}_{\mathcal{M}}$. However if the measuring apparatus is really convenient for measuring the observable

$$A = \sum_i a_i P(a_i) = \sum_i a_i \sum_j^{n_i} P[\varphi_{ij}]$$

with orthogonal states φ_{ij} such that $A\varphi_{ij} = a_i\varphi_{ij}$, then the unitary operator U must fulfill some requirements. Here we will not consider the most general case, but assume that U satisfies the condition $U(\varphi_{ij} \otimes \Phi) = \varphi_{ij} \otimes \Phi_i$ which corresponds to an ideal strong value correlation measurement. [3] Then there exists a *biorthogonal decomposition* of the compound state

$$\psi'(\mathcal{S}+\mathcal{M}) = \sum c_i \varphi^{(a_i)}(\mathcal{S}) \otimes \Phi_i(\mathcal{M})$$

where $\varphi^{(a_i)} = P(a_i)\varphi/c_i$ are normalized projections of φ onto the subspaces $\mathcal{M}(a_i) = P(a_i)\mathcal{H}_{\mathcal{S}}$ and $c_i = \{\mathrm{tr}\,(P(a_i)P[\varphi])\}^{\frac{1}{2}}$ are normalisation constants with $c_i^2 = p_\varphi(a_i)$.[1] In step (III), objectification and reading, the two systems $\mathcal{S}$ and $\mathcal{M}$ are again dynamically independent but still correlated. As subsystems of the total system in the state $\psi'(\mathcal{S}+\mathcal{M})$ they can be described by the reduced (mixed) states

$$W'_{\mathcal{S}} = \sum_i p_\varphi(a_i) P[\varphi^{(a_i)}], \quad W'_{\mathcal{M}} = \sum_i p_\varphi(a_i) P[\Phi_i]$$

respectively. If one assumes, that the mixed state $W'_{\mathcal{M}}$ describes a Gemenge $\Gamma(W'_{\mathcal{M}})$ such that actually one of the states Φ_i pertains to the apparatus $\mathcal{M}$ with probability $p_\varphi(a_i)$ (pointer objectification) then one can obtain the pointer value Z_i simply by

[1] The biorthogonal decomposition of the compound state $\psi'(\mathcal{S}+\mathcal{M})$ was investigated first by J. v. Neumann [12]. In the literature it has also been called "canonical", [4], "normal" [7] or "polar decomposition". [9, 10]

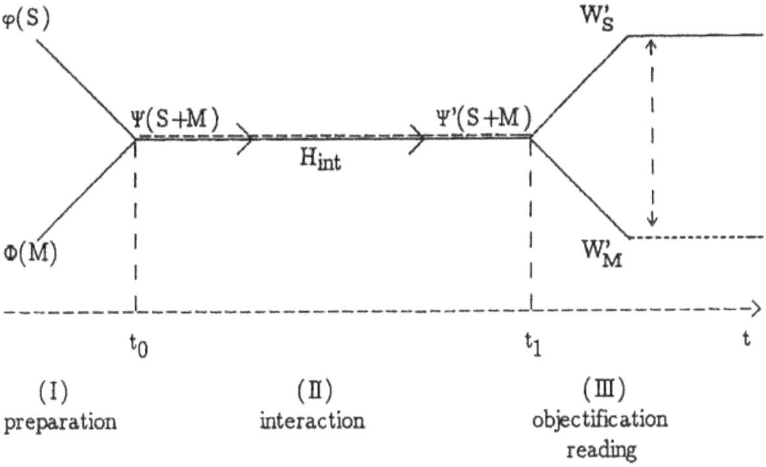

reading. By means of the correlation between the states W'_S and $W'_{\mathcal{M}}$ one finally gets the measuring result, i. e. one of the eigenvalues a_i of A.

We conclude this section with some additional remarks:

i) It is assumed here that in the interval $t_0 \leq t \leq t_1$ the interaction H_{int} induces a state transformation $\psi \rightarrow \psi'$, which can be described by a unitary operator $U(t)$. In order to avoid time-dependent interactions $H_{int}(t)$ and thus difficulties with the group properties of the unitary dynamics, "turning on" and "turning off" the interaction should be understood as an effect of the dynamics, i. e. of the relative motion of $\mathcal{S}$ and $\mathcal{M}$. Hence the measuring process is treated like a scattering process such that the operator U is identified with the respective scattering operator. [3].

ii) There are more general kinds of measurements, which can be described in a similar way. [3]. However in the present paper we consider only unitary measurements of ordinary discrete observables which are repeatable. The reason is, that the considerations of §4 and §5 can only be applied to this restricted class of measurements.

By means of the state $\psi'(\mathcal{S}+\mathcal{M})$ we can give now a more formal expression for the probability reproducibility condition (PRC) mentioned above. The requirement, that the probability $p_\varphi(a_i)$ which is inherent in the state φ is reproduced by the probability of the pointer value Z_i after the interaction H_{int} has been turned off, reads

$$p_\varphi(a_i) = p_{\psi'}(Z_i)$$

$$\text{with} \quad p_{\psi'}(Z_i) = \operatorname{tr}\{P[\psi'(\mathcal{S}+\mathcal{M})] I_{\mathcal{S}} \otimes P[\Phi_i]\}\,.$$

§3 THE PROBLEM OF OBJECTIFICATION

In the description of the measuring process, which was given in §2, there are still some open problems. In step (III) of the measuring process, i. e. in the time interval $t_1 \leq t < \infty$, the two subsystems are dynamically independent and can thus be described by the reduced states W'_S and $W'_{\mathcal{M}}$. These mixed states can be obtained from the state $\psi'(S + \mathcal{M})$ of the compound system $S + \mathcal{M}$ by partial trace operations. In addition we have assumed, that the mixed state $W'_{\mathcal{M}} = \sum_i p_\varphi(a_i)P[\Phi_i]$ of the apparatus $\mathcal{M}$ describes the Gemenge $\Gamma(W'_{\mathcal{M}}) = \{p_\varphi(a_i), \Phi_i\}_i$, i. e. an ensemble $\{p_\varphi(a_i), \Phi_i\}_i$ of pointer states Φ_i and the respective probabilities $p_\varphi(a_i)$. This assumption means, that the system $\mathcal{M}$ with the state $W'_{\mathcal{M}}$ is actually in one of the states Φ_i, but the observer is only partly informed about the state, i. e. he knows only the probability $p_\varphi(a_i)$ of Φ_i.[2] Since in the Gemenge $\Gamma(W'_{\mathcal{M}})$ the pointer observable is *objective*, the transition from the mixed state $W'_{\mathcal{M}}$ to the Gemenge $\Gamma(W'_{\mathcal{M}})$ will be called objectification or more specified "*pointer objectification*". If in addition also the state W'_S of the object system S may be interpreted by the Gemenge $\Gamma(W'_S) = \{p_\varphi(a_i), \varphi^{(a_i)}\}$ then also the system S will be objectively in one of the states $\varphi^{(a_i)}$. The transition from the state W'_S to the Gemenge $\Gamma(W'_S)$ will be called "*system objectification*". It is obvious that the probability reproducibility condition mentioned above only requires the weaker pointer objectification. System objectification means that the measuring process also transforms the initial state φ of the object system into an eigenstate $\varphi^{(a_i)}$ of the measured observable A.

Pointer objectification and system objectification are by no means trivial transitions. In particular, it is not a matter of convention, whether the system S is described by the state W'_S or by the Gemenge $\Gamma(W'_S)$. It is well known, that a mixed state $W_S = \sum w_i P(w_i)$ can be decomposed in many different ways into orthogonal or nonorthogonal states $\chi_i \in \mathcal{H}_S$. [1]. This nonunique decomposition has the consequence, that a mixed state W_S in general corresponds to many different Gemenge $\Gamma^\lambda(W_S)$ $(\lambda = 1, 2, 3, \ldots)$ which are indistinguishable from a statistical point of view. Hence a mixed state W'_S, say, does not determine the states φ^{a_i}, the mixture of which provides the state W'_S. Nevertheless this ambiguity of the Gemenge decomposition of a mixed state W_S, does not exclude the possibility, that the system S with state W_S is actually in one of the many Gemenge $\Gamma^\lambda(W_S) = \{w_i^\lambda, P[\chi_i^\lambda]\}$ and hence in one of the possible states $\chi_i^\lambda \in \mathcal{H}_S$. One could imagine, that the

[2] The subjective ignorance of the observer about the state Φ_i of the apparatus can be expressed according to Shannons formula by the deficiency of information [3].

$$H(W'_{\mathcal{M}}, Z) = -\sum_i p_\varphi(a_i) \ln p_\varphi(a_i)$$

observer knows only the mixed state W_S and that he is not informed about the pure state χ_i^λ. This explanation will be called here the "ignorance interpretation of mixed states". In the following we will investigate the question, whether this ignorance interpretation can be applied to the states W_S' and $W_{\mathcal{M}}'$, which are of particular importance for the measuring process.

1) If the system S is in the pure state $\varphi \in \mathcal{H}_S$ with the decomposition $\varphi = \sum_i (\varphi_i, \varphi)\varphi_i$, where the states φ_i are eigenstates of the discrete nondegenerate observable $A = \sum a_i P[\varphi_i]$, $[A, P[\varphi]]_- \neq 0$, then it is not possible to assume that one of the values a_i pertains to the system. Indeed if S posesses one of the values a_i of A, then S would be in one of the states φ_i with the probability $p_i = |(\varphi_i, \varphi)|^2$. This means, that the state of S would be the mixed state

$$W = \sum |(\varphi_i, \varphi)|^2 P[\varphi_i]$$

and not the pure state $P[\varphi] = \sum_{i,k} (\varphi_k, \varphi)(\varphi, \varphi_i)|\varphi_k\rangle\langle\varphi_i|$.

The states W and $P[\varphi]$ can be distinguished experimentally by the expectation values of any other observable B, which does not commute with A. Indeed the two expectation values

$$\mathrm{tr}\{BW\} = \sum_i |(\varphi_i, \varphi)|^2 \langle\varphi_i B\varphi_i\rangle$$

$$\mathrm{tr}\{BP[\varphi]\} = \sum_{i,k} (\varphi_i, \varphi)(\varphi, \varphi_k)\langle\varphi_k B\varphi_i\rangle$$

differ by the interference term $\sum_{i\neq k} (\varphi_i, \varphi)(\varphi, \varphi_k)\langle\varphi_k B\varphi_i\rangle$ and thus the observable A cannot be objectified in $S(\varphi)$.[3]

2) Next we investigate the question, whether the observable A can be objectified with respect to the mixed state W_S' of S which is generated in step (III) of the measuring process. Here we have to remind ourselves that the state W_S' has been prepared by separating the systems S and $\mathcal{M}$. Starting from the state $\psi'(S+\mathcal{M}) = \sum_i c_i \varphi^{(a_i)}(S) \otimes \Phi_i(\mathcal{M})$ the correlated reduced states $W_S' = \sum_i |c_i|^2 P[\varphi^{(a_i)}]$ and $W_{\mathcal{M}}' = \sum_i |c_i|^2 P[\Phi_i]$ have been obtained by partial trace operations. Let us apply now the ignorance interpretation to the system S in the state W_S'. The system S would then actually be in one of the states φ^{a_i}. On account of the correlation the apparatus $\mathcal{M}$ would then be in the state Φ_i and consequently the compound system in the state $\psi_i' = \varphi^{(a_i)} \otimes \Phi_i$. If $S + \mathcal{M}$ would be in one of the states ψ_i' with probability $|c_i|^2$, then it would be in the mixed state $W' = \sum |c_i|^2 P[\psi_i']$. However the compound system was assumed to be in the pure state $\psi'(S + \mathcal{M})$.

[3]A similar argument holds for degenerate observables A.

Hence the above mentioned argument which excludes the ignorance interpretation of pure states can be applied to $\psi'(\mathcal{S}+\mathcal{M})$ and we conclude that the ignorance interpretation must also be forbidden for the state $W'_{\mathcal{S}}$.

3) It is obvious, that the same argument also applies to the mixed state $W'_{\mathcal{M}}$ of the apparatus. In fact if $\mathcal{M}$ would be in one of the states Φ_i with probability $|c_i|^2$, then the object system $\mathcal{S}$ would be in the state φ^{a_i} and the compound system $\mathcal{S}+\mathcal{M}$ in the state $\Psi'_i = \varphi^{a_i} \otimes \Phi_i$. However if $\mathcal{S}+\mathcal{M}$ would be in one of the states Ψ'_i with probability $|c_i|^2$, then it would be in the mixed state W' and not – as assumed – in the pure state $\Psi'(\mathcal{S}+\mathcal{M})$. On account of the argument (1) mentioned above this cannot be possible. Hence the ignorance interpretation must also be refuted for the mixed state $W'_{\mathcal{M}}$ of the apparatus.

Since the ignorance interpretation cannot be applied to the mixed states $W'_{\mathcal{S}}$ and $W'_{\mathcal{M}}$ we find, that the objectification is not yet contained in our description of the measuring process, neither the "system objectification" nor the weaker "pointer objectification". The situation becomes even more critical on account of the following argument: In order to fulfill the requirement of the pointer objectification for any arbitrary preparation φ of $\mathcal{S}$, the pointer observable $Z_{\mathcal{M}} = \sum Z_i P[\Phi_i]$ must be classical, i. e. it must commute with any other observable of the apparatus $\mathcal{M}$. [3] If the pointer observable $Z_{\mathcal{M}}$ is classical, then $I_{\mathcal{S}} \otimes Z_{\mathcal{M}}$ commutes with any other observable of $\mathcal{H} = \mathcal{H}_{\mathcal{S}} \otimes \mathcal{H}_{\mathcal{M}}$, in particular with the Hamiltonian $H(\mathcal{S}+\mathcal{M})$ of the compound system $\mathcal{S}+\mathcal{M}$ and hence with the unitary operator $U = \exp(\frac{i}{\hbar}Ht)$. From the relation $[I_{\mathcal{S}} \otimes Z_{\mathcal{M}}, U]_- = 0$ it follows for the expectation value of the pointer observable the relation

$$\langle \psi U^+, I_{\mathcal{S}} \otimes Z_{\mathcal{M}} U\psi \rangle = \langle \psi, I_{\mathcal{S}} \otimes Z_{\mathcal{M}} \psi \rangle = \langle \varphi, I_{\mathcal{S}} \varphi \rangle \langle \Phi Z_{\mathcal{M}} \Phi \rangle = \langle \Phi Z_{\mathcal{M}} \Phi \rangle$$

which means that this expectation value does not depend at all on the initial preparation φ of the object system. [3]. Since this result plainly contradicts the probability reproducibility condition, one finds that the assumption of a classical pointer observable is not compatible with the idea of a unitary dynamics of the measuring process.

Concluding this discussion of the objectification problem we arrive at the following result: The quantum theory of the measuring process, as it was described in §2, provides neither the system objectification nor the weaker pointer objectification. The reason is that the correlated reduced states $W'_{\mathcal{S}}$ and $W'_{\mathcal{M}}$ do not admit an ignorance interpretation. – Moreover, the theory of the measuring process presented here, is not only incomplete such that it cannot explain the objectification. The requirement of a classical pointer observable cannot be fulfilled by any Hamiltonian $H(\mathcal{S}+\mathcal{M})$ and the corresponding unitary dynamics. – Hence we should be prepared to much more radical consequences in the theory and the interpretation of the measuring process.

§4 THE PROBABILITY INTERPRETATION

The second problem which should be solved by the theory of the measuring process is the justification of the probability interpretation. If on a system $\mathcal{S}$ with preparation φ the nonobjective observable $A = \sum a_i P(a_i)$ is measured, then the probability to obtain the value a_i as a result is given by $p_\varphi(a_i) = \mathrm{tr}\{P[\varphi]P(a_i)\}$. Within the theory of the measuring process this probability interpretation was expressed by the probability reproducibility condition which postulates, that the inherent probabilities $p_\varphi(a_i)$ of the initial object state φ are reproduced in the statistics of the pointer observable of the apparatus. Within the framework of the measuring process discussed here (§2) we have a strong correlation between the pointer values Z_i and the values a_i of the object system. Hence if the probabilities $p_\varphi(a_i)$ are reproduced in the statistics of the pointer values Z_i, then they are also reproduced in the statistics of the values a_i of the object system.

If $\varphi \in \mathcal{H}_S$ is the preparation of the object system $\mathcal{S}$ and if A is a discrete observable $A = \sum a_i P(a_i)$ with $\{a_i\}_i = X^A$, $A\varphi_{ij} = a_i\varphi_{ij}$, $P(a_i) = \sum_j^{n_i} P[\varphi_{ij}]$ then the decomposition of φ with respect to the orthonormal states φ_{ij} reads $\varphi = \sum_{i,j}(\varphi_{ij},\varphi)\varphi_{ij}$. The real positive numbers

$$p_\varphi(a_i) = \mathrm{tr}\{P[\varphi]P(a_i)\} = \sum_j |(\varphi_{ij},\varphi)|^2$$

are then probabilities in the formal sense, i. e. the mapping

$$p_\varphi^A : B(X^A) \to [0,1], \quad a_i \to p_\varphi(a_i) = \mathrm{tr}\{P[\varphi]P(a_i)\}$$

is a probability measure which satisfies the Kolmogoroff axioms. ($B(X^A)$ is the Borel algebra of X^A) - According to the probability *interpretation* the number $p_\varphi(a_i)$ is the probability that a measurement of the observable A performed on the system $\mathcal{S}$ in the state φ leads to the result a_i. This means, that if one would perform a long series of N measurements of A on equally prepared systems, the relative frequency $f^N(\varphi,a_i)$ of outcomes a_i would approach in some sense the probability $p_\varphi(a_i)$ for $N \to \infty$. It is obvious that this statement has an immediate empirical relevance.

On the other hand the probability interpretation is not an independent assumption, which must be added to the theory of the measuring process in order to relate the theory with experimental results. Since the theory of the quantum mechanical measuring process is understood as a complete and universal theory, it can equally be applied to a single object system $\mathcal{S}$ or to a many body system $\mathcal{S}^{(N)}$ which consists of N equally prepared subsystems $\mathcal{S}_i$. Hence the question whether in a large system $\mathcal{S}^{(N)}$ ($N \gg 1$) the relative frequency $f^N(\varphi,a_i)$ of outcomes a_i

almost equals the value $p_\varphi(a_i)$ should be answered by the theory itself – and not by an additional hypothetical "probability interpretation".

In order to investigate this problem we consider N independent systems $S_1, S_2, \ldots, S_N$ with Hilbert spaces $\mathcal{H}_1, \mathcal{H}_2, \ldots, \mathcal{H}_N$ respectively and the discrete, nondegenerate observable $A = \sum a_i P[\varphi_i]$ with $\{a_i\}_i =: X^A$. If the systems S_i are equally prepared in states $\varphi^{(i)} \in \mathcal{H}_i$ the compound system

$$S^{(N)} = S_1 + S_2 + \ldots + S_N$$

can be described by the state

$$(\varphi)^N = \varphi^{(1)} \otimes \varphi^{(2)} \otimes \ldots \otimes \varphi^{(N)}$$

in the tensor product Hilbert space $\mathcal{H}^{(N)} = \bigotimes_{i=1}^N \mathcal{H}_i$. Let us consider now a set of N systems S_i with states $\varphi_{l_i}^{(i)}(S_i)$ where $\varphi_{l_i}^{(i)} \in \mathcal{H}_i$ are eigenstates of A with $A\varphi_{l_i}^{(i)} = a_{l_i}\varphi_{l_i}^{(i)}$, and $l = \{l_1, l_2, \ldots, l_N\}$ is an arbitrary sequence of N index numbers $l_i : a_{l_i} \in X^A$. For the state of the compound system $S^{(N)} = \sum_{i=1}^N S_i$ we write

$$(\varphi)_l^N = \varphi_{l_1}^{(1)} \otimes \varphi_{l_2}^{(2)} \otimes \ldots \otimes \varphi_{l_N}^{(N)}.$$

Every system S_i possesses a well defined eigenvalue a_{l_i} of the observable A. Hence the relative frequency $f^N(k,l)$ of the particular eigenvalue a_k in the state $(\varphi)_l^N$ with the index sequence l of N elements is given by

$$f^N(k,l) = \frac{1}{N} \sum_{i=1}^N \delta_{l_i,k}$$

Since the states $\{\varphi_i^{(n)}\}_i$ are orthonormal and complete in $\mathcal{H}_n$, we can use the states $(\varphi)_l^N$ as a basis in $\mathcal{H}^{(N)}$. The "relative frequency operator" f_k^N in the Hilbert space $\mathcal{H}^{(N)}$ can then be defined by

$$f_k^N = \sum_l f^N(k,l) P[(\varphi)_l^N]$$

where the sum runs over all sequences $l = \{l_1, l_2, \ldots, l_N\}$ with $l_i : a_{l_i} \in \underline{X}^A$. From this definition one obtains the eigenvalue equation

$$f_k^N (\varphi)_l^N = f^N(k,l)(\varphi)_l^N$$

which shows that the relative frequency of the value $a_k \in X^A$ is an objective property in the state $(\varphi)_l^N$ and given by the value $f^N(k,l)$.

For the following investigations it is more convenient to write the eigenvalue equation in terms of projection operators

$$f_k^N P[(\varphi)_l^N] = f^N(k,l) P[(\varphi)_l^N]$$

and to use the equivalent formulation

$$\mathrm{tr}\left\{P[(\varphi)_l^N]\left(f_k^N - f^N(k,l)\right)^2\right\} = 0.$$

In the Hilbert space $\mathcal{H}^{(\infty)} = \bigotimes_{i=1}^{\infty}\mathcal{H}_i$ we extend the definition of the relative frequency operator by $\hat{f}_k^N = f_k^N \otimes \mathbf{1}_{N+1,\infty}$, where $\mathbf{1}_{N+1,\infty}$ is the unit operator in the subspace $\mathcal{H}_{N+1}^{(\infty)} = \bigotimes_{i=N+1}^{\infty}\mathcal{H}_i$. Clearly the state $(\varphi)^\infty \in \mathcal{H}^{(\infty)}$ is not an eigenstate of the operator $\hat{f}_k^N$. However in the limit $N \to \infty$ we have the

THEOREM 1 (HARTLE) [6].

$$\lim_{N\to\infty} \mathrm{tr}\left\{P[(\varphi)^\infty]\left(\hat{f}_k^N - p_\varphi(a_k)\right)^2\right\} = 0$$

The theorem means that in the limit $N \to \infty$ the state $(\varphi)^\infty$ is an eigenstate of the relative frequency operator with the eigenvalue $p_\varphi(a_k)$.[4] There is, however, no obvious interpretation of this result. At first glance theorem 1 seems to state, that for an infinite number of equally prepared systems S_i the relative frequency of the value $a_k \in X^A$ approaches the probability $p_\varphi(a_k)$. This would mean that at least in the limit the fraction $p_\varphi(a_k)$ of the system S_i with preparation $\varphi^{(i)}$ would posess the value a_k. However, none of the systems $S_i(\varphi^{(i)})$ possesses an A-value $a_i \in X^A$ in an objective sense. According to our assumptions, the observable A is nonobjective with respect to the state φ. Hence the first nonobjectifiability-argument of section (3) can be applied here.

On the other hand the probability interpretation which should be justified here states, that if one would perform a large number of A-measurements on equally

[4]The proof of this theorem can be found in the literature [6] and will thus not be repeated here. We add, however, some comments. It has been pointed out by Ochs [13] that the strong and somewhat unrealistic premises of Hartle's proof (pure state φ, equally prepared and independent systems) can be considerably relaxed and replaced by more realistic assumptions. Furthermore Ochs emphasized, that theorem 1 is essentially a "law of large numbers" which shows, that quantum probabilities fulfill this important requirement. However Ochs' generalized proof makes use of statistical concepts (expectation value, etc.) and thus shows merely the consistency of the probability interpretation with quantum theory but not its deducibility from the theory.

prepared systems, then the relative frequency of outcomes a_i would approach the probability $p_\varphi(a_i)$. Hence probability is related here to a situation *after* the measuring process, i. e. to the mixed states W'_S and $W'_{\mathcal{M}}$ of the system S and the apparatus $\mathcal{M}$, respectively. In the present paper we consider ideal strong value correlation measurements, i. e. the state of S after the interaction (step II) reads $W'_S = \sum_i p_\varphi(a_i)P[\varphi^{(a_i)}]$. Consequently one should expect a result similar to theorem 1, in which the preparation φ is replaced by the post-measurement state W'_S of the system. In fact, if one defines the state $(W'_S)^N = \bigotimes_{i=1}^N W'_{S_i}$ of N independent systems S_i in equal states W'_{S_i} the following theorem holds:

THEOREM 2.

$$\lim_{N\to\infty} \operatorname{tr}\left\{(W'_S)^N \left(f_k^N - p_\varphi(a_k)\right)^2\right\} = 0.$$

proof :

Using the spectral decomposition of f_k^N one obtains for the trace

$$T^N := \operatorname{tr}\{(W'_S)^N(f_k^N - p_\varphi(a_k))^2\} \text{ in the basis } (\varphi)_l^N$$

$$\begin{aligned} T^N &= \sum_l ((\varphi)_l^N (W'_S)^N (f_k^N - p_\varphi(a_k))^2 (\varphi)_l^N) \\ &= \sum_l ((\varphi)_l^N (W'_S)^N (\varphi)_l^N)\left(f^N(k,l) - p_\varphi(a_k)\right)^2 \end{aligned}$$

LEMMA 2.1:. *If the post-measurement state W'_S reads*

$$W'_S = \sum_k P(a_k)P[\varphi]P(a_k) = \sum_k p_\varphi(a_k)P[\varphi^{(a_k)}]$$

then it holds

$$((\varphi)_l^N (W'_S)^N (\varphi)_l^N) = \prod_{i=1}^N p_\varphi^{(i)}(a_{l_i}) =: p_{\{l\}}$$

with $p_\varphi^{(i)}(a_{l_i}) = \left(\varphi_{l_i}^{(i)} P[\varphi^{(i)}]\varphi_{l_i}^{(i)}\right)$.

By means of this Lemma 2.1 one obtains $T^N = \sum_l \left(f^N(k,l) - p_\varphi(a_k)\right) p_{\{l\}}$. Using some properties of the function $f^N(k,l)$ it follows

LEMMA 2.2. (DeWitt [15]): $\sum_l f^N(k,l)p_{\{l\}} = p_\varphi(a_k)$

and furthermore by means of this result

LEMMA 2.3. (DeWitt [15]):

$$\sum_{l} \left(f^N(k,l) - p_\varphi(a_k)\right)^2 p_{\{l\}} = \frac{1}{N}\, p_\varphi(a_k)(1 - p_\varphi(a_k)).$$

Hence one obtains

$$\lim_{N\to\infty} T^N = \lim_{N\to\infty} \frac{1}{N}\, p_\varphi(a_k)(1 - p_\varphi(a_k)) = 0$$

completing the proof of theorem 2.

The proof is based on the assumption that we are dealing here with an ideal strong value correlation measurement of the ordinary discrete observable $A = \sum a_i P(a_i)$, such that the state of S after the interaction reads $W'_S = \sum_i p_\varphi(a_i)\, P[\varphi^{(a_i)}]$. We leave the question open, whether the proof of theorem 2 can also be performed for more general classes of measurements, e. g. non-ideal strong value correlation measurements [3]. However the strong correlation between the values a_i of the object system and the values Z_i of the pointer observable of the apparatus $\mathcal{M}$ seems to be inevitable for the present considerations.

Theorem 2 states, that the relative frequency of the value a_k in an ensemble of N systems $S^{(i)}$ with equal preparation $W'_{S_i} = \sum P(a_k)P[\varphi^{(i)}]P(a_k)$ approaches the probability $p_\varphi(a_k)$ in the limit $N \to \infty$. Provided the mixture can be understood as the description of the Gemenge $\Gamma(W'_{S_i}) = \{p_\varphi(a_i), \varphi^{(a_i)}\}$, then theorem 2 is in fact an important result. It then shows that the number $p_\varphi(a_k) = \mathrm{tr}\,\{P[\varphi]P(a_k)\}$ can be interpreted as the probability of the value a_k in the post-measurement state W'_{S_i} of $S^{(i)}$ and - via the strong value correlation - also as the probability of the corresponding value Z_k of the pointer observable. Hence the interpretation of the number $p_\varphi(a_i)$ as the probability of the pointer value Z_i after the A-measurement follows from the theory itself - and is not an independent hypothesis which must be added as an interpretation.[5]

This interpretation of theorem 2 is essentially based on the assumption, that the post-measurement mixed state W'_{S_i} of the system $S^{(i)}$ can be considered as the description of the Gemenge $\Gamma(W'_{S_i}) = \{p_\varphi(a_i), \varphi^{(a_i)}\}$. From this assumption it follows that also the corresponding pointer value Z_i can be described by a Gemenge with the same weights. However, it follows from the discussion in §3 that neither

---

[5] For more general kinds of measurements which do not allow to prove theorem 2 for the relative frequencies of the values a_k, it is still possible to show that at least the relative frequencies of the pointer values Z_i approach the probabilities $p_\varphi(a_i)$ in the limit of large N. [3]

the system objectification nor the weaker pointer objectification can be achieved in the present case. The mixed state W'_{S_i} is not the preparation of an independent system $S^{(i)}$ but the reduced state of the compound state $\psi'(S^{(i)}+\mathcal{M})$ and strongly correlated to the corresponding reduced state $W'_{\mathcal{M}}$ of the apparatus. Hence the second nonobjectifiability argument of §3 can be applied which claims that neither the state W'_{S_i} nor the state $W'_{\mathcal{M}}$ admits an ignorance interpretation. There is no hope to relax this strong verdict by considering more general kinds of measurements, since the strong value correlations which in the present case exclude the ignorance interpretation are inevitable for the validity of theorem 2. Hence we find that theorem 2 can only be proved under conditions which just exclude its interpretation as a justification of the probability interpretation.

§5 THE MANY WORLDS INTERPRETATION

Unitary strong value correlation measurements of an ordinary discrete observable A do not admit the objectification of A, neither with respect to the preparation $P[\varphi]$ nor with respect to the post measurement mixed states W'_S and $W'_{\mathcal{M}}$. On the other hand the interpretation of the coefficients $p_\varphi(a_i)$ as probabilities in the sense of relative frequencies can be justified for the states $P[\varphi]$ (theorem 1) and for the states W'_S and $W'_{\mathcal{M}}$ (theorem 2) just under the condition of strong value correlation measurements. We will try to investigate the consequences of this controversial situation.

It has been shown in §3, that within the framework of unitary measurements, i. e. within the von Neumann program for the description of the quantum mechanical measuring process the objectification does not take place. Instead the initial preparation φ and the observable A provide a simultaneous description of the complete variety of infinitely many different situations of the system S such that every eigenvalue a_i corresponds to some pointer value Z_i. The final state of the measuring process is completely described by the compound state $\psi'(S+\mathcal{M}) = \sum_i c_i\varphi^{(a_i)}\otimes\Phi(Z_i)$ and the state correlations contained in it. For any definite observer, which is represented by an apparatus $\mathcal{M}$, the probability for a certain value a_i is given by the coefficient $p_\varphi(a_i) = |c_i|^2$ and manifests itself as relative frequency $f^N(i,l)$ in (almost) every memory sequence $l = \{l_1, l_2, \ldots, l_N\}$ of the apparatus. Theorem 1 can immediately be applied to this situation without any additional interpretation.

These remarks can be elaborated somewhat more in detail. In the measuring process the compound system $S^{(N)}$ is coupled to the apparatus $\mathcal{M}$. According to the assumption of ideal strong value correlation measurements the initial state $\psi(S^{(N)}+\mathcal{M}) = (\varphi)^N\otimes\Phi$ of $S^{(N)}+\mathcal{M}$ is transformed into the state $\psi'(S^{(N)}+\mathcal{M})$

with the biorthogonal decomposition

$$\psi'(\mathcal{S}^{(N)} + \mathcal{M}) = \sum_l c_l \, (\varphi)_l^N \otimes \Phi'\{l\}$$

where the sum runs over all series $l = \{l_1, l_2, \ldots, l_N\}$ of N index numbers l_i : $a_{l_i} \in X^A$. The coefficients c_l are given by $c_l = \prod_{i=1}^N c_{l_i}$ with $c_{l_i} = (\varphi_{l_i}^{(i)}, \varphi^{(i)})$ and are related to the probabilities $p_\varphi(a_{l_i}) = |c_{l_i}|^2$ and $p_\varphi(\{l\}) = |c_l|^2 = \prod_{i=1}^N p_\varphi(a_{l_i})$. The state $\Phi'\{l\} = \Phi'\{l_1, l_2, \ldots, l_N\}$ of the apparatus $\mathcal{M}$ registers the index sequence $\{l_1, l_2, \ldots, l_N\}$ of the corresponding pointer values $\{Z_{l_1}, Z_{l_2}, \ldots, Z_{l_N}\}$. Each of these *memory sequences* $\{l_1, l_2, \ldots; l_N\} = l$ possesses a relative frequency $f^N(k, l)$ of the index k and of the pointer value Z_k, respectively. In order to measure the degree to which a given sequence $l = \{l_1, l_2, \ldots, l_N\}$ deviates from a random sequence with weights $p_\varphi(a_{l_i})$ we introduce the function

$$\delta(l_1, l_2, \ldots, l_N) := \sum_k \left[f^N(k, l) - p_\varphi(a_k)\right]^2$$

and call a sequence l *first random* if $\delta(l) < \varepsilon$, where ε is an arbitrarily small positive number, and *nonfirst-random* otherwise. If in the state $\psi'(\mathcal{S}^{(N)} + \mathcal{M})$ all those elements are removed from the superposition, for which the memory sequence l of the apparatus are *nonfirst-random* one obtains the state $\psi'_\varepsilon(\mathcal{S}^{(N)} + \mathcal{M})$. Using some properties of the relative frequency function $f^N(k, l)$ one finds that in the limit $N \to \infty$ the states $\psi'(\mathcal{S}^{(N)} + \mathcal{M})$ and $\psi'_\varepsilon(\mathcal{S}^{(N)} + \mathcal{M})$ coincide and nonrandom memory sequences in the decomposition of ψ' are of the measure zero in the Hilbert space. [15].

For the description of the measuring process the consciousness of the observer is not needed. Automata for the registration of memory sequences are completely sufficient. Several automata can be used for the formation of a von Neumann chain without thereby changing the registered results in the respective memory sequences. The results of the automata persist throughout the whole chain. Moreover two automata are also allowed to communicate with each other since this exchange of information will never lead to a paradoxical situation like Wigner's friend paradox.

Quantum theory in Hilbert space in its present form – does not describe the reality which we usually have in mind but some reality which is composed of many distinct worlds. We are only in one of these worlds. On this reason we are unable to recognize the full determinism of the totality of many worlds which are described by $\psi'(\mathcal{S} + \mathcal{M})$, but only one part of it which is governed by a probability law, – the justification of which is given by theorem 1.

It follows from these arguments, that quantum mechanics can be applied to an individual system. Every automaton, i. e. measuring apparatus cum memory sequence, describes the system correctly according to the probability laws. On this reason quantum theory can be applied even to the universe, its creation and its development. The state of the universe corresponds to a decision tree with an enormous number of branches. Every measuring process splits the world again into many alternative components, but the whole description is completely consistent.

The many worlds interpretation is the correct image of the quantum measurement theory, which we really have. It does not presuppose the pointer-objectification and the system objectification, which cannot be justified by unitary measurements. On the other hand it makes use of the theorem 1 as a legitimation of the probability interpretation, since this theorem need not to be understood here in the sense of an ignorance interpretation. However the price for this consistent interpretation of quantum measurement theory is a very strange ontology: Quantum mechanics does not describe any longer the *one* world, in which we are living but at the same time the totality of all possible worlds.

§6 CONCLUSION

According to the investigation of the preceding sections the rather bizarre "many worlds interpretation" seems to be an inescapable consequence of the quantum theory of the measuring process which was first conceived by J. von Neumann. *However it is not claimed here, that this many worlds interpretation is the correct description of the quantum physical reality.* Instead it is considered rather as a very significant indication of the fact, that the quantum measurement theory mentioned is not yet completely correct. There are several proposals to extend or to modify quantum mechanics such that the difficulties mentioned would disappear. They are represented in the following decision tree

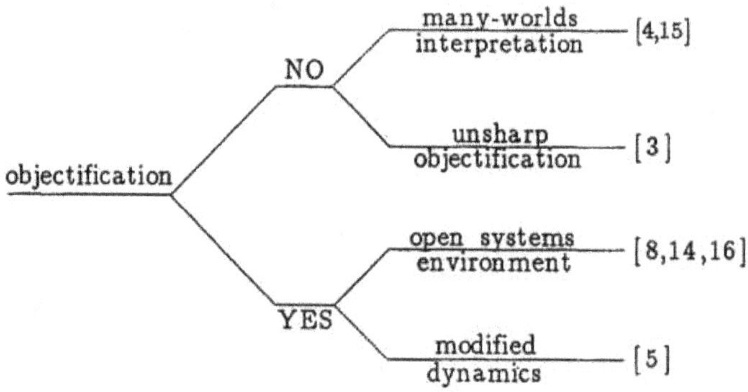

Within the framework of the well known quantum mechanics in Hilbert space the only way out seems to be the relinquishment of sharp measuring results. The results, which can be obtained, would then always be unsharp and the objectification of the pointer observable would take place only approximately – however with an overwhelming probability. It is a consequence of these assumptions that also the preparation of a quantum system could be performed only in an approximate way. It is obvious that this solution of the measuring problem leads to serious difficulties with our traditional concept of reality. The stability of past events and the reliability of our knowledge of facts becomes highly questionable. (cf. the lecture of P. Busch in this volume and [3])

Furthermore there are at least two proposals to modify quantum mechanics in order to obtain a consistent and complete theory of the measuring process:

(i) Quantum mechanics should be generalized such that in addition to closed systems also the environment is taken into account. It is claimed by some authors that a quantum theory of open systems would provide new environment induced superselection rules which could explain the objectification in the measuring process. [8, 14, 16].

(ii) A more radical solution is the modification of the Schrödinger equation by incorporating additional terms which correspond to stochastic corrections. It is argued that this modified dynamics is able to explain the objectification in macroscopic systems like the measuring apparatus. [5]

REFERENCES

1 E. Beltrametti and G. Casinelli (1981), The Logic of Quantum Mechanics, Addison-Wesley, Reading, Mass.

2 E. Beltrametti, G. Casinelli and P. Lahti (1990), Unitary Measurements of Discrete Quantities in Quantum Mechanics, Journal of Mathematical Physics 31, 91-98

3 P. Busch, P. Lahti and P. Mittelstaedt (1990), The Problem of Measurement in Quantum Mechanics (forthcoming)

4 H. Everett (1973), The Theory of the Universal Wave Function, in: The Many-Worlds Interpretation of Quantum Mechanics, B. S. DeWitt and N. Graham, eds., Princeton University Press, Princeton

5 G. C. Ghirardi, A. Rimini and T. Weber (1986), Unified Dynamics for Microscopic and Macroscopic Systems, Phys. Rev. D 34, 470-491

6 J. B. Hartle (1968), Quantum Mechanics of Individual Systems, Am. J. of Phys. 36, 704-712

7 J. M. Jauch (1968), Foundations of Quantum Mechanics, Addison-Wesley Publishing Company, Reading, Mass.

8 E. Joos, H. D. Zeh (1985), The Emergence of Classical Properties Through Interactions with the Environment Z. Physik B59 223-243

9 S. Kochen (1985), A new Interpretation of Quantum Mechanics, in: Symposium on the Foundations of Modern Physics 1985, P. Lahti and P. Mittelstaedt, eds., World Scientific, Singapore, pp. 151-169

10 P. J. Lahti (1990), Quantum Theory of Measurement and the Polar Decomposition of Interaction, Int. J. of Theoretical Physics 29, 339-350

11 P. Mittelstaedt (1989), Philosophische Probleme der Modernen Physik, Bibliographisches Institut, Mannheim

12 J. von Neumann (1932), Mathematische Grundlagen der Quantenmechanik, Springer-Verlag, Berlin

13 W. Ochs (1977), On the Strong Law of Large Numbers in Quantum Probability Theory, J. of Phil. Logic 6, 473-480

14 H. Primas, Contextual Quantum Objects and their Ontic Interpretation, in:

Symposium on the Foundation of Modern Physics 1987, P. Lahti and P. Mittelstaedt, eds., Singapore, p. 251-277

15 B. S. DeWitt (1971), The Many-Universes Interpretation of Quantum Mechanics, Academic Press Inc., New York

16 W. H. Zurek (1982), Environment-induced Superselection Rules, Phys. Rev. D26, 1862-1880

EFFECTIVE INCOHERENCE IN ORTHODOX AND UNORTHODOX THEORIES OF MEASUREMENT

ORESTE NICROSINI
Istituto Nazionale di Fisica Nucleare – Sezione di Pavia
Via Bassi 6, 27100 Pavia, Italy

and

ALBERTO RIMINI
Dipartimento di Fisica Nucleare e Teorica, Università di Pavia
Via Bassi 6, 27100 Pavia, Italy

ABSTRACT

The family of theories of measurement orthodox in the sense that they assume both that the Schrödinger equation is absolutely valid and that the quantum probability law is the only link between the state vector and the experimental results is considered. Such theories are based on the proof that some reasonable assumption guarantees that the state vectors corresponding to the various outcomes of a measurement are effectively incoherent. It is shown that the consideration of the fluctuations of quantum distributions occurring for actual finite ensembles of quantum systems makes effective incoherence insufficient to reconcile, in the orthodox framework, the description of the system object–plus–apparatus with the standard postulates. Examples of unorthodox theories are further considered, showing how the problem related to fluctuations of distributions disappears and discussing the role of effective incoherence.

1. The problem of quantum measurement

In the standard formulation of quantum mechanics, the dynamical evolution of the state vector is given, when the system is left undisturbed, by the Schrödinger equation

$$i\hbar\frac{d}{dt}|\psi\rangle = H|\psi\rangle. \tag{1.1}$$

Alternative postulates are given for the case in which a measurement is performed. First, the probability law states that, when the quantity V represented by the operator $\hat{V} = \sum_v vP_v$ is measured on the system, the probability of getting $V = v$ is given by

$$\Pr(V = v) = \|\phi_v\|^2, \qquad |\phi_v\rangle = P_v|\psi\rangle. \tag{1.2}$$

Furthermore the reduction postulate establishes that, as a consequence of the measurement, the state vector undergoes the sudden transition

$$|\psi\rangle \to |\psi_v\rangle = |\phi_v\rangle/\|\phi_v\|, \qquad |\phi_v\rangle = P_v|\psi\rangle. \tag{1.3}$$

The postulates about measurement deal with the measured system, say S, alone. The measuring apparatus A is supposed to be there but it remains outside the description. The theory of measurement consists in applying the quantum rules to the composite system $S + A$ and in deducing from such a description the usual postulates dealing with S alone (or at least in showing the consistency of the two descriptions).

We use, in describing measurement, the so called ideal scheme. This choice will be discussed in Sect. 6. Referring for simplicity to the case in which the measured variable is nondegenerate, the working of the apparatus requires that the Schrödinger evolution has the characteristic property

$$|\psi_m\rangle|A_0\rangle \quad \longrightarrow \quad |\psi_m\rangle|A_{m,s_m}\rangle, \tag{1.4}$$

where $|A_0\rangle$ is a suitable initial state vector and $|A_{m,s}\rangle$ are the eigenvectors of the pointer variable of A.

If the initial state of S is a superposition $|\psi\rangle = \sum_m a_m|\psi_m\rangle$, then, according to the probability law and the reduction postulate, one of the transitions

$$\begin{array}{ccc} & \xrightarrow{|a_1|^2} & |\psi_1\rangle|A_{1,s_1}\rangle \\ |\psi\rangle|A_0\rangle & \xrightarrow{|a_2|^2} & |\psi_2\rangle|A_{2,s_2}\rangle \\ & & \cdots\cdots\cdots\cdots\cdots\cdots \end{array} \tag{1.5}$$

takes place, with the probabilities indicated over the arrows. On the other hand, the Schrödinger equation applied to $S + A$ gives

$$|\psi\rangle|A_0\rangle \quad \longrightarrow \quad \sum_m a_m|\psi_m\rangle|A_{m,s_m}\rangle. \tag{1.6}$$

The problem of the theory of quantum measurement consists in the fact that the result 1.6, a *pure state* for $S+A$, seems to contradict both the result 1.5, a *mixture*, and common sense. The mixture 1.5 refers to an ensemble which we designate by $\tilde{\mathcal{E}}$ while the pure state 1.6 can refer to a single system or to an ensemble, say $\mathcal{E}$.

2. A family of theories of measurement

We shall now describe the general structure of a very numerous family of theories of measurement. It was initiated, to our knowledge, by the pioneer study of Daneri, Loinger and Prosperi[1] and has continued with a very long list of works. Recently, a sort of compendium of the principles of such an approach to the problem of measurement was presented in written form by Gottfried[2] at Erice in 1989. For reasons which will be clear shortly, we shall refer to this family as orthodox effective–incoherence (OEI) theories.

The logical structure of OEI theories rests on the proof that the Schrödinger equation plus the probability law plus a reasonable assumption are equivalent to the same elements plus the reduction postulate. We shall see in what sense they are equivalent. A preliminary statement is made implicitly or explicitly: that, due to the statistical character of quantum mechanics, "the theory must deal with ensembles, whether pure or not".[2]

Clearly, the first question to be answered by any theory of measurement is: is it possible to distinguish between the ensemble mixture $\tilde{\mathcal{E}}$ and the ensemble pure state $\mathcal{E}$ by further measurements on the system $S+A$? The answer is affirmative, if all Hermitian operators in the Hilbert space of $S+A$ are actually measurable; possibly negative, if not all Hermitian operators do correspond to measurable quantities.[3] Some OEI theories use as an essential ingredient the fact that $S+A$ unavoidably interacts with its environment E; in such a case $S+A$ must be replaced by $S+A+E$ in the above question and answer.

The main argument of OEI theories is based on the remark that a certain class of Hermitian operators in the Hilbert space of the system $S+A$ (or $S+A+E$) does not correspond in practice to measurable quantities. For any actually measurable quantity $\hat{W} = \sum_w wP_w$, one then finds that, in calculating probabilities for the pure state ensemble $\mathcal{E}$, the cross terms can be dropped, so that

$$\Pr_{\tilde{\mathcal{E}}}(W=w) = \Pr_{\mathcal{E}}(W=w). \tag{2.1}$$

The situation can be described by saying that the states in the superposition in $\mathcal{E}$ are *effectively incoherent.* The proof of Eq. 2.1 is based on the recognition of a certain kind of limitation on measurability (this is the reasonable assumption mentioned above) and on the application of the probability law.

According to OEI theories, because of the result 2.1, the ensemble $\mathcal{E}$ can be replaced by $\tilde{\mathcal{E}}$ for all purposes, so that quantum mechanics without reduction gives the same statistical predictions as quantum mechanics with reduction. "This demonstrates that quantum mechanics does describe the 'collapse of the wave function' to precisely the extent required".[2]

The above account is obviously incomplete and oversimplified, but it gives the essence of the argument. The various theories mainly differ from one another in the kind of limitation which is invoked. For example, in the work by Daneri, Loinger and Prosperi quoted above, it is assumed that the measuring apparatus, being macroscopic, is *classical* in the sense that its dynamical variables all commute. This is sufficient to produce effective incoherence at the time immediately after the completion of the measurement. However, since the measurement of commuting variables at different times is in general equivalent to the measurement of noncommuting variables, one must also show that effective incoherence persists in time. This is achieved in Ref. 1 by resorting to the irreversible character of the amplification which accompanies any quantum measurement. Gottfried[2] ascribes effective incoherence to a different origin. We do not enter into details because we are interested here in the general structure of OEI theories. *

Two criticisms are most frequently raised against OEI theories. The first is that the limitation on measurability, which in turn is related to a shifty distinction between microscopic and macroscopic objects, is of a practical nature so that it should not be used in matters of principle. The second criticism is that the contradiction is overcome at the level of ensemble and the particularity of the real world is not described. The first criticism is rejected on the basis of various types of arguments, and the second one because of the previous deliberate choice of dealing only with ensembles.

Two more specific objections can be raised. The first is that OEI theories do not justify the agreement between two successive measurements of the same variable of S. The OEI answer is that the state vector after the second measurement according to the Schrödinger equation is

$$\sum_m a_m|\psi_m\rangle|A^1_{m,s_m}\rangle|A^2_{m,s_m}\rangle \tag{2.2}$$

with obvious meaning of symbols. This state vector clearly exhibits the agreement between the two measurements because of the complete correlation of the index m in each term. One then counterargues that the state vector after the first measurment is

$$\sum_m a_m|\psi_m\rangle|A^1_{m,s_m}\rangle|A^2_0\rangle \tag{2.3}$$

and the probability law, which is the only link between a state vector and the results of a measurement, does not guarantee the agreement. But, it is answered by OEI theories, the application of the probability law is here meaningless because the agreement between two successive measurements concerns single systems. The second objection is that, if one considers a von Neumann chain going back up to the observer's brain, then the state vector 1.6 entails the consideration of superpositions of brain states corresponding to different knowledges. The objection is refused because it does not regard physics and moreover it again involves single systems rather than ensembles.

The final conclusion is: "So I must end with a plea for a crystal-clear explanation of what ails the orthodox theory, and why this affliction is so grave that it warrants rococo defilements of its austere majesty".[2]

We note that the word "orthodox" does not indicate here the original Copenhagen interpretation, according to which no theory of measurement has to be done. Instead, it indicates a set of principles including the Schrödinger equation, the probability law, the limitation on measurability (which all are essential to prove Eq. 2.1) and nothing else. On the basis of such principles, the description of the system $S + A$ is undertaken.

Since we are responsible, together with others, for one of the mentioned "rococo defilements", we will first try to satisfy Gottfried's demand, not searching for inexisting errors, but rather pointing out the insufficiency of the orthodox approach. We will accept both to limit measurability and to deal only with ensembles. We also agree not to call into question brain states. However, we claim the right to deal with actual i.e. finite ensembles. Secondly, we will discuss how our argument pointing out the insufficiency of the orthodox theories fails when applied to typical unorthodox theories and show that, however, the orthodox and unorthodox theories of measurement are more strictly related to one another than it is usually declared, one link among them being the role played by effective incoherence, which after all is a fact.

3. Fluctuations of quantum distributions

For a given ensemble, the distribution of values obtained measuring a certain variable is in general different from the corresponding ideal quantum distribution. This is because actual distributions fluctuate. The fluctuations we are talking about are not the usually considered quantum fluctuations. The latter are related to the deviations of the actual values from the mean value of the measured quantity and are usually calculated by means of the ideal quantum distribution. The former are instead deviations of the actual distribution as a whole from the ideal one. For an ensemle having a definite number of elements and a definite composition, the *distribution of distributions* of a certain dynamical variable is a definite mathematical entity which can be calculated from the theory.

Fig. 1. The ideal quantum distribution (dashed line) and an actual distribution (solid line) of a quantity: an imaginary example.

As it is well known, the correspondence between ensembles and statistical operators is many to one, so that, since the ideal quantum distributions can be calculated from the statistical operator, it happens that different ensembles have the same ideal quantum distributions for all dynamical variables. It has been noted by d'Espagnat[3] that, nevertheless, such ensembles have significantly different distributions of distributions. Of course, determining by measurements a distribution of distributions requires an infinite collection of identical ensembles. There are cases, however, in which measurements even on a single ensemble may be sufficient to draw significant conclusions.

The set, for all quantities, of the distributions of distributions is a significant feature of a given ensemble. d'Espagnat applies these considerations to a discussion of the quantum mechanical concept of state. His conclusion is that we may well decide that a statistical operator represents a state, but then we must remember that observationally different ensembles of physical systems can be in the same state.

We will discuss the consequences of taking into account fluctuations of quantum distributions for the comparison between the mixture and the pure state intervening in the problem of quantum measurement.

4. Application to measurement

We consider again the two competing ensembles $\tilde{\mathcal{E}}$ and $\mathcal{E}$ of Sect. 1. Let $\mathcal{N}$ be the number of systems S on which the quantity V is measured. To simplify notation, we superpose only two states: $|\psi\rangle = a_1|\psi_1\rangle + a_2|\psi_2\rangle$. Then the ensembles $\tilde{\mathcal{E}}$ and $\mathcal{E}$ have the structures

$$\begin{aligned} &\mathcal{N}|a_1|^2 \text{ times:} \quad |\psi_1\rangle|A_{1,s_1}\rangle, \\ &\mathcal{N}|a_2|^2 \text{ times:} \quad |\psi_2\rangle|A_{2,s_2}\rangle \end{aligned} \tag{4.1}$$

and

$$\mathcal{N} \text{ times:} \quad a_1|\psi_1\rangle|A_{1,s_1}\rangle + a_2|\psi_2\rangle|A_{2,s_2}\rangle, \tag{4.2}$$

respectively. If the quantity V is measured again the result $V = v_i$ is obtained $\mathcal{N}|a_i|^2$ times in the case of the ensemble $\tilde{\mathcal{E}}$ and, say, p_i times in that of the ensemble $\mathcal{E}$. The numbers p_i are often nearly equal to $\mathcal{N}|a_i|^2$, but the two sets do not necessarily coincide. So, in spite of effective incoherence, $\tilde{\mathcal{E}}$ and $\mathcal{E}$ are not equivalent.

However, since we are considering fluctuations of quantum distributions, we must also allow for the distribution of the results of the first measurement being different from the ideal one. Then the ensemble after the first measurement according to the reduction postulate is not $\tilde{\mathcal{E}}$, but $\tilde{\mathcal{E}}_{\{n\}}$ having the structure

$$\begin{aligned} &n_1 \text{ times:} \quad |\psi_1\rangle|A_{1,s_1}\rangle, \\ &n_2 \text{ times:} \quad |\psi_2\rangle|A_{2,s_2}\rangle, \end{aligned} \tag{4.3}$$

where n_i is the number of times the result v_i has been obtained in the measurement. Of course, the numbers n_i do not necessarily coincide with the numbers $\mathcal{N}|a_i|^2$. If the ensemble after the first measurement is $\tilde{\mathcal{E}}_{\{n\}}$, the second measurement gives again n_i times the result v_i. If viceversa the ensemble is $\mathcal{E}$, the numbers p_i, describing the outcome of the second measurement, are independent of the numbers n_i because of the structure of $\mathcal{E}$ (in which no trace of the n_i's appears). As before, the numbers p_i are in general different from the numbers n_i, so that $\tilde{\mathcal{E}}_{\{n\}}$ and $\mathcal{E}$ are not equivalent.

The effect we are discussing is not a tiny one. The probabilities to be assigned to the considered fluctuations are easily calculated, being given by the binomial distribution

$$B(\{q\}) = \frac{\mathcal{N}!}{q_1!\, q_2!} |a_1|^{2q_1} |a_2|^{2q_2}, \tag{4.4}$$

where $\{q_1, q_2\}$ is $\{n_1, n_2\}$ or $\{p_1, p_2\}$. The total probability that $\{p\} \neq \{n\}$ is

$$\Pr(\{p\} \neq \{n\}) = \sum_{\{i\}} B(\{i\}) \sum_{\{j\} \neq \{i\}} B(\{j\}). \tag{4.5}$$

The probabilities 4.5 are usually quite large. For example, for $|a_1|^2 = |a_2|^2 = \frac{1}{2}$ and $\mathcal{N} = 10^4$ the result is 0.99. One might object that the probability that $\{p\}$ be simply different from $\{n\}$ is not significant for large $\mathcal{N}$ because of the uncertainties inherent to any experiment. In our opinion the objection is not relevant because we are here discussing two different theoretical descriptions of a given situation. However, probabilities are still large if, by suitably limiting the internal sum in Eq. 4.5, one considers differences significant only when larger than a certain relative amount. E.g. in the same case considered above, the probability that the p_i's be different from the n_i's by at least 1% is 0.48.

There is an obvious relation between the problem we are raising here and that of the agreement between two successive measurements of the same variable of S. But we do not follow this way because we have accepted not to deal with single systems.

Because of the difference between the actual distribution $\{n\}$ of the results of the first measurement and the ideal distribution $\{\mathcal{N}|a|^2\}$, the ensemble $\tilde{\mathcal{E}}$ is replaced by $\tilde{\mathcal{E}}_{\{n\}}$. We note that, for this ensemble, even the ideal quantum distributions are different from those for the ensemble $\mathcal{E}$. In fact, e.g., for the measurement of a quantity $\hat{U} = \sum_i u_i |\phi_i\rangle\langle\phi_i|$ of S, one has

$$\Pr_{\tilde{\mathcal{E}}_{\{n\}}}(U = u_i) = \frac{n_1}{\mathcal{N}} |\langle\phi_i|\psi_1\rangle|^2 + \frac{n_2}{\mathcal{N}} |\langle\phi_i|\psi_2\rangle|^2, \tag{4.6}$$

$$\Pr_{\mathcal{E}}(U = u_i) = |a_1|^2 |\langle\phi_i|\psi_1\rangle|^2 + |a_2|^2 |\langle\phi_i|\psi_2\rangle|^2. \tag{4.7}$$

The difference of the probability distribution 4.6 from 4.7 is significant in spite of fluctuations because a collection of ensembles $\tilde{\mathcal{E}}_{\{n\}}$ can concretely be constructed if wanted. In the limit $\mathcal{N} \to \infty$, the relative frequencies $n_i/\mathcal{N}$ tend to the ideal values $|a_i|^2$, so that the $\tilde{\mathcal{E}}_{\{n\}}$ probability distribution 4.6 tends to the corresponding distribution pertaining to $\mathcal{E}$ given by 4.7. While we insist in claiming our right to deal with finite ensembles, we note that, in the limit $\mathcal{N} \to \infty$, the set $\{p\}$ remains different from the set $\{n\}$, at least if we consider absolute differences as significant.

The ensemble $\tilde{\mathcal{E}}_{\{n\}}$ and the probability associated to it result from the application of the probability law and the reduction postulate to the system S. On the other hand, the ensemble $\mathcal{E}$ is a consequence of the absolute validity of the Schrödinger equation for the system $S + A$. The inequivalence of $\tilde{\mathcal{E}}_{\{n\}}$ and $\mathcal{E}$ comes from the application of the probability law to the system $S + A$ for the second measurement. It is the second application of the probability law which is responsible for the paradox, but we do not see any consistent way to avoid it if the probability law is the only link between the state vector and the outcomes of an experiment.

Our argument is banal and can be summarized as follows. There is something between single systems (forbidden) and infinite ensembles (allowed but unattainable): actual ensembles (allowed, we suppose). For these ensembles the distributions of the results of measurements fluctuate around the ideal distributions. The distributions of distributions are significant features of the considered ensembles. When they are taken into account effective incoherence is no more sufficient to reconcile the Schrödinger equation with the reduction postulate in the orthodox framework.

5. Effective incoherence and unorthodox theories

In this section we shall see how giving up the assumption that the probability law is the only link between state vector and experiment or that of the absolute validity of the Schrödinger equation can solve the problem we raised, remaining true that some kind of effective incoherence is necessary in order that the resulting theory work. We limit ourselves to two typical examples of unorthodox theories, the pilot wave theory and theories in which reduction is described as a real physical process.

A) Pilot wave theory

In the pilot wave theory no reduction of the wave function takes place, so that one is somehow lead to consider the wave function of the universe. However, the original formulation for limited systems is consistent and we will stick to this conceptually simple framework. The description of measurement within the pilot wave theory is due to Bell[4,5] and Bohm and Hiley.[6]

In the pilot wave theory the state of an N-particle system is the pair $(\Psi, \boldsymbol{X})$, where $\boldsymbol{X} = \boldsymbol{x}_1, \ldots, \boldsymbol{x}_N$ is the point in configuration space and $\Psi(\boldsymbol{X})$ is a field in the same space. If the configuration of the system or of a part of it is measured at a certain time, the system or its part are found as specified by the value of $\boldsymbol{X}$ at that time. The configuration $\boldsymbol{X}$ plays the role of "hidden" variable. The evolution of the pair is given by the equations

$$i\hbar\frac{d}{dt}\Psi = H\Psi, \tag{5.1}$$

$$\frac{d}{dt}\boldsymbol{X} = \boldsymbol{J}(\boldsymbol{X},t)/\rho(\boldsymbol{X},t), \tag{5.2}$$

where

$$\rho = |\Psi|^2, \tag{5.3}$$

$$\boldsymbol{J} = \hbar\,\mathrm{Im}\left(\Psi^*\frac{\nabla}{M}\Psi\right), \qquad \frac{\nabla}{M} = \left(\frac{\nabla_1}{m_1},\ldots,\frac{\nabla_N}{m_N}\right). \tag{5.4}$$

It is further supposed that, when an ensemble of systems described by Ψ is prepared, the distribution of $\boldsymbol{X}$ is given by ρ. This is a consistent assumption, i.e. the condition remains true while time is going on, because of the equation

$$\frac{\partial}{\partial t}\rho + \nabla\cdot\boldsymbol{J} = 0, \tag{5.5}$$

which is a consequence of the Schrödinger equation, under the usual hypotheses on H. It may happen that the wave function splits into two parts, $\Psi = \Psi_1 + \Psi_2$, in such a way that the Schrödinger evolution keeps the two terms separated in configuration space during a certain time. Then

$$\begin{aligned} \rho &= \rho_1 + \rho_2, \\ \boldsymbol{J} &= \boldsymbol{J}_1 + \boldsymbol{J}_2 \end{aligned} \tag{5.6}$$

where ρ_1, $\boldsymbol{J}_1$ and ρ_2, $\boldsymbol{J}_2$ are nonzero, respectively, in two domains in configuration space D_1 and D_2 disjoint during the considered time. According to Eq. 5.2, for the purpose of the evolution of a system whose $\boldsymbol{X}$ lays, say, in D_1 the term Ψ_2 can consistently be dropped.

In the pilot wave theory all quantum fluctuations are contained in the distribution of the initial configuration of the system.

Clearly, position has a special role in the pilot wave theory. As a rule, any measurement is, after all, a position measurement; if a variable does not commute with position, its measurement is achieved by a time delayed position measurement. In the pilot wave theory it is so in principle. We will discuss a simple example, in which a coarse measurement of the position of a particle is performed by a pair of detectors A (see Fig. 2). The state variables of the system particle plus detectors are the configuration $\boldsymbol{x}, \boldsymbol{Y}$ and the wave function $\psi(\boldsymbol{x})A(\boldsymbol{Y})$ or $\Psi(\boldsymbol{x},\boldsymbol{Y})$, where $\boldsymbol{x}$ is the position of the particle, $\boldsymbol{Y}$ is the configuration of the

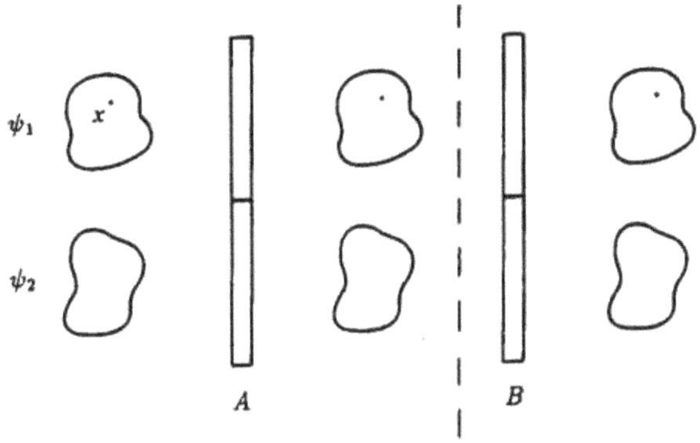

Fig. 2. Coarse position measurement in pilot wave theory. The part of the drawing beyond the dashed line refers to the repetition of the same measurement.

particles in the detectors, $\psi(\boldsymbol{x})$, $A(\boldsymbol{Y})$ and $\Psi(\boldsymbol{x},\boldsymbol{Y})$ are the wave functions of the particle, of the detectors and, when necessary, of the entire system, respectively.

Let the particle wave function before the measurement be the superposition of two packets ψ_1 and ψ_2, each hitting one of the detectors. The evolution of the state variables of the system during the measurement is described by the transitions

$$\begin{aligned} &\boldsymbol{x},\boldsymbol{Y}_0 \longrightarrow \begin{cases} \boldsymbol{x},\ \boldsymbol{Y}_1 \\ \boldsymbol{x},\ \boldsymbol{Y}_2, \end{cases} && \begin{array}{l}\text{depending on whether}\\ \boldsymbol{x} \text{ is within } \psi_1 \text{ or } \psi_2,\end{array} \\ &(\psi_1(\boldsymbol{x})+\psi_2(\boldsymbol{x}))\,A_0(\boldsymbol{Y}) \quad \longrightarrow && \psi_1(\boldsymbol{x})A_1(\boldsymbol{Y})+\psi_2(\boldsymbol{x})A_2(\boldsymbol{Y}), \end{aligned} \tag{5.7}$$

where we have omitted to indicate the trivial time evolution bringing the particle position and wave function through the detectors. $\boldsymbol{Y}_0$, $\boldsymbol{Y}_1$ and $\boldsymbol{Y}_2$ are the configurations of the two detectors before the measurement, after the measurement if $\boldsymbol{x}$ is in ψ_1 and after the measurement if $\boldsymbol{x}$ is in ψ_2, respectively. A_0, A_1 and A_2 are the wave functions of the pair of detectors before the measurement, after the measurement if the particle wave function were ψ_1 alone and after the measurement if the particle wave function were ψ_2 alone, respectively.

If the measurement is repeated by means of a further pair of detectors B (see Fig. 2), the evolution of the state variables during the two measurements is

$$\begin{aligned} &\boldsymbol{x},\boldsymbol{Y}_0,\boldsymbol{Z}_0 \longrightarrow \begin{cases} \boldsymbol{x},\boldsymbol{Y}_1,\boldsymbol{Z}_0 \longrightarrow \boldsymbol{x},\boldsymbol{Y}_1,\boldsymbol{Z}_1 \\ \boldsymbol{x},\boldsymbol{Y}_2,\boldsymbol{Z}_0 \longrightarrow \boldsymbol{x},\boldsymbol{Y}_2,\boldsymbol{Z}_2, \end{cases} \\ &(\psi_1(\boldsymbol{x})+\psi_2(\boldsymbol{x}))\,A_0(\boldsymbol{Y})B_0(\boldsymbol{Z}) \longrightarrow \\ &(\psi_1(\boldsymbol{x})A_1(\boldsymbol{Y})+\psi_2(\boldsymbol{x})A_2(\boldsymbol{Y}))\,B_0(\boldsymbol{Z}) \longrightarrow \\ &\psi_1(\boldsymbol{x})A_1(\boldsymbol{Y})B_1(\boldsymbol{Z})+\psi_2(\boldsymbol{x})A_2(\boldsymbol{Y})B_2(\boldsymbol{Z}), \end{aligned} \tag{5.8}$$

with obvious meaning of symbols. The ensemble after the first measurement is described by the pure state wave function as for the ensemble $\mathcal{E}$ of Sects. 2 and 4, but the results of the second measurement on the various members of the ensemble, rather than being controlled by a second application of the probability law, are now determined by the same values of the variable $\boldsymbol{x}$ which have determined the results of the first one. The distribution of the results of the second measurement is therefore identical to that of the first, as was the case for the ensemble $\tilde{\mathcal{E}}_{\{n\}}$ of Sect. 4. No contradiction of the type discussed in that section arises.

One further problem has, however, to be discussed. We return to the consideration of only one measurement. The original ensemble of systems $S + A$ can be split into two parts, $\mathcal{E}_1$ and $\mathcal{E}_2$, according to having found the particle in one detector or the other. Focusing our attention e.g. on $\mathcal{E}_1$, the systems contained in it are described by the wave function $\Psi_1 + \Psi_2 = \psi_1 A_1 + \psi_2 A_2$, but the distribution of $\boldsymbol{x}, \boldsymbol{Y}$ is that corresponding to Ψ_1 alone. However, immediately after the measurement Ψ_1 and Ψ_2 are obviously separated. Bohm and Hiley[6] correctly argue, using the irreversibility of the amplification which accompanies any quantum measurement, that this condition will remain true for all subsequent times with overwhelming probability, in spite of any possible attempt to rejoin the two terms. Then, as noted above, Ψ_2 can definitively be dropped from the wave function for the systems belonging to the subensemble $\mathcal{E}_1$.

We point out that we have just used effective incoherence and that Bohm and Hiley use, to prove its permanence in time, the same type of argument first used by Daneri, Loinger and Prosperi[1] in the framework of an OEI theory.

B) Reduction as a real physical process

In theories where reduction is described as a real physical process no extra variable is added to the state vector, which remains the only state variable of quantum systems. However, a correction to the Schrödinger equation restricts its absolute validity. Since the purpose of the correction is that of including the reduction process in the evolution equation of the theory and this process is stochastic and nonlinear, the correction will have such features.

In the most intuitive type of reduction models[7] finite stochastic changes of the state vector take place instantaneously. The stochastic process is defined in the following way. Consider a K-dimensional vector of selfadjoint commuting operators $\boldsymbol{A}$ and a real number β. A hit at the point $\boldsymbol{a}$, a K-dimensional vector of real numbers, is the transition

$$\begin{aligned} |\psi\rangle \to |\psi(\boldsymbol{a})\rangle &= |\phi(\boldsymbol{a})\rangle / \|\phi(\boldsymbol{a})\|, \\ |\phi(\boldsymbol{a})\rangle &= \left(\frac{\beta}{\pi}\right)^{K/4} \exp\left(-\tfrac{1}{2}\beta\left(\boldsymbol{A}-\boldsymbol{a}\right)^2\right)|\psi\rangle, \end{aligned} \tag{5.9}$$

from an initial state vector $|\psi\rangle$ to the final state vector $|\psi(\boldsymbol{a})\rangle$. It is assumed that, when a hit occurs, the probability density for its taking place at the point $\boldsymbol{a}$ is

$$\mathcal{P}r(\boldsymbol{a}) = \|\phi(\boldsymbol{a})\|^2. \tag{5.10}$$

It is also assumed that hits occur at random times according to a Poisson process with mean frequency μ and that in the interval between two hits the state vector evolves according to the Schrödinger equation. It is obvious that, if the Schrödinger evolution is frozen, the repeated application of the process (5.9) with the probability law (5.10) produces the reduction of $|\psi\rangle$ onto the common eigenspaces of the operators $\boldsymbol{A}$. If the Schrödinger evolution acts together with the stochastic process, the competition between the two processes can be studied.[7]

In the discontinuous processes just described the way how the reduction mechanism works is quite transparent. The continuous processes[8-11] which are obtained by a proper infinite frequency limit,[12] though less intuitive, avoid the appearance of finite changes taking place instantaneously, express the principle of evolution through a single synthetic equation and allow a simple field theoretic formulation for systems of identical particles. The continuous reduction models are specified by a K-dimensional vector of self adjoint operators $\boldsymbol{A}$ and by a real number γ. Actually, in the physically interesting case[11] the components of the operator vector $\boldsymbol{A}$ are labelled by the point $\boldsymbol{x}$ in the physical space and have the meaning of density of particles around $\boldsymbol{x}$. The state vector $|\psi\rangle$ obeys the norm-conserving nonlinear Itô stochastic differential equation

$$\begin{aligned} d|\psi\rangle = \Big[& -iH\,dt + \int d^3\boldsymbol{x}\,(N(\boldsymbol{x}) - N_\psi(\boldsymbol{x}))\,dB(\boldsymbol{x}) \\ & -\tfrac{1}{2}\gamma \int d^3\boldsymbol{x}\,(N(\boldsymbol{x}) - N_\psi(\boldsymbol{x}))^2\,dt \Big]\,|\psi\rangle, \\ N_\psi(\boldsymbol{x}) = & \langle\psi|N(\boldsymbol{x})|\psi\rangle \end{aligned} \tag{5.11}$$

where the Wiener process $dB(\boldsymbol{x})$ has the properties

$$\begin{aligned} &\overline{dB(\boldsymbol{x})} = 0, \\ &\overline{dB(\boldsymbol{x})dB(\boldsymbol{y})} = \gamma\,\delta^3\,(\boldsymbol{x} - \boldsymbol{y})\,dt. \end{aligned} \tag{5.12}$$

The parameter γ controls the strength of the stochastic process. The density operators $N(\boldsymbol{x})$ are defined by

$$N(\boldsymbol{x}) = \sum\nolimits_s \int d^3\boldsymbol{y}\,g(\boldsymbol{y} - \boldsymbol{x})\,a^\dagger(\boldsymbol{y}, s)a(\boldsymbol{y}, s), \tag{5.13}$$

where $a^\dagger(\boldsymbol{y}, s)$ and $a(\boldsymbol{y}, s)$ are creation and annihilation operators at $\boldsymbol{y}$ with spin component s satisfying canonical commutation or anticommutation relations. The shape function $g(\boldsymbol{x})$ satisfies the condition

$$\int d^3\boldsymbol{x}\,g(\boldsymbol{x}) = 1, \tag{5.14}$$

so that

$$\int d^3\boldsymbol{x}\, N(\boldsymbol{x}) = N, \tag{5.15}$$

N being the total number operator. The function $g(\boldsymbol{x})$ is taken to be

$$g(\boldsymbol{x}) = (\alpha/2\pi)^{3/2} \exp\left(-\tfrac{1}{2}\alpha \boldsymbol{x}^2\right), \tag{5.16}$$

where the range parameter $1/\sqrt{\alpha}$ is intended to be microscopically large and macroscopically small. In Ref. 11 the values

$$1/\sqrt{\alpha} = 10^{-5}\ \text{cm}, \qquad \gamma = 10^{-30}\ \text{cm}^3\ \text{s}^{-1} \tag{5.17}$$

were proposed and discussed. It turns out that the evolution of microsystems is left practically unaltered, while superpositions of state vectors of macrosystems corresponding to different macroscopic densities are quickly reduced to one of their components. The probability of being asymptotically reduced to a certain component is given by the initial value of its squared norm.

The description of measurement within stochastic reduction theories has been studied by Benatti et al.[13] Such theories entail, in the case of the state vector 1.6, its quick reduction to one of the components with probabilities $|a_m|^2$. Since reduction actually takes place, the ensemble after the measurement is $\tilde{\mathcal{E}}_{\{n\}}$ and $\mathcal{E}$ does not represent the physical situation any more. The distribution of the results of a further measurement of the same quantity is $\{n\}$ and the probability distribution for the measurement of another quantity is given by 4.6. No contradiction of the type discussed in Sect. 4 arises.

We want now to show that the classical property, in the sense already discussed in Sect. 2, of macroscopic systems, which is a fact that is believed to follow from their dynamical features independently of the assumption of the stochastic terms in the evolution of the state vector, prevents such systems from exhibiting a stochastic behaviour never experienced. Consider a macroobject A (not necessarily a pointer) and two state vectors $|A_{\text{here}}\rangle$ and $|A_{\text{there}}\rangle$ corresponding to macroscopically distinguishable positions. According to our assumptions, if a superposition of these two state vectors is somehow created it is stochastically reduced to one of its components. Suppose that it is possible to measure a dynamical variable of A having eigenaxes

$$\begin{aligned} |A_1\rangle &= c_{11}|A_{\text{here}}\rangle + c_{12}|A_{\text{there}}\rangle, \\ |A_2\rangle &= c_{21}|A_{\text{here}}\rangle + c_{22}|A_{\text{there}}\rangle. \end{aligned} \tag{5.18}$$

This implies that one can construct an apparatus B with the dynamical property, say,

$$\begin{aligned} |A_1\rangle|B_0\rangle &\longrightarrow |A_1\rangle|B_{\text{above}}\rangle, \\ |A_2\rangle|B_0\rangle &\longrightarrow |A_2\rangle|B_{\text{below}}\rangle, \end{aligned} \tag{5.19}$$

where $|B_{\rm above}\rangle$ and $|B_{\rm below}\rangle$ are macroscopically distinguishable. Then, for the initial state vector, e.g., $|A_{\rm here}\rangle = c_{11}^*|A_1\rangle + c_{21}^*|A_2\rangle$, one has

$$\begin{aligned}
|A_{\rm here}\rangle|B_0\rangle \quad \longrightarrow \quad & c_{11}^*|A_1\rangle|B_{\rm above}\rangle + c_{21}^*|A_2\rangle|B_{\rm below}\rangle \\
& = c_{11}^*c_{11}|A_{\rm here}\rangle|B_{\rm above}\rangle + c_{11}^*c_{12}|A_{\rm there}\rangle|B_{\rm above}\rangle \\
& + c_{21}^*c_{21}|A_{\rm here}\rangle|B_{\rm below}\rangle + c_{21}^*c_{22}|A_{\rm there}\rangle|B_{\rm below}\rangle \\
\longrightarrow \quad & \begin{cases} |A_{\rm here}\rangle|B_{\rm above}\rangle \\ |A_{\rm there}\rangle|B_{\rm above}\rangle \\ |A_{\rm here}\rangle|B_{\rm below}\rangle \\ |A_{\rm there}\rangle|B_{\rm below}\rangle, \end{cases}
\end{aligned} \tag{5.20}$$

where the last arrow indicates the transition due to the stochastic reduction process. Since the measurement we are discussing concerns a macroscopic variable of a macroscopic object, all relevant constituents of the system $A+B$ are macroscopic. According to Eq. 5.20 the system $A+B$ undergoes eventually a stochastic evolution which has never been experienced for such a type of system. The fact is that the evolution 5.19 can never be realized in practice. The impossibility of measuring a dynamical variable having axes different from $|A_{\rm here}\rangle$ and $|A_{\rm there}\rangle$ amounts to the classical property of A. This property, which is not a principle of the theory, it is a dynamical behaviour taking place practically, is essential in order that the theory be in agreement with our experience. Again we find that a property related to effective incoherence and which is a weighty ingredient of an orthodox theory, as discussed in Sect. 2, plays an important role also in an unorthodox theory.

6. Final remarks and conclusions

We have used in our discussion the ideal scheme to describe the dynamics of the measurement process. One could object that the consideration of the unavoidable nonideality of actual measurements would spoil our argument based on fluctuations of quantum distributions. We note, on the other hand, that the orthodox theories we are discussing are formulated within the ideal scheme and that a reformulation of such theories in a nonideal framework should be discussed again from the beginning. At any rate, we observe that our arguments concern the theoretical comparison of two descriptions of a given situation and do not require any resort to experiment. A serious use of nonideality to the aim of the present discussion would require to incorporate it into the principles of quantum theory so deeply as to question in principle that the result of a measurement can be used to foresee the outcome of a successive measurement.

An essential ingredient of the arguments of Sects. 3 and 4 is that the probability associated to a given distribution of the values of a variable can be calculated, so that the distribution of distributions is a quantitatively defined significant feature of the considered ensemble. We think that the calculation of such probabilities requires to ascribe to quantum probabilities a meaning pertinent to the single elements of the ensemble. We admit that in this sense our arguments deal with single systems, but we believe that denying physical significance to numbers which can be calculated from the quantum probability law and in principle be measured amounts to maiming the theory itself.

It is clear that there are deep conceptual differences among theories of measurement like those founded on the orthodox frame in the sense of Sect. 2, those using the idea of the pilot wave and those based on models of real physical reduction. However, we have seen that there are points which are common to different theories. Our first conclusion is that an extensive study of the relations among the different approaches, can shed light on the problem and help in understanding each other.

We have pointed out a serious trouble met by a family of orthodox theories of measurement and we have seen that two different unorthodox theories overcome the difficulty. Our arguments, we think, support the opinion according to which a fully satisfactory description of the system object–plus–apparatus cannot be achieved within the orthodox conceptual framework. To summarize our main conclusion (which is strongly suggested not strictly proved, we admit), we use the words of "the master heretic himself"[2] John Bell: "If, with Schrödinger, we reject extra variables, then we must allow that his equation is not always right. I do not know that he contemplated this conclusion, but it seems to me inescapable".[14] As one can see, our conclusion is not at all new.

References

1. A. Daneri, A. Loinger and G.M. Prosperi, Nuclear Physics **33** (1962), p. 297.
2. K. Gottfried, presented at *Sixty Two Years of Uncertainty: Historical, Philosophical and Physics Inquiries into the Foundations of Quantum Mechanics*, Erice, 1989.
3. B. d'Espagnat, *Conceptual foundations of quantum mechanics*, second edition (W. A. Benjamin, Inc., Reading (Mass.), 1976).
4. J.S. Bell, CERN Preprint TH 1424 (1971); Epistemological Letters (July 1978), p. 1.
5. J.S. Bell, in *Quantum Gravity 2,* edited by C. Isham, R. Penrose and D. Sciama (Clarendon Press, Oxford, 1981), p. 611.
6. D. Bohm and B.J. Hiley, Foundations of Physics **14** (1984), p. 255.
7. G.C. Ghirardi, A. Rimini, T. Weber, Physical Review D **34** (1986), p. 470; Physical Review D **36** (1987), p. 3287.
8. L. Diósi, Physics Letters A **129** (1988), p. 419.
9. P. Pearle, Physical Review A **39** (1989), p. 2277.
10. N. Gisin, Helvetica Physica Acta **62** (1989), p. 363.

11. G.C. Ghirardi, P. Pearle, A. Rimini, Physical Review A **42** (1990), p. 78.
12. O. Nicrosini and A. Rimini, to be published in Foundations of Physics.
13. F. Benatti, G.C. Ghirardi, A. Rimini, T. Weber, Il Nuovo Cimento **100 B** (1987), p. 27.
14. J.S. Bell, in *Schrödinger — Centenary celebration of a polymath,* edited by C.W. Kilmister (Cambridge University Press, Cambridge, 1987), p. 41.

WAVE-PARTICLE EFFECTS IN MODERN MICROELECTRONICS

GÜNTER NIMTZ
II. Physikalisches Institut der Universität zu Köln, D-5000 Köln 41, FRG

and

PETER MARQUARDT
Institut für Oberflächentechnik, Universität-GH Siegen, D-5900 Siegen, FRG

ABSTRACT

Electronic properties of modern microelectronic structures are governed rather by the wave nature than by the corpuscle character of electrons. This novel aspect of transport in minute conductors is exemplified with several wave interference effects observed in recent experiments.

1. Introduction

Modern semiconductor devices have undergone an enormous progress of miniaturization. At the time Bardeen, Brattain, and Shockley (1947) have invented the transistor, the size scale was in the mm range. Nowadays, transistors reach linear dimensions much smaller than 1 μm. In this way, the wave nature of the electrons becomes evident in their transport properties. At low temperatures the wave nature even dominates eventually in small enough metal or semiconductor structures.

The very length we have to compare with the dimensions of a device or with the interdistance of e.g. defect potentials is the *electron wave coherence length* ξ. In the next section we discuss and present typical values of ξ in a semiconductor. In Section 3 we introduce an example of non-classical electron transport which was observed in a standard infrared detector element. Here the coherent electron-electron interaction (EEI) due to random potential scattering contributes up to 10 % of the total element resistance.

In 1988, Dutch and British physicists discovered the transverse quantization of electron transport through a size-tunable electron waveguide. The experiment was carried out with a special split-gate field effect transistor (FET). This transverse size quantization causes discrete jumps of the conductance with the universal value $e^2/\hbar\pi$ (Section 4). The transverse quantization ends eventually in an electron transport cut-off when the de Broglie wavelength λ exceeds twice the waveguide diameter. This cut-off wavelength (COW) property is assumed to become the basis of very small and simple transistors as well as versatile sensors. The concept of such COW devices is introduced in Section 5.

The modern technology of molecular beam epitaxy yields excellent layer structures. The layers are extremely perfect with respect to both lattice defects and chemical impurities. For instance, by means of this technique it is possible to grow different alternating semiconductor layers with less than ten atomic monolayers each. The method allows to taylor potential wells in semiconductor heterostructures as shown in Section 6. These structures are used e.g. for the fabrication of high-quality laser diodes; they may, however, also yield information about the basic understanding of tunneling processes. For instance, the tunneling time can be measured which is still under discussion from the theoretical point of view.

2. The particle-wave transition and the electron coherence length

With decreasing size of a conductor, the classical Drude transport model of electrons as particles fails to describe the observed transport properties. Non-classical contributions due to the electron wave nature become important. The wave nature becomes evident when the coherence length ξ exceeds the size of a conductor in one, two or even three dimensions (D). This corresponds to a 2D, 1D, or zero-D conductor, respectively, the latter often being called a "quantum dot". The wave nature, however, can also be observed when ξ is longer than the average distance between randomly distributed defect potentials in the lattice of the conductor crystal. According to the de Broglie relation, the wavelength is

$$\lambda = \frac{h}{mv} \tag{1}$$

with $h = 2\pi\hbar$ Planck's constant, and m and v the particle mass and velocity, respectively. Due to the high Fermi temperature T_F of a metal (between 10^4 and 10^5 K) the wavelength of an electron in a metal is only some 0.5 nm, i.e. of the order of the interatomic distance. Having in mind that T_F and thus the corresponding velocity v_F depends on the carrier density via the relation

$$v_F = \frac{(3\pi^2 n)^{1/3}\hbar}{m} \tag{2}$$

the wavelength in a semiconductor is usually much longer than in a metal. For instance, the semiconductor $Hg_{0.8}Cd_{0.2}Te$ which is used as infrared photoconductor has a carrier density about nine orders of magnitude smaller than a metal. This low carrier density results in a wavelength of nearly $1\mu m$.

The coherence length of an electron along a diffusive path depends on the diffusion coefficient D and the inelastic scattering time τ_i [1]

$$\xi = (D\tau_i)^{1/2} \tag{3}$$

This relation is equivalent to

$$\xi = (\frac{1}{3}L_e L_i)^{1/2} \tag{4}$$

in a 3D sample with L_e the elastic and L_i the inelastic free path. In general, $L_e \ll L_i$ holds: There are many more elastic than inelastic scattering events. [2] The elastic scattering, changing only the momentum *direction* is mainly due to interactions with lattice defects and lattice vibrations, whereas the inelastic phase-breaking scattering is determined by electron-phonon (EP) and by electron-electron (EE) interactions. The *inelastic* EP scattering rate can be experimentally determined from the energy relaxation of the electron ensemble. [3] The EE scattering time may be estimated from the non-classical conduction contribution. The scattering rates usually decrease markedly at lower temperatures. In a semiconductor, we expect $\xi \approx 0.2\mu m$ near 300 K and $\xi \leq 1\mu m$ at sub-Kelvin temperatures from both experimental and theoretical data. [2] The estimates of the electron wave coherence length show that miniaturization below $1\mu m$ may provoke non-classical transport properties even at 300 K finally. In submicrometer structures, ξ is even enhanced due to the lack of long-wavelength phonons which do not exist in such minuscule systems, which are, however, in charge of important scattering contributions in bulk materials. For propagating elastic excitations it is necessary that $\lambda_{phonon} \ll s$ holds. Thus, in a crystal of $s = 1\mu m$ diameter, only 90 % of the bulk phonon spectrum fulfill this condition. Miniaturization of semiconductor structures allows to study the transition from dominant particle behaviour to dominant wave behaviour. This will be demonstrated with some examples in the following Sections.

3. Coherent Electron-Electron Interaction in a Bulk Semiconductor

In the last decade, many theoretical and experimental investigations were devoted to electron-wave interference effects on the electric resistance. [4] At very low temperatures, when ξ exceeds a characteristic length (e.g. the average distance between defect potentials or a linear dimension of the conductor), the total resistance is given by two contributions

$$R = R_D + \Delta R \tag{5}$$

with R_D the classical Drude resistance and ΔR the non-classical resistance due to electron interference. This is sketched in Fig. 1. Non-classical resistance was infact observed in metal films and semiconductors in the sub-Kelvin range. Essentially there are two non-classical effects, *weak localisation* (WL) and *electron-electron interaction* (EEI). WL is due to the interference of the partial waves of one electron wave packet moving on time-reversed paths, i.e. clock- and counter-clockwise on a loop. [4] EEI is based on a correlated electron interaction at a defect potential. [4]

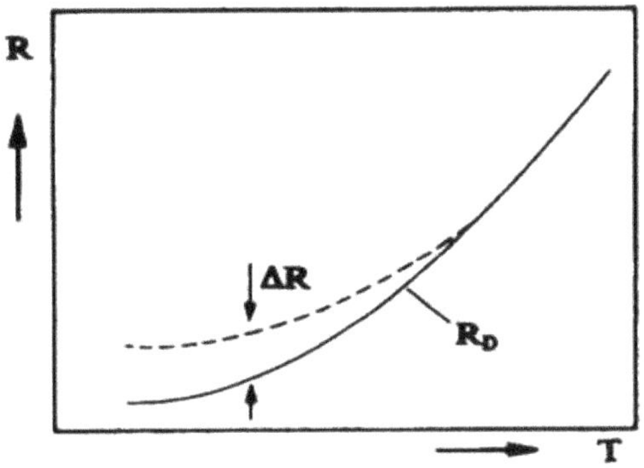

Fig. 1 Qualitative temperature dependence of the total resistance $R = R_D + \Delta R$ given by the classical Drude contribution R_D and the non-classical interference term ΔR.

EEI was predicted by Al'tshuler and Aronov. [5] Here, we present data on EEI observed in an infrared detector element prepared from the narrow-gap semiconductor n-$Hg_{0}.8Cd_{0}.2Te$, having a carrier density about nine orders of magnitude smaller than a metal. Owing to the long mean free path, the small effective mass m (only 0.005 of the free mass m_0) and the low carrier density in this very material, ξ is expected to have unusually large values. Infact, there was observed a non-classical resistance ΔR depending on temperature T and electric field E as predicted for EEI (Figs. 2a and 2b). According to the EEI theory we have

$$\Delta R(T)/R_D = C_1 - C_2\sqrt{T} \tag{6}$$

and

$$\Delta R(E)/R_D = C_3 - C_4\sqrt[3]{E} \tag{7}$$

where $C_{1,2,3,4}$ are constants.

This experiment, recently carried out with a standard device, has reveiled now that electron wave interference can be of importance even at elevated temperatures up to 40 K.[6,7]

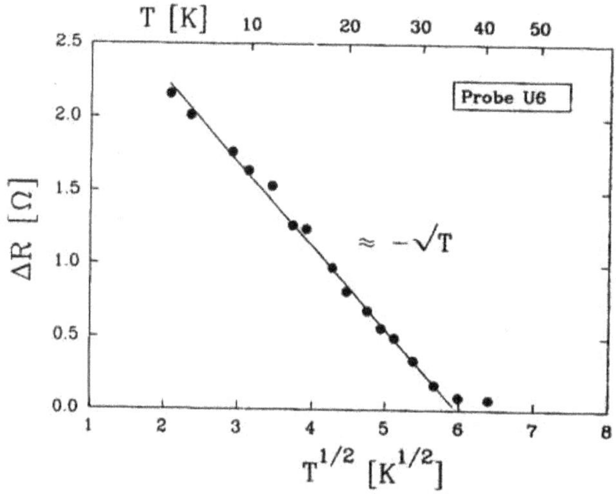

Fig. 2a Non-classical resistance ΔR vs temperature T showing a $\sqrt{T}$ dependence.[6,7]

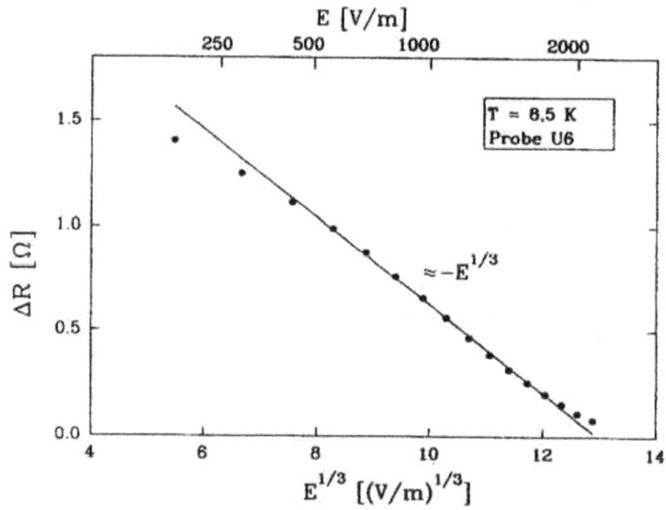

Fig. 2b Non-classical resistance ΔR vs electric field E showing a $\sqrt[3]{E}$ dependence.[6,7]

4. Size-induced quantization of electron transport

EEI as well as WL are interference effects caused by electron scattering at defect potentials in a crystal. Other interference phenomena take place when at least one dimension of the electron confinement becomes smaller than ξ. In this case, we have a 1D, 2D, or 3D size quantization with the following eigenvalues in a cuboidal box of dimensions x, y, and z:

$$E_n = \frac{(n_x h)^2}{8mx^2} + \frac{(n_y h)^2}{8my^2} + \frac{(n_z h)^2}{8mz^2} \tag{8}$$

with $n_x^2 + n_y^2 + n_z^2 = n$ the main quantum numbers for each direction.

Here we present an experiment [8] in which the conductance of a narrow 1D "waveguide" was recently shown to be quantized in multiples of $e^2/\pi\hbar$. The quantization was observed in a field effect transistor (FET) where the electron channel was squeezed like a bottleneck as shown in the inset of Fig. 3. The bottleneck was provoked by a split-gate arrangement on top of the FET, the diameter being proportional to the voltage applied across the split-gate. The exciting result is that the electron transport in small structures becomes quantized independent of the host medium. Finally, at a gate voltage below -2 V the transport is quenched and the resistance becomes infinite. This takes place at the so-called maximum resistance or, correspondingly, at the minimum conductance $e^2/\hbar\pi$.

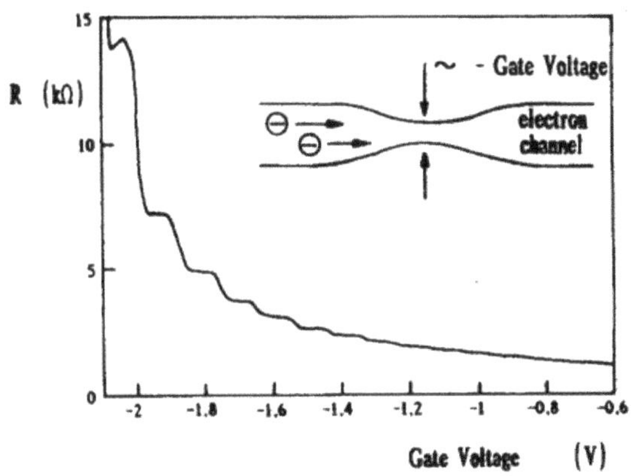

Fig. 3 Resistance R vs gate voltage U. The bottleneck of the electron film is sketched in the inset. [8]

5. Cut-off electron-wave devices and the COW Transistor principle

The experimental results presented in Section 4 can be described also as *transverse* quantization of the electron wave. [9] To some extent, this is analogous to the guided transport of electromagnetic waves. Whenever the diameter of the electron channel was reduced by $\lambda/2$, the resistance in the split- gate experiment increased by a multiple of $\pi\hbar/e^2$ and, finally at a diameter equal to $\lambda/2$, the transport broke down. (In the above experiment, λ was 44 nm). In electrodynamics, this phenomenon is known as the cut-off wavelength (COW) condition of a waveguide. Some time ago, we have proposed a family of novel devices based on this COW principle. [9,10] We have designed a semiconductor bottleneck structure as illustrated in Fig. 4a. In thermal equilibrium, the bottleneck diameter d is smaller than the shortest $\lambda/2$ in the electron ensemble. Accordingly, the bottleneck does not bridge the two reservoirs A and B. The structure, however, can be switched through by heating the electrons in reservoir A. This can be achieved for instance by applying a voltage U_2 between terminals 1 and 3, thus heating the carriers and decreasing their wavelength. They will move through the bottleneck if terminals 1 and 2 are biased properly (U_1). The electrons may also be heated by free carrier photon or phonon absorption. Hence, the COW transistor can also be used for detecting electromagnetic waves or as a temperature sensor. The conduction and valence band edges of the COW structure corresponding to the wave picture are shown in Fig. 4b. Thus, simply by size-tayloring, many device specifications can be achieved.

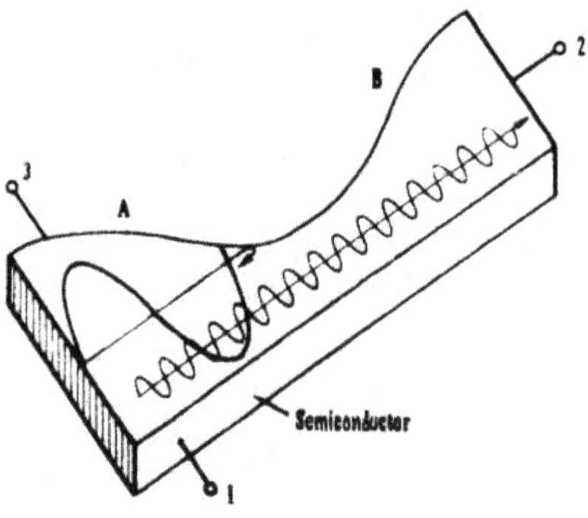

Fig. 4a A COW structure. [9]

We expect these simple devices to operate by the year 1995. Then semiconductor technology will be able to produce on a large scale structures with linear dimensions below 10^{-7} m. The advantages of COW devices are obvious: They do not require any heterostructure nor any special doping profile as needed by present-day devices.

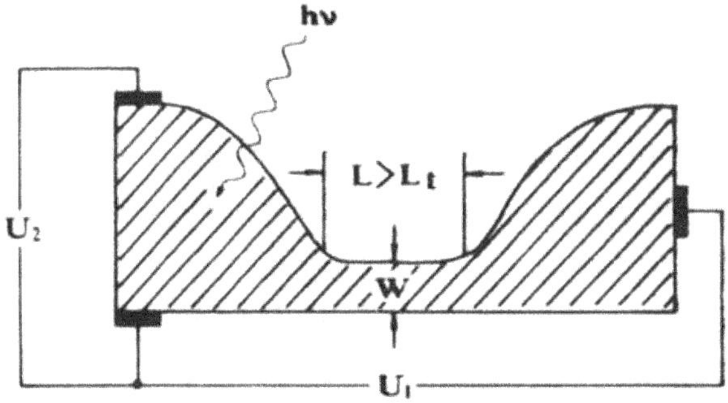

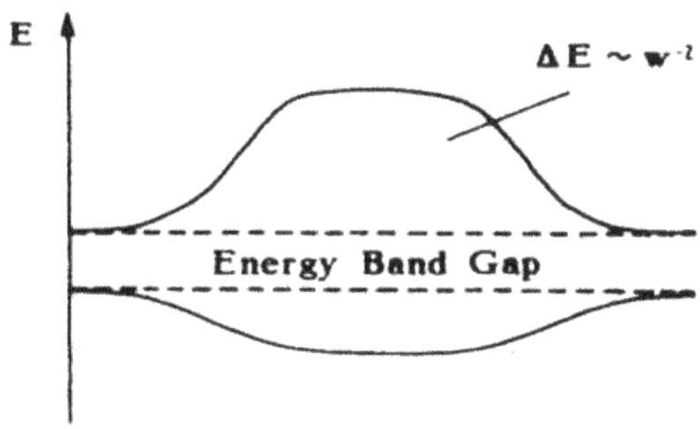

Fig. 4b Energy band edges of the COW structure.

6. Quantum-Well Structures and Electron Tunneling

Modern molecular beam epitaxy (MBE) technology allows to fabricate extremely precise and clean epitaxial layers of semiconductors nearly within one atomic layer. With MBE, heterostructures of semiconductors with different energy band gaps can be prepared. An example of a series of 5 different layers and the correspomding current-voltage characteristics are shown in Figs. 5a and 5b. [11] Within layer thicknesses of several *nm* the electron potential energy can be changed by the proper choice of a semiconducting composition. With semiconductor alloys having different band gaps a quantum well of 5 nm length was prepared. The two potential walls (each 4 nm thick) hinder an electron transport at low electric fields. However,

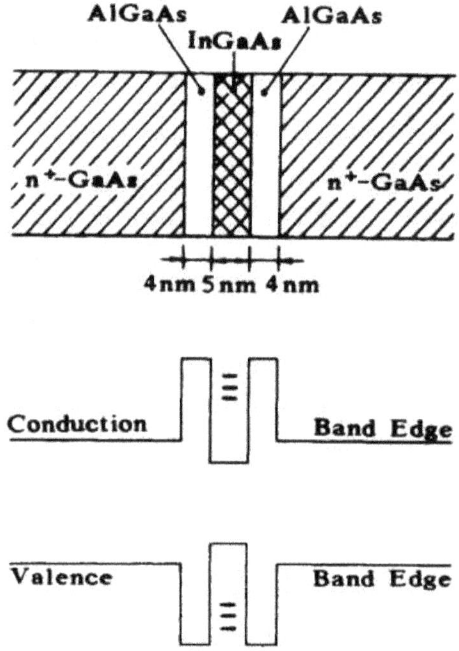

Fig. 5a Semiconductor hetero-structure and corresponding conduction and valence band energies. [11]

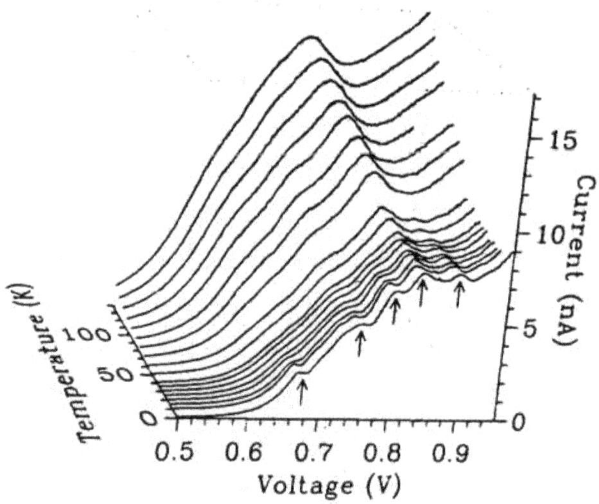

Fig. 5b Current-voltage characteristics of the structure from Fig. 5a with temperature as parameter. [11]

above $5*10^7 Vm^{-1}$ corresponding to a voltage of $0.6V$, electrons start tunneling and a finite current is measured. According to the current-voltage characteristic measured, the electron transport is structured by resonant tunneling. This experiment shows that the widely discussed question of the tunneling time [12] can now be decided. With a structure as investigated by Reed et al. [11] a time-resolved measurement can definitely solve this old problem. The time shift between applied voltage and current with the barrier thickness as parameter should simply and definitely yield the real tunneling time whose definition still is ambiguous in the Copenhagen interpretation of quantum mechanics. [12]

7. Summing-up

Modern microelectronic technology allows to study and at the same time to apply the electron wave nature. The wave nature becomes important whenever the coherence length exceeds a characteristic length, as e.g. a diffusive path length or the linear extension of a confinement. We have shown that in a standard infrared photoconducting detector element coherent electron- electron interaction becomes important at temperatures up to 40 K. The electron conductance in narrow channels is quantized in multiples of $e^2/\pi\hbar$, corresponding to multiples of $\lambda/2$. As a truly important result, this transverse quantization is universal and independent of the host medium. The transverse quantization of electron waves is the basis of a new generation of size-taylored devices. The cut-off wavelength condition can be used to design novel simple devices whenever semiconductor technology allows the production of structures below 0.1 μm. Besides infrared and temperature sensors, the COW principle will lead to memory large scale integration in the 10 nm range within the near future. Recent quantum well MBE structures are expected to allow the measurement of the electron tunneling time and to decide whether it is a real or an imaginary quantity.

The intention of our brief report was to point out that modern microelectronics will give answers to open questions of basic quantum mechanics, particularly those on the wave-particle dualism. We terminate our report by giving some data on the achievements of microelectronics during the past twenty years. In 1968, the silicon single crystal diameter was 32 mm and the integration density was 64 bit; in 1978, 82 mm and 64 kbit were reached. In 1989, the standard diameter of a silicon single crystal is at 200 mm and the integration density of the chips on the production line of IBM is 4 Mbit.

8. Acknowledgement

We gratefully acknowledge the financial support by the Deutsche Forschungsgemeinschaft (Bonn)

9. References

1. D.J. Thouless, *Physics Reports* **13** (1974) 93

2. P. Marquardt and G. Nimtz, *Semicond. Sci. Technol.* **2** (1987) 833

3. G. Nimtz and J.P. Stadler, *Physica* **134B** (1985) 359

4. P. A. Lee and T. V. Ramakrishnan, *Rev. Mod. Phys.* **57** (1985) 287

5. B. L. Al'tshuler and A. G. Aronov, *Solid State Commun.* **30** (1979) 115

6. J. Schilz, J. Lange, L. Mester, and G. Nimtz, *Z. Phys. B* (1990) (in press)

7. J. Lange, Diploma Thesis, Univ. Köln (1990)

8. B. J. van Wees, H. van Houten, C. W. J. Beenakker, J. G. Williamson, L. P. Kouvenhoven, D. van der Marel, and C. T. Foxon, *Phys. Rev. Lett.* **60** (1988) 848

9. G. Nimtz and P. Marquardt, *Appl. Phys.* **A 47** (1988) 317

10. G. Galeczki and G. Nimtz, *Infrared Physics* (1990) (in press)

11. M. A. Reed, J. N. Randall, R. J. Aggarwal, R. J. Matyi, T. M. Moore, and A. E. Wetsel, *Phys. Rev. Lett.* **60** (1988) 535

12. T. F. Spiller, T. D. Clark, R. J. Prance, and H. Prance, *Europhys. Lett.* **12** (1990) 1

A FINITE, RATIONAL MODEL FOR THE EPR-BOHM EXPERIMENT*

H. PIERRE NOYES

Stanford Linear Accelerator Center

Stanford University, Stanford, California 94309, U.S.A.

ABSTRACT

We construct particles and 3+1 space-time from bit-strings of finite length. We show that for 4-events defined by four strings adding to the null string we can define discrete conservation laws invariant under discrete and finite Lorentz transformations. Considering a trajectory for a single particle which "travels" both backward and forward in time and which when read in a causal sense corresponds to pair creation followed by space-like separated spin measurements, we find the usual spin correlations which violate Bell's inequalities. The only "non-local" parameter is our discrete version of the global space-time ordering of special relativity.

1. Introduction

Bell[1] reduced the Bohm[2] version of the EPR[3] *Gedanken*-experiment to the problem of making a physically acceptable model for the calculation of $\left\langle \vec{a} \cdot \vec{b} \right\rangle$. Here $\vec{a}$ and $\vec{b}$ are the spin directions of two space-like separated (massive) spin-$\frac{1}{2}$ particles in a singlet state. Under his (and Einstein's) understanding of "physically acceptable", Bell proved that it is impossible to find such a model compatible with standard quantum mechanics. To my knowledge no model acceptable to Bell (and retrospectively to Einstein) exists which explains the laboratory realizations of many EPR experiments. Our aim in this paper is to model the global correlations predicted by quantum mechanics while preserving the space-like separation and quasi-local character of the two spin-state detections. That such a model exists analogically has been claimed previously.[4] Here the general argument is replaced by a specific physical model.[5]

* Work supported by the Department of Energy, contract DE-AC03-76SF00515.

The basic idea which we use comes from Feynman's image of an electron moving "backward in time" as a positron. From his point of view, the production of an electron–positron pair in a singlet state followed by the subsequent detection of their spins in two space–like separated regions — obviously a Bell–Bohm–EPR situation — can be thought of as a electron which originates in one region, travels backward in time to the pair production event and then forward in time to the other end of its trajectory. The singlet state "production event" when the process is read in causal time order can, when viewed as an event in the middle of a single path connecting two space–like events, be thought of as a "filter" which forces singlet spin–correlation between the particle states on the two sides of the filter when it is traversed in either direction along the single particle path. That its spin is conserved in the process, and hence produces the standard singlet spin correlations when measured, is then no surprise. The underlying symmetry which allows an extended, macroscopic, emission *completed* by an absorption to be read in either space or either time direction is the CPT invariance of relativistic quantum mechanics.

We *could* start from Feynman's path integral approach,[6] were it not for difficulties that "realists" find in his subsequent development of quantum mechanics. He computes the "time evolution" of a single particle system for a *completed* process with specified space–time boundary conditions at the beginning and end of a finite "trajectory" by summing over all allowed "histories". We emphasize that if both space–time endpoints lie in the *past* — which is always the case when they are analysed by a physicist — the replacement of this probability calculation by a differential equation is simply a mathematical convenience. We claim that it is a conceptual mistake to read this (Dirac or Schroedinger) equation as representing the *deterministic* evolution of a complex amplitude. To do so would force the "act of measurement" at the "final" endpoint to "collapse the wave function" leading to many attendant conceptual woes. If one interprets the physical theory as applying only to *past* events, the wave function is "born collapsed", and one neatly avoids falling into this quagmire. Note that this option is also available in classical mechanics and there avoids the criticisms of "determinism" raised by Hume.

Of course this way of cutting the Gordian knot is not novel.[7] I suspect that it is not widely embraced because there are other conceptual difficulties that stand in the path of a "realistic" interpretation of quantum mechanics. One of these is that in order to go from his finite step–length *Zitterbewegung* model for the paths Feynman abandoned the finite and discrete step–lengths by taking a continuum limit, as did others like Jacobson and Schulman,[8] who worked out Feynman's derivation in detail. This opens another can of worms in terms of infinite self–energies, Goedel paradoxes and the like. Karmanov suggested[9] that this step is unnecessary if one starts from Stein's finite and discrete model for relativistic quantum mechanics.[10] A second difficulty for realists is that the step–length used by Feynman is imaginary! In a finite and discrete theory this is simply a "bookkeeping" device that records the non–commutativity between paths introduced by the discrete *Zitterbewegung*, according to McGoveran.[11]

In what follows we hope to show that neither problem occurs in our theory of discrete physics.[12]

2. Bit–String Representation of Lorentz Transformations

Our model uses *bit–strings* of finite length S to represent its fundamental entities; they are defined by

$$\mathbf{a}(S) = (..., b^a_s, ...)_S; \; b^a_s \in 0,1; \; s \in 1,2,....S; \; 0,1,...,S \in \textit{ ordinal integers} \quad (1)$$

They can combine by *discrimination* (XOR) symbolized by "$\oplus$":

$$\mathbf{a} \oplus \mathbf{b} = (..., b^{a\oplus b}_i, ...)_S = \mathbf{b} \oplus \mathbf{a}; \; b^{a\oplus b}_i = (b^a_i - b^b_i)^2 \quad (2)$$

or *concatenation* symbolized by "$\|$":

$$\mathbf{a}(S_a)\|\mathbf{b}(S_b) = (....b_i....)_{S_a}\|(....b^b_j...)_{S_b} = (......, b^{a\|b}_k, ...)_{S_a+S_b}$$

$$b^{a\|b}_k = b^a_i, \; i \in 1,2,...,S_a; \; b^{a\|b}_k = b^b_j, \; j \in 1,2,...,S_b, k = S_a + j \quad (3)$$

To map bit–strings onto one–dimensional integer and half–integer coordinates first note that the Hamming measure $a := \Sigma^S_{s=1} b^a_s$ takes the null string as the "reference ensemble" in McGoveran's definition of *attribute distance*.[13] We restore the symmetry between the symbols "0" and "1" by using for our measure the signed coordinate $q_a(S)$ defined by

$$-\frac{S}{2} \leq q_a := a - \frac{S}{2} \leq +\frac{S}{2} \quad (4)$$

There are $2S+1$ such integrally spaced coordinates for S even and $2S$ for S odd. These integer *or* half–integer coordinates can be related to the usual angular momentum "space quantization" of elementary quantum mechanics by defining

$$J(S)\cos\theta_a := q_a; \; J^2(S) := \frac{S}{2}\left(\frac{S}{2}+1\right) \quad (5)$$

Then integer steps (in cos θ_a, *not* in θ_a) correspond to "rotations" which leave the string length and hence J^2 invariant. Alternatively we can define

$$\tau(S)\cosh\xi_a := q_a; \; \tau^2(S) := \frac{S}{2}\left(\frac{S}{2}+1\right) \quad (6)$$

with $\beta_a := tanh\ \xi_a := \frac{2a}{S} - 1$; "Lorentz transformations" which leave $\tau^2(S)$ invariant for equal steps in $\beta_a S$ imply the usual relativistic addition law for velocities in one dimension. They also imply an absolute reference frame for velocities and directions, which our cosmological construction allows us to identify with the $2.7^{o}K$ background radiation.

In order to define *relative* motion, we note that the constraint

$$\mathbf{a} \oplus \mathbf{b} \oplus \mathbf{c} = \mathbf{0}(S) \tag{7}$$

forces the Hamming measures a, b, c to satisfy the triangle inequalities and hence allows us to define relative angles in a plane by defining the inner product

$$2ab\cos\,\theta_{ab} := \mathbf{a} \odot \mathbf{b} := a^2 + b^2 - |\mathbf{a} \oplus \mathbf{b}|^2 \tag{8}$$

The square of the magnitude of the vector product is then given by

$$|\mathbf{a} \otimes \mathbf{b}|^2 := a^2 b^2 \sin^2 \theta_{ab} \tag{9}$$

and the anti-commutativity by $\theta_{ba} := -\theta_{ab}$. When $c = |a \pm b|$, the three "vectors" lie on a line; our definition of the Lorentz transformation given above is directly applicable. If $|a - b| < |\mathbf{a} \oplus \mathbf{b}| < a + b$, the two strings lie on different lines. The systematic way to compare them is then make a "boost" of each particle along its direction of motion to the coordinate system in which $\beta_a = 0 = \beta_b$, which we already know how to do if S, a and b are even, and then compute the angle. Assuming S even but a odd, we can make the velocity zero, but not the coordinate given by Eq. (4), and if we start from a zero coordinate, the velocity between that and the next position is the limiting velocity (one step in one time interval). We interpret this as the *Zitterbewegung* of a spin-$\frac{1}{2}$ particle, which occurs also in the conventional theories. Then we can interpret the two possibilities $q_{a_0} = \pm\frac{1}{2}$ as the two *helicity states* (i.e. spin quantized along the direction of motion). We separate off a "spin label" string of length two for these cases (i.e. (01) or (10)), and compare only the residual strings of length $2L := S - 2$ (even). Then $L^2 = L(L+1)$, $J = L \pm \frac{1}{2}$ and we obtain the usual angular momentum composition rules. When we discriminate two spin-$\frac{1}{2}$ strings $\mathbf{a}_0$ and $\mathbf{b}_0$ we will obtain the three helicity states of a spin one particle, (11) corresponding to ± 1 and (00) to $\sigma_z = 0$. We can now, with some effort, construct the usual connections between spin and orbital angular momentum at a "Feynman vertex". Our definition of Lorentz transformations (which now must be calculated as boosts along a specified string) and rotations must now leave all discriminations invariant (eg $\mathbf{a}' \oplus \mathbf{b}' \oplus \mathbf{c}' = \mathbf{a} \oplus \mathbf{b} \oplus \mathbf{c}$) as well as the string length S. Extension of these definitions to vector addition would involve us in our interpretation of "second quantization" and is beyond the scope of this paper.

In order to relate this construction of "velocities" and "angles" to laboratory situations, we must specify our *rules of correspondence*.[13] Initially these were given in terms of a *counter paradigm*[14,15] which stated that a bit-string represents a finite step *Zitterbewegung* executed at the speed of light with invariant step-length h/mc which in the laboratory connects two counter firings, or conceptual equivalent, separated by a finite space-time interval. We have recently extended this to an *8-counter paradigm*[5] based on our definition of *4-event*, namely the constraint

$$\mathbf{a} \oplus \mathbf{b} \oplus \mathbf{c} \oplus \mathbf{d} = \mathbf{0}(S) \tag{10}$$

As we explain in more detail in the reference cited, this encompasses all possible connections between four particles (source, sink, $3 \to 1$, $1 \to 3$, $2 \to 2$ (elastic and anelastic scattering) with the usual Mandelstam invariants, crossing symmetry and CPT invariance. To see this briefly, note that if we can assign invariant masses to the four strings, and the process is "on mass shell", four-momentum conservation is equivalent to four-velocity conservation, i.e. $(\Sigma_i\gamma_i, \Sigma_i\gamma_i\vec{\beta}_i)$ is invariant under Lorentz transformations, including rotations. This is in fact a convenient starting point for formulating the quantum mechanical few body problem in a relativistically invariant way.[16] But the Eq. (10) constraint is equivalent (for fixed string length) to saying the number of 1's in any one string is the same as the number of 1's in the string obtained by discriminating the other three. Consequently, with appropriate sign conventions to reverse velocities when particles are changed to anti-particles, the definition of velocity in terms of the number of 1's together with the invariance under rotations and boosts discussed above does indeed insure 4-velocity conservation (*and* 4-momentum conservation!) in 4-events specified by our discrete theory.

The proof of the conservation laws sketched above assumed that we have a unique way to assign mass values to the strings. In order to describe electrons, positrons and γ-rays, this space-time (*content string*) description is concatenated with seven *label* strings. A specific representation which implies the appropriate conservation laws under our definition of the two "events" which specify the space-time boundary conditions for any particular string is

$$\mathbf{e}_L^- := (0101);\ \ \mathbf{e}_R^+ := (1010);\ \ \mathbf{e}_R^- := (0110);\ \ \mathbf{e}_L^+ := (1001)$$

$$\gamma_{LL} := (0011);\ \ \gamma_c := (1111);\ \ \gamma_{RR} := (1100) \tag{11}$$

The labeling scheme itself is derived from the *combinatorial hierarchy*[17] as is explained in more detail in Ref. 12.

In order to *calculate* the electron mass we first calculate the mass of the proton, assumed generated by its gravitational self-interaction,[18] and then calculate the self-energy of the electron due to its electromagnetic interaction with proton-antiproton pairs;[17] this is our way of looking at the original Parker–Rhodes calculation.[19] The string itself can be thought of as emerging from and terminating in two 4–events defined by appropriate equivalents to Eq. (10). This general scheme leads to the conservation of fermion number, weak hypercharge and baryon number across the intervals connecting two scattering events, and to the calculation of the physical and cosmological parameters given in the Table.

3. The EPR–Bohm Experiment in this Discrete Context

Thanks to our Lorentz invariance, and CPT invariance ("crossing symmetry") we can now model the electron-positron EPR experiment using bit-strings in a very simple way. Consider two strings representing, when read in causal order, an electron and a positron starting from a single event and ending in two space-like separated events. For convenience we will take these to be an equal distance from the common event in the coordinate system in which the 3-vector part of the total 4-velocity is zero. This requires us to start with two strings of equal length which discriminate to a single string that can represent the "virtual" (space-like) quantum from which they originate. As in Wheeler-Feynman theory, this is simply the conceptual link between an earlier deflection of a charged particle and the event of interest which is required to conserve energy and momentum between the emission and the materialization process. We cannot read the energy and momentum of the quantum from its "content string" since this has to be either the null (backward light cone) or anti-null (forward light cone) string for (as assumed here) a massless quantum.

So far we have ignored "spin". But, as we outlined above and discuss in more detail elsewhere,[5] in a representation in which the invariant string length S is even, we can distinguish fermions from bosons by whether the number of 1's in the string is odd or even. Then, given odd content strings with $\beta^+ = -\beta^-$ representing the positron and the electron, the odd "1" determines their helicities as opposite. Transformed to the rest system, the angular momentum of the pair then vanishes (as does that of the quantum from which they originate), and we have a bit-string representation of the singlet state. But this correlation persists if we use any arbitrary direction to describe the spin of each separately in space-like separated regions, provided always we use the same *global* definition of direction in comparing the two "measurements". This is a simple property of the strings we have constructed, and holds whether or not the "measurement" is carried out. Of course *if* there is an "interaction" in either of the space-like separated regions, we are dealing with a problem that involves more degrees of freedom than we have included in the model. We would be making a serious error, according to our *modeling methodology*,[12,13] if we did not start our analysis of the situation over again and construct a more complicated model *before* we asked new questions.

Granted our discrete construction and our *rules of correspondence*[13] which con-

Table of Results, June, 1990

General structural results

- 3+1 asymptotic space–time
- combinatorial free particle Dirac wave functions
- supraluminal synchronization and correlation *without* supraluminal signaling
- discrete Lorentz transformations for event–based coordinates
- relativistic Bohr-Sommerfeld quantization
- non–commutativity between position and velocity
- conservation laws for Yukawa vertices and 4–events
- crossing symmetry, CPT, spin and statistics

Gravitation and Cosmology

- the equivalence principle
- electromagnetic and gravitational unification
- the three traditional tests of general relativity
- event horizon
- zero–velocity frame for the cosmic background radiation
- mass of the visible universe: $(2^{127})^2 m_p = 4.84 \times 10^{52}\ gm$
- fireball time: $(2^{127})^2 \hbar / m_p c^2 = 3.5$ million years
- critical density: of $\Omega_{Vis} = \rho/\rho_c = 0.01175$ $[0.005 \leq \Omega_{Vis} \leq 0.02]$
- dark matter = 12.7 times visible matter [10??]
- baryons per photon $= 1/256^4 = 2.328... \times 10^{-10}$ $[2 \times 10^{-10}?]$

Unified theory of elementary particles

- quantum numbers of the standard model for quarks and leptons

with confined quarks and exactly 3 weakly coupled generations

- gravitation: $\hbar c / G m_p^2 = [2^{127} + 136] \times [1 - \frac{1}{3 \cdot 7 \cdot 10}] =$
  $1.70147...[1 - \frac{1}{3 \cdot 7 \cdot 10}] \times 10^{38} = 1.6934... \times 10^{38}$ $[1.6937(10) \times 10^{38}]$
- weak–electromagnetic unification:
  $G_F m_p^2 / \hbar c = (1 - \frac{1}{3 \cdot 7}) / 256^2 \sqrt{2} = 1.02\ 758... \times 10^{-5}$ $[1.02\ 684(2) \times 10^{-5}]$;
  $\sin^2 \theta_{Weak} = 0.25(1 - \frac{1}{3 \cdot 7})^2 = 0.2267...$ $[0.229(4)]$
  $M_W^2 = \pi\alpha / \sqrt{2} G_F \sin^2 \theta_W = (37.3\ Gev/c^2 \sin\ \theta_W)^2$; $M_Z \cos\ \theta_W = M_W$
- the hydrogen atom: $(E/\mu c^2)^2 [1 + (1/137 N_B)^2] = 1$
- the Sommerfeld formula: $(E/\mu c^2)^2 [1 + a^2 / (n + \sqrt{j^2 - a^2})^2] = 1$
- the fine structure constant: $\frac{1}{\alpha} = \frac{137}{1 - \frac{1}{30 \times 127}} = 137.0359\ 674...$ $[137.0359\ 895(61)]$
- $m_p / m_e = \frac{137\pi}{\frac{3}{14}\left(1 + \frac{2}{7} + \frac{4}{49}\right)\frac{4}{5}} = 1836.15\ 1497...$ $[1836.15\ 2701(37)]$
- $m_\pi^\pm / m_e = 275[1 - \frac{2}{2 \cdot 3 \cdot 7 \cdot 7}] = 273.1292...$ $[273.12\ 63(76)]$
- $m_{\pi^0} / m_e = 274[1 - \frac{3}{2 \cdot 3 \cdot 7 \cdot 2}] = 264.2\ 1428..$ $[264.1\ 160(76)]$
- $(G_{\pi N}^2 m_{\pi^0})^2 = (2\hat{m}_p)^2 - m_{\pi^0}^2 = (13.86811 m_{\pi^0})^2$

  [( )] = empirical value (error) or range

nect this model to laboratory experience, we hope to have provided a concrete model for Feynman's imaginative picture of an electron–positron pair as a single particle trajectory connecting two space–like separated spin measurements. For us, granted the physical background we have succeeded in constructing from our theory as represented by the Table of Results, our discrete and finite model reduces the EPR "paradox" to triviality. The space–like connection, granted the Minkowski symmetries and the CPT theorem which are part and parcel of the construction, is just as "real" as a time–like connection. In fact, once one grants our contention that quantum mechanics refers to completed processes that lie in the fixed past, there can be no "collapse of the wave function".

A different conceptual problem with our model is that it allows a *Zitterbewegung* in time as well as in space. This has the consequence that if we imagine a discrete background space in which the possible particle paths are embedded, a particle can go through the "same" space-time "point" many "times", if "time" is represented by an irreversible parameter counting the stepwise evolution of a "path". My contention here is that in any discrete theory, this is a *necessary* consequence of our discrete version of the Lorentz transformation. By making a "boost" referred to some much larger referene mass we can always interject any large number of points between any two points in the original mesh. That this — in general statistical — process is an alternative and completely consistent way of looking at the Lorentz transformation has been argued in detail by McGoveran (Ref. 13, Theorem 41). Of course the same thing happens in a continuum theory, but gets swept under the rug by the assumption that the space–time continuum is structureless. In our opinion, this is one of the obvious sources of the objectionable infinities in second quantized relativistic field theory. We have the conceptual advantage that for us these additional degrees of freedom represent additional external processes that for the moment we can ignore, but which will become relevant when a larger number of particles enter the system. Just as Einstein's coordinates are specified by possible "events" where a light ray could be reflected by a mirror, our coordinates represent potential events where a photon could be emitted or absorbed. The fact that for us the "coordinate string" for a photon is simply the null or the anti-null string,[5] means that we have no way of defining its energy and momentum except by its source and sink (as in Wheeler–Feynman theories).

This brings us to a *physical* interpretation of our *Zitterbewegung* — it is simply the inevitable interaction of any particle with what would be called in second quantized field theory the vacuum fluctuations of the zero–point energy of the massless quantum fields. Casimir had the courage to take this idea seriously, and predicted an attractive force between two conducting plates which has been observed. Others predicted a force in a conducting sphere which is repulsive in the sense that it tends to expand the sphere, and which has also been observed. It is easy to see that our theory predicts the right sign and geometry dependence for these two cases. In fact for us these experiments confirm our prediction that space is 3–dimensional, since they depend only on hc and are independent of the masses and coupling constants involved. A good check on our theory will be whether we can calculate the correct 1/120 factor in the Casimir effect combinatorially.

A second consequence of our finite vacuum fluctuations is that we should be able to make finite combinatorial calculations of all mass ratios and all "mass renormalizations". Our original calculation of the pion mass and the ratio of the proton mass to the Planck mass[20] based on Dyson's limit[21] have both now been found by McGoveran to have well defined combinatorial corrections.[22] The construction presented in this paper allows a firm foundation to be placed under the originally heuristic argument[17] for the Parker-Rhodes calculation[19] of the proton-electron mass ratio. Other mass, mass-ratio and coupling constant calculations are already made (see Summary table) or in the offing.

Whether or not this finite approach holds up in the long run, we trust that we have made a case here for our contention that it allows a new way for us to look at the ERP experiments which avoids some of the old problems.

4. References

1. J. S. Bell, *Physics* (N.Y.), **1** (1964) 195.
2. D. J. Bohm, *Quantum Theory*, Prentice-Hall, Engelwood Cliffs, N.J., (1951) 611-623.
3. A. Einstein, B. Podolsky and N. Rosen, *Phys. Rev.* **47** (1935) 777.
4. D. O. McGoveran, H. P. Noyes, and M. J. Manthey, "On the Computer Simulation of the EPR-Bohm Experiment", in *Bell's Theorem, Quantum Theory and Conceptions of the Universe*, M.Kafatos, ed., Kulwer Academic Publishers, (1989) 153-158 and SLAC-PUB-4729, December 1988.
5. H. P. Noyes, "Bit-String Scattering Theory", *Proc. ANPA 11 (1989)*, F. Abdullah, ed., City University, London EC1V 0HB, and SLAC-PUB-5085, January 1990; "From Bit-Strings to Quaternions", *Proc. ANPA 12* (in preparation).
6. R. P. Feynman and A. R. Hibbs, *Quantum Mechanics and Path Integrals*, McGraw Hill, New York (1965).
7. J. G. Cramer, *Rev. Mod. Phys.*, **58** (1986) 647-87, can be read this way.
8. T. Jacobson and L. S. Schulman, *J. Phys. A, Math. Gen.* **17** (1984) 375-383.
9. V. A. Karmanov, private communication
10. I. Stein, seminars at Stanford, 1978, 1979; —, papers at ANPA 2 and ANPA 3, 1980, 1981; —, *Physics Essays*, **1**, (1988) 155-170; **3** (1990) 66-70.
11. D. O. McGoveran, private communication.
12. H. P. Noyes and D. O. McGoveran, *Physics Essays*, **2** (1989) 76-100.
13. D. O. McGoveran and H. P. Noyes, "Foundations of a Discrete Physics", in *Proc. ANPA 9*, H.P.Noyes, ed, ANPA WEST, 25 Buena Vista Way, Mill Valley, CA 94941, (1988) 37-104 and SLAC-PUB-4526, June, 1989.

14. H. P. Noyes, "A Finite Particle Number Approach to Physics", in *The Wave–Particle Dualism*, S.Diner, et al. eds, Reidel, Dordrecht, (1984) 537–556.
15. Ref. 12, pp 90–91.
16. James V. Lindesay, A. Markevich, H. Pierre Noyes, G. Pastrana, *Phys. Rev.* **D33** (1986) 2339.
17. T. Bastin, H. P. Noyes, J. Amson, C. W. Kilmister, *Int. J. Theor. Phys.* **18** (1979) 455.
18. H. P. Noyes, "Discrete Gravity", in *Physical Interpretations of Relativity Theory. II.*, M.C. Duffy, ed.; conference at Imperial College, Sept. 3–7, 1990 and SLAC–PUB–5218.
19. A. F. Parker–Rhodes, *The Theory of indistinguishables*, (Synthese Library **150**, Reidel, Dordrecht) pp 184–185.
20. H. P. Noyes, "Non–Locality in Particle Physics", SLAC–PUB–1405(rev. Nov. 1975), pp 23–25.
21. F. J. Dyson, *Phys. Rev.*, **85** (1952) 631.
22. D. O. McGoveran, "Advances in the Foundations", *Proc. ANPA 11 (1989)*, F.Abdullah, ed., City University, London EC1V 0HB.

AXIOMATIC QUANTUM PHENOMENOLOGY

ASHER PERES

Department of Physics, Technion—Israel Institute of Technology, Haifa, Israel

ABSTRACT

Quantum phenomena do not occur in a Hilbert space. They occur in a laboratory. The empirical rules obeyed by these phenomena are analyzed. Their natural mathematical expression is in terms of unitary matrices. This leads to the use of Hilbert spaces, for purely phenomenological reasons.

1. Glossary

Let us observe physicists in a laboratory. We see them performing two different kinds of tasks, which can be called *preparations* and *tests*. These are the primitive, undefined notions of quantum theory. Their intuitive meaning can be explained as follows. A *preparation* is an experimental procedure which may be completely specified, or which may involve stochastic processes, such as thermal emission, provided that the statistical properties of these stochastic processes are known, or at least reproducible. For example, the hardware sketched in the left half of Fig. 1 represents a preparation. A *test* starts like a preparation, but it also includes a final step whereby *information* (which was previously unknown) is supplied to an *observer* (namely, the physicist who is performing the experiment). For example, the right half of Fig. 1 represents a sequence of tests. The resulting information is printed on the tape. This information is not trivial because, as seen in Fig. 1, identical tests following identical preparations need not have identical outcomes.

The foregoing statements have only suggestive value. They do not properly define preparations and tests, as this would require prior definitions for the notion of *information* and for related terms such as *known/unknown*, etc. It is obvious that the distinction between preparations and tests involves a direction for the flow of time. The asymmetry between past and future is fundamental in the axiomatic structure of quantum theory. It is similar to the fundamental asymmetry between

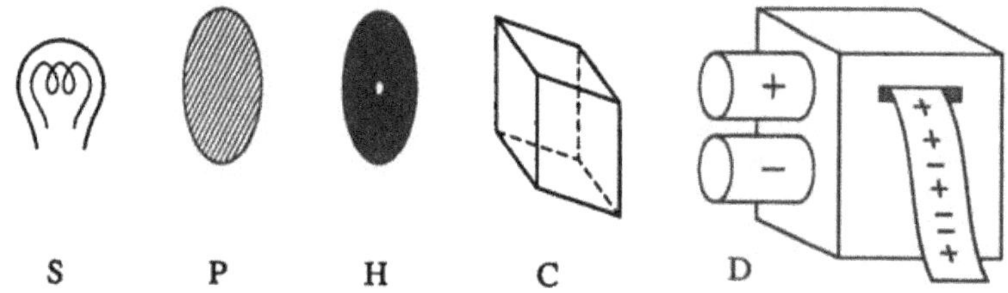

Fig. 1. Light from a thermal source S passes through a polarizer P, a pinhole H, a calcite crystal C, and then it may trigger one of the detectors D. The latter register their output in a device which prints the results.

the past and future light cones in special relativity. It is closely related to the fact that we have a limited amount of free will. We are free to *choose* the preparations and tests that we perform, but our free will stops there. We are not free to choose the future outcome of a nontrivial test.

To continue our glossary, let us try to define a *quantum system.* This is a useful abstraction, frequently met in the literature, but which does not really exist in nature. Ideally, a quantum system would be defined by an *equivalence class of preparations.* For example, there are many equivalent procedures to produce a photon, or a hydrogen atom, etc. The ambiguity of these notions emerges as soon as we think of concrete examples. Is a hydrogen atom in a $2p$ state the same system as one in a $1s$ state? Or is it the same system as a hydrogen atom in a $1s$ state accompanied by a photon? The answer depends on the problem in which we are interested: energy levels or transition rates. As another example, the "quantum system" in a Stern-Gerlach experiment is *not* a silver atom in the atomic beam. It is only the magnetic moment μ of that atom, because the goal of the Stern-Gerlach test is to determine a component of μ. The center of mass of the atom can be treated classically. We must therefore be content with a vague "definition": A quantum system is whatever admits a closed dynamical description within quantum theory.

It is time now to define quantum theory itself: *In a strict sense, quantum theory is a set of rules, giving the probabilities of outcomes of tests that follow specified preparations.*[1] This may not be the way quantum theory is understood by most physicists. They would rather say that quantum theory is used to compute the properties of microscopic objects, for example the energy levels and cross-sections of atoms and nuclei. Despite its uncontested success, the epistemological meaning of quantum theory is fraught with controversy. In the present article, I shall carefully avoid controversial questions, and I shall discuss quantum phenomena in a strictly phenomenological, instrumentalist way.

While quantum systems are somewhat elusive, quantum *states* can be given an unambiguous *operational* definition, based on the notion of *test*. Consider a given preparation and a set of tests, among which some may be mutually incompatible. If these tests are performed many times, after identical preparations, we find that the statistical distribution of outcomes of each test tends to a limit. We can therefore define a state as follows: *A state is characterized by the probabilities of the various outcomes of every conceivable test.* This definition is highly redundant, because these probabilities are not independent. One can specify—in many different ways—a *restricted* set of tests such that the probabilities of the outcomes of these tests determine those of every other test. I shall later return to this point.

As a simple example of definition of a state, suppose that a photon is said to have right-handed polarization. Operationally, this means that if we subject that photon to a specific test (namely, a quarter wave plate followed by a suitably oriented calcite crystal) we can predict *with certainty* that the photon will exit in a particular channel. For any other test, consisting of arbitrarily arranged calcite crystals and optically active media, we can then predict *probabilities* for the various exit channels. (These probabilities are computed in the same way as the classical beam intensities.) Note that the word "state" does not refer to the photon by itself, but to *an entire experimental setup involving macroscopic instruments.*

More generally, we may relate a quantum state to a set of equivalent experimental procedures. The latter may even include purely mental experiments, as long as we are sure that a real experiment is possible *in principle*. For example, it is perfectly legitimate to consider the state of an electron located at the center of the Sun. A measurement of a spin component of that electron is undoubtedly very difficult, and it is ruled out for sure by budgetary constraints; but it is not ruled out by the laws of physics—as they are known today. Therefore it is legitimate to use quantum mechanics to compute the physical properties of a stellar plasma, just as it is used to discuss metallic conduction, or helium superfluidity, that we observe in our laboratory.

The essence of quantum theory is to give a mathematical representation of states (that is, of *preparation procedures*), enabling us to compute the probabilities for the possible outcomes of any test. This article discusses the phenomenology of quantum tests. It starts by listing some basic *empirical* facts. The conceptual implications of these facts are analyzed, and then elevated to the status of "postulates." Although it is not possible to derive the complete formal structure of quantum theory from these empirically based postulates, we shall be able to go a long way toward guessing that quantum states form a complex linear space.

Before we enter into these details, the nature of a quantum test must be clearly understood. A test is more than the mere occurrence of an unpredictable event, such as the blackening of a grain in a photographic plate, or an electric discharge in a particle detector. To be meaningful, these macroscopic events must be accompanied by a *theoretical interpretation*, which always is partly *classical*. For example, the firing of a photodetector in Fig. 1 is interpreted as the arrival of a polarized photon, because we tacitly use the rules of classical electromagnetic theory, according to which a beam of light is split by a calcite crystal into two beams with opposite polarizations. Likewise, the Stern-Gerlach experiment is interpreted as the measurement of a magnetic moment, because it could indeed be such a measurement if we sent little compass needles through the Stern-Gerlach magnet, instead of sending silver atoms. When nuclear physicists measure cross sections, they assume that the nuclear fragment trajectories are classical straight lines between the target and the various detectors. Without this assumption, the macroscopic positions of the detectors could not be converted into angles for the differential nuclear cross sections. Quantum theory appears only at the *next* stage, to explain, or predict, the possible values of the magnetic moment, the cross sections, the wavelengths, etc.

2. Repeatable and maximal tests

Consider two consecutive identical tests, following each other with a negligible time interval between them, so that we may neglect any dynamical evolution of the quantum system, between these tests. If the tests *always* yield identical outcomes, for any preparation, they are called *repeatable*. For example, consider two identical calcite crystals, arranged for testing the linear polarizations of incoming photons, as in Fig. 2. There are three detectors. It is found empirically that only the upper and lower ones may be excited; the central one never is. This is indeed what classical electromagnetic theory would predict for light rays: any trajectory different from those indicated by the dotted lines is impossible to achieve. Note that we *must* tacitly imagine the existence of quasi-classical paths, as indicated by the dotted lines, because this is the only way to interpret the outcome of the experiment. Without some kind of interpretation, experiments are meaningless.

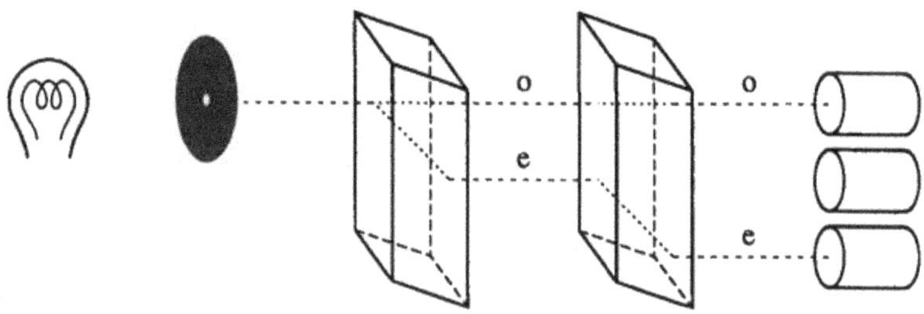

Fig. 2. A repeatable test: the second calcite crystal always confirms the result given by the first one.

There is, however, a delicate point here: We also tacitly assume that the route followed by a photon, when it is tested by the first calcite crystal, does not depend on the existence of a second calcite crystal that we may place between the first one and the detectors. We believe that this photon would have followed the same route even if the second crystal had not been present. Such an assumption is needed in order to be able to say that there are *two* consecutive tests here, and to compare their results, in spite of the fact that the result of the first test is not recorded, but only inferred. This assumption is natural, because of our deeply rooted classical prejudices—we imagine that each photon follows a well defined trajectory. However, this assumption is obviously counterfactual, and it is not verifiable.

In real life, most quantum tests are *not* repeatable. For example, if we measure the momentum of a neutron by observing the recoil of a proton in a photographic emulsion or in a bubble chamber, it is obvious that the momentum of the neutron after the measurement cannot be the same as before it. Truly repeatable tests exist only in the imagination of theorists. They are idealizations, like rigid bodies, or Carnot engines; and like them, indeed, they are very useful in theoretical discussions. In this article, I shall assume that tests have been designed so as to be repeatable, and the word "test" will mean a *repeatable test*, unless specified otherwise.

Let N be the maximum number of different outcomes obtainable in such a test, for a given quantum system. Then, any test that has exactly N different outcomes is called *maximal* or *complete*. For example, a Stern-Gerlach experiment is a complete test for the value of a component of a magnetic moment. It always has, irrespective of the orientation of the magnet, $(2s + 1)$ different outcomes for atoms of spin s. An *incomplete test* is one where some outcomes are lumped together, for example, because the experimental equipment has insufficient resolution.

The adjective *maximal* or *complete* should not be misunderstood. A linear polarization test, such as the one sketched in Fig. 1, is complete only with respect to the polarization of the photon. It yields no information about other properties which the photon may have, such as its position or momentum. Likewise, the Stern-Gerlach experiment is a complete test for spin, while other degrees of freedom are ignored. The result of each one of these tests is observed by *correlating* the value of the internal degree of freedom (polarization or spin) which is tested, to the *position* of the outgoing particle, which can then be detected by macroscopic means.

The notion of completeness of a quantum test is radically different from its counterpart in classical physics. For example, in classical mechanics, one can specify *all* the components of the angular momentum $\mathbf{J}$ of a rotating body. A complete description of the body *must* therefore include all of them. However, when we attempt to measure the components of $\mathbf{J}$ of very small systems such as atoms, we find *empirically* that a very precise measurement of one of the components of $\mathbf{J}$ (namely, with an accuracy better than $\sqrt{J\hbar}$) not only precludes the measurement of the other components, but it even alters the expected values of these other components in an uncontrollable way. We thus see that we have to *choose* between different, *mutually incompatible* tests. For example, we may orient the Stern-Gerlach magnet in any direction we please. Why then is such a Stern-Gerlach test called *complete*? The reason can be stated as the following postulate:

> A. **Statistical determinism.** *If a quantum system is prepared in such a way that it certainly yields a predictable and definite outcome in a specified maximal test, then the various outcomes of* **any** *other test have definite* **probabilities**. *In particular, these probabilities do not depend on the details of the procedure that was used to prepare the quantum system, to make it yield the predictable outcome in the given maximal test. Such a system is said to be in a* **pure** *state.*

The simplest way to obtain quantum systems in a given pure state is to subject them to a complete test, and to discard all the systems that did not yield the desired outcome. When this has been done, *all the past history of the selected quantum systems becomes irrelevant.* The fact that a quantum system in a pure state is bound to bring a definite outcome, if subjected to a specific maximal test, is the *most complete description* that can be given of that system.

Another important notion is that of *equivalent* tests:

> B. **Equivalence of maximal tests.** *Two maximal tests are equivalent if every preparation that yields a* **definite** *outcome for one of these tests also yields a* **definite** *outcome for the other one. In that case, any other, arbitrary preparation (namely one which does not yield a predictable outcome for these tests) will yield the same* **probabilities** *for corresponding outcomes of both tests.*

For example, a Stern-Gerlach experiment measuring J_z (for arbitrary spin) is equivalent to an experiment measuring $(J_z)^3$, or any other single valued function of J_z. Note that postulate B demands that N different preparations yield *definite and different* outcomes for each one of the maximal tests. Equivalence is not guaranteed for tests that merely yield, in some cases, identical *probabilities*.

In real life, most preparations do not yield pure states, but *mixed* ones. After an imperfect preparation, no maximal test has a predictable outcome. For example, photons originating from an incandescent lamp are in a mixed state of polarization. In this particular case, their polarization is completely *random*. Any test for polarization should yield approximately equal numbers of photons with opposite polarizations, if the apparatus has the same efficiency for both outcomes of that test. Such an apparatus is called *unbiased*, and we shall henceforth consider only unbiased tests. This example suggests the following generalization:

C. **Random mixtures.** *It is possible to prepare quantum systems with N states in such a way that* **every** *unbiased maximal test has the same probability, N^{-1}, for each one of its outcomes.*

A random mixture is the state that corresponds to a complete lack of knowledge of the past history of the quantum system. Postulate C seems innocuous, but it has far reaching consequences. First, we note that, for each quantum system, the state which is a random mixture is *unique* (there cannot be several distinct types of random mixtures). This follows from the very definition of a state, namely the set of probabilities for the various outcomes of every conceivable test. All these probabilities are equal to N^{-1}. Moreover, this unique random mixture is *dynamically invariant.* This can be shown as follows: An unbiased maximal test may include doing nothing but waiting, for a finite time. In that case, a quantum system, initially prepared as a random mixture, and allowed to evolve according to its internal dynamical properties, must remain in the state which is a random mixture. If it weren't so, the "idle test" would yield probabilities different from N^{-1}, contrary to the definition of a random mixture. Postulate C may therefore be called the *law of conservation of ignorance.*

3. Consecutive tests and interference effects

Let us consider now *correlations* between the outcomes of *different* consecutive tests, such as the Stern-Gerlach experiments illustrated in Fig. 3. Assume that the collimation of the atomic beam ejected by the oven is very sharp. We can then imagine that three narrow beams leave the first magnet. By adjusting the field intensity in that magnet, the angular separation of these beams can be made large enough, so that *they do not overlap* when they enter the second magnet, and yet not too large, so that the second magnet performs essentially the same test for each one of the three beams that impinge on it. If these conditions are satisfied, it becomes possible to imagine the existence of quasi-classical trajectories—as we did for Fig. 2—in order to give a meaning to the experiment. We can consider the setup shown by Fig. 3 as *two* consecutive tests, rather than a single test with nine possible outcomes (for which we would have no direct interpretation). The location of the impact point reveals not only the outcome of the final test, but also the outcome of the test performed with the first magnet. (Moreover, we assume that *if* the second magnet had not been there, the trajectory through the first magnet *would* have remained the same. As explained above, this is a natural, but unverifiable, counterfactual assumption.)

The nine spots on the detector plate can therefore be unambiguously identified as corresponding to outcomes a, b, and c of the first test, and outcomes α, β, and γ of the second test (Latin and Greek letters will be used to label the outcomes of

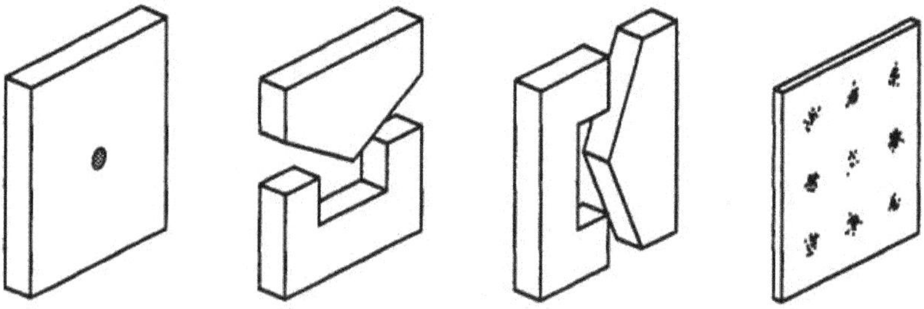

Fig. 3. Two consecutive Stern-Gerlach experiments for particles of spin 1. The drawing has been compressed by a factor 10 in the longitudinal direction. From left to right: an oven, a vertical magnet, a horizontal one, and a detector.

the first and second test, respectively). Let $I_{\mu m}$ be the intensities of these spots. Define $P_{\mu m} = I_{\mu m}/\sum_\nu I_{\nu m}$. This is the probability for the observation of outcome μ, following the preparation of pure state m. Obviously,

$$\sum_\mu P_{\mu m} = 1. \tag{1}$$

The probability matrix $P_{\mu m}$ depends solely on the two tests that are involved in the experimental setup. It does not depend on the source of particles.

It will now be shown that $P_{\mu m}$ satisfies not only (1), but also

$$\sum_m P_{\mu m} = 1. \tag{2}$$

Matrices with nonnegative elements satisfying both (1) and (2) are called "doubly stochastic" in probability theory. The proof of Eq. (2) is based on the obvious fact that any combination of consecutive tests can be considered as a single test. In general, this combined test may be biased, even if the individual tests are not. For example, in Fig. 2, we could *select* one of the outcomes of the first test before performing the second one, by placing an absorber in one of the beams, between the two crystals. Likewise, in Fig. 3, we could bias the final result by performing a selection among the beams which leave the first magnet or, more gently, by subjecting them to an additional inhomogeneous magnetic field, even a weak one, in order to cause different precessions of the magnetic moments of the atoms. Suppose however that we are careful not to introduce any bias by treating differently the beams leaving the first test. Then, if the quantum systems being tested are prepared as a *random* mixture (see postulate C), the probability for each one of the N distinct outcomes of the second test is $p'_\mu = N^{-1}$, just as $p_m = N^{-1}$ was the probability for the mth outcome, in the first part of the combined test. On the other hand, we have $p'_\mu = \sum_m P_{\mu m} p_m$, because $P_{\mu m}$ is the probability that outcome μ follows preparation m. Comparing these results, we readily obtain Eq. (2).

Finally, consider the same two complete, repeatable tests executed in *reverse order.* We likewise define a probability table $\Pi_{m\mu}$, for the observation of outcome m, following a preparation of pure state μ. It is found *empirically* that

$$\Pi_{m\mu} = P_{\mu m}. \tag{3}$$

This can be stated as the following law:

> D. **Law of reciprocity.** *Let ϕ and ψ denote pure states. Then the probability of observing outcome ϕ in a maximal test following a preparation of state ψ, is equal to the probability of observing outcome ψ in a maximal test following a preparation of state ϕ.*

This reciprocity law has no classical analogue. In some instances, it is possible to derive it from symmetry arguments, for example in the case of tests for the linear polarization of photons along different directions: The probability that a photon with one of the polarizations will pass a test for the other polarization can depend only on the angle between the two directions. However, in general, different complete tests are not related by symmetry operations. An unsuccessful attempt to derive the reciprocity law from thermodynamic arguments was made by Landé.[2] Here, we accept it as an empirical fact.

One further step will now bring us to the heart of quantum physics. Consider *three* consecutive repeatable maximal tests, as illustrated in Fig. 4. The first and last tests are identical. As before, it is assumed that the beams which leave each magnet are well separated, so that they do not overlap when they pass through the next magnet, but nevertheless their separation is not too large, so that the next magnet performs essentially the same test for each one of the beams that impinge on it. If these conditions are satisfied, and if we use spin $\frac{1}{2}$ particles for simplicity, eight clusters of points appear on the detector plate (there would be 27

Fig. 4. Three consecutive Stern-Gerlach experiments for spin $\frac{1}{2}$ particles. Eight clusters of points appear on the detector plate, corresponding to the two possible outcomes of each test.

such clusters if particles of spin 1 were used, as in Fig. 3). Each cluster can be labelled by three indices, such as $m\mu n$, indicating that the three tests successively performed on each particle gave outcomes m, μ, and n, repectively (the same set of Latin indices can be used to label the outcomes of the first and last tests, since these tests are identical). If the particles are initially unpolarized (e.g., the atomic beam is produced by evaporation in an oven) the intensity of cluster $m\mu n$ is proportional to

$$\Pi_{n\mu}\, P_{\mu m} = P_{\mu n}\, P_{\mu m}\,, \tag{4}$$

where the right hand side follows from the reciprocity law (3).

Let us now gradually turn off the field of the *second* magnet of Fig. 4. The horizontal separation of the clusters on the detector plate will decrease but, as long as these clusters are distinguishable, their intensities remain constant. As the field continues to decrease, each one of the four pairs of clusters begins to coalesce. One could then naively expect to get, when the second magnet is completely turned off, four clusters with intensities proportional to $\sum_\mu P_{\mu n} P_{\mu m}$. On the other hand, we know that it cannot be so, because this would violate the *repeatability* property of the first and third tests, which are identical: If the second magnet is inactive, all the particles hitting the detector plate must be concentrated in *two* clusters only, those corresponding to identical outcomes of the initial and final tests.

This contradiction, between the naive prediction $\sum_\mu P_{\mu n} P_{\mu m}$ and the correct result, can be explained as follows: When the second magnet is gradually turned off, it ceases to satisfy the requirement that the beams which leave it are well separated. As they come to overlap, two of the pairs of beams interfere constructively and reinforce each other, giving the *double* of the sum of their separate intensities, while the two other pairs interfere destructively and annihilate each other (see Fig. 5). This phenomenon is one of the cornerstones of quantum theory:

> E. **Principle of interference.** *If a quantum system can follow several possible paths from a given preparation to a given test, the probability for each outcome of that test is* **not** *in general the sum of the separate probabilities pertaining to the various paths.*

An analogy with the case of polarized photons is helpful, because statistical properties of photons such as the total intensity of a light beam are adequately handled by classical electromagnetic theory. Instead of the three magnets, let there be three calcite crystals. The first and last crystals test vertical versus horizontal polarization, and the middle crystal tests polarization at $\pm 45°$. If we gradually reduce the thickness of the middle crystal, the eight resulting beams will behave as shown in Fig. 5. Whenever two beams overlap, there is either constructive interference (the amplitude is double and therefore the intensity quadruple of that of a single beam) or destructive interference (the amplitude is zero).

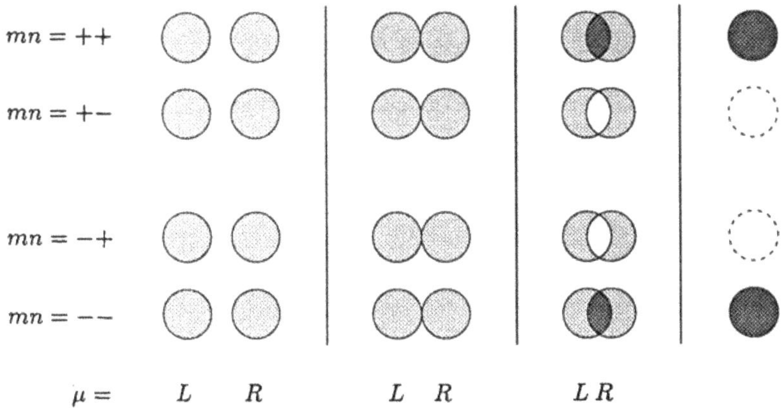

Fig. 5. From left to right: Behavior of the eight beams resulting from the experiment sketched in Fig. 4, when the field in the middle magnet is gradually turned off. Interference effects double the amplitude of the beams with $m=n$, and annihilate those with $m \neq n$.

It is interesting to note that the principle of interference E could be made plausible by simple qualitative arguments, involving only crude mental experiments, such as the one sketched in Fig. 4. (On the other hand, the law of reciprocity D had to be accepted as an empirical fact.) How can such a far reaching principle, which grossly violates our classical intuition, be obtained without any detailed knowledge of the dynamical laws underlying quantum phenomena? The novel feature of quantum physics which inexorably led us to the principle of interference can be succintly stated as follows: *Quantum tests may depend on classical parameters which can be varied* **continuously**, *and nevertheless these tests have fixed,* **discrete**, *outcomes.*

4. Transition amplitudes

Interference effects are not peculiar to quantum physics. They also occur in acoustics, optics, and other types of classical wave motion. The amplitudes of these waves usually satisfy a set of *linear* differential equations. Therefore, when several paths are available for the propagation of a wave, the *amplitudes* combine linearly. On the other hand, the observed *intensity* of these wave phenomena is given by the energy flux, and the latter is *quadratic* in the field amplitudes. Therefore, the intensities—contrary to the amplitudes—are not additive.

In a quantum description of optical interference, the energy flux is proportional to the number of photons per unit area, and therefore to the *probability* of arrival of these photons. The principle of interference E, which applies to all quantum systems (photons, electrons, atoms, ...), suggests that transition probabilities such as $P_{\mu m}$ are the squares of *transition amplitudes*, and that the latter combine in a linear fashion. Moreover, we know from classical optics that phase relationships are essential, and that polarization states are conveniently described by complex numbers. We are thus led to postulate the existence of a *complex* transition amplitude, $C_{\mu m}$, for obtaining outcome μ in a test that follows preparation m. We shall soon see even more compelling reasons for the use of complex numbers.

The postulated transition amplitude $C_{\mu m}$ satisfies

$$|C_{\mu m}|^2 = P_{\mu m} \,. \tag{5}$$

Likewise, we define, for the inverse transition, an amplitude $\Gamma_{m\mu}$, satisfying

$$|\Gamma_{m\mu}|^2 = \Pi_{m\mu} \,. \tag{6}$$

We know from the reciprocity law, Eq. (3), that $|\Gamma_{m\mu}| = |C_{\mu m}|$, but we do not know the phases, as yet. An appropriate choice of these phases is our next problem.

Suppose that we have several consecutive tests, and assume for the moment that a *single* path can lead from the initial preparation to the final outcome. The probability for that path is the product of the consecutive probabilities for each step, as in Eq. (4). It is therefore natural to define the transition amplitude for a sequence of maximal tests as the *product of the consecutive amplitudes of each step*. For example, in the triple Stern-Gerlach experiment (Fig. 4), the amplitude for the path labelled $m\mu n$ is $\Gamma_{n\mu}C_{\mu m}$. Up to this point, nothing was assumed that had any physical consequence: The phases of the transition amplitudes $C_{\mu m}$ still are irrelevant, and we could as well use the probabilities $P_{\mu m}$. It is only now that we introduce a new physical hypothesis (borrowed from classical wave theory):

> F. **Law of composition of transition amplitudes.** *The phases of the transition amplitudes can be chosen in such a way that, if several paths are available from the initial state to the final outcome, the complete amplitude for that outcome is the algebraic* **sum** *of the amplitudes corresponding to the various paths.*

Consider again the triple Stern-Gerlach experiment of Fig. 4. If the central magnet is switched off, the amplitudes $\Gamma_{n\mu}C_{\mu m}$ have to be summed over μ, which is the unresolved intermediate outcome. On the other hand, all we really have in this case is a pair of *identical* maximal tests, whose results must always agree. Therefore the overall transition probability, from the pure state m (prepared by the first Stern-Gerlach magnet) to the pure state n (obtained from the last magnet) is

δ_{nm} . Let us assume that the complete transition *amplitude* is δ_{nm} too, without any extra phase factor. We then obtain

$$\sum_{\mu} \Gamma_{n\mu} C_{\mu m} = \delta_{nm} . \tag{7}$$

The matrices C and Γ thus are the inverses of each other. On the other hand, we must have $|\Gamma_{n\mu}| = |C_{\mu n}|$. Thus, if we were able to choose phases in such a way that

$$\Gamma_{n\mu} = C^*_{\mu n} , \tag{8}$$

we would obtain the *unitarity* relation

$$\sum_{\mu} C^*_{\mu n} C_{\mu m} = \delta_{nm} . \tag{9}$$

We are now faced with a well defined algebraic problem: Given $P_{\mu m}$, can we find a unitary matrix $C_{\mu m}$ satisfying Eq. (5) ? Remarkably, it turns out that this problem has solutions, for a *dense* set of $P_{\mu m}$, provided that these $P_{\mu m}$ are *doubly stochastic* matrices, namely they satisfy both (1) and (2). This condition is necessary, but it is not sufficient. As we shall presently see, the probabilities $P_{\mu m}$ must satisfy not only (1) and (2), but also some complicated inequalities. Doubly stochastic matrices that satisfy Eq. (5) are called *orthostochastic.*[3]

The absolute values $|C_{\mu m}|$ being already determined, the problem is to assign *phases* to the complex numbers $C_{\mu m}$, in such a way that (9) is valid. Let us count the number of independent equations in (9). It is enough to consider the case $m > n$, because $m < n$ corresponds to the complex conjugate equations, and, if $m = n$, Eq. (9) is automatically satisfied by virtue of (1). Thus, counting separately the real and imaginary parts, there are $N(N-1)$ equations (9) to be satisfied. These equations, however, are not independent, because the $|C_{\mu m}|$, which are given by (5), are themselves not independent. They must satisfy the N constraints (2), of which only $(N-1)$ are independent, because the sum of these constraints is automatically satisfied, by virtue of (1). The final count thus is $(N-1)^2$ independent algebraic conditions imposed on the phases of $C_{\mu m}$.

This is also the number of unknown *nonarbitrary phases* that have to be determined. Indeed, we may, without affecting the validity of the unitary conditions (9), choose arbitrarily the phases of an entire row and an entire column of the matrix $C_{\mu m}$. We are thus left with $(N-1)^2$ nontrivial phases, to be determined by $(N-1)^2$ independent nonlinear algebraic conditions. It is therefore plausible that there is some dense domain of values of the $P_{\mu m}$, in which our problem has a *finite* number of solutions. Let us see this in detail.

For $N = 2$, the problem is trivial. The case $N = 3$ can be handled as follows.[3] Choose phases such that $C_{\mu 1}$ and C_{1m} are real. Eq. (9) for $nm = 12$ then gives

$$|C_{11}C_{12}| + e^{i\beta}|C_{21}C_{22}| + e^{i\gamma}|C_{31}C_{32}| = 0, \tag{10}$$

where $e^{i\beta} = C_{22}/|C_{22}|$ and $e^{i\gamma} = C_{32}/|C_{32}|$. This is a complex equation for the two unknown phases β and γ. A simple graphical solution is to draw a triangle with sides $|C_{\mu1}C_{\mu2}|$. This is possible if, and only if, all the triangle inequalities such as

$$|C_{11}C_{12}| \leq |C_{21}C_{22}| + |C_{31}C_{32}|, \tag{11}$$

are satisfied by the given values of $|C_{\mu m}| = \sqrt{P_{\mu m}}$. Obviously, if β and γ are a solution, $-\beta$ and $-\gamma$ are another one. This is the only ambiguity in this case.

The solution of Eqs. (5) and (9) for the case $N \geq 4$ is much more complicated.[4] The space of the allowed values of $P_{\mu m}$ has a subspace of dimension $< (N-1)^2$, in which the various algebraic constraints are not independent, and therefore in which there is a *continuous* family of solutions for the phases of $C_{\mu m}$. This difficulty indicates that we ought to reverse our approach. The amplitudes $C_{\mu m}$ should be considered as the *primary*, fundamental objects, and the probabilities $P_{\mu m}$ derived from them, in spite of the fact that only the $P_{\mu m}$ are directly observable.

The same conclusion can also be reached by considering a modified triple Stern-Gerlach experiment. Let us orient the last magnet in a direction that is *not* parallel to the first one. The third magnet then provides a *new* test, whose outcomes we are going to label by boldface letters, **r**, **s**, etc. When, in the modified experiment, the central magnet is turned on, all the paths are distinguishable, and the probability for path $m\mu\mathbf{r}$ is $\Pi_{\mathbf{r}\mu}P_{\mu m}$. We therefore write the transition amplitude for that path as $\Gamma_{\mathbf{r}\mu}C_{\mu m}$, where $|\Gamma_{\mathbf{r}\mu}|^2 = \Pi_{\mathbf{r}\mu}$, as usual. Let us now gradually turn off the central magnet, so that the various μ become indistinguishable. The natural generalization of the sum over paths (7) obviously is

$$\sum_{\mu} \Gamma_{\mathbf{r}\mu} C_{\mu m} = \mathbf{C}_{\mathbf{r}m}, \tag{12}$$

where $\mathbf{C} = \Gamma C$, again is a unitary matrix. Its element $\mathbf{C}_{\mathbf{r}m}$ is the transition amplitude from preparation m to outcome **r**. Note that the matrices Γ and C in (12) are time ordered: the earliest is on the right, the most recent on the left. This property is readily generalized to the case of several intermediate states. It suggests that, in a fully developed formalism for quantum theory, the dynamical evolution of a quantum system must be represented by a product of unitary matrices.

From here on, it is easy to guess the correct route. The characteristic property of unitary matrices is that they preserve the norm of a vector. Complex vectors thereby appear on the quantum theoretical stage. Their obvious role will be to represent quantum *states*. In particular, the N outcomes of a maximal test shall correspond to an *orthogonal basis*, and the $C_{\mu m}$ matrix shall be the transformation matrix from one basis to the other. One more postulate is needed:

> **G. Principle of superposition.** *Any complex vector, except the null vector, represents a realizable pure state.*

This raise a difficult question: how to design an experimental setup which prepares the pure state represented by a given vector. Unfortunately, this question cannot be answered by quantum theory. As explained above, quantum theory allows to compute probabilities for the outcomes of tests following specified preparations. On the other hand, quantum theory does not supply any instructions for setting up the laboratory procedures that are used as preparations and tests (just as Euclidean geometry does not instruct us how to manufacture rulers and compasses). These procedures are freely designed by physicists, and then their design can be analyzed, *a posteriori*, with the help of quantum theory.

We still have to define a *scalar product* for a pair of vectors. Eq. (8) indicates that this product is Hermitian; the complex vector space describing quantum phenomena is a Hilbert space. The generic case of scalar product is that of a pair of vectors which are neither parallel nor orthogonal. These vectors correspond to physical states that are neither identical, nor totally different, but rather "partly alike."

Let us therefore define a "test for state ψ" as one which always succeeds for quantum systems prepared in state ψ, and always fails for those in a state orthogonal to ψ. This need not be a maximal test, but only one that singles out ψ from all the states orthogonal to ψ. A last postulate is the following:

> H. **Projection postulate.** *Let ϕ and ψ be normalized vectors. The transition amplitude for a quantum system prepared in state ϕ to pass successfully a test for state ψ is the scalar product of ϕ and ψ.*

The law of reciprocity D then becomes a trivial consequence of the Hermitian nature of the scalar product.

In summary, the empirical, macroscopic rules obeyed by quantum phenomena are most naturally expressed in terms of unitary matrices. This justifies the use of Hilbert spaces, on purely phenomenological grounds.

Acknowledgments

This work was supported by the Gerard Swope Fund, and by the Fund for Encouragement of Research.

References

1. H. P. Stapp, *Am. J. Phys.* **40** (1972) 1098.
2. A. Landé, *Foundations of Quantum Theory: A Study in Continuity and Symmetry*, (Yale Univ. Press, New Haven, 1955).
3. Y. H. Au-Yeung and Y. T. Poon, *Linear Algebra and Appl.* **27** (1979) 69.
4. G. Auberson, *Phys. Letters B* **216** (1989) 167.

NECESSARY AND SUFFICIENT CONDITIONS FOR AN INDIVIDUAL DESCRIPTION OF THE MEASUREMENT PROCESS

HANS PRIMAS

Laboratory of Physical Chemistry, ETH
ETH-Zentrum, CH-8092 Zürich, Switzerland

ABSTRACT

Von Neumann's Hilbert-space formulation of quantum mechanics is restricted to locally compact phase spaces and therefore of rather limited validity. It excludes a proper description of genuinely irreversible stochastic processes, symmetry breakings, phase transitions and the classical properties of the quantum world. These restrictions prevent a physically realistic formulation of the measuring problem. In the more general codification by algebraic quantum mechanics, these limitations do no longer exist and the well-known no-go theorems do not apply. An individual ontic interpretation of C*-algebraic quantum mechanics is not only possible, but in fact much simpler and more informative than epistemic ensemble interpretations. It follows that the effective dynamics of an individual open quantum object is given by a nonlinear stochastic integro-differential equation for the state vector. Allowing for the irreducible contextuality of any observing situation, the individual formulation of the measurement problem is well-posed, and the statistical ensemble interpretation can be derived in a conceptually well-defined manner.

1. What is the measurement problem?

In the pioneer days of quantum mechanics, microscopic systems were described by wave functions, while the experimental set-ups were considered as macroscopic and were described in terms of traditional classical physics. However, *the distinction between "microscopic" and "macroscopic" has become obsolete*; as we know today, there are small molecular systems having classical properties, and there are macroscopic systems showing also quantal properties. In order to pose the measurement problem properly, it is important not to confuse microscopic and quantal, macroscopic and classical. Furthermore, we have to distinguish carefully between individual and statistical descriptions, and between ontic and epistemic interpretations. The interrelations are delicate and nontrivial. To make matters worse, the established terminology is unwise and troublesome.

First of all, what is called a *measurement* in the literature on the foundations of quantum mechanics is a fiction, for which we always should use Pauli's [1] expression *measurement of the first kind*. In any case, a measurement of the first kind is a poor, inadequate and misleading model for the link between theory and experiment. In the laboratory, *there are no measurements of the first kind*. No measurement gives a sharp value, no measurement is instantaneous.

The heart and essence of the so-called measurement problem is the question whether the measuring tools can be included in the quantum-mechanical description. The Copenhagen view does *not* present quantum mechanics as a *universal theory*, it *presupposes* observational tools, but does not describe them. Certainly one should not try to make a theory of the measuring process on the basis of a theory which is unable to explain engineering physics. The only consistent course is to show, first of all, that quantum theory is indeed capable to describe the well-known phenomena of classical analytical point mechanics, classical continuum mechanics, the theory of nonlinear dynamical systems, stochastic processes, thermostatics, phase transitions, and so on.

One of the central theoretical problems in molecular quantum mechanics is the analysis of the interaction of an individual, small, non-isolated quantum object with its environment. The "measurement process" of quantum mechanics is just a *special case* of such an interaction. Its description requires a *theory of interactions* between quantal and classical degrees of freedom of a general quantum system. Such an enterprise is mathematically difficult, but *all conceptual tools are available* in the codification of quantum mechanics in terms of C*- and W*-algebras. The remaining open problems are of purely physical and mathematical nature, *the inclusion of a human observer, or even his consciousness, is completely unnecessary*. The philosophical problems associated with quantum-mechanical measurements are the very same as in classical physics. That is not to say that the problem of the observer in physical theories is nonexistent or trivial, but only that *quantum theory does not pose new problems not already present in classical science*.

2. On the universal validity of quantum mechanics

The first subject of investigation must be the question of the *domain of validity* of the basic theory. If we posit hypothetically the *universal validity of quantum mechanics*, we have to avoid paradoxes of self-reference. If we want to speak *about* an object, it is compulsory to introduce a *metalanguage* which is not part of the «universally valid theory». This state of affairs has nothing to do with quantum mechanics, but is a *logical* necessity, similar to the semantic incompleteness theorem by Tarski.

In the following, I adopt the *working hypothesis* that – *in a proper codification* – quantum mechanics has universal and unlimited validity at the level of atoms, molecules and non-cosmological macroscopic bodies. Subatomic, genuinely Lorentz-relativistic phenomena, and problems related to men's free will and consciousness will, however, be excluded. Neither the projection postulate nor the expectation-value postulate are presupposed, *but they have to be derived in an appropriate form from first principles of a universally valid quantum theory*. This task is the *measurement problem.*

A measuring instrument is necessarily an *open system* but it has not to be macroscopic, the essential feature is its *classical* behavior in the sense of a *Boolean* logical structure. Amplification devices may be desirable but they are not necessary. As an example, one may consider *molecular coding* and the subsequent phenotypic expression of the genetic material. The emergence of irreversibility has nothing to do with an additional amplification process but requires a breaking of the time-inversion symmetry. Evidently, *symmetry breakings* are the most common features of our everyday experience, they show up not only in the macroscopic but already in the *mesoscopic and molecular domain*. For example, there are *circular DNA-molecules*, which possess at the same time purely quantal properties, like their photochemical characteristica, and strictly classical properties, like their knot-type and knot-number. *Enzymes* act as molecular measuring devices and require a classical behavior for their function. The *immune system* represents a molecular organization with memory, which obtains information about its outside environment and whose main task is to maintain the molecular identity of the living system. In order to understand such systems, one needs a theory of matter which is valid in the whole molecular domain, including systems of mesoscopic and macroscopic dimension, which can describe both the quantal and the classical properties, and of a single individual object in an objective way. For this task, neither the Hilbert-space formalism of traditional quantum mechanics, nor the Copenhagen view are sufficient.

3. Classical quantum systems

Every experiment ever performed in physics, chemistry and biology has a classical description, that is, a description in terms of a Boolean language. A quantum-theoretical description of such an experiment has to be in terms of *classical quantum systems*. A quantum system with a commutative algebra of observables has a *Boolean* logical structure and behaves like a system in classical physics. On the other hand, the formalism of algebraic quantum mechanics allows the description of every dynamical system of classical physics as a dynamical C*- or W*-system with a *commutative* algebra of observables. That is, the characteristic element of classical physics is its *Boolean structure*, and not the fact that Planck's constant of action is a hidden parameter. Therefore we adopt the following definition[2]:

> *DEFINITION : CLASSICAL QUANTUM SYSTEMS*
>
> *A quantum system is said to be classical if its algebra of observables is commutative.*

Note that every classical quantum system depends on Planck's constant. That is, the limit $\hbar \to 0$ is in itself not relevant for the discussion of classical quantum systems. As counterexamples show, the classical behavior of quantum systems is, as a rule, also *not related* to a large number of particles involved.

The formulation of quantum theory in terms of algebraic quantum mechanics puts in evidence that the existence of classical quantum subsystems is an intrinsic feature of every realistic quantum system. That is, *the existence of classical quantum systems has not to be postulated, but is a consequence of a proper mathematical codification of quantum mechanics*. The fact that such classical quantum systems are unknown in traditional quantum mechanics is due to the unrealistic irreducibility postulate, introduced in the Hilbert-space codification by von Neumann without any physical reasons.

The idea of a measurement *presupposes* that we are able to discriminate between the observing and the observed system. *Our ability to describe the world cannot go farther than our ability to isolate objects*. The separation of the universe into two parts, a part *which sees* and a part *which is seen*, is a conceptual necessity, which requires that the object system is not Einstein–Podolsky–Rosen-correlated with the observing system. Bohr's requirement that "the functioning of the measuring instrument must be described within the framework of classical physical ideas" [3] has not to be prerequired, but *follows* from a theorem of algebraic quantum mechanics [2,4] by a much weaker assumption:

> *RESULT OF ALGEBRAIC QUANTUM MECHANICS :*
>
> *Two interacting dynamical quantum systems are never Einstein–Podolsky–Rosen-correlated if and only if one of the systems is a classical quantum system.*

A quantum system qualifies as a measuring tool if and only if it is not Einstein–Podolsky–Rosen-correlated with the rest of the world. If this is not factually true, but corresponds to abstractions the experimenter makes, *then we have to make the very same abstractions in our theoretical discussion.*

4. What do we have to prove?

As a starting point for an individual description of the measurement process, we need a mathematically rigorous codification of the first principles of quantum mechanics which is internally consistent and empirically valid for the whole molecular and engineering domain. On this basis, we have to prove

a) the existence of classical quantum systems,
b) the feasibility of an individual and ontic interpretation,
c) the possibility of symmetry breakings, in particular the breakdown of the symmetries related to time-inversion and the so-called "collapse",
d) the existence of truly irreversible processes (i.e. backward deterministic and forward completely nondeterministic stochastic processes),
e) the existence of memories,
f) the causal behavior of measurement systems ("the cause must be prior to the effect"),
g) the admissibility of engineering science,
h) the existence of measuring processes and the approximate validity of the expectation-value postulate,
i) the compatibility of the physical principles with the human free will, which is constitutive for experimental science.

To solve the measurement problem, it is not sufficient to derive only the conditional probabilities for their *postulated* occurrence, we have *to prove* from first principles that quantum events actually occur. This implies that one has to demonstrate that the pointer observable belongs to the center of the algebra of observables and that pointer states are represented by *disjoint* states, i.e. by pure states which have different dispersionfree values for the pointer observable. In order to prove the disjointness of the pointer states, one has to describe the measuring tools in quantum-mechanical terms. *A reduced description of the object system is not sufficient.* For example, in a statistical description it is not sufficient to show that the final reduced density operator is diagonal with respect to the measured observable. In the individual description, it is not sufficient to show that the effective state vector is governed by a nonlinear stochastic Schrödinger equation with different possible final states.

5. First principles of quantum mechanics and their codifications

In order to make use of the powerful tools of modern mathematics, we need a systematic and comprehensive mathematical formalization of the physical principles of quantum theory. However, we have to be on our guard since no mathematical codification does fully justice to physical first principles, and every codification adds something which cannot be defended by physical principles. The popular no-go theorems of measurement theory are important because they throw light on the limitations of a particular *codification*, but usually their *physical content* is meager. *There is no reason to identify quantum mechanics with any of its codifications.* If we have difficulties with a particular codification, we have to check the physical validity of the codification before we change or reject the physical principles of quantum mechanics.

The kinematical first principles are the easiest to codify. In quantum mechanics, Heisenberg's commutation relations characterize quantum kinematics. There seems to be a broad agreement that Weyl's group-theoretical rephrasing of Heisenberg's commutation relations is the proper thing to do. If we adopt this point of view, the codification of quantum kinematics is straightforward: *it is a representation theory of Weyl's canonical commutation relations over a given phase space.*

Much more problematic is the codification of *quantum dynamics*. Usually, the dynamics of a *free system* is given group-theoretically and can be represented by a one-parameter group of *automorphisms*. However, in general automorphic time evolutions do not correspond to reality because they do not take the actual physical *interactions* into account. The first principles of physics require that the time evolution of a nonelementary system is uniquely determined by the time evolution of the free systems and their *interactions*. The only two fundamental interactions which are relevant in the Galilei-relativistic domain are *electromagnetism* and *gravitation*. It would be science fiction to adopt the admittedly elegant and convenient hypothesis that the time evolutions of closed systems are always automorphic. Sometimes they are, sometimes they are not – that is not a matter of belief or axiomatization, but a matter of *proof* starting from the free evolutions and the known physical interactions.

Von Neumann's Hilbert-space codification presupposes that the phase space (necessary for the formulation of Weyl's canonical commutation relations) is *locally compact*. Since the phase space of an actual measuring instrument never is locally compact, von Neumann's codification excludes without necessity the representations required for the kinematical description of measurement instruments. Its generalization – the *Fock-space quantum field theory* – is theoretically inconsistent. A universal theory of matter must be able to describe matter in its interaction with the radiation field and other environmental interactions. *Von Neumann's Hilbert-space codification of quantum mechanics, the Fock-space formalism, or any known version of quantum logics are not such universally valid theories.* These codifications are unable to describe without ad hoc assumptions experimental facts like the emergence of symmetry breakings, classical observables, and truly irreversible phenomena.

This limitation of the traditional formulations is not a weakness of quantum mechanics. It is due to the use of an improper mathematical codification of the basic ideas of quantum mechanics. A universally valid formulation of quantum mechanics has to be *representation-independent*. Such formulations are available since many decades. The best presently available codification of the principles of quantum theory is *algebraic quantum mechanics*. Algebraic quantum mechanics is nothing else but a precise and complete codification of the heuristic ideas of quantum kinematics of the pioneer days. *No ad hoc modifications, no hidden variables, no quantization procedures, no many-worlds interpretations are necessary*. Moreover, algebraic quantum mechanics encompasses all kinds of physical systems, it is a straightforward generalization both of classical point mechanics and traditional quantum mechanics. In the domains proper of these theories, nothing is changed. Nevertheless, both theories are treated in the same framework. In contradistinction to von Neumann's codification, the algebraic formulation avoids the limitations of the Stone-von Neumann uniqueness theorem, and is also appropriate for quantum fields, infinite-dimensional dynamical systems, Hamiltonian descriptions of nondeterministic stochastic processes, or for a truly quantum-theoretical description of classical systems.

The uniqueness theorem by Stone und von Neumann implies that all Hilbert-space representations of a quantum system are physically equivalent if and only if the corresponding phase space is locally compact. In the generic case – and this is the crucial point – there are many *physically inequivalent representations*. That is, Hilbert-space representations of quantum mechanics are possible, but they are not universal but *context-dependent*. For systems whose phase spaces are not locally compact there always exist contextual superselection rules and contextual classical observables. *There is a dramatic qualitative difference between the behavior of finite and infinite systems*. For finite systems, the phase space is locally compact and there is only one W*-representation (uniqueness theorem by Stone and von Neumann). For infinite systems the phase space is not locally compact, there are infinitely many physically inequivalent W*-representations, and symmetry breakings are the rule.

Sometimes it is argued that "*measuring apparatuses as well as human observers are finite systems*"[5]. This remark is, however, not appropriate since measuring apparatuses and human observers are necessarily *open* systems whose interactions with the rest of the world cannot be neglected. They have no *robust* description in terms of traditional quantum mechanics as *Hamiltonian* systems with *finitely many* degrees of freedom. Any Hamiltonian description of dissipative open systems has to include electromagnetic interactions and requires infinitely many degrees of freedom and, therefore, goes beyond the limits of traditional quantum mechanics.

The existence of physically inequivalent faithful representations for infinite systems implies the existence of *classical observables* in most representations. That is, classical observables have not to be postulated, they can be rigorously *derived* from algebraic quantum mechanics. Usually, the C*-algebra $\mathcal{A}$ of intrinsic observables is assumed to be simple, so that *there are no intrinsic classsical observables*. Nevertheless, the structure of the C*-algebra $\mathcal{A}$ completely determines all transition probabilities und all superselection rules. The corresponding classical observables become manifest only in the weak closure of appropriate representations. Since the choice of an appropriate faithful representation π depends on the physical context, we speak of *contextual classical observables*. A faithful representation $\pi(\mathcal{A})$ of a C*-algebra $\mathcal{A}$ is

C*-isomorphic to $\mathscr{A}$, so that at the first glance nothing is changed. However, the representation $\pi(\mathscr{A})$ acts as an algebra of bounded operators on a Hilbert space $\mathscr{H}_\pi$, and the weak closure of $\pi(\mathscr{A})$ in the algebra of all bounded operators acting on $\mathscr{H}_\pi$ is a W*-algebra $\mathscr{M}_\pi$ which characterizes the particular representation π chosen. The contextual classical observables associated with the representation π are elements of the center $\mathscr{Z}_\pi$ of the W*-algebra $\mathscr{M}_\pi$,

$$\mathscr{Z}_\pi := \{Z \mid Z \in \mathscr{M}_\pi, ZM = MZ \text{ for all } M \in \mathscr{M}_\pi\} \quad .$$

Contextual classical observables are *emergent* in the sense that they are generated by the algebra $\mathscr{A}$ of intrinsic observables together with a contextual topology, nevertheless they are not functions of the elements of $\pi(\mathscr{A})$. A typical contextual observable is *temperature*, it is manifest in all global representations which are relevant for the description of systems in thermal equilibrium.

6. Endophysics and exophysical descriptions

If we adopt the working hypothesis that quantum mechanics is universally valid, we have to cut a Gordian knot. All *fundamental* first principles we know refer to *strictly closed systems*. Since strictly closed systems cannot be observed from outside, such descriptions are not operational. On the other hand, an observer cannot be part of the object system itself, since a description must not be self-referential, but has to be formulated in terms of some metalanguage.

A helpful distinction has been made by Otto Rössler [6] and David Finkelstein [7, 8] by introducing the notions of *endophysics* and *exophysics*. Probably misusing their ideas, I adopt nevertheless their way of speaking:

DEFINITIONS:

A strictly closed physical system without any concept of an observer is called an *endoystem*.

If the endoworld is divided into an observing and an observed part, we speak of an *exophysical description*.

The world of the observers with their communication tools is called an *exosystem*.

All fundamental universally valid first principles we know refer to strictly closed systems, hence belong to deterministic endophysics. They are universally valid but not operational. Strictly speaking, there is nothing outside an endosystem. The endophysical description is a view without perspective, it is God's panorama, the "*view from nowhere*", as modern philosophers say.

On the other hand, *experimental* physics belongs to *exophysics*. Every exophysical description requires a division of the holistic quantum world. Yet, the endoworld does not present itself already divided. *We have to divide it.* The measurement problem is an *interface problem between an endophysical and an exophysical description.* Endophysics has universal laws, but does not refer to measurements. Exophysics can be operational, but has no universal laws. The Copenhagen view bypassed the issue by some ad hoc precepts, like the "collapse of the wave function", or the projection postulate.

Endophysics describes ontic aspects of the material reality in a non-operational way, *yet its first principles are not sufficient for a characterization of exosystems*. This fact does not imply that we cannot go from endophysics to an exophysical description, but for such an enterprise we need additional *regulative principles*. Every operationalization of an endophysical theory must be formulated in another, essentially richer language, an appropriate *exophysical metalanguage*. The very possibility of experimental and engineering science *presupposes* the existence of a subject–object distinction and the possibility of a free choice of the initial conditions. It requires an exophysical domain in which phenomena can be experienced, and in which irreversibility can become manifest. While endophysics is about being, exophysics deals with observations, knowledge, and measurements which *reflect* the objectively real.

Endophysics is different from exophysics[6], so that two modes of descriptions are required: a formalism for the universally valid quantum endophysics, and a formalism for a context-dependent quantum exophysics. In algebraic quantum mechanics, quantum endophysics can be characterized by a C*-algebra $\mathscr{A}$ of intrinsic observables, while every exophysical description – which is necessarily context-dependent – is characterized by a W*-algebra $\mathscr{M}$, which can be obtained via an appropriate faithful representation of the algebra $\mathscr{A}$ of intrinsic observables[9–13]. The distinction between the endophysical *context-independent C*-algebraic formalism* and the exophysical *context-dependent W*-algebraic descriptions* is of crucial importance. Many of the conceptual difficulties and alleged paradoxes of quantum mechanics are due to the failure to distinguish properly between endophysical and exophysical descriptions.

Endophysical algebraic quantum mechanics describes the fundamental *intrinsic observables* by a C*-algebra $\mathscr{A}$ which has to be chosen as small as possible, for example as a separable, simple and nuclear C*-algebra. The *potential intrinsic properties* are characterized by the selfadjoint elements of $\mathscr{A}$, the possible *intrinsic states* of the system by extremal positive linear functionals on $\mathscr{A}$. Exophysics depends on regulative principles which are context-dependent, hence not universally valid. In the algebraic codification, the algebra $\mathscr{M}$ of contextual observables is a W*-algebra with a separable predual, statistical states are represented by *normal* linear functionals on $\mathscr{M}$, while the contextual observables are given by *positive operator-valued measures* in $\mathscr{M}$.

Quantum *endo*physics does not predict what happens in a physical experiment, since in an endoworld there is not yet any concept of observers or observing tools. It is a strictly deterministic theory which describes what is supposed to be objectively real, that is, the reality existing independently of human observations. A statistical interpretation of quantum endophysics does not make sense, but it is possible and consistent to adopt a non-operational and non-probabilistic *individual and ontic interpretation* for quantum endophysics. Since such an endophysical interpretation makes no reference to the abilities of human observers, it is *strongly objective* in the sense of d'Espagnat[14].

7. The construction of Baconian exophysics from endophysics

In order to go from the hypothetically universally valid endophysical description to a contextual exophysical description, one has to introduce *regulative principles* reflecting the view of the observer. Till today, all well-confirmed applications of quantum mechanics belong to Baconian exophysics. In *experiments* – in contradistinction to *observations* – one *prepares* systems in initial states, *controls* some of the variables, and finally *measures* a particular variable. The regulative principles of experimental scientific research and engineering science are to a large extent characterized by Francis Bacons's motto *dissecare naturam.* Baconian science requires power to create initial conditions, entertains the idea that we can learn from the past for the future, and rejects teleological considerations. That is, its regulative principles stress *the facticity of the past and the probabilistic predictability of the future*. Since Baconian science is not identical with empirical science, we have to introduce explicitly the following regulative principles for contemporary engineering exophysics :

a) It is possible to dissect nature in such a way that the exosystem (including the experimenter) is not Einstein–Podolsky–Rosen-correlated with the quantum object under investigation.
b) All processes manifest in the exosystem are backward deterministic, and there are some processes which are forward purely nondeterministic. That is, teleological processes are explicitly excluded.
c) In the exosystem, all hardware is realizable by non-anticipatory input-output systems, i.e. by systems whose present states are determined by the past history.
d) Nature can be manipulated and the initial conditions required by experiments can be brought about by exophysical interventions.
e) There exists an additional level which can influence the exosystem without being influenced by it, and which is not subject matter of the basic physical theory.

If we would like to derive the results of laboratory experiments from the first principles of quantum endophysics, we have to adopt in the theoretical discussion the very same regulative principles as the experimenter does. Since the Baconian regulative principles are widely adopted in present-day experimental science, the challenge of the measurement problem is to prove that the Baconian normative principles are indeed *compatible* with the first principle's of quantum endophysics. This is not a trivial affair. *The measurement problem is solved if and only if we can construct a representation of the C*-algebra of the universally valid but nonoperational quantum-endophysics which fulfills the additional requirements of Baconian science*. The first condition refers to the separability of the external world from the quantum object, it is the prerequisite for the existence of any exophysical description. The second and third point require a breaking of the time-inversion symmetry, (which poses no difficulty) and the rejection of advanced solutions (which is the important part). The requirement of a dissipative and causal (in the sense "the stimulus precedes the response") character of the exosystem requires that in the classical part of the exosystem there are no forward deterministic and backward nondeterministic processes. This can be achieved by breaking the time-inversion symmetry, together with the rejection of all advanced solutions. The last two items are certainly problematic but already present in classical physics and in engineering science. Since I do not believe that consciousness or free will can be reduced to physical properties of the organism such as brain states, I have to include all ideas of choice and purpose in appropriate *regulative principles* which are not derivable from first principles of endophysics.

The Baconian world view has no absolute or a priori validity, it may be compatible with but is certainly not implied by quantum endophysics. There is the possibility of alternative forms of knowledge. The Baconian mode of explication sets its own well established normative standards of judgement we often tacitly adopt for our research. The dissipativity and the backward determinism of Baconian science are posited, not derived. They are derivable from the basic endophysical laws only if we add appropriate *boundary conditions*. If the principles of quantum endophysics are invariant under time-inversion, and if the endophysical dynamics is bidirectional deterministic, then a *teleological* exophysical description is consistent with the basic endophysical laws if and only if a *Baconian* exophysical description is consistent[13].

8. The individual description of quantum objects

The time evolution of a quantum object can be described *individually* or *statistically*. However, the appropriate mathematical formulations are fundamentally different: The individual description requires C*-algebraic formalism whereby the selfadjoint elements of the C*-algebra $\mathcal{A}$ represent the *potential properties* of the system. A statistical ensemble description has to be in terms of a contextual W*-algebra $\mathcal{M}$, whereby the positive operator-valued measures in $\mathcal{M}$ stand for the *experimentally accessible quantities*. In the traditional Hilbert-space formalism of quantum mechanics, the distinction between the C*-algebraic and the W*-algebraic description is hidden since in this exceptional case, there is a one-to-one correspondence between the C*-formulation (in terms of the C*-algebra $\mathcal{A} = \mathcal{B}^{\infty}(\mathcal{H})$ of compact operators acting on the Hilbert space $\mathcal{H}$) and the W*-formulation (in terms of the W* algebra $\mathcal{M} = \mathcal{B}(\mathcal{H})$ of all bounded operators acting on $\mathcal{H}$). This case is exceptional because it is the only case in which the bidual $\mathcal{A}^{**}$ of $\mathcal{A}$ equals $\mathcal{M}$, $\mathcal{M} = \mathcal{A}^{**}$, while in general one has $\mathcal{M} \subset \mathcal{A}^{**}$. This typical situation is well known from classical Hamiltonian point mechanics on a phase space Ω, where the individual dynamics is given by a trajectory in the phase space Ω, and where the dynamical variables are real elements of the C*-algebra of continuous functions on Ω vanishing at infinity, $\mathcal{A} = \mathcal{C}^{\infty}(\Omega)$, while the statistical description is in terms of the W*-algebra of bounded Borel-measurable functions on Ω, $\mathcal{M} = L^{\infty}(\Omega)$ and probability densities in $L^{1}(\Omega)$. In this case we have the typical situation $\mathcal{M} \subset \mathcal{A}^{**}$, so that the statistical description can be derived (under additional assignments) from the individual one, *but not the other way round*.

In classical mechanics, a pure state is characterized by a point of the phase space Ω. If the phase space is uncountable (e.g. $\Omega = \mathbb{R}^{2n}$), then there is clearly no way to measure the state with absolute precision, or to prepare a system in such a way as to know in which pure state the system is. That is, *in classical mechanics the concept of a pure state is not operational*. In the framework of algebraic quantum mechanics, one describes the experimentally accessible states by *statistical states*, which are characterized mathematically by *normal* positive linear functionals on the W*-algebra $\mathcal{M}$ of contextual observables. For classical mechanics over the phase space Ω this algebra is given by $\mathcal{M} = L^{\infty}(\Omega)$, while the normal states are probability densities, i.e.

positive elements of $L^1(\Omega)$. In this formalization, we can say that the pure states of classical point mechanics are not operational since there are no *normal* pure states[15].

The situation in quantum mechanics is quite different. In contrast to classical mechanics, for quantum systems with finitely many degrees of freedom there *are* normal pure states. But there is a pivotal difference between classical and quantal systems: An open quantal system can be entangled with its environment by Einstein–Podolsky–Rosen correlations. That is, if a quantum system is in a pure state, its subsystems are in general not in pure states. As opposed to the classical world view, the quantum world cannot be cut arbitrarily into pieces. Quantum systems like electrons, atoms or molecules do in fact not exist in an absolute sense since they are entangled with their environments. In the mathematical description, this predicament is reflected by the fact that such an entangled system cannot be in a pure state, so we must indeed renounce ascribing physical attributes to it in an absolute way. Hence, the difficulty does not lie in the existence of external disturbances, but in the possibility of Einstein–Podolsky–Rosen correlations.

A realistic description of quantum systems is possible if and only if there are no Einstein–Podolsky–Rosen correlations between the object system and its environment. Therefore I adopt the following definition of an object[2, 9, 12, 15, 16]:

DEFINITION : OBJECTS

An object is defined to be an open quantum system, interacting but not Einstein–Podolsky–Rosen-correlated with the environment.

It follows that objects are exactly those quantum systems for which at every instant a maximal description in terms of pure states is possible. An object is something having individuality and properties, so that we can interpret a pure quantum state of an object as an ontic state. Here the notion of an "ontic state" refers to a mode of being, describing characteristics existing independently of any observation, while the notion of a "pure state" refers to a merely mathematical concept, meaning an extremal positive linear functional on the algebra of observables.

It is perfectly legitimate to base a realistic interpretation of individual quantum objects on a nonoperational and nonprobabilistic description in terms of *potential* and *actualized* properties, where the actualized properties correspond to the elements of reality in the sense of Einstein. The selfadjoint elements of the C*-algebra $\mathcal{A}$ of intrinsic observables represent the *potential properties* of an individual quantum object. Those properties which are *actualized* at the instant t characterize the ontic *state* ρ_t. Ontic states are supposed to describe exhaustively the physical reality at a particular time. A property $A \in \mathcal{A}$ is actualized at time t if $\rho_t(A^2) = \rho_t(A)^2$.

These definitions apply to classical systems as well as to quantum systems. Classical systems are characterized by the fact that the set of *potential properties* equals the set of *actualized properties*. In other words, *all* potentially possible properties (described by *intrinsic observables*) have always a dispersion-free value. In quantum mechanics, however, the set of actualized properties always is a *proper* subset of the set of potential properties, so we have to distinguish carefully between properties actualized at some instant, and potentially possible properties. It is assumed that every potential property can be actualized under appropriate circumstances, but in quantum mechanics it is impossible to actualize *all* potential properties at the *same instant.*

9. The dynamics of quantum objects in its individual and statistical description

In the usual statistical description of a measurement of the first kind, a pure initial statistical state is supposed to change into a mixture, the mixture being that of eigenstates of the measured observable with weights equal to the probability of finding the respective eigenvalues as the measured value of the observable. In an individual account, the notion of a "mixed state" does not exist. In the individual description, the dynamics maps ontic states into ontic states, or in mathematical terms, pure states into pure states. There is no *immediate* correspondence between the properties of individuals and the properties of ensembles, but starting with the individual description, the necessary ergodicity conditions for the feasibility of a Gibbsian ensemble description can be given.

Consider a *quantum system* with finitely many degrees of freedom which interacts strongly with a nontrivial quantal environment. The endophysical description is then based on a C*-algebra $\mathscr{A} = \mathscr{A}_{\mathrm{b}} \otimes \mathscr{A}_{\mathrm{f}}$, where $\mathscr{A}_{\mathrm{b}}$ is the C*-algebra of the intrinsic observables of the *bare* system, and $\mathscr{A}_{\mathrm{f}}$ is the C*-algebra of the intrinsic observables of the *free* environment. Typically $\mathscr{A}_{\mathrm{b}}$ is C*-isomorphic to the algebra of compact operators acting on some Hilbert space, while $\mathscr{A}_{\mathrm{f}}$ is typically a separable antiliminal C*-algebra. The characterization of the dynamics on the C*-algebraic level is a notoriously difficult problem. If we know the interactions between the bare object system and the free environment, we know in principle the intrinsic dynamics of the composed system. Since transition probabilities (which are intrinsically defined on the C*-algebraic level) are invariant under automorphisms, it follows at once that *a measuring-type dynamics cannot be automorphic*. An important effect of the interaction of the bare system with its environment (e.g. the electromagnetic radiation field) is that the object system becomes *dressed* (e.g. by a cloud of virtual photons) and more or less *disentangled* from the new effective environment – unfortunately, this is a very complicated dynamical process and the main stumbling block for a rigorous discussion of the measuring process. It is important to recognize that *actual laboratory experiments refer always to dressed objects*, never to bare systems.

The measurement problem can be solved if we can find a faithful representation π of the C*-algebra $\mathscr{A}$, such that the W*-algebra $\mathscr{M}_\pi$ generated by this representation fulfils all the regulative principles required for an exophysical description of a laboratory experiment. The center $\mathscr{Z}_\pi$ of the algebra $\mathscr{M}_\pi$ of contextual observables represents the classical part of the measuring apparatus, it is always a classical, intrinsically chaotic dynamical system. In the simplest case, the W*-algebra is a W* tensor product $\mathscr{M}_\pi = \mathscr{M}_{\mathrm{d}} \overline{\otimes} \mathscr{M}_{\mathrm{e}}$, where $\mathscr{M}_{\mathrm{d}}$ is a factor of type I, representing the *dressed object* which is no longer entangled with the *exosystem*, described by the commutative W*-algebra $\mathscr{M}_{\mathrm{e}} = \mathscr{Z}_\pi$.

Since $\mathscr{M}_{\mathrm{e}}$ is a commutative algebra, every ontic state ρ_t on the tensor product $\mathscr{M}_{\mathrm{d}} \overline{\otimes} \mathscr{M}_{\mathrm{e}}$ necessarily is a *product state*, $\rho_t = \alpha_t \otimes \beta_t$. This fact implies that all expectation values $\rho_t(AB)$ of product operators $AB \in \mathscr{M}_{\mathrm{d}} \overline{\otimes} \mathscr{M}$, $A \in \mathscr{M}_{\mathrm{d}}$, $B \in \mathscr{M}_{\mathrm{e}}$, factorize with respect to ontic states, i.e. $\rho_t(AB) = \alpha_t(A)\beta_t(B)$. If the W*-algebra $\mathcal{M}_d$ of contextual observables of the dressed object system is a factor of type I, then $\mathcal{M}_d$ is W*-isomorphic to the algebra of all bounded operators acting on some Hilbert space $\mathcal{H}$, $\mathcal{M}_d \cong \mathcal{B}(\mathcal{H})$, so that the ontic state of the object system at time t can be represented by a state vector $\Psi_t \in \mathcal{H}$, such that $\alpha_t(A) = \langle \Psi_t | A \Psi_t \rangle$, and the time dependence $t \mapsto \Psi_t$ of the ontic state of the object system is given by a dynamical map $\Psi \mapsto \Psi_t$ which in general is neither an automorphism nor a group. If the intrinsic time-inversion symmetry is broken, then both backward deterministic and forward completely nondeterministic, and forward deterministic and backward completely nondeterministic realizations are possible.

The effective ontic state of the dressed object depends also on the initial condition of the ontic state of the effective environment, which is a classical system with an atomfree phase space Ω, $\mathcal{X}_\pi \cong L^\infty(\Omega)$. The initial value of the individual state of the effective environment can be characterized by a point ω of the phase space Ω, so that the state vector Ψ_t of the dressed object is also a function $\omega \mapsto \Psi_t(\omega)$ of the initial values $\omega \in \Omega$. For a precisely fixed initial value ω, the time evolution $t \mapsto \Psi_t$ of the state vector of the dressed object is *strictly deterministic*. It can be shown that the evolution $\omega \mapsto \omega_t$ of the *free* environment is of crucial importance for the nature of the dynamics of the *dressed* object. Let $f: \Omega \to \mathbb{C}$ be a smooth complex-valued function, and define f_t by $f_t(\omega) := f(\omega_t)$. The *individual autocorrelation function*[17]

$$\phi(t|\omega) := \lim_{T\to\infty} \frac{1}{2T} \int_{-\infty}^{\infty} f_{s+t}(\omega)\, f_s^*(\omega)\, ds \quad , \quad t \in \mathbb{R} \quad ,$$

characterizes to a large extent the *intrinsic chaotic nature* of the environment, where we use the notion *chaos* in the original sense of Norbert Wiener[18]. If ϕ is continuous at zero and if the spectral distribution of f is absolutely continuous, then the dynamical system loses the memory of the initial conditions, $\lim_{t\to\infty} \phi(t|\omega) = 0$, and the individual function $t \mapsto f_t(\omega)$, with ω fixed, represents a truly chaotic motion which is *crypto-deterministic* in the sense of Whittaker[19].

Provided the conditions for the validity of the Birkhoff[20] individual ergodic theorem are fulfilled, one can turn from the individual chaotic motions in the sense of Wiener to an *ensemble description* in terms of a Kolmogoroff probability space (Ω, Σ, μ), and consider the family $\{f_t(\omega) | t \in \mathbb{R}, \omega \in \Omega\}$ as a stochastic process. The individual crypto-deterministic motion $t \mapsto f_t(\omega)$ is extremely sensitive to the initial value ω, two close-by trajectories diverge exponentially with time, and a small uncertainty in the initial condition grows rapidly. Since the phase space Ω is atomfree and not locally compact, say of the type $\mathbb{R}^\infty$, it is impossible – *even in principle* – to know the *precise* initial state ω of the free exosystem so that exact predictions or retrodictions are impossible. In Baconian experimental science we are interested in *predictions* (as opposed to retrodictions), so we need a statistical estimation of the *initial conditions* (as opposed to final conditions). If we know the epistemic statistical initial state of the *classical* environment, then we know the relevant probability density $p: \Omega \to \mathbb{R}^+$, so that we can evaluate an appropriate classical ensemble average $\langle A \rangle_t$ for an observable A of the dressed object,

$$\langle A \rangle_t = \int_{\Omega} \langle \Psi_t(\omega) \,|\, A\Psi_t(\omega)\rangle \, p(\omega)\, \mu(d\omega) \quad .$$

Of course, one can write this average also in terms of a density operator D_t

$$\langle A \rangle_t = \mathrm{tr}(D_t A) \quad ,$$

but in this reduced description we lose the information that the actual mixture has a *unique decomposition* into disjoint states. *An ignorance interpretation of a mixture is possible if and only if its components are disjoint states.* In the traditional Hilbert-space formalism, a mixture has no unique decomposition into pure states, so that a naive ensemble interpretation proceeds along shaky grounds.

In the statistical ensemble description, the effective dynamics of the dressed object is given by a nonlinear stochastic integro-differential equation for the state vector. In particularly simple models, the effective dynamics $\Psi \to \Psi_t$ can be shown to be given by a *nonlinear stochastic Schrödinger equation* which in some models may even be an Itô-differential equation. This nonlinear stochastic Schrödinger equation is a *derived effective* equation of motion, and should not be confused with similarly looking phenomenologically postulated equations, or with context-independent nonlinear modifications of quantum mechanics. The *nonlinear* behavior of the dressed quantum object is a *secondary* phenomenon due to *feedback effects*, that is, due to the polarization of the classical environment by the object and its backreaction on the quantum object. The *stochastic nature* of the dynamics of the dressed object reflects the influence of the environment. If the environment is intrinsically chaotic, the retarded description of the object dynamics is genuinely irreversible.

A measurement process is characterized by the fact that in a crypto-deterministic individual description, the final value of the pointer observable depends not only on the initial state of the object system but also in an extremly sensitive way on the individual initial state of the environment, so that the individual outcome of the measurement is not predictable, while the relative frequencies can be estimated. A *good* measurement instrument has to fulfill many additional conditions, for example that after a sufficiently long time T the expectation-value postulate of quantum mechanics is fulfilled in a good approximation,

$$\langle \Psi_0 \,|\, M\Psi_0 \rangle \approx \int_{\Omega} \langle \Psi_T(\omega) \,|\, M\Psi_T(\omega)\rangle \, p(\omega)\, \mu(d\omega) \quad ,$$

where M is the measured object observable and Ψ_0 the initial state vector. Note that in an individual description, the measuring process transforms the initial state vector crypto-deterministically into the final state vector $\Psi_T(\omega)$. The "actualization of the potentialities" is nothing but the *continuous yet chaotic* dynamical evolution $\Psi_0 \mapsto \Psi_T(\omega)$ which transforms non-actualized potential properties into actualized ones, and actualized properties into non-actualized ones. The associated probabilities of a statistical description are inevitable and irreducible, and not a matter of ignorance.

References

1. W. Pauli: *Die allgemeinen Prinzipien der Wellenmechanik*. In: *Handbuch der Physik*. Ed. by H. Geiger and K. Scheel. Second ed.. Berlin. Springer. 1933.
2. H. Primas: *Chemistry, Quantum Mechanics and Reductionism*. Berlin. Springer, 1981, second edition 1983.
3. N. Bohr: *Quantum physics and philosophy, causality and complementarity*. In: *Philosophy in the Mid-Century. A Survey*. Ed. by R. Klibansy. Florence. La Nuova Italia Editrice. 1958. Pp. 308–314.
4. G. A. Raggio: *A remark on Bell's inequality and decomposable normal states*. Letters in Mathematical Physics **15**, 27–29 (1988).
5. G. C. Ghirardi, A. Rimini and T. Weber: *Unified dynamics for microscopic and macroscopic systems*. Phys. Rev. **D 34**, 470–491 (1986).
6. O. E. Rössler: *Endophysics*. In: *Real Brains, Artificial Minds*. Ed. by J. L. Casti and A. Karlqvist. New York. North-Holland. 1987. Pp. 25–46.
7. D. Finkelstein and S. R. Finkelstein: *Computational complementarity*. Int. J. Theor. Phys. **22**, 753–779 (1983).
8. D. Finkelstein: *Finite physics*. In: *The Universal Turing Machine. A Half-Century Survey*. Ed. by R. Herken. Hamburg. Kammerer & Unverzagt. 1988. Pp. 349–376.
9. H. Primas: *Mathematical and philosophical questions in the theory of open and macroscopic quantum systems*. In: *Sixty-Two Years of Uncertainty: Historical, Philosophical and Physical Inquiries into the Foundations of Quantum Mechanics*. Ed. by A. I. Miller. New York. Plenum. 1990. Pp. 233–257.
10. H. Primas: *Induced nonlinear time evolution of open quantum objects*. In: *Sixty-Two Years of Uncertainty: Historical, Philosophical and Physical Inquiries into the Foundations of Quantum Mechanics*. Ed. by A. I. Miller. New York. Plenum. 1990. Pp. 259–280.
11. H. Primas: *The measurement process in the individual interpretation of quantum mechanics*. In: *Quantum Theory Without Reduction*. Ed. by M. Cini and J. M. Lévy-Leblond. Bristol. Adam Hilger. 1990. Pp. 49–68.
12. H. Primas: *Realistic interpretation of the quantum theory for individual objects*. La Nuova Critica. In Press (1990).
13. H. Primas: *Time-asymmetric phenomena in biology. Complementary exophysical descriptions arising from deterministic quantum endophysics*. In: *Proceedings of the International Workshop "Information Biothermodynamics"*. Torun. In press.
14. B. d'Espagnat: *Reality and the Physicist. Knowledge, Duration and the Quantum World*. Cambridge. Cambridge University Press. 1989.
15. H. Primas: *Foundations of theoretical chemistry*. In: *Quantum Dynamics of Molecules. The New Experimental Challenge to Theorists*. Ed. by R. G. Woolley. New York. Plenum Press. 1980. Pp. 39–113.
16. H. Primas: *Contextual quantum objects and their ontic interpretation*. In: *Symposium on the Foundations of Modern Physics, 1987. The Copenhagen Interpretation 60 Years after the Como Lecture*. Ed. by P. Lahti and P. Mittelstaedt. Singapore. World Scientific. 1987. Pp. 251–275.
17. N. Wiener: *Generalized harmonic analysis*. Acta Math. **55**, 117–258 (1930).
18. N. Wiener: *The homogeneous chaos*. Amer. J. Math. **60**, 897-936 (1938).
19. E. T. Whittaker: *Chance, freewill and necessity in the scientific conception of the universe*. Proc. Phys. Soc. **55**, 459–471 (1943).
20. G. D. Birkhoff: *Proof of the ergodic theorem*. Proc. Nat. Acad. Sci. U.S. **17**, 656–660 (1931).

SOME ASPECTS OF NEUTRON QUANTUM OPTICS

H. RAUCH
Institut für Kernphysik, Technische Universität Wien,
Wiedner Hauptstraße 8-10, A-1040 Wien, Austria

ABSTRACT

An attempt is made to describe neutron interferometry experiments in terms of quantum optics. Several aspects of spin-state interferometry are discussed in relation to the wave-particle behaviour and to the measuring problem of quantum mechanics. It is shown that a previously performed double resonance experiment can be interpreted as a magnetic analogue of the superconducting Josephson effect. A new version of a double resonance experiment is proposed which is inspired by the Bohr-Einstein contraversy concerning the existence of a particle trajectory.

1. Introduction

All neutron optical effects are described by the wave properties which exist also for massive particles. Totalreflection, single-slit and double-slit diffraction, diffraction from various gratings and crystals as well as from different surface layers have been observed in the pat[1]. Interferometer experiments exhibit for any kind of radiation the most direct evidence of the wave-particle dualism and they are, therefore, closely related to interpretational questions of quantum mechanics.

Many textbook experiments of quantum mechanics have been realized since the invention of neutron interferometry by Rauch, Treimer and Bonse in 1974[2]. The macroscopic coherence of the strictly periodic arrangement of atoms inside a perfect silicon crystal provides the basis for the production of widely separated coherent neutron beams by dynamical Bragg diffraction. Up till now, such interferometers are designed monolithically to preserve the parallelism of the lattice planes between the beam splitter, the mirror and the analyzer crystal. Figure 1 shows the standard three-plate interferometer and a skew symmetrically cut interferometer which provides even more space for sample insertion and for beam handling. Perfect crystal interferometers are cut from high quality silicon pieces used in advanced semiconductor technology. The characteristic dimensions of such interferometer crystals

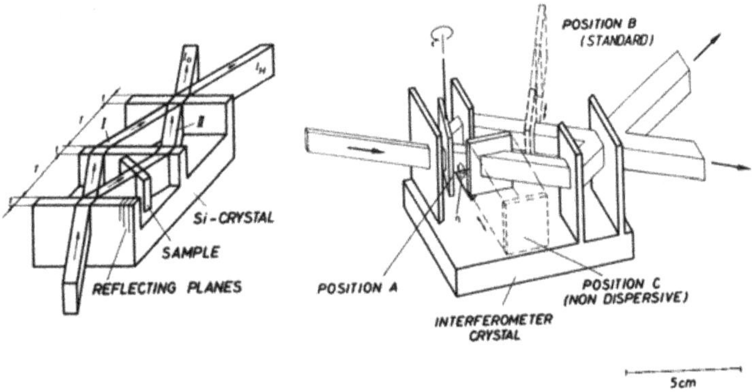

Fig. 1: Sketch of a symmetic and of a skew symmetric perfect crystal neutron interferometer with an indication of different measuring methods.

are about 10x10x70 cm, and the beam separation, for thermal neutrons with wavelengths of about 1.8 Å, is of the order of 5 cm. Therefore, the neutron interferometer represents a real macroscopic quantum system which makes possible the realization of many fundamental experiments of quantum mechanics.

An incident beam is split coherently at the first crystal plate, reflected at the middle plate and coherently superposed at the third plate (Fig.1a). It follows immediately from general symmetry considerations that the wave functions in both beam paths, which compose the beam in the forward direction behind the interferometer, are equal ($\psi_0^I = \psi_0^{II}$), because they are transmitted-reflected-reflected (TRR) and reflected-reflected-transmitted (RRT), respectively.

The whole theoretical treatment of the diffraction process is based on the dynamical diffraction theory, which describes the diffraction of plane or spherical waves by a strictly periodic potential. To preserve the interference properties over the length of the interferometer, the dimensions of the momolithic system have to be accurate on a scale comparable to the so-called Pendellösung length Δ_0, which is for an ordinary silicon reflection of the order of 50µm. The whole interferometer crystal has to be placed on a stable goniometer table under conditions avoiding temperature gradients and vibrations.

A phase shift between to two coherent beams can be produced by nuclear, magnetic or gravitational interactions. In the first case, the phase shift is most easily calculated using the index of refraction which reads for weakly absorbing materials as

$$n = \frac{K}{k} = \sqrt{1 - \frac{\bar{V}}{E}} \simeq 1 - \lambda^2 \frac{Nb_c}{2\pi} \qquad (1)$$

where b_c is the coherent scattering length and N is the particle density of the phase shifting material. The mean interaction potential due to nuclear interaction inside a sample can be obtained by using the Fermi-pseudopotential

$$\bar{V} = \frac{2\pi\hbar^2}{m} \int b_c\, \delta(\vec{r} - \vec{r}_i)\, d\vec{r} = \frac{2\pi\hbar^2}{m} Nb_c \qquad (2)$$

As in ordinary light optics the change of the wave function is obtained as follows:

$$\psi \rightarrow \psi_0\, e^{i(n-1)kD} = \psi_0\, e^{-iNb_c\lambda D} = \psi_0\, e^{i\chi} . \qquad (3)$$

Therefore, the intensity behind the interferometer is given by

$$I_0 \propto |\, \psi_0^I + \psi_0^{II}\, |^2 \propto (1 + \cos\chi) \qquad (4)$$

The intensity of the beam in the deviated direction (I_H) follows from particle conservation ($I_0 + I_H$ = const). The phase shift χ follows from equ. (2) and (3) as $\chi = -Nb_c\lambda D$. The wavelength dependence of the phase shift causes a reduction of the visibility of the interference pattern at high orders due to the finite coherence length of the beam in longitudinal direction ($\Delta x \simeq \lambda^2/\Delta\lambda$). There exists a nondispersive sample position (C) in which the phase shift becomes independent from the wavelength ($\chi = -2d_{hkl}Nb_cD_o$) because the wavelength is given as $\lambda = 2d_{hkl} \sin \Theta_B$ and the effective thickness is given as $D = D_o/\sin \Theta_B$. In this case higher orders of interferences can be observed due to the much larger coherence length of the beam perpendicular to the reflecting lattice planes[3].

Thus, the intensities behind the interferometer vary as a function of the thickness D of the phase shifter, the particle density N or the neutron wavelength λ indicating a pure wave-like behaviour of the neutrons inside the interferometer.

All the results of interferometric measurements, obtained till now can be explained well in terms of the wave picture of quantum mechanics and the complementarity principle of standard quantum mechanics. Nevertheless, one should bear in mind that the neutron also carries well defined particle properties and has to be considered as a composed system consisting of one "up-" and two "down"-Quarks (Tab. 1).

Tab. 1: Properties of the Neutron

Particle Properties

mass:	$m_o = 1.6749286(10).10^{-24}$ g
spin:	$s = \frac{1}{2}\hbar$
magnetic moment:	$\mu = -1.91304275(45)\ \mu_K$
live time:	$\tau = 888.6(3.5)$ s
confinement radius:	$R = 0.7$ fm
electrical polarizability:	$\alpha_e = 1.1\ {}^{+0.4}_{-0.6}\ .10^{-3}\ fm^{-3}$
nuclear structure:	u - d - d

Wave Properties

Compton wavelength:	$\lambda_c = \frac{h}{m_o c} = 1.3x10^{-13}$
phase difference $0 < \chi < 2\pi$	

for thermal neutrons
$\lambda_B = 1.8A, \quad v = 2.2x10^5$ cm/s

deBroglie wavelength:	$\lambda_B = \frac{h}{mv} = 1.8x10^{-8}$ cm
coherence length:	$\lambda_c = \frac{\lambda^2}{\Delta\lambda} = 1.x10^{-5}$ cm
packet length:	$\lambda_p = v.\Delta t = 1$ cm
decay length:	$\lambda_d = v.\tau = 2x10^8$ cm

All neutron interferometric experiments pertain to the case of self-interference, where during a certain time interval, only one neutron is inside the interferometer, if at all. Usually, at that time the next neutron has not yet been born and is still contained in the uranium nuclei of the reactor fuel. Although there is no interaction between different neutrons, they have a certain common history within predetermined limits which are defined, e.g., by the neutron moderation process, by their movement along the neutron guide tube, by the monochromator crystal and by the special interferometer set-up. Therefore, any real interferometer pattern contains single particle and ensemble properties together.

In the follwing chapter we discuss the coupling of the neutron magnetic moment to static and time-dependent magnetic fields and different aspects of coherence and statistic properties of coherent neutron beams. Quantum optical questions come into play both from the quantization of resonance flipper fields and from the neutron beam itself.

2. Spin state interferometry

The related interaction is caused by the dipole coupling of the magnetic moment of the neutron $\vec{\mu}$ to a magnetic field $\vec{B}$ ($H = -\vec{\mu}\vec{B}$). Therefore, the propagation of the wavefunction is given by

$$\psi \rightarrow \psi_0\, e^{-i(Ht/\hbar)} = \psi_0\, e^{-i(\vec{\mu}\vec{B}t/\hbar)} = \psi_0\, e^{-i\vec{\sigma}\vec{\alpha}/2} = \psi(\alpha) \tag{5}$$

where $\vec{\alpha}$ represents a formal description of the Larmor rotation angle around the field $\vec{B}$ ($\alpha = (2\mu/\hbar)\int B dt = (2\mu/\hbar v)\int B ds$). This wave function shows the typical 4π-symmetry of a spinor

$$\begin{aligned} \psi(2\pi) &= -\psi(0) \\ \psi(4\pi) &= \psi(0) \end{aligned} \tag{6}$$

wheras 2π-symmetry exists only for the expectation values.

The 4π-periodicity becomes visible in interferometer experiments, and it has been verified experimentally in early neutron interferometric experiments[4,5], where the intensity for unpolarized incident neutrons was found to be

$$I_0 = |\, \psi(0) + \psi(\alpha)\, |^2 \propto \left(1 + \cos\frac{\alpha}{2}\right) \tag{7}$$

It should be mentioned, that in the case of fermions this 4π-symmetry can always be attributed to real rotations[6,7]. Today, the most precise value for the periodicity factor is $\alpha_0 = 715.87 \pm 3.8$ degrees[8]. The 4π-periodicity effect has been observed for unpolarized as well as for polarized neutrons, which demonstrates the intrinsic feature of this phenomenon and the self-interference properties involved in these kind of experiments. These experiments have been continued by using polarized incident neutrons and rotating the polarization vector inside the interferometer[9,10]. In this connection the quantum mechanical spin superposition law has been verified in two different manners.

In the first case (static flipper) the wave function is changed by the flipper according to equation (5), which now has to be applied for polarized incident neutrons:

$$\psi \to e^{i\chi}\, e^{-i\vec{\sigma}\vec{\alpha}/2}\, |z\rangle = e^{i\chi}\, e^{-i\sigma_y \pi/2} |z\rangle = -i\sigma_y\, e^{i\chi} |z\rangle = e^{i\chi} |-z\rangle \tag{8}$$

Thus, two wave functions with opposite spin directions are superposed at the third plate

$$\psi \;\alpha\; (\, |z\rangle + e^{i\chi}\, |-z\rangle \,) \tag{9}$$

which corresponds to the situation proposed by Wigner[11] in 1963 to verify the quantum mechanical spin superposition law. In this case the intensities in the O- and H-beams are equal, and the beams are polarized in the (x,y)-plane, i.e. perpendicular to both the inital spin directions. The angle of the polarization in the (x,y)-plane is given by the nuclear phase shift.

$$\vec{P}_0 = \frac{\langle \psi \,|\vec{\sigma}|\, \psi \rangle}{I_0} = \begin{pmatrix} \cos\chi \\ \sin\chi \\ 0 \end{pmatrix} \tag{10}$$

This demonstrates that a pure initial state is transfered to a pure final state which is different from the states before superposition. The interference pattern appears only, if a polarization analysis is performed in the $|x\rangle$ - or in the $|y\rangle$ - direction. The experimental results agreed with the theoretical predictions completely (see[9]).

The second version of the spin superposition experiment was performed with a Rabi-type resonance flipper which is also commonly used in polarized neutron physics. This kind of interaction is time-dependent and, in addition to the spin-inversion, an exchange of the resonance energy $E_{HF} = \hbar\omega_r = 2\mu B_0$ occurs between the neutron and the resonator system, which has to be considered in the interferometric experiment. This energy exchange was observed in a separate experiment, with a very strong guide field B_o, where the energy resolution ΔE of the apparatus was better than the Zeeman energy splitting $(\Delta E < E_{HF})$[12].

Thus, the wave function of the beam with the flipper changes according to

$$\psi \to e^{i\chi}\, e^{i(\omega-\omega_r)t}\, |-z\rangle. \tag{11}$$

Therefore, again a spin-up and a spin-down state are superposed at the position of the third plate. The final polarization of the beam in the forward direction is now given by

$$\vec{p} = \begin{pmatrix} \cos(\chi+\omega_r t) \\ \sin(\chi+\omega_r t) \\ 0 \end{pmatrix} \tag{12}$$

and lies again in the (x,y)-plane but now rotates within this plane with the resonance-(Larmor) frequency without being driven by a magnetic field. A stroboscopic method was needed for the observation of this effect. The direction of the polarization in the (x, y)-plane depends on the status (phase) of the resonance field and, therefore, has to be measured synchronously with this phase[10].

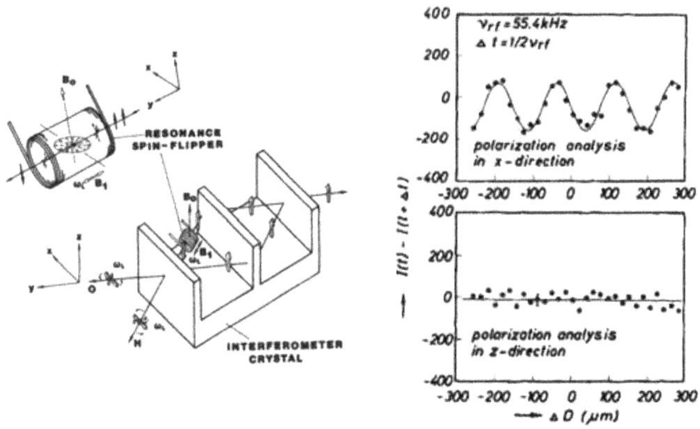

Fig. 2: Experimental results of the spin-superposition experiment using a time-dependent resonance flipper[10].

The obsered interference pattern (Fig.2) demonstrates that coherence persists, although a well defined energy exchange between the neutron and the apparatus exists. Thus, an energy exchange is not automatically a measuring process. The exchanged photon cannot be used for a quantum nondemolition measurement. In our experiment, the following argument based on different uncertainty relations can be used: Firstly, one single absorbed or emitted photon of the resonator cannot be detected because of the photon number - phase uncertainty relation, which can be used in its simplest form ($\Delta N \Delta \phi > 2\pi$). A second version of the beam path detection could be based on the observation of the energy change of the neutron. This can only be achieved, if the energy resolution of the instrument fulfills the relation $\Delta E \leq 2\mu B_o$. On the other hand, the stroboscopic measuring method requires time channels $\Delta t \leq 1/2\nu_{HF} = h/4|\mu|B_o$, which provides another constraint on the experiment. Both conditions cannot be fulfilled with respect to the energy-time uncertainty relation concerning the beam parameters $\Delta E \Delta t \geq \hbar/2$. Therefore, we conclude that a simultaneous detection of the beam path through the interferometer and of the interference pattern remains impossible. An analogeous conclusion has been drawn from an analysis of this experiment in terms of a coherent- and a number state behaviour of the resonance flipper system[13].

3. Double coil experiment

In chapter 2 we have shown that a simultaneous measurement of the beam path and of interference can not be achieved due to various uncertainty relations. The question remains whether there exists a trajectory of the particle et all. In this connection a double coil experiment was proposed where a spin-flip and an energy exchange occures in both coherent beam paths[14] (Fig.3). In this case the final polarization lies in the $|-z\rangle$ direction and the energy transfer $\hbar\omega_r$ can be smaller or larger than the energy resolution ΔE because this information can not be in any way associated with a beam path detection. According to our previous considerations the change of the wave functions with the resonance flippers turned to the resonance frequency can be written for polarized incident neutrons ($|z\rangle$) and for different modes of operation as follows.

a) Both flippers are operated synchronously without a phase shift between the flipper fields:

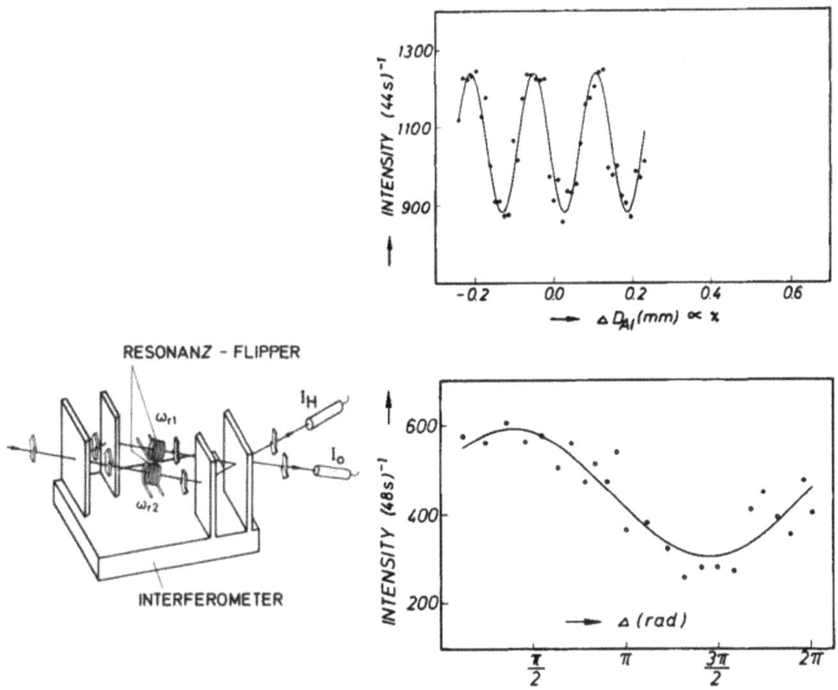

Fig. 3: Sketch of the experimental arrangement (left) and typical results of the double coil experiment (right) as a function of a phase shift of the neutron beams (above) and of the flipper fields (below) (after[15]).

$$\psi \rightarrow e^{i\chi}\, e^{i(\omega-\omega_r)t}\, | -z > + e^{i\chi}\, e^{i(\omega-\omega_r)t}\, |-z > \qquad (13)$$

This results in an intensity modulation

$$I_0 \;\alpha\; 1 + \cos\chi \qquad (14)$$

which is independent of the flipper fields.

b) Both flippers are operated synchronously with a distinct phase relation Δ:

$$\psi_0 \rightarrow e^{i\chi}\, e^{i(\omega-\omega_r)t}\, |-z > + e^{i\chi}\, e^{i\Delta}\, e^{i(\omega-\omega_r)t}\, |-z >. \qquad (15)$$

In this case, the intensity modulation is given by

$$I_0 \propto 1 + \cos(\chi+\Delta). \tag{16}$$

i.e. the interference patters becomes sensitive to the relative phase difference of the flipper fields. Both effects have been verified experimentally[15] (Fig.3). From these experimental results follows that the interference properties are preserved, although an energy exchange $\hbar\omega_r$ certainly has taken place. Only quanta within a narrow energy band around $\hbar\omega_r$ and no others are excited inside the flipper resonator. Therefore, one could believe that the spin flip and the energy transfer process to the neutron occured inside one of the two coils, which would demonstrate that the neutron has chosen one of the two possible paths or energy and momentum conservation is broken in individual processes.

This kind of experiments can be extended to make this situation even more clear. Suppose one produces a pulsed beam with burst length shorter than the dimension of the interferometer and shorter than the distance of the two coils which are now arranged spatially shifted as shown in Fig. 4[16]. Furthermore let us apply an energy shift $\Delta E = 2\mu B_0$

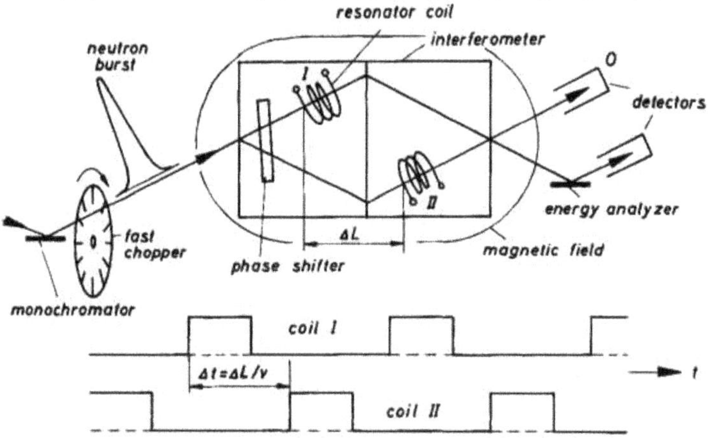

Fig. 4: Sketch of a proposed double coil experiment where the energy exchange in the discontinuosly operating flipper coils and the interference pattern can be measured simultaneosly (after[16]).

larger than the energy width of the beam, i.e. the energy transfer becomes measurable and let us operate the flipper coils only if the neutron burst passes through the coils. In this case the wave function for the time interval when the burst is just in between the two coils read as

$$| \psi > = e^{i(\omega-\omega_r)t} \; |-z > + e^{i\chi} \; |+z > \tag{17}$$

which exhibits energy conservation for the quantum exchange in flipper I. At the location of superposition at the third interferometer plate the wave function has to be written as

$$| \psi > = e^{i(\omega-\omega_r)t} \; |-z > + e^{i\chi} \, e^{i(\omega-\omega_r)t} \; |-z > \tag{18}$$

which yields an interference pattern

$$I(\chi,\Delta E) = \int | A(E+\Delta E)|^2 \; (1 + \cos\chi) \; dE \tag{19}$$

where the energy change and the interference pattern can be measured simultaneously. The whole situation and the temporal evaluation suggest an individual neutron an energy exchange in one coil only, which raises again questions of the Bohr-Einstein contraversy concerning the meaning of the wave function. The experiment shown in Fig. 4 seems to be feasible in any respect, if it is worthwhile for interpretational questions of quantum mechanics should be discussed before within the physical community and during this conference.

4. Neutron Josephson-effect analogue

The double coil experiment discussed in chapter 3 was also performed operating the two coils at shightly difference frequencies. Now, the wave functions change according to

$$\psi \rightarrow e^{i(\omega-\omega_{r1})t} \; |-z > + e^{i\chi} \, e^{i(\omega-\omega_{r2})t} \; |-z >. \tag{20}$$

Therefore, the intensity behind the interferometer exhibit a typical quantum beat effect, given by

$$I \;\alpha\; 1 + \cos[\chi+(\omega_{r1}-\omega_{r2})t] \tag{21}$$

Thus, the intensity behind the interferometer oscillates between the forward and deviated beam without any apparent change inside the interferometer[15]. The time constant of this modulation can reach a macroscopic scale which is again correlated to an uncertainty relation $\Delta E \Delta t \geq \hbar/2$. Fig. 5 shows the result of an experiment, where the periodicity of the intensity modulation, $T = 2\pi/(\omega_{r1}-\omega_{r2})$, amounts to $T = (47.90\pm0.15)$s caused by a frequency difference of about 0.02 Hz. This corresponds to a mean energy transfer difference between the two beams of $\Delta E = 8.6\times10^{-17}$ eV, and to an energy sensitivity of 2.7×10^{-19} eV, which is higher by many orders of magnitudes than that of other advanced spectroscopic methods. There exists a kind of memory effect in the system which guides the neutron to the predicted direction behind the interferometer.

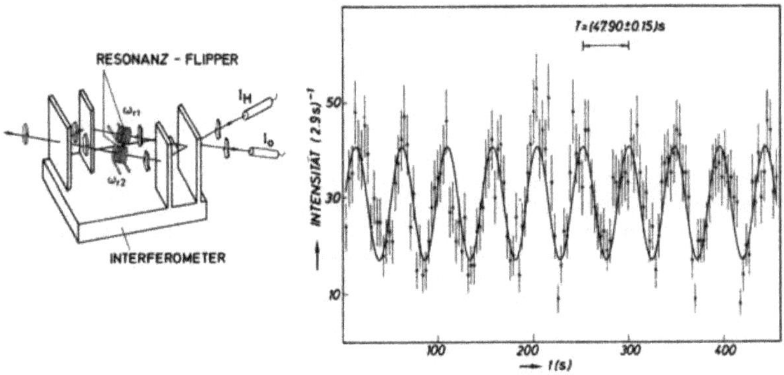

Fig. 5: Quantum beat effect observed when the frequencies of the two flipper coils are slightly different[15].

A comparison of equations (15), (20) and (21) show that this quantum beat effect (Fig. 5) can also be described by a time dependent phase

$$\Delta = \Delta_2 - \Delta_1 = (\omega_{r2} - \omega_{r1}) \quad t = \frac{2\mu B_o}{\hbar} t \tag{22}$$

which follows from

$$\frac{\delta}{\delta t} (\Delta_2 - \Delta_1) = \omega_{r2} - \omega_{r1} = \frac{1}{\hbar} 2\mu B_o \tag{23}$$

and yields to an intensity modulation

$$I \propto (1 + \cos\Delta(t)) \tag{24}$$

These equations are magnetic analogues to the well-known Josephson-effect in superconducting tunnel junctions[17] (Fig. 6). In this case the wave functions of the superconductors 1 and 2 are given by the density of the Cooper pairs n_s and their common Boson-phase ϕ

$$| \psi_s >_{1,2} = \sqrt{n_s^{(1,2)}}\ e^{i\phi_{1,2}} \tag{25}$$

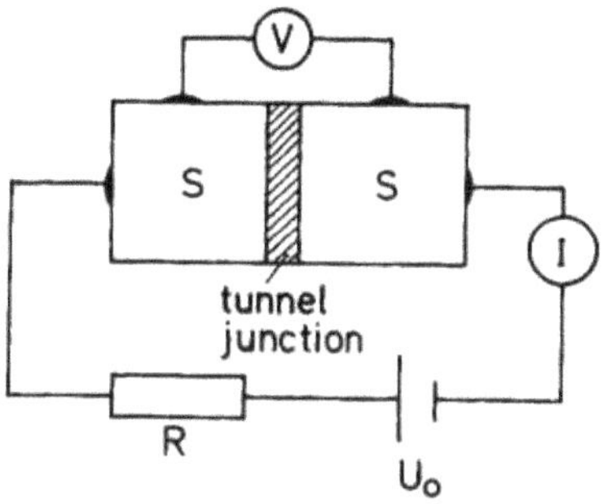

Fig. 6: Arrangement for the observation of the Josephson effect in a superconducting tunnel junction.

which varies in the case of an applied electrical field as

$$\frac{\delta}{\delta t}\ (\phi_2 - \phi_1) = \frac{1}{\hbar}\ (E_1 - E_2) = \frac{1}{\hbar}\ 2eV \tag{26}$$

and gives

$$\phi = \phi_2 - \phi_1 = \frac{2eV}{\hbar}\ t \tag{27}$$

The related oscillation of the electrical current through the tunnel junction becomes

$$I_s = I_{smax} \sin \phi \tag{28}$$

The equivalence of equations (23) and (26) is evident and even the factor 2 appears in both cases, which is in the superconducter Josephson case related to the tunneling of two electrons (Cooper pairs). The electric and the magnetic Josephson-effects are both phase-sensitive phenomena and they are therefore extremely sensitive concerning any additional driving force.

5. Coherence measurements

The coherence function is defined as the auto-correlation function of the wave function

$$\Gamma(\vec{\Delta}, t) = < \psi^*(0,0) . \psi(\vec{\Delta}, t) > \tag{29}$$

where $\Gamma(\vec{\Delta}, 0) = \Gamma(\vec{\Delta})$ determines the spatial and $\Gamma(0,t) = \Gamma(t)$ the temporal coherence function. $\Delta = (1-n)D$ is the optical path difference or the spatial shift of the overlapping wave trains and is related to the real part of the phase shift $(\chi'=\vec{k}\vec{\Delta})$. The interference pattern can be written in this terms as:

$$I \propto 1 + < \frac{\Gamma(\vec{\Delta})}{\Gamma(0)} > \cos \chi' \tag{30}$$

where the average has to be taken over the beam cross section. For gaussian wavepackets with momentum widths δk it is easy to show that

$$\frac{\Gamma(\Delta)}{\Gamma(0)} = \prod_i \exp\,[-(\Delta.\delta k_i)^2/2] \tag{31}$$

which has been verified experimentally for all three directions[3]. The reduction of the visibility of the interference pattern at high order determines the coherence length in the direction where the phase shift has been applied.

A reduction of the visibility of the interference pattern can also be achieved by using an absorbing phase shifter in one beam path. In this case a complex phase shift exists

$$\chi = \chi' + i\,\chi'' \tag{32}$$

where $\chi' = -Nb_c\lambda D$ and $\chi'' = N\sigma_t D/2$ (σ_t ... total attenuation cross section). The visibility now varies as

$$\frac{\Gamma(\Delta)}{\Gamma(0)} = \sqrt{a} \tag{33}$$

where $\underline{a} = I/I_0 = \exp[-\sigma_t ND]$ denotes the general beam attenuation, which is in this case a stochastic process for every neutron. If the beam attenuation is achieved in a deterministic manner by a the insertion of a smaller slit or by a slowly rotating shutter defining a certan open to close ratio the coherence function varies as

$$\frac{\Gamma(\Delta)}{\Gamma(0)} = \underline{a} \tag{34}$$

which causes a remarkable difference especially at small transmission probabilities as it is shown in Fig. 7[18].

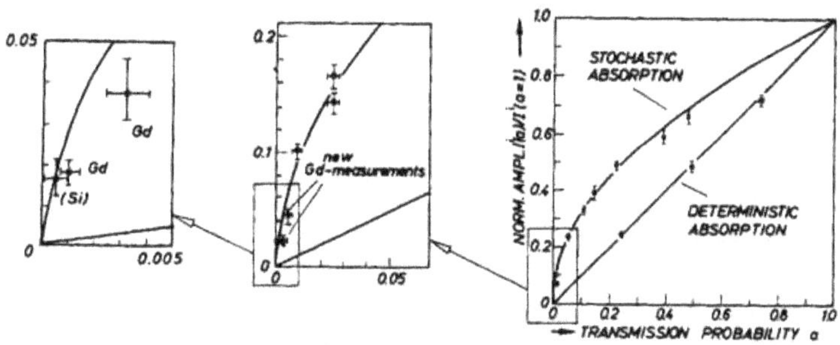

Fig. 7: Normalized amplitudes of the interference pattern in the case of stochastic and deterministic absorption[18].

All these coherence measurements can be put together if one defines new parameters for the wave-like (W) and particle-like (P) behaviour of a system of interfering beams I and II. These quantities are related to two measurable quantities, to the general transparency ($T = I/I_0$) and to the visibility of the interference pattern.

$$W = TV = \langle \mid \psi^{I} \psi^{II} \mid \rangle \tag{35}$$

$$P = 1 - TV = 1 - \langle \mid \psi^{I} \psi^{II} \mid \rangle$$

which yields the conservation law

$$W + P = 1 \tag{36}$$

All performed interferometer experiments are described within the hatched area of Fig. 8.

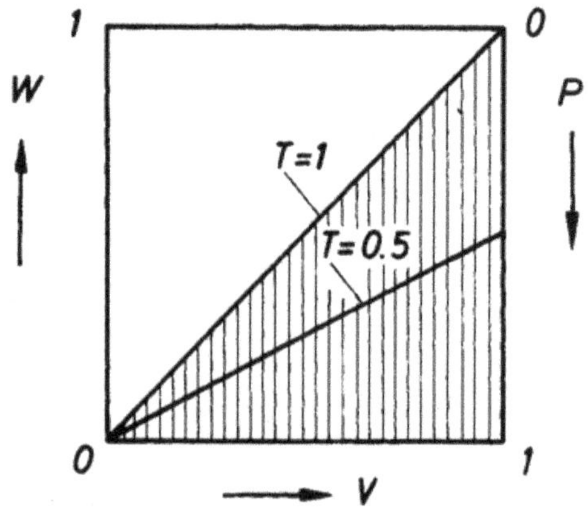

Fig. 8: Synopsis of wave-like and particle-like character of neutrons in different neutron optical experiments[18].

In an other experiment the internal statistics of the interfering beams has been investigated[18]. It has been shown that the count rate of the interfering beams fulfil Poissonian statistics if the measurement intervals are fixed. Taking interferograms with a few counts only the particle-number-phase uncertainty relation has been verified. This relation reads in the case V ≠ 1 as

$$(\Delta\chi)^2(\Delta N)^2 = [1 - (1-V^2)^{1/2}]^{-1} \xrightarrow[V \to 1]{} 1 \tag{37}$$

with $\Delta N = \sqrt{N}$. Taking this aspects into account one can define a coherent state behaviour for the two coherently coupled beams in the interferometer. The coherence lengths, which are determined by the momentum distribution functions, are analogeous to the uniform wave-packet phenomena in coherent atomic states and a time average has to be taken instead of an ensemble average. This justifies the interpretation of many neutron optical phenomena in terms of quantum optical effects.

Most of the experimental work was performed at the High Flux Reactor of the Institut Lane-Langevin in Grenoble. The author likes to thank his coworkers for the substantial support during the course of the experiments. The financial support due to the Fonds zur Förderung der wissenschaftlichen Forschung (project S4201) in acknowledged as well.

References

1. V. F. Sears, *Neutron Optics*, (Oxford University Press, 1989)
2. H. Rauch, W. Treimer, U. Bonse, *Phys. Lett.* **47A** (1974) 369
3. H. Rauch, E. Seidl, D. Tuppinger, D. Petraschek, R. Scherm, *Z. Physik* **B69** (1987) 313
4. H. Rauch, A. Zeilinger, G. Badurek, A. Wilfing, W. Baunspiess, U. Bonse, *Phys. Lett.* **54A** (1975) 425
5. S. A. Werner, R. Colella, A. W. Overhauser, C. F. Eagen, *Phys. Rev. Lett.* **35** (1974) 1053
6. A. Zeilinger, *Nature* **294** (1981) 544
7. H. J. Bernstein, *Nature* **315** (1985) 42
8. H. Rauch, A. Wilfing, W. Bauspiess, U. Bonse, *Z. Physik* **B29** (1978) 281
9. J. Summhammer, G. Badurek, H. Rauch, U. Kischko, A. Zeilinger, *Phys. Rev.* **A27** (1983) 2523
10. G. Badurek, H. Rauch, J. Summhammer, *Phys. Rev. Lett.* **51** (1983) 1015
11. E. P. Wigner, *Am. J. Phys.* **31** (1963) 6
12. B. Alefeld, G. Badurek, H. Rauch, *Z. Physik* **B41** (1981) 231
13. M. D. Scully, H. Walther, *Phys. Rev.* **A39** (1989) 5229
14. J. P. Vigier, *Pramana* **25** (1985) 397
15. G. Badurek, H. Rauch, D. Tuppinger, *Phys. Rev.* **A34** (1986) 2600
16. H. Rauch, J. P. Vigier, *Phys. Lett.* (in print)
17. B. D. Josephson, *Rev. Mod. Phys.* **46** (1974) 251
18. H. Rauch, J. Summhammer, M. Zawisky, E. Jericha, *Phys. Rev.* **A42** (1990) No. 7 (in print)

LOCATING THE MANY WORLDS

JOE ROSEN*
Department of Physics
The Catholic University of America
Washington, DC 20064, USA

ABSTRACT

A novel, compact scheme for the origin and subsequent evolution of the Universe is proposed, whereby the Universe is identified with a baby universe occurring within itself. This picture indicates an intimate meshing between gravitation, space-time and the quantum and offers a framework for the many worlds interpretation of quantum theory.

I would like to present a novel world picture -- not a theory but rather a conceptual scheme -- that contains an ensemble of universes, yet avoids the pitfall of laws of behavior for universes.

The first step in setting up the picture is to consider the idea of baby universes. Although we do not (yet) know the laws of quantum gravity, by the general nature of quantum physics it is reasonable to assume that at the Planck scale (about 10^{-35}m and 10^{-43}s) space-time is a foam of fluctuations, quantum gravitational fluctuations, forming and disappearing incessantly and spontaneously, much like bubbles in a boiling liquid. Of these fluctuations there are those that develop into baby universes, becoming disconnected from the Universe, so that their contents are inaccessible to us.

The second step in setting up the picture is to consider the idea, which is rather in vogue these days, of a quantum origin to the Universe. According to this idea the Universe developed from a quantum fluctuation in some unspecified substrate. Thus the Universe, itself originating in a quantum fluctuation, gives rise through quantum fluctuations to baby universes within itself. Presumably there are

---

*On leave from School of Physics and Astronomy, Tel-Aviv University, 69978 Tel-Aviv, Israel

those baby universes that similarly give rise within themselves to grandchild universes, and so on ad infinitum, baby universes within baby universes within baby universes ....

The third step in setting up the picture is to identify the Universe with a baby universe occurring within itself, thereby imposing a closed topology, as it were, on the scheme. This is the self-generating Universe picture. Note that no laws of behavior for universes have entered the picture. The behavior of the Universe is governed by the laws of physics that are valid within itself, i.e., the Universe controls its own behavior, even its own coming into existence.

Which one of all the gravitational fluctuations in the Universe is the Universe? We cannot know. First, it is physically impossible to look inside a baby universe. Second, even if we imagined we could peer into baby universes and see what is going on there, we still would not be able to choose the one that is our reality. More precisely, just as our reality consists of our past and present and its emanating quantum possibilities for the future, so we would find an unimaginable number of baby universes containing the same past and present as ours with all possible futures. Thus, over and above the physical impossibility of looking inside a baby universe, we have this uncertainty in labeling the one that is the Universe, the uncertainty that is just the quantum uncertainty of our future.

What about the unimaginable number of baby universes that are not the Universe? For self-consistency of the picture these other baby universes must also be universes, and physics has a perfect role for such a multitude of universes: the plethora of quantum possibilities. We thus obtain a framework for quantum theory in its many worlds interpretation. Here lie all the "discarded" universes, all those failed quantum possibilities, all possible realities that are not ours. They are not our reality, but they are part of it, are contained in it in an essential way. They are contained in our reality just as our reality is contained in itself. They are generated by our reality just as our reality is generated by itself, through quantum gravitational fluctuations producing baby universes.

This self-generating Universe picture has beautiful compactness and completeness: The Universe is the cause and effect of its own existence (as well as of all its quantum alternatives), a self-generating unity, the ultimate bootstrap.

We are used to the idea that quantum effects are responsible for the stability of atoms, for chemical bonding, for the blackbody spectrum, etc. If the Universe behaved classically it would be a very different place; for example, we would not be around to consider what a quantum universe might be like. According to the self-generating Universe picture, without quantum effects the Universe would not even exist, since a quantum effect, specifically, gravitational fluctuation, the quantum fluctuation of space-time, lies at the root of its existence. On the other hand, according to this picture, without space-time there would be no quantum effects, since it is the fluctuation of space-time that brings about the many universes in which quantum possibilities are realized.

The electroweak and strong interactions are aspects of the

Universe that seem to be only weakly coupled to space-time and to the quantum aspect of the Universe. They seem to be playing their roles against the space-time background, so to speak, without affecting space-time to any significant degree. In a similar vein, they seem to be passively obeying the quantum laws of the Universe without themselves affecting those laws. Our more or less successful theoretical treatment of these interactions -- quantum field theories on space-time -- corresponds to that picture.

In contrast, the gravitational interaction is intimately involved with space-time, as Einstein tells us, and with the quantum aspect of the Universe, as the self-generating Universe picture tells us. As for our theoretical treatment of gravitation, we have known for some time that it is not just an ordinary field theory on a space-time background; the general theory of relativity describes a strong mutual influence between gravitation and space-time. And a quantization of gravitation has not yet been achieved. The self-generating Universe picture predicts that indeed we will not successfully quantize it along the lines of quantum electrodynamics or quantum chromodynamics (except perhaps to an approximation). A successful theory will have to include an intimate meshing, with mutual influence, among gravitation, space-time, and the quantum. In certain of its limiting cases it will give us the general theory of relativity and quantum theory.

THEORY OF MACROSCOPIC QUANTUM TUNNELING IN THIN SUPERCONDUCTING WIRES AND COMPARISON WITH EXPERIMENTS

SEIJI SAITO and YOSHIMASA MURAYAMA
Advanced Research Laboratory, Hitachi, Ltd.
Hatoyama, Saitama 350-03, Japan

ABSTRACT

Voltage generation due to macroscopic quantum tunneling is discussed within the framework of a time dependent Ginzburg-Landau equation. The theory predicts that charge neutrality approximately holds, greatly simplifying the theoretical expression of the tunneling rate. It is shown that the derived temperature dependence of resistivity is qualitatively consistent with recently observed phenomena.

1. Introduction

Recent advances in microfabrication technology have led to discoveries of several macroscopic quantum phenomena.[1] A typical example is the observation of macroscopic quantum tunneling (MQT) which demonstrates that the macroscopic objects can obey the quantum mechanical principle. The MQT effect has been theoretically investigated[2] and confirmed in experiments using miniaturized Josephson junctions.[3] This means that the quantum-mechanical measurement process will become essential in such micro-circuits.

Superconducting states correspond to the local minima of the free energy of Josephson junctions. As is well known, thermal activations cause hopping from one local minimum to another so that the superconducting phase difference across the junction varies with time. The time variations of the phase difference generate voltages across the junction according to the AC Josephson relation. The thermal activation effects are large near the critical temperature and decrease exponentially at lower temperatures. Therefore, voltage generation due to the MQT can be observed in this temperature range. The MQT effects can not be observed in the usual Josephson junction, because the free energy barrier is large, making the tunneling probability small. The miniaturization of the junction has enabled us to observe the MQT effects by reducing the tunneling barriers.

Recently, electron beam lithography has allowed fabrications of metal wires with a width of several hundred angstroms, which is close to or less than the magnetic field penetration depth. Thermal activations allow the magnetic flux quantum to easily pass through such thin wires. Thus, similar phenomena can occur in such thin wires without the Josephson junction. In fact, Langer et al. proved that thermal activations cause the decay of currents.[4] This was confirmed in several experiments.[5] Here, again, the thermal activation effects decrease exponentially with the temperature so that voltage generation due to the MQT can be expected.

This problem was formulated based on a time dependent Ginzburg-Landau (TDGL) equation in our first paper[6] in which the crossover from the thermal activation to the MQT was shown to occur at a temperature close to the critical temperature. Giordano[7] independently showed that the same phenomena occurred in thin indium wires produced by electron beam lithography. However, our theory could not be directly compared to Giordano's experiments since they dealt with the zero-temperature model by neglecting the dissipation effects. Later, we reported a model which includes a diffusion-like dissipation term in the TDGL equation that is relevant to the experimental conditions.[8] The theory proves a Giordano's experimental formula for resistivity. This report summarizes our two previous reports[6,8] and shows that the theory is consistent with the experiments performed by Giordano.[7,9,10]

2. Basic Equations

The starting point of our discussion is the TDGL equation derived by Abraham and Tsuneto,[11] which may offer a phenomenological model for Type-I superconductors near the critical temperature. The TDGL equation is a straightforward extension of the GL equation in the thermal equilibrium, including first-order and second-order time derivatives as follows:

$$\frac{\hbar^2}{2m^*}\left[(\frac{1}{i}\nabla-\frac{e^*}{\hbar c}\mathbf{A})^2+\frac{1}{c_q^2}(\frac{\partial}{\partial t}+\frac{ie^*}{\hbar}\Phi)^2\right]\Psi-\alpha\Psi+\beta|\Psi|^2\Psi+\gamma(\frac{\partial}{\partial t}+\frac{ie^*}{\hbar}\Phi)\Psi=0. \quad (1)$$

The first-order time derivative describes a diffusion-like time evolution and represents the dissipation effect. Due to the existence of this term, it is not possible to obtain an appropriate Lagrangian. Hence, we employ here a rather microscopic approach suitable for a direct quantization scheme, obviating the diffusion-like term. Couplings between the collective mode and the reservoir modes cause dissipations. The former mode in this case is described by the wave function of the condensed Cooper pairs given by the GL amplitude, and the latter modes are the quasi-electrons which do not form Cooper pairs. Such quasi-particles as phonons may be another cause of dissipation. Here, the present interaction seems the most suited to the present formulation. Here, we do not rely on an elaborate formulation based on BCS theory but simply assume that the reservoir modes can be described by an infinite number of spatially distributed harmonic oscillators. We also assume that the linear couplings transfer the energy from the collective mode to the reservoir modes. The Lagrangian for this system is given by

$$\begin{aligned}L=&\frac{\hbar^2}{2m^*}\left\{\frac{1}{c_q^2}|(\frac{\partial}{\partial t}+\frac{ie^*}{\hbar}\Phi)\Psi|^2-|(\nabla-\frac{ie^*}{\hbar c}\mathbf{A})\Psi|^2\right\}+\alpha|\Psi|^2-\frac{1}{2}\beta|\Psi|^4\\&+\Sigma_j m_j\left\{|(\frac{\partial}{\partial t}+\frac{ie^*}{\hbar}\Phi)\xi_j|^2-\omega_j^2|\xi_j|^2\right\}\\&-\left\{\Psi^*\Sigma_j C_j\xi_j+\Psi\Sigma_j C_j\xi_j^*\right\}-\Omega|\Psi|^2\\&+\frac{1}{8\pi}\left[(\frac{1}{c}\frac{\partial\mathbf{A}}{\partial t}+\nabla\Phi)^2-(\nabla\times\mathbf{A})^2\right].\end{aligned} \quad (2)$$

The fact that the Euler-Lagrange equation can reproduce the original TDGL equation when a certain assumption is made about the coupling constants justifies using this Lagrangian. The following equation can be derived from the Euler-Lagrange equations when the reservoir oscillator variables are eliminated:

$$\frac{\hbar^2}{2m^*}\left[(\frac{1}{i}\nabla-\frac{e^*}{\hbar c}\mathbf{A})^2+\frac{1}{c_q^2}(\frac{\partial}{\partial t}+\frac{ie^*}{\hbar}\Phi)^2\right]\Psi$$

$$-\alpha\Psi+\beta|\Psi|^2\Psi+K(-i\frac{\partial}{\partial t}+\frac{e^*}{\hbar}\Phi)\Psi=0. \tag{3}$$

First, the equilibrium solution must not be shifted by couplings with the reservoir oscillators. In the equilibrium situation, the time derivative and the scalar potential are equal to zero, so that $K(0) = 0$ must be satisfied. This condition determines the unknown coefficient Ω in Eq.2 as $\Omega=\Sigma_j C_j^2/m_j\omega_j^2$. The resultant expression for $K(\omega)$ is given by

$$K(\omega)=\frac{2\omega^2}{\pi}\int_0^\infty\frac{Z(\omega')d\omega'}{\omega'(\omega'^2-\omega^2)},\qquad Z(\omega)=\frac{\pi}{2}\sum_j\frac{C_j^2}{m_j\omega_j}\delta(\omega-\omega_j). \tag{4}$$

If a linear function of ω is assumed for the spectral function $Z(\omega)$, that is $Z(\omega)=\gamma\omega$, the TDGL equation is reproduced by the Euler-Lagrange equation because $K(\omega)=i\gamma\omega$ is obtained by integrating Eq.4. This case has been called "the strictly linear" case by Caldeira-Leggett.[2]

3. Partition Function and Effective Action

The quantum-mechanical partition function can be derived by using the usual procedure in quantized field theory as follows:

$$Z=\mathrm{tr}\{\exp[-\frac{H}{k_BT}]\}=\iint\mathbf{D}\Psi(\mathbf{r}\tau)\mathbf{D}\Psi^*(\mathbf{r}\tau)\mathbf{D}\Phi(\mathbf{r}\tau)\exp[-\frac{1}{\hbar}S_{eff}], \tag{5a}$$

$$S_{eff}=\int_{-\tau_\beta/2}^{\tau_\beta/2}d\tau\int d^3\mathbf{r}\left[\frac{\hbar^2}{2m^*c_q^2}|(\frac{\partial}{\partial\tau}+\frac{ie^*}{\hbar}\Phi)\Psi|^2\right.$$

$$+\frac{\hbar^2}{2m^*}|\nabla\Psi|^2-\alpha|\Psi|^2+\frac{1}{2}\beta|\Psi|^4$$

$$\left.+\Psi^*Q(-i\frac{\partial}{\partial\tau}+\frac{e^*}{\hbar}\Phi)\Psi+\frac{1}{8\pi}(\nabla\Phi)^2\right], \tag{5b}$$

where the electrostatic approximation is used because the superconducting channel is narrow and the magnetic field produced by the current is small and the reservoir oscillator variables have been integrated. The nonlinear time derivative represents the contributions from the coupling with reservoirs. The function $Q(\omega)$ is related to the function $K(\omega)$ defined in Eq.4 by $Q(\omega)=K(i\omega)$. Therefore, the real time action integral is transformed into the Euclidian action integral by the transformations $it\to\tau$, $i\Phi(\mathbf{r}t)\to-\Phi(\mathbf{r}\tau)$. The Euclidian action integral can be considered an analytic continuation of the real time action integral.

The rest of this paper discusses a thin superconducting wire whose width is much smaller than the penetration depth, making one-dimensional approximation valid. The equations for the stationary points can be derived by functionally differentiating the effective action S_{eff}. Using the amplitude and phase representation $\Psi=fe^{i\phi}\sqrt{(\alpha/\beta)}$, where the amplitude is expressed in a dimensionless form, the following equations are derived by the functional differentiations with respect to the amplitude, phase and scalar potential, respectively:

$$\tau_0^2\frac{\partial^2 f}{\partial\tau^2}+\xi^2\frac{\partial^2 f}{\partial x^2}-\tau_0^2(\frac{\partial\phi}{\partial\tau}+\frac{e^*}{\hbar}\Phi)^2 f-\xi^2(\frac{\partial\phi}{\partial x})^2 f$$

$$+f-f^3-\alpha^{-1}Q_r(-i\frac{\partial}{\partial\tau}+\frac{\partial\phi}{\partial\tau}+\frac{e^*}{\hbar}\Phi)f=0, \tag{6a}$$

$$\frac{1}{c}\frac{\partial\rho_\tau}{\partial\tau}-\frac{\partial j}{\partial x}=0, \tag{6b}$$

$$\frac{\partial^2\Phi}{\partial x^2}=-4\pi\rho_\tau, \tag{6c}$$

where ξ is the GL coherence length given by $\xi(T)=[\hbar^2/4m\alpha]^{1/2}$, τ_0 is a time constant equal to ξ/c_q, and Q_r is the real part of the operator Q. The second and third equations are the continuity equation and the Poisson equation which define the charge and current densities, respectively, as follows:

$$\rho_\tau=-\frac{1}{4\pi\lambda_D^2}\left[f^2(\Phi+\frac{\hbar}{e^*}\frac{\partial\phi}{\partial\tau})+\frac{\tau_c\tau_G}{2\tau_0^2}\frac{\delta\Lambda_{eff}}{\delta\Phi}\right], \tag{7a}$$

$$j=\frac{1}{4\pi\lambda^2}(\frac{\hbar c}{e^*})f^2\frac{\partial\phi}{\partial x}, \tag{7b}$$

where τ_G is the relaxation time defined by $\tau_G=\gamma/\alpha$. The functional Λ_{eff} is defined by

$$\Lambda_{eff}=\int_{-\tau_\beta/2}^{\tau_\beta/2}d\tau\int_{-l/2}^{l/2}dx fQ(-i\frac{\partial}{\partial\tau}+\frac{\partial\phi}{\partial\tau}+\frac{e^*}{\hbar}\Phi)f. \tag{8}$$

In Eq.8, the bold-face $\mathbf{Q}$ is related to Q by $Q=\tau_c\gamma(e^*/\hbar)^2\mathbf{Q}$ where some dimensional factors have been factorized. In the denominator of the charge density, λ_D is defined by $\lambda_D=\lambda(c_q/c)$ where c_q is the phase velocity in Eq.1 which has the order of the Fermi velocity. The Fermi velocity is much smaller than the light velocity, so that λ_D is much smaller than the penetration depth λ. Therefore, it can be safely assumed that λ_D goes to zero since the characteristic length of the space variation in Type-I superconductor is the penetration depth.

In the approximation that λ_D goes to zero, the numerator of the charge density in Eq.7 must also go to zero since the charge density itself should be finite. This condition simply becomes the Josephson phase relation given by

$$\Phi+\frac{\hbar}{e^*}\frac{\partial\phi}{\partial\tau}=0 \tag{9}$$

in the small dissipation limit where the second term of the charge density Eq.7 can be neglected. It also holds even when dissipation occurs. $\mathbf{Q}(\omega)$ is real and an even function of ω, and $f(x,\tau)$ is a real function of x and τ. Thus, Λ_{eff} is an even functional of the function $\partial\phi/\partial\tau+(e^*/\hbar)\Phi$. Therefore, $\delta\Lambda_{eff}/\delta\Phi$ is an odd functional of this function, and, hence, it is equal to zero if the Josephson phase relation holds.

The Josephson phase relation tells us that the scalar potential does not completely disappear, but is proportional to the time derivative of the phase so that a nonvanishing electrostatic energy exists. This electrostatic energy is shown to be much smaller than the kinetic energy term which is proportional to $(\partial f/\partial\tau)^2$ in the effective action. The electrostatic energy can be rewritten in a form that is also proportional to $(\partial f/\partial\tau)^2$ as follows:

$$\frac{\sigma}{8\pi}\int_{-l/2}^{l/2} dx(\frac{\partial\Phi}{\partial x})^2 = \frac{\sigma}{2\pi}(\frac{\hbar}{e^*})^2\frac{J_0^2}{\xi^2}\int_{-l/2}^{l/2} dx\frac{1}{f^6}(\frac{\partial f}{\partial\tau})^2, \tag{10}$$

where σ is the cross-section of wire. The ratio of both the coefficients can be given in terms of the ratio λ/ξ and the ratio v_F/c as follows:

$$r = \frac{4}{3}(\frac{\lambda}{\xi})^2(\frac{v_F}{c})^2\frac{J_0^2}{f^6}, \tag{11}$$

where $c_q=v_F/\sqrt{3}$ (v_F:Fermi velocity) is used.[11] The ratio r is proved to be much smaller than unity in Type-I superconductors because J_0 and f are dimensionless factors and, thus, are the order of unity. It means that the charge neutrality approximately holds, greatly simplifying the tunneling rate expression.

4. Tunneling Rate and its Comparison with Experiments

Finally the following expression for the effective action is derived:

$$S_{eff} = \frac{H_c^2}{4\pi}\xi\sigma\tau_0\int_{-\tau_\beta/2}^{\tau_\beta/2} d\tau\int_{-l/2}^{l/2} dx\left[(\frac{\partial f}{\partial\tau})^2 + (\frac{\partial f}{\partial x})^2 - \frac{J_0^2}{f^2} - f^2 + \frac{1}{2}f^4\right]$$

$$+ \frac{H_c^2}{4\pi}\xi\sigma\tau_G\int_{-\infty}^{\infty} d\tau\int_{-\tau_\beta/2}^{\tau_\beta/2} d\tau'\int_{-l/2}^{l/2} dx\frac{(f(x\tau)-f(x\tau'))^2}{2\pi|\tau-\tau'|^2}, \tag{12}$$

where H_c is the critical magnetic field defined as $H_c^2/4\pi=\alpha^2/\beta$. The coordinate x and imaginary time τ are measured in units of ξ and τ_0. When deriving Eq.12, the nonlinear time derivative Q in Eq.5b that represents the dissipation has been rewritten by the nonlocal interaction term given by the double integral with respect to τ in the last line in Eq.12.

The WKB instanton approximation[12] gives the tunneling rate as follows:

$$\Gamma = A\exp[-B/\hbar],$$

$$B = S_{eff}[f_b] - S_{eff}[f_0], \tag{13}$$

where f_0 is the local minimum solution and f_b is the saddle-point solution for the effective action. The dominant contribution comes only from the exponent B which will be discussed in the rest of this paper. The integrals are expressed in the dimensionless form so that they have the order of unity. Therefore the prefactors determine the order of the exponent B. The only difference between the first and second prefactor is the definition of their time constants τ_0 and τ_G. In the large dissipation limit, the second term dominates the first one so that the exponent B has the order of the second prefactor as follows:

$$B_L \sim \frac{H_c^2}{4\pi}\xi\sigma\tau_G. \tag{14}$$

In the small dissipation limit, the first term dominates so that the exponent B has the order of the first prefactor as follows:

$$B_S \sim \frac{H_c^2}{4\pi}\xi\sigma\tau_0. \tag{15}$$

Using the GL coefficients given by the BCS-Gorkov theory gives the temperature dependences of $B_L \propto \Delta T^{1/2}$ and $B_S \propto \Delta T (\Delta T = T_c - T)$ and their following ratio:

$$\frac{B_S}{B_L} \sim \frac{\tau_0}{\tau_G} = 0.57\left(\frac{\Delta T}{T_c}\right)^{1/2}. \tag{16}$$

The resistivity is proportional to the tunneling rate by $R=(\Phi_0/J)\Gamma$ (Φ_0: magnetic flux quantum).[4] Giordano[7] has shown that the slope of resistivity-vs.-temperature graph in logarithmic scale suddenly changes at a crossover temperature below the critical temperature. The resistivity was consistent with the thermal activation theory[4] above the crossover temperature. Below this temperature, no theory could explain this enhanced exponent. Giordano has intuitively derived the following parametrized form for the exponent B and determined the coefficient from the experimental fitting:

$$B \sim 0.2\frac{H_c^2}{4\pi}\xi\sigma\tau_G. \tag{17}$$

He conjectured that the height of the GL free energy corresponds to that of the potential energy in the Josephson junction, and, that the inverse of the relaxation time corresponds to the frequency at the bottom of the potential. However this intuitive correspondence has not yet been theoretically justified. The experiment was performed between 0.2 and 0.7 K. In this temperature range, τ_0/τ_G has a value from 0.14 to 0.26, as can be estimated using Eq.16. Thus, the theory shows that the dissipation has a significant effect supporting the experimental formula Eq.17. However, both the exponents B_L and B_S still have the same order of magnitude, so that the dissipation might be unimportant. According to private communication from Giordano,[10] the succeeding experiments show that the exponent B is proportional to ΔT^{α} where α has a value from 0.25 to 0.5, strongly indicating that the dissipation dominates. Thus, our theory seems to be consistent with the experiments, supporting Giordano's intuitive derivation of the tunneling rate.

5. Quantitative Evaluation of Small Dissipation Limit

The dimensionless integral can be estimated in the small dissipation limit using a trial function method.[6] The trial function for the suddle-point solution is chosen as an extension of the bubble solution in the thermal equilibrium as follows:[6]

$$f^2 = f_0^2 - \Delta\mathrm{sech}^2[\rho\sqrt{(p\Delta)}], \tag{18}$$

where $\rho=\sqrt{(x^2+\tau^2)}$. In Eq.18, the parameters p and Δ are determined from the variational problem that minimizes the exponent B. The analytic evaluation for the dimensionless integral can be performed by using the expansion around the critical current. The minimization of B gives

$$B \simeq 3.45\frac{H_c^2}{4\pi}\xi\sigma\tau_0(1-J_0^2/J_c^2). \tag{19}$$

On the other hand, the thermal activation theory[4] yields the following activation rate using the expansion around J_c:

$$\Gamma = A_{th}\exp(-F_{th}/k_BT), \quad F_{th} = 0.810\frac{H_c^2}{4\pi}\xi\sigma(1-J_0^2/J_c^2)^{5/4}. \tag{20}$$

Rough estimation of the crossover temperature can be obtained by equating both the exponents, $B/\hbar=F_{th}/k_BT$, and solving with respect to the temperature. It should be noted that the trial function method gives an upper bound for the exponent and, therefore, a lower limit for the crossover temperature. Using the temperature dependence of the GL-coefficients given by the BCS-Gorkov theory yields

$$T_{cross} > T_c C(J_0),$$
$$C(J_0) = (-a + \sqrt{(a^2+4a)})/2, \tag{21}$$
$$a = (1-J_0^2/J_c^2)^{1/2}.$$

The constant $C(J_0)$ is weakly dependent on the current J_0 and has a value of 0.5 to 0.6 except in the very close vicinity of J_c. Thus, the crossover temperature has the order of the critical temperature. It should be compared to the crossover temperature in the Josephson junction case in the small dissipation limit which is around several 100 mK.[3] The crossover temperature is expected to be much higher in thin superconducting wires than in the Josephson junction. Our rough estimation of the lower limit of the crossover temperature is dependent on the selection of the trial function. Therefore, the crossover temperature closer to the critical temperature can be expected to be obtained by using the more relevant trial function or the computer simulation.

The dependence of the tunneling rate on current has been experimentally investigated[9] near the critical current. It was shown that the slope of the exponent changes suddenly at a current lower than the critical current, suggesting a crossover from the thermal activation to the MQT. The exponent B was shown to have approximately linear dependence on the current above and below the crossover current. Our theory showed that the exponent is proportional to $(1-J_0^2/J_c^2)^{5/4}$ above the crossover current and $(1-J_0^2/J_c^2)$ below the crossover current. Both the dependences are close to being linear dependences near the critical current, thus, conforming with the experiments. We must include the dissipation effect in the estimation given in this section. However, the overall feature might not be changed since the dissipation term could be renormalized in the kinetic energy term just as it can be renormalized in the capacitance term in the Josephson junction case.

6. Conclusions

The resistivity arising from the MQT was evaluated based on the one-dimensional TDGL equation that may be a phenomenological model for a thin superconducting wire. This theory showed that a fitting formula for the resistivity can be reproduced. Use of the GL-coefficients given by the BCS-Gorkov theory showed that the dissipation is dominant in the temperature range of the experiments.

Both the theory and experiments showed that the crossover temperature is near the critical temperature and the effect of the dissipation is not so strong compared to the Josephson junction case where the large dissipation effect lowers the crossover temperature down to several 10 mK. This fact enables us to observe another phenomena, the macroscopic quantum coherence (MQC) effects which are an important problem in the measurement theory. The MQC effects were not observed in the Josephson junction because strong dissipation prevents the quantum transition. The thin superconducting circle has been theoretically proven to behave like a SQUID although there is no Josephson junction.[13] Therefore, it is possible to observe the MQC effect using a thin superconducting circle.

Finally, we should note that another related phenomenon observed by Giordano[14] showed that the tunneling rate could be enhanced at several currents. This is thought to be the result of the resonant tunneling which was discussed in the Josephson junction experiment.[15] The resonance tunneling phenomena have not yet been observed in the Josephson junction experiments. We conjecture that the reason for this is that the resonant peaks are broadened due to dissipation. This fact suggests that the MQC effects in the thin superconducting circle, where the effect of dissipation may be smaller than in the usual SQUID, should be observed.

7. Acknowledgements

We would like to express sincere gratitude to Drs. E. Maruyama and A. Fukuhara for their continuous support through this entire work. Thanks are also due to Professors M. Namiki and I. Ohba for their helpful discussions. Useful discussions and comments given during the symposium by Professors F. E. Schroeck, A. Peres, and H. P. Stapp are also greatly appreciated.

8. References

1. R. A. Webb et al., in *Proc. 2nd Int. Symp. Foundations of Quantum Mechanics*, eds. M. Namiki et al. (The Physical Society of Japan, Tokyo, 1987) p. 193.
2. A. O. Caldeira and A. J. Leggett, Ann. Phys. (NY) **149** (1983) 374.
3. R. F. Voss and R. A. Webb, Phys. Rev. Lett. **47** (1981) 265.
4. J. S. Langer and V. Ambegaokar, Phys. Rev. **164** (1967) 498; D. E. McCumber and B. I. Halperin, Phys. Rev. **B1** (1970) 1054; I. H. Duru and H. Kleinert, J. Low Temp. Phys. **42** (1981) 137.
5. J. E. Lukens, R. J. Warburton and W. W. Webb, Phys. Rev. Lett. **25** (1970) 1180; R. S. Newbower, M. R. Beasley and M. Tinkham, Phys. Rev. **B5** (1972) 864; J. R. Miller and J. M. Pierce, Phys. Rev. **B8** (1973) 4164.
6. S. Saito and Y. Murayama, Phys. Lett. **A135** (1989) 55.
7. N. Giordano, Phys. Rev. Lett. **61** (1988) 2137.
8. S. Saito and Y. Murayama, Phys. Lett. **A139** (1989) 85.
9. N. Giordano, in *Proc. Int. Symp. on Nanostructure Physics and Fabrication*, 1989, eds. W. P. Kirk and M. Reed (Academic, New York, to be published).
10. N. Giordano, private communication.
11. E. Abrahams and T. Tsuneto, Phys. Rev. **152** (1966) 416; N. R. Werthamer, *Superconductivity*, ed. R. D. Parks (Dekker, New York, 1969) Chap. 6.
12. S. Coleman, Phys. Rev. **D15** (1977) 2929; C. G. Callan and S. Coleman, Phys. Rev. **D16** (1977) 1762.
13. H. J. Fink, V. Grunfeld and A. Lopez, Phys. Rev. **B35** (1987) 35; H. J. Fink and V. Grunfeld, Phys. Rev. **B33** (1986) 6088.
14. N. Giordano and E. R. Schuler, Phys. Rev. Lett. **63** (1989) 2417.
15. N. Hatakenaka, H. Takayanagi and S. Kurihara, in *Proc. 19th. Int. Conf. on Low Temp. Phys., to be published.*

UNSHARPNESS IN MEASUREMENT YIELDS INFORMATIONAL COMPLETENESS

Implementations in Quantum Signal Processing and Natural Biological Systems

FRANKLIN E. SCHROECK, JR.

Department of Mathematics

Florida Atlantic University

Boca Raton, Florida 33431

U.S.A.

ABSTRACT:

The distinction between unsharp and inaccurate measurements is given, and contrasted by a theorem that unsharp measurements may be informationally complete. Comparison with the standard but informationally incomplete sharp momentum and sharp position observables, as well as time and frequency observables of signals, is made theoretically and in demonstration. We give a method of implementing unsharp measurements with illustrations from experiment and natural biological systems. Information content in these implementations is analyzed, and implications for signal processing and pattern recognition are given.

I. Introduction

"Physics" at one time encompassed more than what is now included in the traditional physics curriculum. This has occurred in two ways:

(i) Areas split off from physics, and as they evolved independently, cross-communication decreased;

(ii) The formalism of physics seemed ineffective in application to some areas.

We shall outline an extension of the formalism for quantum mechanics to the positive operator formalism, which

(i′) establishes ties with the evolved status of other disciplines,

(ii′) broadens the range of applicability of quantum theory.

To obtain this extension, we begin with a look at an unconscious but unnecessarily restrictive mode of thought we carry into quantum measurement theory. We begin with a simple classical example illustrating the same misconception:

Suppose that I fire a gun at a target a number of times, obtaining the following histogram (Figure 1) for the x-coordinates of the impact points (in cm from left edge of the target):

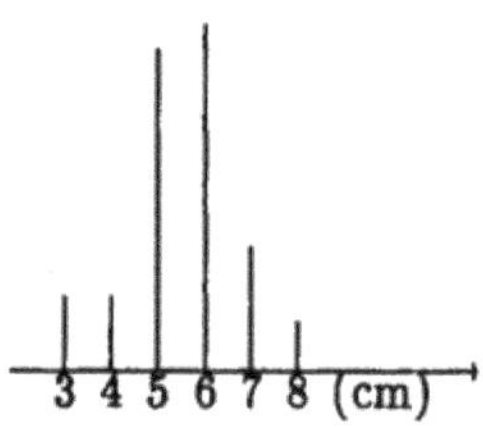

Figure 1

Does this indicate that I'm not a very accurate marksman? If I use this as the distribution for the x–coordinate, and use it to predict the outcome of firing one more round, is this experiment inaccurate? unsharp?

On the other hand, if I show you the joint distribution in both x– and y–coordinates (Figure 2), you will see that the x–distribution is unsharp, but the data is not at all inaccurrate. You might even be able to predict the location of the x coordinate of the next shot.

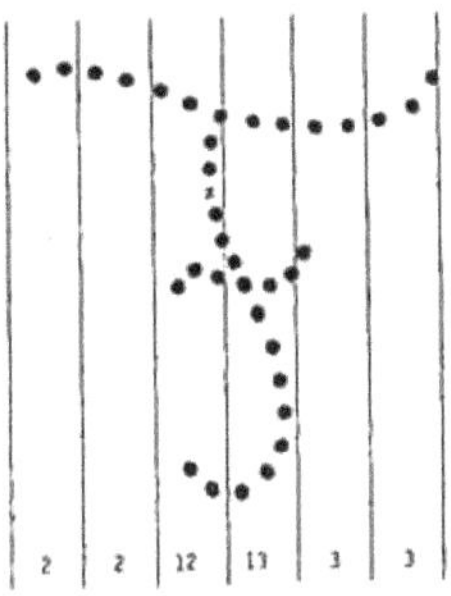

Figure 2

Even if I had given you the histograms for both of the marginal coordinate distributions, still you might have thought that the experiment was inaccurate.

Conclusion 1: The failure to record joint distributions may lead to a misinterpretation of experimental data.

As another example, consider the following frequency spectrum: (see Bennett, 1976, pg 401). Can you identify the pattern? It is the power spectrum of the complete playing of "The Rite of Spring." Had you been given the joint time/frequency data (the musical score) it would be easy to identify the pattern.

Conclusion 2: There is more information in joint distributions than in the collection of marginal distributions.

For pattern recognition, this is an essential point. For an example from the biological realm, consider the problem of bat echo–location. A bat must (a) locate the prey [position], (b) determine the trajectory of the prey [momentum], (c) identify the prey as to size, texture, and shape in order to avoid flying into windows or cliffs, eating falling rocks, tennis balls, etc. [pattern recognition], (d) avoid mistaking signals from nearby bats as reflections off prey [multiple signal discrimination], (e) overcome other problems of a noisy environment [noise surpression].

To bring points (a) and (b) to the level of quantum mechanics, we must face the theorem of Wigner (1971):

Theorem: In the usual Hilbert space framework for quantum mechanics, there is no family of observable projection operators $\{F(q,p)\}$ with marginals giving the spectra of position Q and momentum P respectively; i.e., such that for a state with wavefunction φ,

$$\int <\varphi, F(q,p)\varphi> dp = <\varphi, E^Q(q)\varphi> = |\varphi(q)|^2,$$
$$\int <\varphi, F(q,p)\varphi> dq = <\varphi, E^P(p)\varphi> = |\tilde{\varphi}(p)|^2. \qquad (1)$$

Here $\{E^Q(q)\}$ and $\{E^P(p)\}$ are the spectral families for Q and P respectively.

To make matters worse, there are many publications showing that complete knowledge of the marginal distribution densities $|\varphi(q)|^2$ and $|\tilde{\varphi}(p)|^2$ does not determine φ uniquely in any non-trivial sense.

Let us rephrase this:

Definition: A collection of observables (operators) $\{A_\alpha | \alpha \in I\}$ for a system described in a Hilbert space $\mathscr{H}$ is said to be informationally complete iff the set of quantum expectations $\{Tr(\rho A_\alpha) | \alpha \in I\}$ uniquely determines the quantum density ρ.

In this language, $\{E^Q(q) | q \in \mathbb{R}^3\} \cup \{E^P(p) | p \in \mathbb{R}^3\}$, whose expectations correspond to the complete position and momentum marginal distributions, is informationally incomplete always, whether measured accurately or not.

Nonetheless, there are positive operator valued measures (POVM's) [as opposed to projection valued measures (PVM's)], with density $T(q,p)$ such that

$$\int \langle \varphi, T(q,p)\varphi \rangle \, dp = \text{a classical density convolved with } \langle \varphi, E^Q(q)\varphi \rangle,$$

$$\int \langle \varphi, T(q,p)\varphi \rangle \, dq = \text{a classical density convolved with } \langle \varphi, E^P(p)\varphi \rangle. \quad (2)$$

Furthermore, from these $\langle \varphi, T(q,p)\varphi \rangle$, one can reconstruct not only the $|\varphi(q)|^2$ and the $|\tilde{\varphi}(p)|^2$, but also all information about the state given by φ; that is, the family $\{T(q,p)\}$ is informationally complete. We will indicate a natural condition on a commonly encountered class of POVM's sufficient to guarantee informational completeness.

It is a mathematical fact that if the marginals of a POVM can describe non-commuting operators, then the POVM cannot be a PVM. It is also a fact (Busch and Lahti, 1989) that no commuting set of projections can be informationally complete. Since PVM's correspond to sharp measurements while POVM's include unsharp measurements, we see that obtaining the desirable property of informational completeness necessitates the introduction of experimentally measurable POVM's.

So far, we have focussed on the variables of position and momentum, which together constitute phase space $\mathbb{R}^6$. We could formulate our discussion far more generally to include spin, angular orientation, dilation, etc. This analysis may be found in the literature. We present the discussion in phase space solely to simplify the notation.

We establish the notation, that for any Borel set, Δ, in $\mathbb{R}^6$,

$$A(\Delta) \equiv \int_\Delta T(q,p) \, dq dp. \quad (3)$$

These $A(\Delta)$ form our POVM. The experimental and mathematical meaning for $A(\Delta)$ and $T(q,p)$ will be developed presently.

The topics for this lecture may now be listed :

#1 How can one perform laboratory procedures to obtain the $\langle \varphi, A(\Delta)\varphi \rangle$ or the $\langle \varphi, T(q,p)\varphi \rangle$? Notice that these may correspond to unsharp (but hopefully not inaccurate) measurements of the marginal observables.

#2 How much information is contained in these "unsharp" measurements, compared to the sharp measurements of both $\{E^Q(q)\}$ and $\{E^P(p)\}$ independently?

#3 Are these "marginally unsharp" measurements in fact "accurate"?

II. Experimental Origins of the $A(\Delta)$ and $T(q,p)$

Consider a screen for detecting particles. Partition the screen into a collection of boxes $\{\Delta_i\}$ in configuration space; within each box the recording is done in channels Δ^j according to momentum (or kinetic energy). [See Figure 3.] Set $\Delta_{ij} = \Delta_i \times \Delta^j$ and call the resulting observable for recording a result in Δ_{ij}

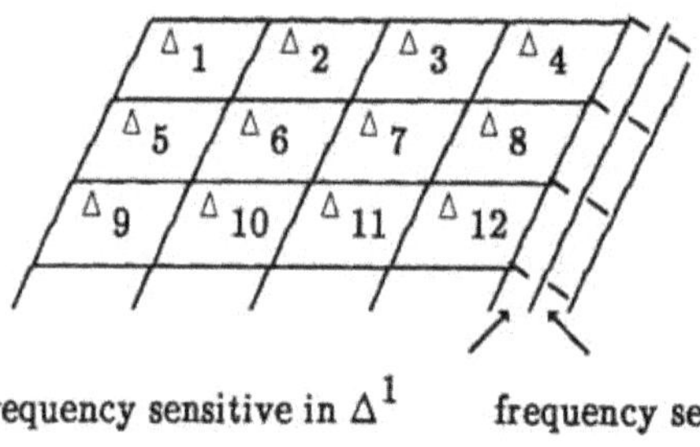

Figure 3

"$A(\Delta_{ij})$." The collection $\{A(\Delta_{ij})\}$ forms a POVM (Foulis & Schroeck, 1990). [In this last reference it is proven that, in general, starting from an operational manual, quantum measurement always leads to a POVM.]

If the screen is constructed uniformly in space, then the unitary translation operators $W(q,0)$ interrelate with the $A(\Delta_{ij})$ by

$$W(q,0)^\dagger\, A(\Delta_{ij})\, W(q,0) = A(U(q,0)\Delta_{ij}), \qquad (4a)$$

where $U(q,0)\Delta_{ij}$ is a simple translation of Δ_{ij} in the configuration part Δ_i. If the screen also is equally efficient at measuring all momenta, then

$$W(q,p)^{\dagger} A(\Delta_{ij}) W(q,p) = A(U(q,p)\Delta_{ij}), \tag{4b}$$

where the $U(q,p)$ represents a translation in phase space by (q,p). For the screen in Figure 2, as partitioned, only a lattice subset of the $W(q,p)$ has this property. The $\{W(q,p)\}$ form the standard representation of the Heisenberg group, and any POVM $\{A(\Delta)\}$ satisfying (4b) is said to be "covariant with respect to the Heisenberg group." This same structure abstracted to the level of a general group and general POVM gives a "system of covariance." Analysis of this general mathematical structure has led to many results of physical importance both foundationally (Prugovečki, 1984; Ali, 1985; Ali and Prugovečki, 1986; Ali, Brooke, Busch, Gagnon, and Schroeck, 1987; Brooke and Schroeck, 1989) and in proposing and analyzing crucial experiments (Mittelstaedt, Prieur, and Schieder, 1987; Busch, 1987; Daubechies, 1988; Busch and Schroeck, 1989).

This same structure may be derived from a different argument: One supposes that, in detecting a particle in Δ_{ij}, the screen forces the particle into a state with expected position in Δ_i and momentum in Δ^j. If the particle, when in a state with expected location at the origin of phase space, has state vector η, then $W(q,p)\eta$ has expected phase space position (q,p). Hence, the probability of recording the original particle "at (q,p)" is given by the transition probability

$$|\langle W(q,p)\eta,\varphi\rangle|^2 = \langle\varphi,T(q,p)\varphi\rangle, \tag{5}$$

where

$$T(q,p) = |W(q,p)\eta\rangle\langle W(q,p)\eta|. \tag{6}$$

Integrating over Δ_{ij} yields the probability $\langle\varphi,A(\Delta_{ij})\varphi\rangle$ of detection in Δ_{ij}.

In this derivation, one can see the connection with the quantum formalism. A large number of fundamental quantum mechanical experiments have been reanalyzed in recent years, showing that the POVM formalism is present everywhere; and covariance is usually satisfied, sometimes requiring only a compensation for non–uniform or non–continuous measuring apparati. This analysis, in turn, shows the feasibility of measuring unsharply but simultaneously, informationally complete sets such as all spin components, all components of helicity, etc. (For example, see Busch, 1987; Busch & Schroeck, 1989).

For a second example, consider radar or sonar and the pattern recognition problems present in this field. On the surface, the fact that time and frequency of the signals form a Fourier pair may seem to be a mathematical accident. On closer examination, the time delay of a signal is directly proportional to the distance between emitter and reflector, and the (Doppler) frequecy shift is proportional to the relative velocity. The Heisenberg group acts naturally on such systems (as well as the affine group), and it is no mathematical accident; the Fourier relationship follows. Identifying time with position, frequency with momentum, is proper in this context.

To analyze a signal, s, one can take a test signal, η, delay it by time τ, shift the frequency by ν to obtain $U(\tau,\nu)\eta$, mix the two signals and record the intensity (energy) of the mixed signals obtaining:

$$I(\tau,\nu) = \|s + U(\tau,\nu)\eta\|^2 = \|s\|^2 + \|\eta\|^2 + 2\mathrm{Re}\langle s,U(\tau,\nu)\eta\rangle. \quad (7)$$

By computing the contrast (visibility) of the resulting interference pattern obtained by making small shifts in ν, one recovers $\langle s,T(\tau,\nu)s\rangle$ (Brooke & Schroeck, 1988), where

$$T(\tau,\nu) = |U(\tau,\nu)\eta\rangle\langle U(\tau,\nu)\eta|. \quad (8)$$

The terms $\langle s,T(\tau,\nu)s\rangle$ are cross–correlations between η and s. We shall show the informational completeness of these $T(\tau,\nu)$ (Klauder, 1964; Healy and Schroeck, 1990). Therefore, the energy in the mixed signals, as a function of τ and ν, must also be informationally complete. Some bats, in their echolocation processes, seem to perform such cross-correlations by first obtaining an overlap of their emitted signal with the reflected signal, and then shifting frequency and comparing again in a part of their sensory system that parallel computes (in a neural network) with different time delays between emitted and received signal (Simmons, Howell, & Suga, 1975; Suga, 1990). [See Figure 4.]

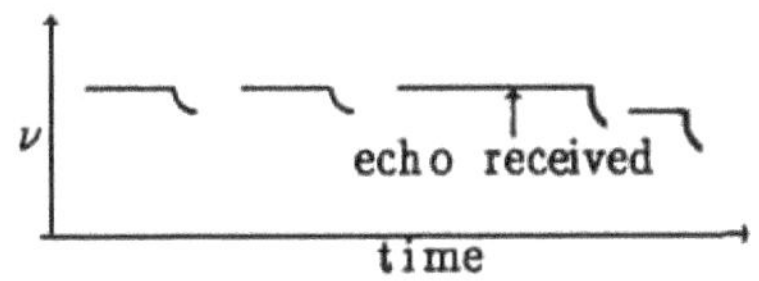

Figure 4

More impressively, with this system, bats can measure time and frequency with a precision at the quantum limit of uncertainty! (Graphically illustrated in Simmons, Howell, & Suga, 1975, pg 204.) This is a necessity, since bats of the same family share a small range of available frequencies, and the bats need to avoid cross-talk.

Closely related to this analysis is the subject of speech recognition. This topic is important both for artificial hearing devices for the hearing impared, and for voice activated computers. Biologists use joint time/frequency analysis (speech spectrograms) for recognizing vocalization patterns in birds, primates (Bauers, 1989), and perhaps other species. The joint time/frequency spectrograms of the human voice show patterns that are characteristic of words and are fairly speaker independent. [See Figures 5, 6, 7, courtesy of Marchette and Priebe, 1987.]

Analysis based on only marginal time and/or marginal frequency versus amplitude (the usual analysis) have so far failed to yield working

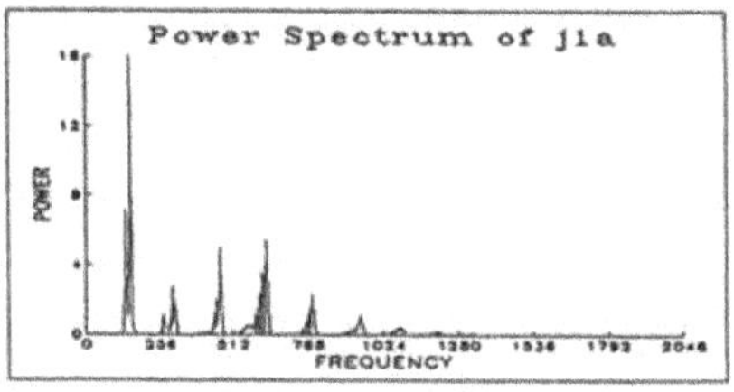

Figure 5

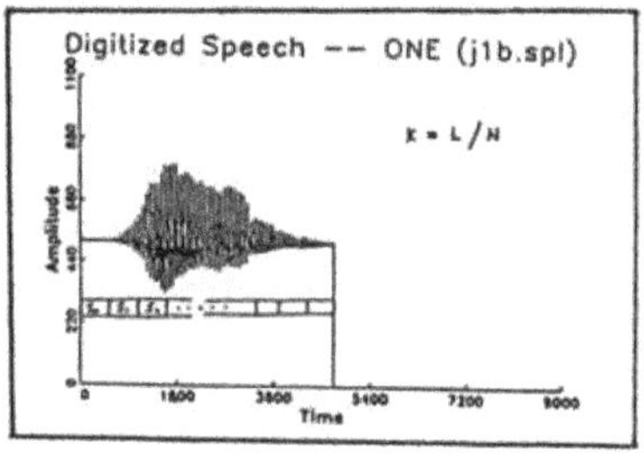

Figure 6

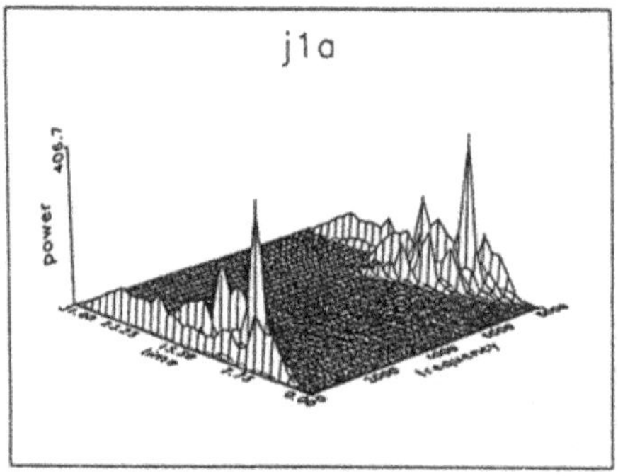

Figure 7

speaker-independent speech recognition systems. Perhaps, that is due to analyzing only informationally incomplete sets of data.

III The Main Result

<u>Theorem</u>: (Healy and Schroeck, 1990) Let G be a locally compact topological group . Let H be a closed subgroup of G. Suppose that the homogeneous space G/H of G has left invariant measure μ. Let σ:G/H $\rightarrow$ G be a Borel section. Let U denote a unitary, strongly continuous, irreducible representation of G on

Hilbert space $\mathscr{H}$. Suppose there exists $\eta \in \mathscr{H}$ such that $<\eta,U(\sigma(x))\eta> \neq 0$ for almost all x in G/H, $U(h)\eta = \eta$ for all $h \in H$, $\int_{G/H}|<\eta,U(\sigma(x))\eta>|^2 d\mu(g)$ is finite, and there exists $B:G/H \times G/H \to \mathbb{C}$ such that, for all x,y in G/H, $U(\sigma(x)^{-1}\circ\sigma(y)^{-1}\circ\sigma(x)\circ\sigma(y))\eta = B(x,y)\eta$. Then $\{T(x) = |U(x)\eta><U(x)\eta|$ such that $x \in G/H\}$ is informationally complete in $\mathscr{H}$, and the POVM defined by

$$A(\Delta) = \int_\Delta T(x)\, d\mu(x), \quad \Delta \in \text{Borel}(G/H), \tag{9}$$

is informationally complete and covariant with respect to G/H.

Combining this with the theorem (Busch, 1989) that the linear span of an informationally complete set of operators is dense in the set of all bounded operators on $\mathscr{H}$, we see that all bounded observables may be written as (limits of) integrals of measurable functions with respect to this POVM.

None of these theorems would be of any use if they were only abstract theorems. The beauty of the results lies in the explicit formulas in the proof of the recent theorem of the author with D. Healy, in which the state ρ of the system is reconstructed exactly from the data $\{Tr(\rho T(x))\}$ and the operators are written as integrals of the set $\{T(x)\}$ over G/H with uniquely determined weight function. This latter result may be viewed as a generalization of the spectral theorem into the realm of non-commuting operators, and providing a powerful new functional analysis. **With these explicit reconstructions in hand, it is clear that these unsharp measurements are in no way inaccurate.**

Result (9) was known for the Heisenberg group, in which case G/H $\cong$ phase space (Klauder, 1964). Application of the present theorem shows that in both the Heisenberg and Gallilei groups, the joint localization operator in phase space is informationally complete. Aspects of informational completeness have also been proven for the affine and Poincaré groups (Healy & Schroeck, 1990). Thus by recording only the cross-correlations (interferences) $|<U(\sigma(x))\eta,\varphi>|^2$ for all x

ϵ G/H, one completely characterizes φ, and one may efficiently analyze this data for pattern recognition and other purposes.

By the continuity of these cross–correlations in x, one need only collect the data at discrete points in G/H to obtain approximations to the terms $<\varphi,A(\Delta)\varphi>$ in precisely the same way Riemann sums approximate Riemann integrals. Hence, with discrete data one may still obtain (approximate) informational completeness <u>since one has started with it</u>. In specific situations, one might even use a finite set.

Amazingly enough, the data is also efficiently stored in fewer bits of this discrete data, leading to a method of data compression. For example, one may take a picture, reexpress the data by these cross–correlations, and then visually compare the picture reconstructed from this new data. Doing this with "Lena," the standard benchmark, one obtains (Daugman, 1988) a very satisfactory picture with significant data compression.

Further analysis of the explicit reconstruction formula shows that this method of analysis is numerically stable, stable under high frequency noise, and is amenable to subtraction of structured noise (a consequence of the informational completeness property when applied to the noise). Moreover, the entropy analysis of the next section shows that by exploiting the ability to change to a new η, one can build adaptive processors. This last flexibility is also exploited by some types of bats that search for prey with one wave form, and change the wave form (to a "chirp") after locating the prey, so as to identify the object, judge the velocity, etc. [See Figure 4.]

Since the rotation group is contained in the groups above, informationally complete measurements of spin and angular momentum are subsumed in these results. Similarly for the problem of pattern recognition when the pattern is simply rotated. Dilations, used for edge detection and other purposes are contained in the results for the affine group, yielding a so-called "mathematical microscope."

IV Entropy Considerations

In the proof of informational completeness, one shows that

$$\int_{G/H} T(x)\, d\mu(x) = \lambda 1,\ \lambda > 0. \tag{10}$$

For η normalized, this λ is independent of η. Therefore, we may normalize μ by dividing by λ; i.e., we may take $\lambda = 1$. Then, for any quantum density operator ρ on $\mathscr{H}$, defining

$$\rho_c(x) = \mathrm{Tr}(\rho T(x)), \tag{11}$$

we see that ρ_c is a classical Kolmogorov probability density on G/H (Schroeck, 1985). Given ρ, the form of ρ_c depends on the choice of η. Suppose $\rho_c^{\,i}$ is based on the choice η^i, i = 1,2. Then we may compare the Boltzmann–Gibbs–Shannon entropies

$$S(\rho_c^{\,i}) = \int \rho_c^{\,i}(x) \ln(\rho_c^{\,i}(x))\, d\mu(x) \tag{12}$$

for i = 1,2. Choosing $\rho_c^{\,i}$ on the basis of which gives the lower uncertainty, one achieves better data compression. On the other hand, from the analysis of Ruch (1975) and Ruch and Mead (1976), convex functions other than *xlnx* may be used for obtaining estimates of data compression, since they are all "mixing homomorphic transformations." Thus, in the case that the Boltzmann–Gibbs–Shannon entropies coincide accidentally, switching to another convex function could single out the better process. In signal analysis, one may use a wave form generator to change test signal. In quantum measurement, one either changes test particles or changes instrumentation.

IV. Summary

We have indicated that in many experimental situations, including but not limited to signal processing, neuro–biological systems, and standard interferometry, data may be taken in the form of cross–correlations with a standardized state. Rather than reducing this information to the form of position distributions, or momentum distributions, one may use a new form of analysis to obtain all information directly from the cross–correlations. We therefore propose that we all change our mode of thinking concerning data taking, and rather than worrying incorrectly that unsharp measurements are inaccurate, seek out ways to take data in this cross–correlated way.

References

1. S. T. Ali, "Stochastic Localization, Quantum Mechanics on Phase Space and Quantum Space–Time," *Nuovo Cimento* 11 (1985) 1–128.

2. S. T. Ali, J. A. Brooke, P. Busch, R. Gagnon, and F. E. Schroeck, Jr., "Current Conservation as a Geometric Property of Space–Time," *Can. J. Phys.* 66 (1988) 238–244.

3. S. T. Ali and E. Prugovečki, three consecutive papers on harmonic analysis, *Acta Appl. Math.* 6 (1986) 1–18, 19–46, 47–62.

4. K. Bauers, "The Role of Vocal Communication in the Intra–Group Social Dynamics of Stumptailed Macaques," Ph.D. Thesis, Univ. of Wisconsin at Madison, 1989.

5. W. R. Bennett, Jr., **Scientific and Engineering Problem–Solving with the Computer**, (Prentice–Hall, New York, 1976), §8,10, p. 401.

6. J. A. Brooke and F. E. Schroeck, Jr., "Half Integer Spin Representations of the Poincaré Group on Phase Space," *Nucl. Phys. B (Proc. Suppl.)* 6 (1989) 104–106.

7. P. Busch, "A Characterization of Informationally Complete Sets of Physical Quantities," preprint, 1988.

8. P. Busch, "Some Realizable Joint Measurements of Complementary Observables," *Found. Phys.* 17 (1987) 905–937.

9. P. Busch and P. J. Lahti, "The Determination of the Past and Future of a Physical System in Quantum Mechanics," *Found. Phys.* 19 (1989) 633–678.

10. P. Busch, and F. E. Schroeck, Jr., "On the Reality of Spin and Helicity," *Found. Phys.* **19** (1989) 807–872.

11. D. Chester, F. J. Taylor, and M. Doyle, "The Wigner Distribution in Speech Processing Applications," *J. Franklin Inst. (U.S.A.)* **318**, no. 6 (1984) 415–430.

12. I. Daubechies, "Time–Frequency Localization Operators: A Geometric Phase Space Approach," *IEEE Trans. Info. Th.* **34** (1988) 909–996.

13. J. G. Daugman, "Relaxation Neural Network for Non–Orthogonal Image Transforms," *Proc. IEEE Int'l Conf. Neural Net. 1988*, pp. I547–I560.

14. D. Foulis and F. E. Schroeck, Jr., "Stochastic Quantum Mechanics Viewed from the Language of Manuals," *Found. Phys.*, 1990, to appear.

15. D. M. Healy and F. E. Schroeck, Jr., "On Informational Completeness of Covariant Observables and Wigner Coefficients," preprint, Dartmouth Coll., 1990.

16. J. Klauder, "Continuous–Representation Theory. III. On Functional Quantization of Classical Systems," *J. Math. Phys.* **17** (1964) 177–187.

17. D. Marchette and C. Priebe, 1987, private communication.

18. P. Mittelstaedt, A. Prieur, and R. Schieder, "Unsharp Particle–Wave Duality in a Photon Split Beam," *Found. Phys.* **17** (1987) 891–904.

19. E. Prugovečki, **Stochastic Quantum Mechanics and Quantum Spacetime**, D. Reidel Pub. Co., 1984.

20. E. Ruch, "The Diagram Lattice as Structural Principle," *Theor. Chim. Acta (Berl.)* **38** (1975) 167–183.

21. E. Ruch and A. Mead, "The Principle of Increasing Mixing Character and Some of Its Consequences," *Theor. Chim. Acta (Berl.)* **41** (1976) 95–117.

22. F. E. Schroeck, Jr., "Entropy and Measurement in Stochastic Quantum Mechanics," in **Symposium on the Foundations of Modern Physics**, P. Lahti and P. Mittelstaedt, eds., World Scientific Pub. Co., 1985, 573–590.

23. J. A. Simmons, D. J. Howell, and N. Suga, "Information Content of Bat Sonar Echoes," *Am. Scientist* **63** (1975) 204–215.

24. N. Suga, "Biosonar and Neural Computation in Bats," *Sci. Am.*, June, 1990, 60–68.

25. E. P. Wigner, "Quantum Mechanical Distribution Functions Revisited," in **Perspectives in Quantum Theory**, W. Yourgrau and A. van der Merwe, eds., M.I.T. Press, Cambridge, 1971, pp. 25–36.

HOW TO TEST FOR CARTESIAN DUALISM BY QUANTUM EXPERIMENTS

Euan J. Squires
Dept. of Mathematical Science, University of Durham,
Durham City, DH1 3LE, England

ABSTRACT

We show how it is possible to construct a realistic, local and Lorentz invariant, version of quantum theory. The model gives different predictions to those of the orthodox theory in certain EPR-like situations. We discuss whether such differences should have been seen in presently available experimental data, and conclude that it is probably reasonable to suppose that they would not, although the situation is marginal.

If improved experiments fail to show any deviations from the predictions of orthodox quantum theory, then we conclude that realism requires the existence of something outside "physics", where we define this term to imply locality and Lorentz invariance. It is then natural to associate this "something" with the non-physical ingredient of a Descartes-like dualism, e.g. with consciousness. Some possible models are described, and it is noted that they may suggest solutions to some of the long standing problems regarding the relation between physics and conscious mind.

1. Introduction

In order to obtain useful information from the result of an experiment, that is, information that goes beyond the result itself, it is necessary to use some theoretical input. One of the reasons behind our lack of progress in understanding quantum theory, is that we have not been able to find suitable input that has allowed testable predictions to be made. The supreme counter-example to this is, of course, Bell's theorem. In this article I shall endeavour to suggest a set of assumptions and definitions which allow us to decide experimentally whether Descartes was correct to postulate the need for "something" (e.g. conscious mind) outside of physics. In the last section I shall discuss alternative possibilities for the way this non-physical property operates.

First, it is necessary to remind ourselves of the basic measurement problem of quantum theory. This is the topic of the next section.

2. The Measurement Problem of Quantum Theory.

We see this in its simplest form by the standard Stern-Gerlach thought experiment for the measurement of an electron spin in the z-direction. If this spin is initially $+1/2$ then the particle will be deflected, say, upwards, where it will pass through a detector, and I will then observe it. This will constitute my measurement of the spin. The *relevant* part of the wavefunction will have the form $|+, Me^+>$, where the Me^+ refers to a particular state of my brain (corresponding to my knowing that the electron had spin $+1/2$). Similarly, if the initial spin of the electron had been $-1/2$, then the wavefunction after the measurement would have been $|-, Me^->$.

So far all is well; the wavefunction accurately reflects my experience. However, the linearity of quantum theory now leaves us no freedom. If, initially, the electron had a definite spin component in a direction different to that of the z-axis, then the state after the measurement would be

$$|\Psi>= \alpha|+, Me^+> +\beta\,|-, Me^->. \tag{2.1}$$

This contains both results, although I will be aware of only one. The measurement problem of quantum theory is how to restore the correlation between the result of which I am aware, and the description given by physics. Of course, in most textbooks on quantum theory the state of the brain is not included. Then it is possible to regard the problem as being one of interpretation. Here we see that the calculation simply gives the wrong answer. This is the real measurement problem!

Note that everything in the above has been grossly oversimplified; I am not a simple two-state system, there is a large detection apparatus which has been ignored, as has the fact that everything is continually interacting with an unknown environment, with a microwave background, maybe with wormholes, and who knows what else. What these things mean, of course, is that I cannot *in practice* do an interference experiment to verify the above wavefunction. However, such an experiment, real or imagined, plays no role in our discussion. We obtained the wavefunction from the linearity of quantum theory, not from experiment. (We should not, of course, forget the role of interference. It is because this is seen so clearly in microscopic systems that we are led to use the standard linear formalism of quantum theory.)

3. The First Assumption: Realism.

We now make the first key assumption of this article. This is what I shall call the assumption of **Realism** (though in no sense will I enter into a technical discusion of what might be meant by this word). For my purpose what this will mean is that the wavefunction in Eq.2.1 cannot be the complete description of reality because there is nothing in it that corresponds to my experience of one result.

Since this is an assumption, I am free to make it, and no defence is needed. However, I believe that such an assumption is an essential requirement if we are to do proper physics. To illustrate what I mean here I will compare two opposing positions. First, let me take the alternative view, as expressed for example by Dieks[1] , where the assumption is not made. Instead, it is postulated that whenever a unique decomposition of the form

$$|\Psi>=\sum \alpha_i|\chi_i>|\phi_i>, \tag{3.1}$$

where the $|\chi>$ and the $|\phi>$ are orthogonal states, exists, then the individual sub-systems are really in one of the states $|\chi_i>$ and $|\phi_i>$, respectively. Thus, although the calculation does not give the right answer, i.e. the answer corresponding to our experience, we simply say that it *means* the right answer. We decree that, regardless of the result of the calculation, the answer really means the answer we observe. I do not believe that this is the correct way to do physics; when we obtain the wrong answer from a calculation we should rather deduce that it is based on an incorrect theory. (See also the articles in this volume by D.Albert, who has a more specific objection to this procedure, and by B.van Fraassen, who suggests a similar model to that of Dieks.)

I am sure that I am not being entirely fair here, but it is of interest to contrast the verbal contortions that are necessary in ideas of this sort, with genuine realistic models[2] [3] [4] [5] [6] , in which the physics properly reflects experience. Here we make the obvious response to the measurement problem noted in the last section, and accept that the failure of the expression for the wavefunction given in Eq.2.1 to agree with experiment is due to the fact that it is wrong. The true quantum theory is not linear. The various ways in which the Schrödinger evolution equation can be modified, so that there is rapid collapse to position eigenstates when macroscopic systems are involved, without upsetting the accurate predictions for microscopic systems, have been well presented elsewhere, so they need not be explained here (see also the article by Ghirardi in this book). I will however express the opinion that the existence of such models should remove for ever the need for anti-realistic treatments of quantum physics. Serious physicists who do not wish to get involved in consciousness (and some would say that this last qualification is redundant), and are unwilling to admit hidden variables, now have an honest way of doing quantum theory.

The difficulty with models of this sort is that they are non-relativistic, and that, since they are explicit, they reveal in a stark fashion the essential non-locality of quantum phenomena. It is very hard to see how we can reconcile this non-locality with the requirements of special relativity. Before we discuss this further, however, we need to introduce our second key assuption.

4. The second assumption: choice

An important ingredient in the proof of Bell's theorem is that the settings of the two polarisers (or, in the actual Aspect experiments, the decision as to which of the polarisers the two beams are sent) is *effectively* **free**, i.e. that it can be made just prior to the photon reaching it, and that the choice at one side is independent of the choice at the other side and of the nature of the decay. Of course, in a fully deterministic theory, such free choices do not exist and it is therefore not possible to give a logically sound proof of Bell's theorem. This is why I have inserted the word "effectively" in the above statement. John Bell[7] refers to this as the assumption of "deterministic chaos" or "pseudo randomness". Theories without such a property he calls "superdeterministic".

What is being forbidden here is that the whole experiment, decay of the atom, settings of the polarisers, etc., all have a common cause, and that there is a clever conspiracy to ensure that the results of quantum theory hold. It is only when we exclude such things that the idea of local causality can be given meaning; otherwise nothing can be regarded as the cause of anything else; the world just "is". In orthodox quantum theory, where there are genuine random events, this particular assumption is certainly correct (even without the "effectively").

In passing we note that if this assumption is not valid for the actual Aspect experiments then we could try to make it true by using signals from distant galaxies, which have never been in causal contact, to set the polarisers. Alternatively[8] we might try the possibility that human mind acts are (in this sense) free, and use them explicitly to make the settings.

Here, however, we remain with the standard assumption. In its support I would suggest the following scenario. I always imagine that what appears on my television screen is a consequence of what happens in the studio (or on the sports field or whatever). If this is not a proper way of understanding the world, but if they should rather be thought of as having a common origin, then it seems to me that it would be quite possible, and would indeed occur occasionally, for the result of the World Cup, as seen on my screen, to be different from that which happens on the playing field.

We have now made the necessary assumptions. To see how we can test for something beyond physics we need to know what physics is. This we discover in the next section.

5. A definition of physics

If we are to find criteria for measuring whether there is something in the world beyond physics then clearly we must have an understanding of what we mean by "physics", i.e. we must define the word. (I find it strange that many people have very positive views about duality, for and against, without seeming to worry about this detail). I shall not of course attempt to give, nor shall I need, a complete definition of physics. Only two properties will be required: ***Physics is relativistically invariant, and is, at least approximately, local.*** That relativistic invariance is a reasonable property of physics we might justify by asking any well-taught physics student throughout the world. More seriously, we really do not know how to do physics without relativity. Some will object that people managed very well until 1905, but we have learned a lot since then, and it is not easy to return to the idea of a preferred inertial frame. With regard to locality, it is clear that some sort of locality is necessary if we are to do physics. Indeed this is one of the reasons why Einstein insisted upon it. The idea that something physical can happen, simultaneously (on a particular space-like hypersurface) thoughout the whole universe, is so far removed from the physics to which we are accustomed, that it seems reasonable to call it something else.

We can explain more clearly what we shall mean by this definition if we combine it with the second assumption above. We then require that there are free events which only affect physics within their forward light cones (here we are explicitly excluding the occurence of any faster-than-light signalling). The possibility of a contradiction with the results of quantum theory predictions is now evident, and indeed has been well-known, in some form or other, since the early days of quantum theory (detailed discussion is given by Aharonov and Albert[9]). Suppose that we do not allow the existence of anything outside of physics. Then, from our first assumption, we know that something must happen within physics that corresponds to the perceived polarisation measurements in an EPR-type of experiment. For definiteness it is convenient to think of this as a collapse of the wavefunction, but this is not really essential. The information as to which particular collapse has occurred can only reach points within the forward light cone. Thus, if two collapses occur, one in the left-hand beam and one in the right-hand beam, and if these lie outside of their respective forward light cones, then they must be independent. In particular, if the measurements are in the same direction there would be nothing to prevent both measured polarisations being the same, in violation of angular-momentum-conservation. (I am implicitly using Bell's theorem here in order to remove the possibility that each particle carries with it a hidden variable which determines the outcome.) Similar violations of other conservation rules, e.g. charge, would also be expected in individual events. The conservation rules would hold only for averages. The fact that this does not seem to be the way the world is has led to the idea that some departure from Lorentz invariance is inevitable[10] [11] (from our point of view this would imply duality, in the sense that something beyond physics, as we have defined it, would be required). As we shall see below, however, such a conclusion is not forced upon us.

When Descartes introduced a distinction between the material, or physical, body and the immaterial mind, it is clear that he could not have defined physics as we have done. Perhaps, however, he came as close to it as possible, since he recognised that locality was an important feature of physical things, in contrast to the fact that non-physical things do not have positions.

6. A wavefunction in space

The discussion of locality in quantum theory is obscured by the well-known fact that a many-particle wavefunction does not live in our familiar 3+1 dimensional space-time, but in configuration space of 3N space dimensions (plus one, or N, time dimensions[3]). It does not represent a distribution of "something" over all space-time.

We can remove this problem by a simple trick (this seems to be the appropriate word), which was first suggested to me by a Durham research student[8] , and which has also been used by Eberhard[10] . The idea here is that the complete wavefunction $|\Psi>_x$ should be regarded as existing at all points of space and time. For example, in a universe of one scalar particle, there would be a complex function $<\mathbf{r}|\Psi>_x$ at each space-time point x. We should contrast this with the idea of a field, which we think of as a (complex) *number* $<\mathbf{r}|\Psi>$ existing at each position $\mathbf{r}$.

Apart, possibly, from trivial phase-factors, the wavefunction might be expected to be independent of the position $\mathbf{x}$, i.e. all points should possess the same information. This, however, is not quite true because of the wavefunction collapses (or, more generally, the *free* events of section 4), which happen at specific space-time points. As we noted in section 5, the information about a particular collapse travels to other points along the forward light-cone. In other words, the information $|\Psi>_x$ is continually being updated by news of free events elsewhere, but it will always be unaware of those events outside of its past light-cone. It is not important here for us to know *how* this updating occurs; for this we would need specific equations. In the case of a single spin-half particle, it is possible[12] to find a modified Dirac equation which generalises the collapsing Schrödinger equation of Ghirardi, Pearle and Rimini[13] . The propagation of information along the light cone then arises through the normalisation integrals which appear in these equations, and which are defined along the light cone, rather than at a specific time as in the non-relativistic models. An interesting toy-model, showing the effects of finite velocity of transmission, is obtained by considering playing Pearle's "gamblers ruin" game[4] with the players in, say, California and Italy communicating through the postal system.

In the usual applications of non-relativistic quantum theory, it is assumed that the information is available at all points simultaneously, and it is this assumption that gives the familiar results. We will now obtain different results, so our next task is to investigate how these diferences will show up in experiments.

7. Experimental tests of the effects of finite velocity signals

We consider an EPR-like situation where we measure spins along the same axis. Angular momentum conservation then predicts that we should obtain opposite (say) spins for the L and R particles. Suppose that the spin projection of the L particles is a free event (in the sense of section 4) at the point $\mathbf{x}_L$ at time t_L (which will be a time after the L-particle has reached the L-detector). We will obtain the expected result of quantum theory provided the information regarding this collapse has reached the right detector, at $\mathbf{x}_R$, before the spin of the R-particle is measured. In the work of Eberhard[10] it is assumed that the measurement is made as soon as the particles reach the detector. The experimental agreement with normal quantum theory then requires that signals must pass from $\mathbf{x}_L$ to $\mathbf{x}_R$ (or *vice versa*) in less time than the difference between the times of the L and R particles reaching the detectors. Clearly this requires signals faster than light, which we have forbidden.

The above condition, however, is too strong. The measurements are not made when the particles reach the detectors, but when a collapse occurs, and this may be a considerable time later (see below). Thus the minimal requirement for agreement with quantum theory is that, after a collapse has occurred at one side, no collapse occurs at the other within the time $|\mathbf{x}_L - \mathbf{x}_R|/c$ required for a light signal to travel between the two detectors. Since the collapses presumably occur at random times, this can never hold strictly. Rather we should say that the probability for a "wrong" result is given approximately by

$$P = \frac{1}{2}(1 - e^{-\lambda}), \tag{7.1}$$

where

$$\lambda = \frac{|\mathbf{x}_L - \mathbf{x}_R|}{c}\tau_c^{-1}. \tag{7.2}$$

Here τ_c is the average time for collapse. For example, with a distance between the detectors of 13m (the maximum used in the Aspect experiments[14] we find

$$P \sim \frac{1}{2}\frac{10^{-7}}{\tau_c}, \tag{7.3}$$

with τ_c measured in seconds. In the original paper of Ghirardi et al.[2] a collapse time of the order of 10^{15} seconds for a single particle was suggested. For a reasonable size detector, containing about 10^{22} relevant particles, this would imply $P \sim 1/2$, in contrast to the fact that the experiments agree with quantum theory. It seems therefore as though collapse times as rapid as this can already be ruled out by the experiments. On the other hand, collapse times longer by about two orders of magnitude would give very little observable effect.

It is clear that we are in an interesting, marginal, situation. This means that good experiments to look for effects of this nature should seriously be considered. It would be necessary to perform the above calculations in more detail and in the framework of specific collapse mechanisms and particular experimental arrangements. We note that the effects are more likely to occur with large detectors, separated by large distances.

An important ingredient in the above estimate is the average time taken for a macroscopic measuring apparatus to collapse (τ_c). Unfortunately it is not easy to know what this has to be, in particular, to put a lower limit on it. We tend to believe that a macroscopic apparatus records a result "immediately"; but just what does this mean: is, say, 1sec., sufficiently fast? I have suggested elsewhere[15] an experiment that would allow us to find a very crude limit, with an accuracy that might conceivably go down to about 10^{-3}sec.

I believe experiments of this type are worth doing. If positive results are found, a new area of physics will have been opened up. I would guess, however, that negative results, confirming the expectations of quantum theory, are more likely. Whether such results will ever be able to eliminate completely the possibility of a collapse mechanism that is consistent with relativity is not clear. It may well be that there will always be a small "window" of possibilty that cannot be closed, i.e. that it will be possible to choose the parameters to have a sufficiently rapid collapse, and which nevertheless gives rise to an undetectable number of "wrong" events caused by limitations of the finite velocity of signals. The record of ingenuity of experimentalists, when suitably motivated, suggests that such a window can be made very small; so small in fact that a new problem will arise, namely, why has nature apparently been so cunning as to choose the parameters sufficiently carefully to hide interesting quantum effects for ever from our view? Given that, at least in most models, the relevant collapse time is seemingly totally unconnected to the parameters that determine the size of apparatus, the maximum distance to the detectors, etc., such a conspiracy would be hard to believe. Indeed, even with the information we have at present, it must be regarded as surprising.

In the next section we shall take seriously the possibilty that in fact nature is telling us that the above approach is wrong, and that rather we should look for the bridge between the wavefunction of quantum theory, Eq.2.1, and our experience, in something that is outside physics.

8. Many worlds and conscious mind

We shall now assume that physics is completely described by the wavefunction, i.e. by an object $|\Psi>$, existing at all points of space and time, and evolving with time according to a linear equation (in a non-relativistic approximation this will be the Schrödinger equation). Such an assumption, which is the key to the so-called *Many Worlds* interpretation of quantum theory, is in accord with the fact that quantum theory correctly predicts interference effects in all cases where it has been tested. The assumption also has the advantage that it allows us to use the Anthropic Principle to explain the otherwise amazing coincidences that appear to be necessary for the emergence of intelligent life in the universe.

It now follows from our previous discussion (section 3) that we must have something outside of what we have called physics. This something is the extra ingredient that relates the wavefunction to our experience, so it seems reasonable that we should consider the possibility that conscious mind itself is the non-physical thing. Such an idea seems to involve the minimum of extra assumptions, since we know that consciousness certainly exists and that there is no natural place for it within physics. Although the separation we are making here resembles in many respects the traditional dualism of Decartes, there is an interesting change of perspective. Descartes regarded the physical world as something *material and concrete*, and he added to this the rather nebulous "soul", which was very immaterial. Here, in contrast, physics is represented by a wavefunction defined in configuration space (in quantum cosmology, the space includes the set of all 3-manifolds, etc.), and the *material* things (particles) arise through the influence of consciousness. Rather than being *one* aspect of a dualistic reality, matter is a consequence of the interplay between two non-materialistic concepts, a physical wavefunction and a conscious mind.

Now it might well be argued that little is gained by making this sort of assumption, and that we are really not going beyond the "anti-realist" ideas that we tried to reject in section three. Certainly it must be admitted that we cannot now write any nice equations, or do any new calculations to quantify what is happening. Instead we have to rely on very vague words. Nevertheless, I believe that we are saying things of real interest, and that it is even possible to go a little further and discuss several possible theories about the nature of consciousness and of its relation to the wavefunction, i.e. to the physical world. In this way quantum theory might help to solve one of the oldest problems of thinking people, the relation between mind and matter. In particular, we are here claiming that, when an observer becomes aware of a result, following some experiment, then something, having to do with his consciousness, happens with a probability $|\alpha|^2$ or $|\beta|^2$, respectively. (Here, for simplicity, we are are keeping to the two-state situation described by a wavefunction like that in Eq.2.1). But, *what* happens? Two answers to this question seem to have been discussed in the literature. (Some writers on the Many-Worlds model, maybe even Everett, would say that nothing happens. For reasons already stated I do not regard this as an acceptable answer. See also refs.[16] [17])

What appears to be the most popular option (see, for example, Albert and Loewer[16] , Stapp[18] and Lockwood[19]) requires the existence of an infinite number of minds belonging to every person. Each of these minds becomes aware of one result of an experiment, i.e. it selects one of the elements of the superposition of the brain states. The choice is made in a random way; this being defined in terms of the only weight function that is available, namely, $|\alpha|^2$ and $|\beta|^2$ in the simple case. Thus the usual probability rule of quantum theory is guaranteed, in the sense that the proportion of minds in each state is equal to the quantum probability. Also, because we have an infinite number of minds, it is certain that all the branches, i.e. all the elements of the superposition, will be occupied by conscious minds. Thus all results of the experiment exist and have identical ontological status. Here we see the origin of the "many-worlds" name, although, since there really is only one physical world, described be the single wavefunction, other names, e.g. "many views of one world"[20], are more accurate descriptions.

Perhaps the main problem with this idea is that it is very uneconomic. Somehow the idea of an *infinite* number of minds, all calling themselves "me", is hard to accept. Similarly it seems unreasonable that all possible results of an experiment (consistent with the wavefunction) are actually observed. On this picture, for example, there exists a me who does not believe in quantum theory; this me has only become aware of the possible, though statistically highly unlikely, result that no interferences have ever been seen. As I sit here writing this article, this other me, who probably has no particular worries about reality, is perhaps struggling to explain the energy levels of hydrogen. We can sympathise!

A slight modification of this idea seems, at first sight, to be more promising. Here we replace the infinite set of minds by one mind which is associated with some continuous parameter, say, μ, ranging between 0 and 1. Then, in a measurement like that associated with Eq.2.1, the part with $0 < \mu < |\alpha|^2$ sees the + spin, and the remainder sees the − spin. A similar splitting occurs at each subsequent measurement. The advantage of this over the previous idea is that now only one mind goes to each branch of the wavefunction. However, this fact kills the model. The reason is that, unless we can find some significance for a particular parameterisation, we have lost the quantum probability rule: both channels are identical and $|\alpha|^2$ has played no role. To put this point another way, if instead of the paremeter μ we used ν which is defined to be some function of μ, then, on splitting the mind as suggested above, we would obtain a different result.

I prefer an alternative explanation of what it is that "happens" when I make a measurement[17]. Here we have *one* mind which observes *one* world; the conscious mind simply selects one of the results, randomly, again with the only available quantities $|\alpha|^2$ and $|\beta|^2$ giving the weights. To me this is conceptually easier to accept than the idea of splitting of consciousness. However, it does have a big problem. After *I* have observed a particular result of an experiment, how can *you* be sure that if you observe the same experiment then you will obtain the same result (as you must if I can assume that all the people I meet are conscious). It was to ensure that there would always be one mind in each branch that in the above we required an infinite number of minds. Here, instead, we seek the answer in some sort of universality of consciousness. Another way of seeing that something like this is required, is to note that here only consciousness can carry the non-locality which we know from the Bell's theorem to be implicit in quantum theory, provided, as is true in this theory, that experiments ultimately yield unique results.

The fact that microphysics is here suggesting (even if only in a very loose sort of way) the existence of a universal conscious mind is a striking fact. Many people (including Schrödinger) have supported such an idea. It removes one unsatisfactory aspect of all "dual" models, because we do not have to suppose that there is a set of individual minds waiting to be dropped into babies (and baby animals?) as they are born. Just as there is only one wavefunction, so there is only one mind. An individual person is defined by his or her physical body. "I" am a particular window between conscious mind and the world of physics. My experiences are the picture of the physical state of my brain selected by conscious mind. Presumably, as we move from human brains, through those of animals, insects, amoeba, thermostats(?), and etc., the windows become more opaque, and the pictures less clear.

The other major problem with dual models of mind and matter: *how can the non-physical conscious mind act on physics?*, is also solved through these ideas. There is no such action; there is only *selection*. We can imagine that, in some circumstances, my choice of a particular behaviour might depend on the selection of one of a set of branches of the wavefunction associated with some part of my brain. Normally this choice would be made at random, in accordance with the usual probability laws. However the experience of free-will might occur when I (that is, conscious mind acting on my body) make a *deliberate* choice of an outcome. The mind does not affect physics, it simply selects the physics that is experienced.

I am very grateful to many people, in particular, D.Albert, Y.Ben-Dov, M.Bitbol and H.Stapp for discussion of the above ideas. Further relevant material is given by Bitbol and Stapp in their articles in this book.

REFERENCES

1. D.Dieks, Phys.Letters **A142** (1989) 439
2. G.C.Ghirardi, A.Rimini and T.Webber, Phys.Rev. **D34** (1988) 470
3. J.S.Bell, *Speakable and Unspeakable in Quantum Mechanics* (Cambridge, 1987) p.201
4. P.Pearle, Int.J.Th.Phys. **18** (1979) 489
5. L.Diosi, Phys.Rev. **A40** (1989) 1165
6. N.Gisin, Helv. Phys. Acta **62** (1989) 363
7. J.S.Bell, *La Nouvelle Cuisine*, to appear in *Between Science and Technology*, ed. A.Sarlemiju and P.A.Kroes (Elsevier, 1989)
8. L.Hardy, private communication
9. Y.Aharonov and D.Albert, Phys.Rev.**D21** (1980) 3316 and **D24** (1980) 359
10. P.Eberhard, in *Quantum Theory and Pictures of Reality*, ed. W.Schommers (Springer-Verlag, 1989) p.169
11. J.S.Bell, private communication
12. E.J.Squires, *Wavefunction localisation and Lorentz invariance*, Durham preprint DTP 89/47 (1989)
13. G.C.Ghirardi, P.Pearle and A.Rimini, *Markov Processes in Hilbert space and continuous spontaneous localisation*, Univ. of Trieste preprint (1989)
14. A.Aspect, P.Grangier and G.Roger, Phys.Rev.Letters **47** (1981) 460
15. E.J.Squires, *How long does it take to make a measurement?*, Durham preprint, DTP 90/21, to be published in Phys.Letters **A**
16. D.Albert and B.Loewer, Synthese, **77** (1988) 195
17. E.J.Squires, Found.of Phys.Letters, **1** (1987) 13 and in *Quantum Theory without Reduction*, ed.Cini and Levy-Leblond (Adam Hilger, 1990) p.151
18. H.P.Stapp, *Quantum Theory and the Emergence of Patterns in the Universe*, Lawrence Berkeley preprint LBL-27862 (1989)
19. M.Lockwood, *Mind, Brain and the Quantum* (Blackwell, Oxford, 1989)
20. E.J.Squires, Eur.J.Phys.**8** (1987) 171

Quantum Measurement and the Mind–Brain Connection

Henry P. Stapp

University of California
Lawrence Berkeley Laboratory
Berkeley, California 94720

Abstract

It is argued that quantum measurements do pose a problem, within the context created by the fundamental aim of science, which is identified as the construction of a cohesive, comprehensive, and rationally coherent idea of the nature of the world in which we live. Models of nature are divided into two classes: (1), those in which there is a selection process that, for any possible measurement, would, if that measurement were to be performed, pick out one single outcome, and, (2), all others. It is proved that any model of class (1) that reproduces the predictions of quantum theory must violate the condition that there be no faster-than-light influences of any kind. This result is used to motivate the study of models in which unitary evolution is maintained and there is no selection of unique outcomes. A consideration of ontic probabilities, historical records, and the form of the mind-brain connection leads to an elaboration of the Everett many-worlds interpretation that appears to provide the basis of satisfactory solution of the measurement problem.

1. The Problem

An essential part of the problem of quantum measurements is to correctly identify the problem. Indeed, many scientists believe that quantum measurements pose no scientific problem at all, and that the subject of this conference lies, therefore, outside the realm of science. Even among scientists who profess to see a problem there is considerable divergence about precisely what this problem is: certainly many of the differences of opinion expressed at this conference stem directly from differing perceptions of the nature of the problem.

van Fraassen[1] has identified the problem as the reconciliation of the truism that quantum theory describes an indeterministic world with the fact that quantum theory describes the evolution of an isolated system deterministically. This comes close to the mark, but needs elaboration.

The basic work in this field is that of von Neumann.[2] Within von Neumann's framework the problem, if there be one, is connected to his postulate that the quantum system evolves in two fundamentally different ways, either normally in accordance with a unitary transformation (type II process), or, during the course of a measurement process, in accordance with a (type I) process of the form

$$\rho \rightarrow \sum_i P_i \rho P_i Tr P_i \rho, \tag{1.1}$$

where the set of P_i are a set of disjoint projectors ($P_i P_j = P_i \delta_{ij}$) that sum to unity ($\sum P_i = I$).

A projector P_i might be, for example, the operator defined by the requirement that it take the value *one* when acting on states in which the position of the center of mass of a certain pointer of some measuring devices lies within some specified domain, say to the right of the resting point, and takes the value *zero* otherwise. Or P_i might project onto the "yes" state of a memory unit on a magnetic tape, or onto the states of a laboratory notebook in which appear the statement "the pointer moved to the right".

The von Neumann reduction is compatible with an ignorance interpretation. It is logically completely different from the similar–looking reductions that can occur in the *reduced* density matrices that can be formed by performing partial traces over selected degrees of freedom: the von Neumann reduction is a reduction of the density matrix of the entire universe, or of the largest system that can be described as a quantum system.

Zeh[3], Zurek[4], and Joos[5] have examined the reductions that can occur in the reduced matrices formed by taking partial traces over unobserved environmental degrees of freedom. As those authors have repeatedly emphasized, such considerations leave essentially untouched the basic problem associated with the von Neumann reduction. Various contributions[6,7] to this conference have examined reductions associated with certain other reduced density matrices, and have arrived at the similar conclusion that the problem of the von Neumann reduction is left untouched.

But what is this problem? In fact, there is no problem if von Neumann's quantum formalism is ascribed the significance specified by the Copenhagen interpretation. According to this interpretation the sole correct use of the quantum formalism is merely to make calculations of expectations pertaining to observations obtained under classically described conditions.[8,9] Within this framework the "measurement process" of the von Neumann description must be identified as part of a process whereby some human observer is trying to acquire knowledge about a quantum system: "measurements" do not occur in the physical world considered alone; they are tied specifically to the acquisition of knowledge by human observers, and the reduction postulate is simple part of the computations scheme by which those observers calculate expectations pertaining to their observations.

Within this orthodox framework a question of *consistency* can arise in a situation involving a chain of measuring devices linking the quantum system to the brain of the observer. This question concerns the dependence of the calculated expectation upon *which* of the devices in this chain are considered to be "classically describable", i.e., upon *which* of the devices the operator P_i act.

von Neumann's analysis shows that for good (nondemolition) measurements the calculated results will not depend significantly upon where one places the 'Heisenberg cut' that divides the classically described world from the part of the world that is described quantum mechanically.

The clarity of von Neumann's presentation invites one to consider von Neumann's reduction process to be a process that occurs physically within nature herself, rather than in the computations of scientists. Indeed, Ghirardi, Rimini, and Weber,[10] and Pearle[11] have accepted this invitation to construe the von Neumann process as a real physical process. However, the orthodox way to solve the problem of the reduction is to accept the Copenhagen view that the quantum formalism is merely a set of rules for calculating expectations pertaining to our observations. These rules are considered to be adequately justified by their practical success: their origin, or physical basis, is asserted to be in no need of explanation.

There are, in my opinion, two major objections to this pragmatic approach. The first is that it is overly restrictive. Most physical systems have properties that can be understood as consequences of the quantum nature of these systems, in spite of the fact that no one has ever prepared these systems, or will ever assigned a wave function to then. The stability of planets and rocks, and the physical properties of metals, inorganic compounds, and organic compounds, have presumably existed long before any quantum physicist assigned a wave function to them. Thus there is evidently some element of truth in the quantum description that is independent of the mental acts of quantum physicists: the mental acts of quantum physicists depend upon the existence of quantum properties, not vice versa. Hence the pragmatically construed theory is, in some sense, incomplete.

The second objection is related to the first. The task of science is not simply to provide a foundation for engineering practice, or to make predictions about how laboratory experiments will turn out. We do laboratory experiments in order to find out something about the nature of the world, not merely to perfect and test prescriptions about how to compute expectations pertaining to the results of laboratory experiments. To restrict the aims of science in the way suggested by a narrowly construed Copenhagen interpretation is to confuse the means of science with its ends. The proper goal of science is to provide, by means of empirical investigations and theoretical analysis, a cohesive, comprehensive, and rationally coherent idea of the nature of the world in which we live. To be diverted from this goal by philosophy is to become confused by philosophy, not illuminated by it. From this perspective the EPR-Bell analysis establishes not the futility of pursuing the real, but rather the inadequacy of classical conceptions of the real.

The possible models of nature are of two fundamentally different kinds: (1), those in which nature would, in any measurement situation that we might set up, make some absolute selection between the alternative possible observable outcomes, and, (2), those in which this property does not hold. The models of Ghirardi, Rimini, Weber, and Pearle lie in the first class, as does the pilot-wave model of Bohm.[12] The many-minds model of Everett,[13] to the extent that it can be formulated as a truly coherent, rational, and well-defined model of nature, lies in the second class. Section two of this paper gives a proof that models in the first class must necessarily involve some kind of faster-than-light influence. Certain earlier proofs of similar results have involved auxiliary assumptions such as determinism, counterfactual definiteness, the EPR criterion of physical reality, or hidden-variable factorization. The proof given in section 2 employs no such assumption, but is based, rather, on the assumption that if a measurement were to be performed then the outcome would be determinate. The proof is based on predictions of quantum theory pertaining to the outcomes of the three-particle spin-correlation experiments discussed by Greenberger, Horne, and Zeilinger.[14]

The result obtained in section two is used to motivate the work of section three, in which an effort is made to meet the objection to the many-minds approach that it is not sufficiently well formulated to be regarded as a rationally coherent model of nature.

2. Significance of Bell's Theorem: Local Selection Entails Superluminal Influence

Bell's original theorem[15] is usually formulated as the assertion that 'No local deterministic hidden-variable theory is compatible with the predictions of quantum theory'. This result is useful in the context of an effort to construct a rationally coherent model of nature that is compatible with the predictions of quantum theory: it rules out a large class of theories. This class consists of theories that satisfy the following two premises:

1. *Reality Premise.* The outcomes that a certain set of mutually exclusive alternative possible measurements 'would have if they were to be performed' is determined by a set of simultaneously well-defined functions of a set of hidden variables.

2. *Locality Premise.* The outcome that 'would appear' in any spacetime region R_i, under any condition that might be set up there, is independent of which measurement is chosen and performed in any region R_j that is spacelike separated from the first.

This reality premise contradicts two basic quantum precepts, namely the precept that nature is indeterministic, and the precept that observables corresponding to mutually exclusive alternative possible measurements cannot have simultaneously determinate values. Hence the theorem appears merely to confirm the inadequacy, in the quantum domain, of classical ideas about reality.

The theorem can, however, be reformulated so that: (1), the reality premise expresses ideas widely accepted among quantum physicists, principally the idea of *macroscopic* physical realism; and, (2), the locality premise is a clean expression of the condition that there be no faster-than-light influence of any kind. Then the theorem no longer simply reaffirms the inadequacy of classical concepts. Rather, it allows the no-faster-than-light-influence condition to discriminate between two fundamentally different conceptions of nature that are both compatible with the usual quantum precepts.

The reformulated version is this: 'No general selection process compatible with the predictions of quantum theory can be local.'

A *selection process* is a process that will, for any local measurement, under the condition that this measurement be performed, select a single outcome that, in principle, will appear to all observers who examine the result of this measurement.

A selection process is operative in both the collapse model of Ghirardi, Rimini, and Weber,[10] and the pilot-wave model of Bohm[12]: in both models nature selects a single outcome for any performed measurement, and this outcome will, in principle, appear to all observers who examine the result of this measurement. However, no selection process is operative in the many–minds model of Everett[13]: in this model *each* of the alternative possible outcomes will, in principle, appear in the consciousness of some corresponding observer.

A *general* selection process is a selection process in which: (1), the decisions as to which measurements are to be performed in each of several spacetime regions are indeterminate until the moment that these choices are made; and, (2), for each of these regions the selection process will act no matter which of the alternative possible measurements is chosen and performed there.

The collapse model of Ghirardi, Rimini, and Weber and the pilot-wave model of Bohm are both deterministic, as originally formulated. However, the choices among possible measurements can be made indeterministic without disturbing these models in any essential way. One way to do this is simply to imbed the quantum world described in these models in a classical external field that acts only in certain small spacetime regions, and that can shift the choices of measurements in arbitrary ways. Alternatively, one can, for each such choice, add a small quantum mechanical 'choice system' that is dynamically isolated from the rest of the system until the moment of choice. The extra classical variables that are, according to these models, associated with these choice systems can be kept indeterminate until the moment of choice. At this moment of choice one particular set of values for the set of classical choice parameters is chosen, and one particular measurement will consequently be chosen. But there are an infinite number of possible values that were not chosen. Thus 'what would have happened in the other cases' remains indeterminate.

This construction allows the theories of Ghirardi, Rimini, and Weber, and of Bohm, to provide models of general selection processes.

The *reality premise* of the reformulated theorem is simply that a general selection process is operative. This premise is essentially the demand that macroscopic physical realism hold: the active presence of a general selection process entails that those observable properties of measuring devices that characterize outcomes of measurements become determinate whether or not they are observed by someone; these macroscopic observable properties are not fundamentally observer dependant, as they are in the many-minds model.

It is this close connection to macroscopic physical realism that provides the motivation for the study of general selection processes. Interest in the possibility of reconciling quantum theory with physical realism motivated also another generalization of Bell's theorem, namely the one formulated as the assertion that 'No theory can be compatible with both local realism and the predictions of quantum theory'. However, the reality premise of that theorem[16] is far stronger than macroscopic physical realism.[17]

The locality premise of the reformulated theorem will be defined in due course. First, however, a model of a general selection process more general than either of the two models described above will be constructed. To make this model concrete, and closely connected to the proof of the theorem, it will be formulated within the context of an experimental arrangement discussed by Greenberger Horne, and Zeilinger.[14] This GHZ experiment is a generalization of the EPR-Bohm spin-correlation experiment considered by Bell. It has certain advantages, which will soon become evident.

The GHZ experiment involves three spin-$\frac{1}{2}$ particles, which are measured in three mutually spacelike separated spacetime regions $R_i, i\epsilon\{1,2,3\}$. The initial spin state is

$$\psi = \frac{1}{\sqrt{2}}\left(\begin{pmatrix}1\\0\end{pmatrix}_1\begin{pmatrix}1\\0\end{pmatrix}_2\begin{pmatrix}1\\0\end{pmatrix}_3 - \begin{pmatrix}0\\1\end{pmatrix}_1\begin{pmatrix}0\\1\end{pmatrix}_2\begin{pmatrix}0\\1\end{pmatrix}_3\right), \tag{2.1}$$

where the eigenstates of σ_z^i are used. In region R_i a choice is to be made to measure either σ_x^i or σ_y^i, and the chosen measurement is then to be performed in R_i.

Let X_i represent the condition that a measurement of σ_x^i be chosen and performed in region R_i, and let Y_i represent the condition that a measurement of σ_y^i be chosen and performed in region R_i. Let $x_i = +1$ and $x_i = -1$ represent the two alternative possible outcomes that might appear under condition X_i, and let $y_i = +1$ and $y_i = -1$ represent the two alternative possible outcomes that might appear under condition Y_i. Then quantum theory makes the following predictions:

$$\text{If } X_1X_2X_3 \text{ then } x_1x_2x_3 = -1, \tag{2.2}$$

$$\text{If } X_1Y_2Y_3 \text{ then } x_1y_2y_3 = +1, \tag{2.3}$$

$$\text{If } Y_1X_2Y_3 \text{ then } y_1x_2y_3 = +1, \tag{2.4}$$

and

$$\text{If } Y_1Y_2X_3 \text{ then } y_1y_2x_3 = +1. \tag{2.5}$$

The first prediction is that if the measurements of σ_x^1, σ_x^2, and σ_x^3 were to be chosen and performed in the regions R_1, R_2, and R_3, respectively, then the appearing triad of values x_1, x_2, and x_3 would have product minus one. The three other predictions are defined analogously.

A random-variable model of a general selection process, operating in the GHZ context, is constructed in the following way. Let the six functions

$$x_i(\rho(X_i; ...), ...) \qquad i\epsilon\{1,2,3\} \tag{2.6}$$

and

$$y_i(\rho(Y_i; ...), ...) \qquad i\epsilon\{1,2,3\} \tag{2.7}$$

be defined. The quantity $\rho(X_1; ...)$ is, in general, a random variable (i.e., a *set* of values with a weight assigned to each value) that will, however, if σ_x^1 is measured, be transformed into a determinate value. The five other ρ's are defined analogously.

These six quantities ρ are six among a host of similar quantities. The basic selection process of the universe is described in terms of these quantities ρ. Initially these ρ's are all indeterminate (e.g., random variables), but some of them will eventually become, under the action of the fundamental process of the universe, transformed into determinate values.

This fundamental process, which is the generation of determinate properties, consists of a sequence of events, each of which is a transformation of some ρ from its original indeterminate status into a determinate value. Each ρ is associated with a spacetime point, called its location.

The dots in $\rho(X_1; ...)$ represent the set of ρ's upon which it might depend. In principle these ρ's could be located at points scattered over all of spacetime. But this sort of scattering would create the danger of a 'gridlock', in which no ρ could become determinate because it would depend functionally upon other ρ's that are indeterminate.

A gridlock can be prevented by, for example, introducing a preferred coordinate system, and hence a preferred temporal ordering, and requiring each ρ to be functionally independent of all ρ's having temporally later locations, and specifying also that the sequence of events be ordered in accordance with this same temporal ordering. This latter condition would yield a fundamental process that acts in accordance with the intuitive idea that what happens today becomes fixed prior to the fixing of what happens tomorrow.

An alternative possibility, more in line with ideas from the theory of relativity, is to introduce no preferred coordinate system, but to allow no ρ to depend upon any ρ that is not located in its own backward light cone, which is the backward light cone whose apex lies at its own location.

The final three dots in the expression $x_1(\rho(X_1; ...), ...)$ represent the other ρ's upon which this function can depend. One demand imposed by the requirement that a general selection process be operative is that if the measurement of σ_x^1 were to be chosen and performed in R_1 then expression $x_1(\rho(X_1;...),...)$ would yield some determinate value, +1 or −1. One way to ensure this property would be to demand: (1), that *both* triads of dots in this expression be restricted to sets of ρ's whose locations lie in the backward light cone of $\rho(X_i;...)$; and, (2), that all of these ρ's become transformed into determinate values *prior* to the event at which $\rho(X_i;...)$ becomes determinate. (The word 'prior' used here refers to *process time*, which is the time variable associated with the fundamental selection process: it increases by one unit at the occurrence of each event in the fundamental sequence of selection events.)

The quantity $\rho(X_i;...)$ is directly associated with the selection of the outcome of the measurement of σ_x^1, and its location lies in R_1. Another important ρ whose location lies in R_1 is the quantity ρ_1; it is the becoming determinate of ρ_1 that fixes which measurement will be performed in R_1. The location of ρ_1 lies in the backward light-cone of $\rho(X_i;...)$. The quantities ρ_2 and ρ_3 are defined analogously.

Within this random-variable model the locality (i.e., no-faster-than-light-influence) premise is expressed as the set of conditions: "Under any condition X_1,

$$\begin{aligned}&[x_1(\rho(X_i;..),...)]_{X_2X_3}\\&=[x_1(\rho(X_i;...),...)]_{Y_2Y_3}\\&=[x_1(\rho(X_i;...),...]_{X_2Y_3}\\&=[x_1(\rho(X_i;...),...)]_{Y_2X_3},\end{aligned}\tag{2.8a}$$

together with the analogous conditions for x_2, x_3, y_1, y_2 and y_3. The subscript X_2X_3 represents the condition that ρ_2 and ρ_3 are assigned determinate values, and these values are allowed to be *any* values such that σ_x^2 is chosen and performed in R_2 and σ_x^3 is chosen and performed in R_3. The other subscripts are defined analogously.

Conditions (2.8a) would be satisfied if the light-cone conditions mentioned above were to be satisfied: the determinate value appearing under condition X_1 would then be independent of whether X_2 or Y_2 would be chosen by the becoming determinate of ρ_2, and of whether X_3 or Y_3 would be chosen by the becoming determinate of ρ_3.

The locality premise of the reformulated theorem is similar to (2.8a). It asserts that: "Under any condition X_1

$$[x_1]_{X_2X_3}=[x_1]_{Y_2Y_3}=[x_1]_{X_2Y_3}=[x_1]_{Y_2X_3}.\text{"}\tag{2.8b}$$

The analogs for $x_2,...,y_3$ are also demanded. The meaning and rationale of these conditions will now be described.

What is under consideration here is one single experimental situation in which choices are about to be made that will determine which measurements are to be performed in the three regions. The condition "Under any condition X_1" stands for: "If an experimental procedure for measuring σ_x^1 were to be chosen in R_1, and this procedure were to be carried out". Under this condition, according to our reality premise, some determinate outcome, either +1 or −1, must be selected in R_1. This outcome is not predetermined, and it could logically depend upon which measurements will later, in some frame of reference, be chosen and performed in the other two regions. However, the condition that there be no faster-than-light influence of any kind can be taken to mean that, as far as processes in R_1 are concerned, it is as if these choices to be made in R_2 and R_3 do not exist. Thus, in the realm of theoretically allowed possibilities, there are not several alternative possible selections in R_1, corresponding to the various alternative possible choices that might later be made in R_2 and R_3, and hence a logical possibility that these various selections could be different. There is, logically, insofar as these differences in future choices are concerned, just one selection of one outcome.

The subscript X_2X_3 on the first term in (2.8b) represent the conditions: "If a procedure for measuring σ_x^2 were to be chosen and carried out in R_2, and if a procedure for measuring σ_x^3 were to be chosen and carried out in R_3". The other subscripts have analogous meanings.The various subscripts in (2.8b) refer, therefore, to alternative possible measurments that cannot be simultaneously performed.

Quantum theory warns us to be wary of such situations. However, it does not inveigh against the idea that if a measurement of σ_x^1 were to be performed then some determinate outcome would be selected, and that this selection cannot depend upon choices that have not yet been made, and hence do not yet exist.

Deterministic theories often entail rigid connections among outcomes associated with alternative possible future conditions. Thus the logical situation engendered by the no–faster–than–light influence assumption is somewhat similar to the one encountered in deterministic theories. However, there is a crucial difference: the demand (2.8b) specifies that the outcomes selected under the various alternative possible conditions indicated there must be the *same* outcome, without entailing that this single outcome be predetermined.

To formalize the conditions represented by (2.8b) we shall appropriate the machinery normally employed within the context of deterministic theories, where one can contemplate rigid theoretical connections among outcomes associated with alternative possible measurements. But we will use this machinery in a way that does not entail predetermined outcomes. Thus the first part of (2.8b) will be expressed, in part, by the assertion:

"Under any condition X_1, if $[(X_2X_3)$ and $(x_1 = +1]$ then [if (Y_2Y_3) then

$$(x_1 = +1)]" \tag{2.8c}$$

In words this says: "If X_1 were to be chosen then: if [X_2 and X_3 were to be chosen and x_1 were to be +1] then [if, *instead*, Y_2 and Y_3 were to be chosen then x_1 would be +1]".

This assertion imposes the theoretical demand that the outcomes selected in R_1 under the alternative possible conditions in $R_2 \cup R_3$ be the *same* outcome, without implying that this single outcome be predetermined. The word "would" that occurs in the verbal equivalent of (2.8c) is not a consequence of any use of physical determinism. It imposes the no–faster–than–light–influence condition that the selection in R_1 be made without reference to future free choices. The *lack of dependence* of a stochastically made selection upon future free choices is not determinism.

A compact expression of (2.8c), generalized to include also the possible outcome -1, is this:

$$((X_1X_2X_3)\text{and}(x_1 = p)) \Rightarrow ((X_1Y_2Y_3) \Rightarrow (x_1 = p)). \tag{2.8d}$$

Some other parts of (2.8b) that will be used in the ensuing proof are, in this compact notation,

$$((Y_1X_2X_3)\text{and}(x_2x_3 = -p)) \Rightarrow ((X_1X_2X_3) \Rightarrow (x_2x_3 = -p)) \tag{2.8e}$$

and

$$((X_1Y_2Y_3)\text{and}(y_2y_3 = p)) \Rightarrow ((Y_1Y_2Y_3) \Rightarrow (y_2y_3 = p)). \tag{2.8f}$$

We turn now to the proof. The first two predictions of quantum theory, (2.2) and (2.3), assert that

$$(X_1X_2X_3) \Rightarrow (x_1x_2x_3 = -1) \tag{2.9}$$

and

$$(X_1Y_2Y_3) \Rightarrow (x_1y_2y_3 = +1). \tag{2.10}$$

Accepting these two conditions, and also (2.8d), we obtain the following sequence of syllogisms:

$$\begin{aligned}
&((X_1X_2X_3) \text{ and } (x_2x_3 = -p)) \\
&\quad \Rightarrow ((X_1X_2X_3) \text{ and } (x_1x_2x_3 = -1) \text{ and } (x_2x_3 = -p)) \\
&\quad \Rightarrow ((X_1X_2X_3) \text{ and } (x_1 = p)) \\
&\quad \Rightarrow ((X_1Y_2Y_3) \Rightarrow ((x_1y_2y_3 = 1) \text{ and } (x_1 = p))) \\
&\quad \Rightarrow ((X_1Y_2Y_3) \Rightarrow (y_2y_3 = p)).
\end{aligned} \tag{2.11}$$

Thus our premises (2.9), (2.10), and (2.8d) entail that

$$((X_1X_2X_3) \text{ and } (x_2x_3 = -p)) \Rightarrow ((X_1Y_2Y_3) \Rightarrow (y_2y_3 = p)). \tag{2.12}$$

In words, this result asserts that:

Under any condition X_1, if X_2 and X_3 were to be chosen and the product of the then-determinate values x_2 and x_3 were to be $-p$ then, if, *instead*, Y_2 and Y_3

were to be chosen, the product of the then-determinate values y_2 and y_3 would be p.

This theoretical condition, which relates outcomes allowed under alternative possible conditions, is *directly entailed* by our premises: if it were to fail then the determinate value x_1 would be *forced to depend* upon which measurements are later to be chosen in $R_2 \cup R_3$.

The no-faster-than-light influence conditions entail that the condition X_1 can be omitted from (2.12). Indeed, by using (2.8e), (2.12), and (2.8f) one obtains the sequence of syllogisms

$$\begin{aligned}
&((\,Y_1X_2X_3) \text{ and } (x_2x_3 = -p)) \\
&\quad \Rightarrow ((X_1X_2X_3) \Rightarrow (x_2x_3 = -p)) \\
&\quad \Rightarrow ((X_1X_2X_3) \Rightarrow ((X_1X_2X_3) \text{ and } (x_2x_3 = -p))) \\
&\quad \Rightarrow ((X_1X_2X_3) \Rightarrow ((X_1Y_2Y_3) \Rightarrow (y_2y_3 = p))) \\
&\quad \Rightarrow ((X_1X_2X_3) \Rightarrow ((X_1Y_2Y_3) \Rightarrow ((X_1Y_2Y_3) \text{ and } (y_2y_3 = p)))) \\
&\quad \Rightarrow ((X_1X_2X_3) \Rightarrow ((X_1Y_2Y_3) \Rightarrow ((Y_1'Y_2Y_3) \Rightarrow (y_2y_3 = p)))).
\end{aligned} \tag{2.13}$$

The sequence, in the last line, "then if X_1 were to be chosen instead of Y_1, then if Y_2Y_3 were to be chosen instead of X_2X_3, then if Y_1' were to be chosen instead of X_1" can be shortened to "then if $Y_1'Y_2Y_3$ were to be chosen instead of $Y_1X_2X_3$". One special case is that in which Y_1' is the same as the original Y_1. In this case (2.13) becomes

$$((Y_1X_2X_3) \text{ and } (x_2x_3 = -p)) \Rightarrow ((Y_1Y_2Y_3) \Rightarrow (y_2y_3 = p)), \tag{2.14}$$

which is (2.12) with X_1 replaced by Y_1. Thus the two results together have the precondition "If (X_1 or Y_1)". But X_1 and Y_1 are the only possibilities considered in R_1. Thus this precondition can be omitted. This yields

$$((X_2X_3) \text{ and } (x_2x_3 = -p)) \Rightarrow ((Y_2Y_3) \Rightarrow (y_2y_3 = p)). \tag{2.15}$$

The second two predictions of quantum theory, (2.4) and (2.5), yield, in the same way,

$$((Y_2X_3) \text{ and } (y_2x_3 = q)) \Rightarrow ((X_2Y_3) \Rightarrow (x_2y_3 = q)). \tag{2.16}$$

Condition (2.15) entails that

$$\begin{aligned}
&[(X_2 \text{ and } (x_2 = m)) \text{ and } (X_3 \text{ and } (x_3 = n))] \\
&\quad \Rightarrow [(Y_2Y_3) \Rightarrow (y_2y_3 = -mn)] \\
&\quad \Rightarrow [(Y_2 \text{ and } (y_2 = r)) \Rightarrow (Y_3 \Rightarrow (y_3 = -mnr)).
\end{aligned} \tag{2.17}$$

This entails that

$$\begin{aligned}[X_2 \ \text{and}\ (x_2 = m))] \\ &\Rightarrow [((X_3 \ \text{and}\ (x_3 = n)) \ \text{and}\ (Y_2 \ \text{and}\ (y_2 = r))) \\ &\Rightarrow (Y_3 \Rightarrow (y_3 = -mnr))]\end{aligned} \tag{2.18}$$

On the other hand, (2.16) entails that

$$\begin{aligned}[(X_3 \ \text{and}(x_3 = n)) \ \text{and}\ (Y_2 \ \text{and}\ (y_2 = r))] \\ &\Rightarrow [(Y_3 \ \text{and}\ (y_3 = -mnr)) \Rightarrow (X_2 \Rightarrow (x_2 = -m))].\end{aligned} \tag{2.19}$$

Insertion of this relationship into (2.18) entails that

$$\begin{aligned}[X_2 \ \text{and}\ (x_2 = m)] \\ &\Rightarrow [((X_3 \ \text{and}\ (x_3 = n)) \ \text{and}\ (Y_2 \ \text{and}\ (y_2 = r))) \\ &\Rightarrow (Y_3 \Rightarrow (X_2' \Rightarrow (x_2' = -m)))].\end{aligned} \tag{2.20}$$

The numbers n and r are arbitrary, and do not appear elsewhere. Thus, by virtue of our reality premise, which demands that if X_3 were to be chosen then some outcome $x_2 = \pm 1$ must be selected, and that if Y_2 were to be chosen then some outcome $y_2 = \pm 1$ must be selected, we may omit the conditions "and $(x_3 = n)$" and " and $(y_2 = r)$". This simplification reduces (2.20) to

$$\begin{aligned}[X_2 \ \text{and}\ (x_2 = m)] \\ &\Rightarrow [(X_3 Y_2) \Rightarrow (X_2' \Rightarrow (x_2' = -m))]\end{aligned} \tag{2.21a}$$

In words, this asserts that:

"If some X_2 were to be chosen and the selected outcome were to be $x_2 = m$, then if some X_3 were to be chosen, and if, instead of X_2, *any* Y_2 were to be chosen, then if, instead of X_3, *any* Y_3 were to be chosen, then if, instead of Y_2, *any* X_2' were to be chosen then the selected outcome must be $x_2' = -m$" (2.21b)

The primes on the final X_2' and x_2' are inserted to distinguish them from the original X_2 and x_2. The words "*any*" have been inserted to emphasize that this result is required to hold no matter which of the conceivable alternative possible procedures for measuring the indicated quantity is chosen. For example, in our models, different determinate values of ρ_i, or its analogs in the GRW and Bohm models, could create nonidentical experimental conditions corresponding to a measurement of, say, σ_y^i, and these nonidentical conditions might lead to different outcomes: this is allowed, *a priori*.

The construct "if *any* Y_2 were to be chosen instead of X_2", and the other similar ones in (2.21b), were introduced as parts of a formalization of the delicate concept that certain outcomes are nonpredetermined yet independent of choices to be made later. This purely logical construct involves no physical sequence of choices of alternative possibilities: the alternative possibilities all exist on a par. No possibility can be destroyed by drawing from it a conclusion about what would, by virtue of a certain relationship of sameness, hold if, instead, some other possibility were to be chosen. Consequently, the fact that under some condition X_2 the outcome $x_2 = m$ could appear, which is ensured by the initial condition in (2.21), cannot be destroyed by the logical acts of considering some alternative measurements. Thus the *possibility* that X_2' *could be* the same as X_2, and that x_2' *could be* the same as x_2, cannot be destroyed by the intermediate logical steps in (2.21).

We now make explicit the final assumption, which is that the choices of which measurement are to be performed are fixed by variables that are initially indeterminate, so that all of the alternative possibilities are allowed. This assumption was, in fact, part of the requirement for a general selection process. This assumption means that the initial assumption in (2.21) can be satisfied for some m, and that the intermediate conditions can be satisfied. But then (2.21) yields a contradiction. For the logical acts of considering the alternative possibilities cannot destroy the possibility, that the final X_2' could be the same as the initial X_2, and that the outcome selected under that final condition X_2' could be $x_2' = m$. This contradicts the conclusion, asserted by (2.21), that x_2' must be $-m$.

The conclusion is that the following four conditions cannot be simultaneously enforced:

1. Indeterminate Choices
2. Macroscopic Physical Realism
3. No Faster-Than-Light Influence of any Kind.
4. Validity of the Predictions of Quantum Theory

The first and second assumptions provide the variable causes and determinate effects need to formulate clearly the no-faster-than-light- influence assumption. The operative parts of these first two assumptions, in the context of the proof, are contained in the premise that a general selection process is acting.

This conclusion is fundamentally different from the conclusion of either Bell's original theorem, or arguments based on the EPR criterion of physical reality. Those conclusions are essentially that certain classical conceptions of reality must fail. The present conclusion is that some property that is not contrary to the quantum precepts, and that is widely accepted by quantum physicists, must fail

The result derived here is completely understandable. The proof merely confirms a fact that could have been anticipated from the beginning, namely that one cannot mutilate the integrity of the basic mathematical structure of quantum theory, as represented in its unitary covariant law of evolution, by superposing the alien "collapse of the wave function", without disrupting the property of no faster–that–light influence. To retain this property one must retain the parallel observers demanded by the unitary law of evolution. If this is done, and attention is paid to the central fact of the interpretation of quantum theory, which is that the probabilities it defines are probabilities for the formation of *combined records* of all the facts being correlated, and that these combined records can be formed only by bringing the records of the separate facts together in the intersection of the forward light–cones of the regions in which these facts are generated, then the predictions of quantum theory are readily seen to arise from superposition of amplitudes without recourse to faster–than–light influences.

In particular, if one writes x_i^+ and x_i^- as normalized solutions of $\sigma_x^i x_i^\pm = \pm x_i^\pm$, and expresses the spin state ψ in the forms appropriate to the four experiments, namely as

$$\frac{1}{4}\Big[(x_1^+ + x_1^-)(x_2^+ + x_2^-)(x_3^+ + x_3^-) - (x_1^+ - x_1^-)(x_2^+ - x_2^-)(x_3^+ - x_3^-)\Big] = \frac{1}{2}\Big[x_1^- x_2^+ x_3^+ + x_1^+ x_2^- x_3^+ + x_1^+ x_2^+ x_3^- + x_1^- x_2^- x_3^-\Big]$$

and

$$\frac{1}{4}\Big[(x_1^+ + x_1^-)(y_2^+ + y_2^-)(y_3^+ + y_3^-) + (x_1^+ - x_1^-)(y_2^+ - y_2^-)(y_3^+ - y_3^-)\Big] = \frac{1}{2}\Big[x_1^+ y_2^+ y_3^+ + x_1^+ y_2^- y_3^- + x_1^- y_2^+ y_3^- + x_1^- y_2^- y_3^+\Big],$$

and its two cyclic counterparts, and recognizes that in the unitary evolution each spin state will be converted into the corresponding macroscopic state of the observer, then one sees that the quantum predictions arise from a superposition–principle cancellation, completely incomprehensible from the classical point of view, between certain combinations of the *macroscopic* states of the parallel observers. Even though these states are macroscopic their cancellations will not be disrupted by the uncontrollable phases associated with macroscopic states, for these phases, though uncontrolled, are common.

3. Ontic Probabilities, Historical Records, and Many–Minds

The conclusion of the foregoing section is that, within an indeterministic setting in which choices among possible measurements are indeterminate, one cannot reconcile the idea that the predictions of quantum theory are valid with the ideas that: (1), there are no faster-than-light influences of any kind; and, (2), the outcome of any local measurement would become determinate if the measurement were to be performed. Any rationally coherent model of nature must violate at least one of these four conditions. Each of the four options contradicts an intuition derived either from our direct experience of the world, or from our experience with earlier partially successful theories. The intuitions derived from earlier partially successful theories are probably the more likely to prevail: much of the history of philosophy is a warning to be wary of our direct intuitions, whereas in the progression of physical theories, from classical theory, to relativity theory, to quantum theory, each successive theory is partially incorporated into its successors, and hence presumably captures certain essential features of an eventual full scientific understanding of nature. This consideration suggests that a cohesive, comprehensive, and rationally coherent model of nature is more likely to emerge from adherence to the basic principle of relativistic quantum theory, namely covariant unitary evolution, than from adherence to the principle of macroscopic physical realism.

Relativistic quantum theory is fundamentally probabilistic: it has probabilities built into its basic structure. This elevation, by quantum theory, of probability to fundamental status highlights the severe difficulties that have plagued the efforts of mathematicians and philosophers to make the concept of probability absolutely clear within the context of classical ideas about nature.

What does it mean to say that the outcome of coin flip has a 51% probability to be heads and a 49% probability to be tails? It means, of course, that, if one were to do a large number, n, of identical coin flips, then the probability that there would be m heads and $n-m$ tails would be $(.51)^m(.49)^{n-m}n!/m!(n-m)!$. But this answer just shifts the question to that of the meaning of this latter probability.

One can assign *epistic* meanings in these cases of large numbers of identical processes by introducing the vague idea that if the relative probability of some possibility in a situation that will occur only once is "almost zero" then we can be "almost certain" that this possibility will not occur. Alternatively, one can define an epistic probability by *imagining* the unrealizable possibility of an infinite number of repetitions of the stochastic process. But the question is then: What is the *ontic*, or ontological, basis of the concordance of our usual experiences with these epistic concepts?

One approach to this question is illustrated by the Bohm pilot-wave model. At some initial time, the beginning of the universe, the wave function is suppose to have some definite form, and then *the single classical world* (which in this model is assumed to exist) is supposed to be fixed at some *single initial position* in accordance with the probability density specified by the wave function. But it makes no sense, ontologically, to say that the one single world is placed at one single position in accordance with a smeared-out probability distribution. To make ontic sense of this idea one would have to introduce an infinite number of alternative possible worlds, and distribute them in accordance with this rule. But that would bring back the many worlds that the theory was supposed to eliminate. The GRW model is faced with a similar difficulty.

The logical situation, therefore, is that quantum theory itself, to the extent that the quantum state is assigned ontic status, has ontic probabilities built into it. But if one disrupts the theory by introducing a selection process, in order to bring the model into line with our intuitions about macroscopic physical realism, then, although epistic meanings can be introduced by contemplating, in our imagination, an infinite number of copies of the system, the ontic foundation of probability becomes lost: the original indeterministic-probabilistic theory is replaced by a quasi-classical deterministic theory, supplemented by an irrational probability idea in which the ontic probabilities are expressed in terms of things that do not exist. However, if one foregoes macroscopic physical realism then one can retain a rational concept of ontic probability, and, at the same time, exclude the need for faster-than-light influences.

In quantum theory the natural carriers of ontic probabilities are historical records. To explain the meaning of this statement we begin by considering the concept of a record. The importance of records to the understanding of quantum theory has often been emphasized.

A proto-type record is a magnetic tape with a sequence of locations, labelled by an index i, such that each location is either magnetized or unmagnetized; or a book with a sequence of locations, specified by an index i, such that each location is either blank, or contains a symbol 1. Naturally occurring records can also be considered, but we focus on these man-made ones.

Let P_i be the projection operator that projects onto states of the universe in which:

1. A record, say a tape or book, is present in a nondestructive environment, and

2. Location i is magnetized, or contains the symbol 1.

The operator $\overline{P}_i$ projects onto states where this same tape or book is present in the nondestruative environment, but the location i is unmagnatized, or does not contain the symbol 1. Thus $P_i\overline{P}_i = 0$, and $[P_i, \overline{P}_i] = 0$. The symbol R_i will stand for either P_i or $\overline{P}_i$.

We employ the Heisenberg representation in which the state of the universe is a fixed state Ψ, and each operator P_i refers to some particular time.

The defining properties of records are:

A. Quasi-Stability:

$$P_i\Psi = P_i'\Psi$$

B. Copiability:

$$P_i\Psi = P_{j(i)}'\Psi$$

C. Combinability:

$$P'_i\Psi = P_{k(i)}\Psi$$

$$P''_i\Psi = P_{j(i)}\Psi$$

In the quasi-stability property the operators P_i and P'_i are supposed to refer to perceptibly different times, and to be, at those different times, projectors onto the states in which the specified physical property (e.g., a symbol 1 in position i) holds.

The copiability property specifies that there are physical mechanisms that can create new records that are the same, insofar as information content is concerned, as the original record.

The combinability property specifies that there are physical mechanisms that can combine the information content of the two records into a single record.

An essential point, here, is that, due to interactions with the environment, the projector P^S_{12} onto a superposition $\varphi_1 + e^{i\psi}\varphi_2$ of eigenstates of two projectors P_1 and P_2 of the proto-type kind would not specify a record: the quasi-stability property of such a projector would be violated. However, the sum $P_{12} = P_1 + P_2$ satisfies the defining characteristics of the projection operators that define records.

An *historical record* is a record existing at one time that is a combination of records created at different times. An example is a history book, or a scientist's laboratory notebook, containing replications of records made at different times during the course of an experiment, and containing, therefore, information about preparations and outcomes originally represented by records that may no longer exist.

The ontic probability postulate is this:

Suppose P_I is a projector such that $P_I\Psi \neq 0$ represents a part of the universe in which there is an historical record R that records the fact that some system was prepared in a certain way, and that a certain measurement was subsequently performed upon it, and that a record of the experiment and its outcome would be formed regardless of which outcome of the measurement eventually turns up. Let P_F be the projector on the record of one of the possible outcomes of the measurement. Then

$$\text{Prob}(P_F : P_I) = \langle P_F P_I \rangle / \langle P_I \rangle$$

is the probability that, under the conditions specified by $P_I = 1$, the historical record R contains the record of the outcome specified by $P_F = 1$. Here $\langle\ \rangle$ represents expectation value in the state Ψ.

This postulate refers to nothing that does not exist: it refers only to existing historical records; it refers to nothing existing only earlier, or only later, or in anyone's imagination. It refers to an *ontic* probability.

In accordance with the meaning of probability, if $\text{Prob}(P_F : P_I) = 0$ then the historical record R cannot contain a record of the result specified by $P_F = 1$.

To develop the meaning of ontic probability we examine some of its further properties.

Consider a typical Stern-Gerlach type of measurement. Suppose the preparation corresponds to preparing the spin in the plus x direction. Then the condition on the associated historical record is

$$P_I\Psi = \frac{1}{2}(\sigma_x + 1)P_I\Psi.$$

The operator P_I acts directly on the historical record, but the operator $\frac{1}{2}(\sigma_x + 1)$ acts on the spin state of the particle just after the preparation. The above equation represents the fact that the historical record records the fact that the initial spin state was the one for which $\sigma_x = +1$.

Let P_I represent also the condition that the component in the z direction was later measured, but no assertion about the outcome of that later measurement. Then the two possibilities for the record of this outcome are specified by projectors P_1 and P_2, where

$$P_1P_I\Psi = \frac{1}{2}(\sigma_z + 1)P_I\Psi \qquad P_1\epsilon\{R_i\}$$

and

$$P_2P_I\Psi = \frac{1}{2}(\sigma_z - 1)P_I\Psi \qquad P_2\epsilon\{R_i\}$$

One may then compute

$$\begin{aligned}
&\text{Prob}(P_1 : P_I)\\
&\quad = \langle P_1P_I\rangle/\langle P_I\rangle\\
&\quad = \langle P_IP_1P_1P_I\rangle/\langle P_I\rangle\\
&\quad = \frac{\langle P_I\frac{1}{2}(\sigma_x + 1)\frac{1}{2}(\sigma_z + 1)\frac{1}{2}(\sigma_z + 1)\frac{1}{2}(\sigma_x + 1)P_I\rangle}{\langle P_I\frac{1}{2}(\sigma_x + 1)P_I\rangle}\\
&\quad = \frac{1}{2}
\end{aligned}$$

and, likewise,

$$\text{Prob}(P_2 : P_I) = \tfrac{1}{2}.$$

One may also compute

$$\begin{aligned}
&\text{Prob}(P_1P_2 : P_I)\\
&\quad = \langle P_1P_2P_I\rangle/\langle P_I\rangle\\
&\quad = \langle\frac{1}{2}(\sigma_z + 1)\rangle\frac{1}{2}(\sigma_z - 1)P_I\rangle/\langle P_I\rangle\\
&\quad = 0
\end{aligned}$$

This result implies that the historical record cannot contain both the record that affirms that the outcome of the σ_z measurement was +1 and the record that affirms that the outcome of the σ_z measurement was −1.

The results obtained above show that the historical record of the experiment has the following three properties: (1), the historical record does not contain both a record of the fact that outcome $\sigma_z = +1$ occurred and a record of the fact that the outcome $\sigma_z = -1$ occurred; (2), the probability is $\frac{1}{2}$ that the historical record contains a record of the fact that outcome $\sigma_z = +1$ occurred; and (3), the probability is $\frac{1}{2}$ that the historical record contains a record of the fact that the outcome $\sigma_z = -1$ occurred.

By deriving these properties we have, for our simple example, *derived the von Neumann reduction postulate* in an ontic, or ontological, framework in which the unitary law of evolution is strictly maintained. In particular, even though the unitary law of evolution is maintained, the historical record contains a record of the fact that either one *or* the other outcome occurred, *not both*, and each of the two alternative possibilities has probability $\frac{1}{2}$.

How is this result connected to our experience? To answer this question one needs a theory of the mind-brain connection, since the theory presented above deals only with the physical aspects of nature.

A quantum theory of the mind-brain connection has been developed elsewhere.[18] The essential point of that theory is the feature that conscious mental process is an *isomorphic image* of the process in the brain that creates the physical basis of memory. That brain process consists precisely of the creation of a sequence of historical records, and each conscious experience is an isomorphic image within the psychological domain of an historical record created in the brain.

Given this theory of the mind-brain connection, the relationship of the foregoing physical theory to experience is direct: the perception by the human observer of the external historical record creates within the brain an historical record that is an image of the external historical record. This brain process is accompanied by an experience that is an isomorphic image of that latter historical record. But in this case the results obtained earlier at the physical level carry directly over to experience: the experience will correspond to a recording of one or the other of the two alternative possible outcomes, not both, and each possibility will have weight $\frac{1}{2}$.

The unitary law of evolution is retained here. Thus there will be two historical records generated, one with a recording of the outcome $\sigma_z = +1$, and the other with a recording of the outcome $\sigma_z = -1$. But what does it mean to say that each of these two alternatives has probability $\frac{1}{2}$ if both alternatives "occur", in an absolute sense.

Actually, the real logical problem is with the "classical" case, in which only one of the two alternatives occurs: in that classical case there are severe logical difficulties. But if *both* of the alternatives have ontic status then it is no longer necessary to explain what *does* exist in terms of what *does not exist*, and the question of the meaning of ontic probably becomes tractable.

Major problems in science are often resolved together. There was already in classical physics a problem with "time": the existing world was conceived to occupy a zero-duration time slice separating a nonexistent past from a nonexistent future. This ontology was troubling to philosophers: Bergson, for example, upheld the idea that the existing present moment has finite "duration". Classical relativity theory went to the other extreme: it had no natural place at all for the present "now"; the contents of entire spacetime continuum was spread out in an ontologically uniform way.

Relativistic quantum theory, as described in the Heisenberg representation, again laid out they contents of the entire spacetime continuum in a uniform way: there was no special place "now".

This difficulty was resolved in the Copenhagen epistemological interpretation by bringing in the experiences of observers, and, in models having a selection process, by bringing in the sequence of selection events, each of which is represented within the theory by a change in the otherwise-fixed Heisenberg state of the universe. Here, however, we seek an ontic solution with no selection of unique determinate outcomes.

In the context of the probability issue there is the related question of "occurrences" or "happenings": probabilities are generally probabilities for something to "occur". But in relativistic quantum theory, without the Copenhagen addendum, or a selection process of some sort, there is no notion of any occurrence: the whole spacetime structure of probabilities is simply laid out in a uniform way, with no indication of what the probabilities are probabilities *of*.

There is a further related problem: a comprehensive model of nature must have counterparts of all things that are known to exist. But the only things really known to exist are human experiences. This aspect of reality is left out of the wave function of quantum theory, but cannot be left out of a comprehensive model of nature.

These various problems are resolved together if one identifies "occurrences" or "happenings" with mental events, and specifies that it is *these* occurrences that are governed by the ontic probabilities. In this case, if unitary evolution is maintained, there will be a mental event corresponding to the formation in the brain of each of the alternative possible memory–type historical records, and the relative probabilities that quantum theory assigns to these alternative possible historical records can be interpreted in terms of the idea of a splitting of experiential reality into separately experienced branches. Indeed, as described and explained in detail in ref. 18, each "present experience" is directly experienced as a temporal process imbedded in a context that can cover long periods of time, and there is an intuitive feeling that the occurring experience is picked from among many alternative possibilities . This intuition is, from the point of view of unitary quantum theory, valid: any experienced succession of events is just one experienced succession from among many that co-exist with it in an absolute sense. The relative probability that quantum theory ascribes to any specified one of these alternative experiential sucessions, within a specified collection of such successions, is just the relative probability that a single experiential succession, picked randomly from this collection, will be that specified one.

The above remarks are, in my opinion, completely in line with the views set forth by Everett, and constitute merely a slight elaboration upon them. I have sought, here, by providing philosophic perspective, and focus on key points, to defend the thesis that those ideas, taken in conjunction with the description of the mind-brain connection set forth in ref. 18, provide a natural and reasonable basis for both a satisfactory resolution of the measurement problem in quantum theory and a cohesive, comprehensive, and rationally coherent conception of the nature of the world in which we live.

Acknowledgments

Discussions of the topic of section 2 with Robert Clifton, Bas van Fraassen, Jim Cushing, Don Bedford, and Jerry Finkelstein were very helpful. Discussions of the topic of section 3 with David Albert, Y. Ben-Dov, Euan Squires, and Michel Bitbol were also very helpful.

This work was supported in part by the Director, Office of Energy Research, Office of High Energy and Nuclear Physics, Division of High Energy Physics of the U.S. Department of Energy under Contract DE-AC03-76SF00098 and in part by the National Science Foundation under grant PHY85-15857.

References

1. B.C. van Fraassen, *The model interpretation of quantum theory*, in this volume.

2. J. von Neumann, *Mathematical Foundations of Quantum Mechanics*, (Princeton U.P., Princeton N.J., 1955) Ch. V and VI.

3. H.D. Zeh, *Foundations of Physics* **1** (1970) 69.

4. W.H. Zurek, *Annals of the New York Academy of Sciences* **480** (1986) 89.

5. E. Joos, *Annals of the New York Academy of Sciences 480* (1986) 6, and (with H.D. Zeh), *Z. Phys.* **B59** (1985) 223.

6. P. Mittelstaedt, *The objectification in the measuring process and the many-worlds interpretation*, in this volume.

7. P. Busch, *Macroscopic quantum systems and the objectification problem*, in this volume.

8. N. Bohr, *Atomic Physics and Human Knowledge*, (Wiley, New York, 1958) pp.64, 71, 90; *Essays 1958-1962 on Atomic Physics and Human Knowledge*, (Wiley, New York, 1963) p.60.

9. H.P. Stapp, *Amer. J. Phys.* **40** (1972) 1098.

10. G. Ghirardi, A. Rimini, and T. Weber, *Phys. Rev.* **D 34** (1986) 470.

11. P. Pearle, *Phys. Rev.* **A39** (1989) 2277.

12. D. Bohm, *Phys. Rev.* **85** (1952) 166.

13. H. Everett, *Rev. Mod. Phys.* **29** (1957) 452.

14. D.M. Greenberger, M. Horne, and A. Zeilinger, in *Bell's Theorem, Quantum Theory, and Conceptions of the Universe*, ed. M. Kafatos, (Klumer Academic Press, Dordrecht, 1989). N. David Mermin, *Physics Today*, June 1990, p.9.

15. J.S. Bell, *Physics* **1** (1964) 195.

16. J.F. Clauser and A. Shimony, *Rept. Prog. Phys.* **41** (1978) 1881.

17. J.P. Jarrett, *Nous*, *18* (1984) 569.

18. H.P. Stapp, *A Quantum Theory of the Mind-Brain Interface*, Lawrence Berkeley Laboratory Report LBL-28574 EXPANDED.

WOLFGANG PAULI AND THE SPIRIT OF MATTER

HERBERT VAN ERKELENS

Research group 'History, philosophy and social aspects of science', Faculty of Physics and Astronomy
Free University, De Boelelaan 1081
1081 HV Amsterdam, the Netherlands

ABSTRACT

After World War II Wolfgang Pauli tried to find a unified framework for quantum mechanics and depth psychology. It is shown that a kind of alchemical process motivated him.

1. Introduction

A curious silence has surrounded the person of Wolfgang Pauli for more than thirty years. Since his death in 1958 only few biographical details have come to light about the later years of Pauli. In 1988 a serious study by the physicist K.V. Laurikainen appeared which sheds light on Pauli's later philosophical thought. This book entitled *Beyond the Atom*[1] is based on the correspondence between Pauli and the physicist Markus Fierz, who had been Pauli's research assistant at the Polytechnic in Zurich and, at the time of their correspondence, was professor of theoretical physics at the University of Basle.

In a letter of Pauli to Fierz, dated January 19, 1953, Pauli refers to the fierce controversy between the seventeenth-century physicist Johannes Kepler and his contemporary, the alchemist and physician Robert Fludd. As outlined by Pauli himself in his essay on Kepler[2], Fludd defended, in opposition to the new quantitative thinking of Kepler, a view of nature that includes the inner experience of the natural philosopher. Remarkable is Pauli's own comment on this controversy. He writes to Fierz[3]: 'I myself am not only Kepler, but also Fludd.' Apparently the controversy between Kepler and Fludd reflected a deep inner conflict in himself.

In the same letter to Fierz Pauli writes about the way

the unconscious tried to reconcile the two conflicting standpoints of physics and alchemy by producing images that could mediate between the two[3]: 'My *search* is for a process of conjunction (unification of opposites), but I have only partially succeeded in this. Nevertheless first an exotic woman (slit-eyed Chinese) appeared in my dreams. Later also a strange, light-dark man who seemed to know something about the unification of opposites which I sought.'

It is my intention here to follow some of the steps in Pauli's search for a unification of the physical and the alchemical standpoints. To that end I will concentrate my attention on the two dream-figures Pauli mentions above: the slit-eyed woman and the stranger. As may be concluded from the unpublished correspondence of Pauli with Carl Gustav Jung and two of his co-workers, Aniela Jaffé and Marie-Louise von Franz, both dream-figures opened up for Pauli a hidden dimension of reality which for three hundred years had been ignored by the followers of Kepler and Newton: the spirit of matter. The dialogue of the modern physicist Wolfgang Pauli with this spirit belongs to the most puzzling, fascinating, and moving documents of our time. In the mind of Pauli the spirit of matter, for centuries seemingly dead, struggled for its own resurrection.

2. The Persian

When after World War II Pauli returns from America to Switzerland, he is an unhappy man. He has lost some of his fascination for the world of modern physics. And inwardly he is teased by a figure who in his dreams wants to be admitted to the Polytechnic and causes a lot of tumult there. Pauli calls this figure "the Persian." In a dream of December 1947 Pauli has a fascinating discussion with the Persian. In the days preceding the dream Pauli was very depressive. Apparently his energy had flown to the unconscious where it activated the figure of the Persian. The interesting part of the dream from the stand-point of physics runs as follows[4]:

I arrive at my former house. I see how a dark-skinned young man in whom I recognize the Persian is putting objects into the house through the window, I make out a circular *piece of wood* and several letters. Then he approaches me in a friendly manner and I begin a conversation with him:

I: "You are not allowed to study?"
He: "No, therefore I study in secret."
I: "What subject are you studying?"
He: "Yourself!"
I: "You speak to me in a very sharp voice!"
He: "I speak as someone to whom everything else is forbidden."
I: "Are you my shadow?"
He: "I am between you and the Light, so you are my shadow, not the reverse."
I: "Do you study physics?"
He: "There your language is too difficult for me, but in *my* language *you* do *not* understand physics!"

The significance of this conversation becomes clear if we take an earlier dream of Pauli into account from the period when Albert Einstein and Niels Bohr were discussing the issue of the *incompleteness of quantum mechanics*. In that time Pauli dreamt about a man looking like Einstein who showed him that quantum mechanics only describes a one-dimensional intersection of a two-dimensional more meaningful reality.[5] When he awoke, Pauli immediately realized that the second dimension lacking in the quantum mechanical description of nature could only be the unconscious with its archetypal contents. Apparently the archetypes - defined by Jung as hypothetical ordering factors in the unconscious with the capacity to structure mental images - can equally well structure material processes.

In the years following this dream Pauli tried to find a new language that could make the hidden dimension in nature accessible for our intellect. Since this dimension is neither physical nor psychic, Pauli concluded that this language should be *neutral* with respect to the distinction between psyche and matter. In his correspondence with Jung we actually find a number of attempts by Pauli to construct this *neutral language* from physical and mathematical symbols which frequently occurred in his dreams.[6]

But in the dialogue with the Persian we see something that is far more fascinating and in a sense also more baffling than these rather academic attempts of Pauli to go beyond quantum mechanics. In the dream the second dimension is no longer an area of psychophysical connections waiting to be exploited by some new scientific method. It has become a person who makes a very definite ethical appeal to Pauli. The hidden dimension of nature reveals itself as a spirit.

Moreover the Persian considers himself as someone who knows about the secret workings of nature. He is unable to understand the difficult language of modern physics. But in *his* language *Pauli* would not understand nature. And now he wants to be admitted to the Polytechnic. So he is of the opinion that his view or his knowledge of nature must no longer be ignored by contemporary science. And therefore he puts pressure upon the famous theoretical physicist Pauli to become his bridge to the world of the Polytechnic.

3. The stranger

In November 1948 the dark-skinned Persian merges with another dream-figure who has blond hair and a light skin, "the blond man." Out of this fusion a new inner partner arises. Pauli calls this man frequently "the light-dark stranger" or "the stranger," because everything this dream-figure embodies is so unfamiliar to him. In a dream of October 1949 the stranger tries to drive Pauli by means of fire out of his familiar field of mathematics and physics[7]:

> I am with colleagues on one of the upper floors of a house where a local conference on mathematics and physics is being held. I see that under my name a

course of cookery is announced: "Start: December 15." Surprised I ask a young man near me why the course begins so late in the year. He answers: "Because then the Nobel prize will be granted."

Now I notice that a *fire* has started in the adjacent room. I take fright (*affect*), run down a a staircase along many floors (hurrying - *panic*). Finally I succeed in getting outside. Looking back I see that two floors of the house, where the colleagues were gathered, are burnt down. I walk across the level ground and enter a *garage*. I see that a taxi is waiting for me and that the taxidriver fills the tank with petrol. I look more closely, I recognize "him", *the* light-dark "stranger". Immediately I feel secure. "Probably *he* has lighted the fire", I think, without saying it aloud. He says to me quietly: "Now we can refuel, because upstairs there has been a fire. I will take you where you belong!" Then he drives me off.

The dream indicates a transformation. First Pauli attends a conference on mathematics and physics. This conference is held high above ground level in one of the upper floors of a house. It is a world that has lost contact with daily reality, a world in which higher mathematics has become the sole admitted means of imagining matter. Pauli has to give a course of cookery, that is, a course in which matter is taken as concrete and tangible raw material that with the help of fire will be transformed into more worthwile, but equally concrete products.

But suddenly the fire directs itself at Pauli. He has to fly from the conference and runs down out into the open where a taxi-driver is waiting him. He feels secure as soon as he has recognized the taxi-driver as the stranger. In the dream Pauli changes from a theoretical physicist into a follower of the stranger. That is also the immediate impression Pauli himself had of the dream[7]: 'I awoke with a feeling of great relief and it seemed to me that essential progress had been made.' A further transformation is suggested by the motif of the car-drive. The taxi-driver will bring Pauli 'where he belongs,' to his new spiritual homeland.

In a sense this homeland is the way the stranger perceives the world. Pauli's task is now to enter this homeland, to experience the world from this new perspective and to become familiar with the stand-point of the stranger. But in this stage of his development the contrast between the stand-point of the stranger and Pauli's own view of the world is still too great for fruitful cooperation. This appears from a letter Pauli wrote in November 1950 to Jung's wife Emma. In it he remarks about the stranger[8]:

> 'My dream-figure is "ambivalent": on the one hand he is a spiritual light of superior knowledge, on the other a chtonic spirit of nature. But his knowledge brings him again and again back to nature and his chtonic origin is in fact the source of this knowledge. So in the end both sides turned out to be two aspects of the same personality. (...) His actions are always decisive, his words conclusive, though often incomprehensible. Women and children follow him gladly and he tries frequently to teach them. In general he regards his surroundings (in particular me) as wholly ignorant and uneducated in comparison with himself. He does not reject the old manuscripts about magic, but he considers them as merely a popular preliminary for uneducated people (like me).'

Here Pauli has acknowledged the stranger as the alchemical *lumen naturae*, the light of nature to be found in the unconscious, but closely related to the world of human beings and of nature. Erich Neumann, a pupil of Jung, calls this light an Eros light 'springing from strong emotion, from essentially feeling-toned constellations of the unconscious.' This wisdom always applies to relations between living things and contrasts with 'the objective, abstract wisdom of the patriarchal logos which looks at things from a distance.'[9]

Accordingly, this light of nature is so different from the light of science that Pauli is being confronted now with a deep rift in his own personality. Although he is being considered by many of his colleagues as the 'conscience of physics', the stranger considers him wholly uneducated. In this stalemate a woman comes to the rescue. It is the eternal feminine who as a dark unknown or slit-eyed Chinese woman takes the lead in Pauli's imaginative world and brings about the transformation needed to make him feel at one with himself again.

4. Eros

In January 1951 Pauli falls in love with Marie-Louise von Franz, a co-worker of Jung. When Pauli expresses his feelings for her for the first time in a letter, she is celebrating her 36th birthday. But despite her age she had developed a deep understanding of the unconscious, having worked with Jung for more than fifteen years. Accordingly, her influence on Pauli's inner development is great. Probably for the first time in his life Pauli could share his own deeper self with a woman.

For years Pauli had known that it was his destiny to develop a unified framework for modern physics and depth psychology. Before the encounter with Von Franz Pauli tried to work out this framework through the creation of a neutral language in which physical processes are described in such a way that the new formulation also applies to psychological and parapsychological processes. Now he discovers that this approach is too abstract and too intellectual. Besides physics, psychology and neutral language a fourth element is needed. In a letter to Jung Pauli calls this element: Eros.[10]

One may wonder why love is required at all to bridge the rift between physics and psychology. Since both Pauli as a physicist and Von Franz as a psychologist are interested in unifying physics and psychology, their common interest would have been enough to bring the two together. Indeed, Von Franz had already assisted Pauli in translating Latin fragments of Kepler and Fludd.

Even so Pauli himself regards Eros as essential for his undertaking and his love for Von Franz seems to have a deeper spiritual meaning. Again Pauli's inner dialogue with the stranger gives us a clue. What the stranger wants of Pauli is to be taken seriously. But who is the stranger? He is a modern incarnation of the alchemical spirit Mercurius, a modern parallel to the figure of Mephistopheles in Goethe's *Faust*. And like Mercurius in medieval alchemy the stranger ties Pauli down to his inner destiny through the experience of love. According to Jung Mercurius 'tempts us in the world of sense; he is the *benedicta viriditas* (blessed greenness) and the *multi flores* (many flowers) of spring (...) He is at the same time a Hermes Chtonios and an Eros, yet it is from him that there issues the "light surpassing all lights," the *lux moderna*, (...) the light veiled in matter.'[11]

Consequently, Pauli gets into serious trouble. In a letter to Von Franz, dated April 18, 1951, Pauli writes 'that we should more entertain the *amor coelestis* than the *amor vulgaris*.'[12] Here Pauli uses the terminology of Marsilio Ficino to discriminate between heavenly, spiritual love and ordinary love. In his opinion the heavenly love consists of a 'a meditative spirituality expressive of feeling.' Presumably Pauli tries to save his skin from his love-affair. This is completely understandable, but in a footnote Pauli mentions a negative motif for doing so. He complains that Professor Jung intended his book *The Psychology of the Transference* for doctors, i.e. for therapists. Pauli remarks then: 'I defend the thesis that the future of Jung's psychology does not lie in therapy and in the hands of doctors, but in natural philosophy, that is to say in any case leads to the Faculty of Philosophy.'

What Pauli writes here is rather strange. From a psychological point of view his relation with Von Franz is a transference relationship. And an important element of such a relationship is that it can heal deep inner conflicts. But Pauli does not seem to like this therapeutic effect at all. He wants to conceive a nice piece of natural philosophy. This might be one of the reasons why Von Franz has certain reservations with respect to Pauli's love. In the end Pauli is not willing to surrender himself completely to his love and to suffer the consequences.

Nevertheless, during the first three years of their relationship Pauli is loyal enough to his feeling side to enter more and more deeply into one of the most pressing problems of our time: the lack of soul in the modern scientific conception of the world. In 1952 he discovers that the meditative spirituality mentioned above is better suited to express the soul in matter than his neutral language. In October he writes to Von Franz[13]: 'For a long time I have wanted to write to you in order to communicate a certain mood in which a kind of unified view begins to assert itself in me. In order to give it expression I have to write in a curious way, half fantastic, half rational. In this way a kind of "meditation" is developing which also contains two dreams you do not know yet.'

A year later this meditation is completed by Pauli. He calls it *Die Klavierstunde - Eine aktive Phantasie über das Unbewusste*[14] (The Piano-lesson - An Active Fantasy about the Unconscious). It is dedicated in friendship to Marie-Louise von Franz and contains in a purely meditative and imaginative manner the answer to all the problems Pauli has wrestled with over the years. The figure of the stranger is now called "the master," and instead of the former sharp collisions with this inner authority a woman piano-teacher now mediates between Pauli and his own higher self.

5. The Piano-lesson

In *Die Klavierstunde* Pauli moves into a dimension beyond space and time where he meets a woman piano-teacher who is dependent on the master and is accustomed to follow his authority blindly. But thanks to the arrival of Pauli she now gets the opportunity to develop herself more independently and to act like a bridge between Pauli and the master.

The piano symbolizes the one world of creation. This one world is split into two. Scientists understand with their higher mathematics the *words* of creation, but not the *meaning* symbolized by the piano-playing of the woman. The depth psychologists around Carl Gustav Jung acknowledge that meaning, but they in their turn fail to understand the words Pauli uses as a physicist. And Pauli suffers from the fact that he cannot bring these two schools together. The unity of the world seems lost. Therefore Pauli has to take a piano-lesson, that is, he must develop his feeling side before he can experience the world again as a unity.

The piano-teacher is a personification of Pauli's feminine soul, his *anima*, who can help him to bring about this transformation of his personality. In his dreams she appears as a dark unknown woman who sometimes looks like Marie-Louise von Franz. But more frequently she is slit-eyed and embodies a view of the world which is exotic from the point of view of Pauli's rational ego-consciousness. Her Chinese face signifies further to Pauli that she holds the key to that experience of the world that in China has been crystallized in the ancient concept of Tao. Frequently Pauli calls China 'the realm of the middle.' Here middle means the hidden central point of the psyche denoted by Jung as the Self[15]: 'The Self is not only the centre, but also the whole circumference which embraces both conscious and unconscious; it is the centre of this totality, just as the ego is the centre of consciousness.'

In the room of this woman Pauli no longer feels himself torn between the two schools that have put him in distress. He calls her a lady and during the discussion that develops they both try to understand how the world from one point of view appears as the incarnation of the impersonal and abstract language of higher mathematics and from another is filled with purpose and meaning.

At the end of their discussion Pauli again feels sadness. In his imagination he sees his master's homeland, the new land that has risen from the sea of the unconscious and in the future may feed the people who are hungry for a new spirituality. Pauli is afraid that he will never enter this homeland. But he promises his lady that he will try to reconcile the master with the world of science. Then the voice of the master sounds more friendly than before: 'That is what I have long waited for.' Thereupon Pauli bows deeply for his lady and wants to leave the room. But the master says: 'Wait. Transformation of the centre of evolution.' And Pauli remarks: 'In earlier times one said: lead transforms into gold.'

The text of *Die Klavierstunde* then ends as follows[16]:

> At that moment the lady slipped a ring from her finger which I had not seen yet. She let it float in the air and taught me:
> "I suppose you know the ring from your school of mathematics. It is the *ring i*."
> I nodded and I spoke the words:
> "The i makes the void and the unit into a couple. At the same time it is the operation of rotating a quarter of the whole ring."
> *She*: "It makes the instinctive or impulsive, the intellectual or rational, the spiritual or supernatural, of which you spoke, into the unified or monadic whole that the numbers without the i cannot represent."
> *I*: "The ring with the i is the unity beyond particle and wave, and at the same time the operation that generates either of these."
> *She*: "It is the atom, the indivisible, in Latin ..."
> With these words she looked at me significantly, but it seemed to me unnecessary to speak Cicero's word for the atom aloud."
> *I*: "It turns time into a static image."
> *She*: "It is the marriage and it is at the same time the realm of the middle, which you can never reach alone but only in pairs."
> There was a pause, we waited for something.
> Then: The *voice of the master* speaks, transformed, from the centre of the ring to the lady:
> "Remain merciful."
> Now I knew I could go, out of the room into normal time and normal everyday space.
> When I was outside, I noticed that I was wearing my coat and hat. From afar I heared a C major chord of four tones CEGC, apparently played by the lady herself when she was on her own again.

6. The ring i

Here we encounter the culminating-point of Pauli's alchemical journey. At the moment he surrenders to his inner calling, the reconcilement of the master with the spirit of science, his beloved Lady presents him with the symbol that actually may bring a new orientation to the world of science. As a mathematical element the complex unit circle is simplicity itself, but as a symbol it is as profound as the Cross in Christianity. Its subtlety consists primarily in that it represents a marriage of two dimensions, the real and the imaginary. As such the complex unit circle bridges the gulf between the world of ordinary macroscopic matter and the world of atoms. When Werner Heisenberg and Erwin Schrödinger introduced the imaginary unit as a fundamental element into our description of matter, suddenly all the seemingly contradictory appearences of quantum particles could be harmoniously integrated in the concept of the complex probability amplitude.

This ability of quantum physics to reconcile the wave and particle aspects of matter had impressed Pauli so much that he hoped that this branch of physics might also serve as a model for the more encompassing undertaking of reconciling the physical and the spiritual aspects of matter.[17] Here the unconscious presents Pauli with exactly the same symbol that helped modern physics to unlock the closed doors of atomic and subatomic matter. So there really seems to be a close connection between quantum physics and other fields of experience.

Pauli's teacher in physics and friend Niels Bohr had stressed this connection from an epistemological point of view. For Bohr the *complementarity* of wave and particle aspects of matter was but one example of a wider class of phenomena in which seemingly irreconcilable pairs of opposites need not be contradictory. The statements about the ring i in Pauli's *Klavierstunde* strongly support Bohr's philosophy of complementarity. We can even look upon the ring i as the symbolic, mathematical expression for complementarity, as has been proposed recently by depth psychologist J. Marvin Spiegelman and physicist

Victor Mansfield.[18]

The ring i, however, has a wider meaning than complementarity. Being a symbol and not a philosophical concept the ring i can express feeling tones that are absent in Bohr's philosophy. Since Pauli himself never published anything related to this final chord of 'the piano-lesson', it is our task to encircle this additional richness of the ring i in order to benefit from it. In the first place we should notice that the ring i symbolizes Eros and relationship. But genuine relationship presupposes individuation. Or in the words of Von Franz[19]: 'Eros is in its essence only meaningful if it is completely, uniquely individual.' For Pauli this probably meant that he had to take his love for Von Franz seriously. Only with her could he arrive in the 'realm of the middle', that is, in the middle position between the instinctive and the spiritual, the physical and the psychic, the scientific and the religious.

Seen from this perspective, the ring i also symbolizes a relationship with nature which is no longer dominated by the tendency to objectify nature as a reality outside the observer. We cannot establish a proper relationship with matter, as long as we ignore our inner motive to penetrate into its mysteries. Matter is not only outside if its spirit can speak to us from inside. Ignoring this inner relationship may be dangerous, as Pauli himself noted in a letter to Fierz[20]: 'It appears to me that the ignored "after-effects" of observation will *yet* enter into the picture (as atomic bombs, general anxiety, "the Oppenheimer case", for instance), but in an *undesirable form*. The famous "incompleteness" of quantum mechanics (Einstein) is still factually present somehow-somewhere (...) This has to do with *integral relationships between "inside and outside" which are lacking in contemporary science* (but which alchemy had suspected, and which may also be detected in the symbolism of my dreams, symbols that I believe to be typical for contemporary physicists).'

Finally we should notice that the ring i has a transforming effect on the master-figure who had frightened Pauli so much through the years. Entering into a genuine, understanding relationship with the spirit of matter can transform its fiery nature. Suddenly the master speaks from the centre of the ring i and seems to have lost his threatening power. The homeless, wandering messenger of God is transformed into Pauli's own innermost living centre. As a true alchemist he has realized the philosophers' stone. Now he possesses the great treasure that could bring him the peace of mind and the meaning of life he was looking for.

7. REFERENCES

First I want to express my gratitude to the Swiss analyst Eva Wertenschlag who helped me to understand the significance of Pauli's struggle with the unconscious.

I also wish to thank Dr. Marie-Louise von Franz, Mrs. Aniela Jaffé, Dr. Franz Jung and Dr. Carl Alfred Meier. They kindly gave me permission to quote from or to paraphrase passages in the unpublished letters and manuscripts of Pauli to be found in the archives of the ETH in Zurich, section 'Wissenschaftshistorische Sammlungen' (WHS).

1. K.V. Laurikainen, *Beyond the Atom, The Philosophical Thought of Wolfgang Pauli* (Springer-Verlag, Berlin Heidelberg, 1988), p. 89/90.
2. W. Pauli, 'Der Einfluss archetypischer Vorstellungen auf die Bildung naturwissenschaftlicher Theorien bei Kepler', in: C.G. Jung/W. Pauli, *Naturerklärung und Psyche*, Studien aus dem C.G. Jung-Institut Zürich IV (Rascher Verlag, Zurich, 1952).
3. K.V. Laurikainen, *Beyond the Atom*, p. 89/90.
4. Dreams of Pauli in the archive of Aniela Jaffé, ETH Zurich, WHS, Hs 1090:71.
5. Letter of Pauli to Jung, May 27, 1953, ETH Zurich, WHS, Hs 1056:30867.
6. For instance, in a letter to Jung, dated December 12, 1950, Pauli gives a wider, *neutral* meaning to the term radioactivity.
7. Dreams of Pauli in the archive of Aniela Jaffé, ETH Zurich, WHS, Hs 1090:71.

8. Letter of Pauli to Emma Jung, ETH Zurich, WHS, Hs 1091:401.
9. Erich Neumann, *The Child, Structure and Dynamics of the Nascent Personality* (Harper & Row, New York, 1973), p. 53.
10. Letter of Pauli to Jung, May 27, 1953, ETH Zurich, WHS, HS 1056:30867.
11. C.G. Jung, *Alchemical Studies, Collected Works*, Vol. 13 (Routledge & Kegan Paul, London, 1983), par. 299.
12. Letter of Pauli to Von Franz, April 18, 1951, ETH Zurich, WHS, Hs 176:14.
13. Letter of Pauli to Von Franz, October 12, 1952, ETH Zurich, WHS, Hs 176:52.
14. 'Die Klavierstunde', unpublished manuscript by Pauli, ETH Zurich, WHS, Hs 176:85.
15. C.G. Jung, *Psychology and Alchemy, Collected Works*, Vol. 12, (Princeton University Press, Princeton, 1980), par. 44.
16. 'Die Klavierstunde', p. 20/21.
17. Letter of Pauli to Jung, February 27, 1953, ETH Zurich, WHS, Hs 1056:19392.
18. J. Marvin Spiegelman and Victor Mansfield, 'Complex Numbers in the Psyche and Matter', *Harvest* (Analytical Psychology Club, London), Fall 1990.
19. Marie-Louise von Franz, *A Psychological Interpretation of the Golden Ass of Apuleius* (Spring Publications, Zurich, 1970), Chapter XIII, p. 1.
20. K.V. Laurikainen, *Beyond the Atom*, p. 144/45.

THE MODAL INTERPRETATION OF QUANTUM MECHANICS

Bas C. van Fraassen
Philosophy Dept., Princeton University
Princeton, NJ 08540 USA

ABSTRACT

It was crucial to von Neumann's interpretation to equate : (observable M pertaining to system X has value r) = (system X is in an r-eigenstate of M) = (the probability = 1 that a measurement of M on X will/would yield value r). The modal interpretation equates the second with the third, but not with the first; the probability in the third is attached to a proposition of the first (value attribution) type.

1. INDETERMINISM AND DETERMINISM IN QUANTUM MECHANICS

Quantum physics describes an indeterministic world; yet applied to an isolated system, describes its state as evolving deterministically. How can the two be reconciled? There are three main alternatives. The <u>first</u> is that in certain isolated systems, the state does after all not develop deterministically. This is von Neumann's alternative, which introduces the Projection Postulate to describe moments when the evolution in accordance with Schroedinger's equation is interrupted.[3] It is also, more recently, the alternative of Ghirardi and his co-workers. The <u>second</u> denies not the determinism but the apparent indeterminism. It says that a measurement is not a process characterizable as the evolution of an isolated system, ever: a measurement is an interaction <u>incompletely described</u>, by leaving out something or other. Perhaps the most radical interpretation along this line is the idea that quantum theory was devised to describe only situations in which an observer (or at least, measuring environment) is involved, while leaving that part out of the description. John Wheeler[4] noted this as a major alternative; it is certainly reminiscent of some early Copenhagen texts. Equally radical, but in a different direction, is Everett's many-world interpretation: indeterminism is an illusion, and disappears if we describe all the worlds there are besides our own.

The <u>third</u> alternative is to deny neither the determinism of the total system evolution, nor the indeterminism of outcomes, but to say that the two are different aspects of the total situation. The modal interpretation takes this third road, via a denial of the identification of value attributions to observables with attributions of states -- the state can then develop deterministically, with only statistical constraints on changes in the values of the observables. There are several variants of the modal interpretation, but only one ("Copenhagen variant") will be discussed here. In comparison with the first two options listed above, it allows for isolated systems in the scope of quantum theory, but classifies the indeterminism as reality and not mere appearance.

2. THE MODAL INTERPRETATION

On von Neumann's interpretation, attributions of values to observables are classifications of states:

(1) Observable B has value b _if and only if_ a B-measurement is certain to have outcome b.

This looks classical, for in classical world-picture, measuring was just looking, without disturbing, as if by a disembodied intelligence. It was also assumed there that for any observable, at any time, there _is_ a _unique_ value which is _certain_ to be shown by a correct measurement. But what happens to this picture when we admit states in which measurement has uncertain outcomes?

The term "state" is familiar from classical mechanics, thermodynamics, chemistry, engineering, and so on. Classically we would assume (a) that a state is given if we know the value of all observables, and (b) that if we knew the state, then we would know all there is to know about how the system will develop if left alone and how it will react if acted upon. Such action-upon includes measurement, of course. If observables simply have values, there to be seen if we look, and the temporal evolution is determined entirely by what those values are, then there is no need to keep (a) and (b) separate. But to be cautious we must distinguish two concepts of state: the _value state_, specified by describing the values of the observables, and the _dynamic state_, specified by stating how the system will develop if isolated, and how if it is acted upon in any fashion. Measurement is an interaction, so the prediction of measurement outcome probabilities belongs to the role of the dynamic state.

Another way to present this distinction is to distinquish between two sorts of propositions. A _value-attributing proposition_, symbolized <m,E>, says that observable m actually has a value in E, and a _state-attributing proposition_, symbolized [m,E] says that the state is such that a measurement of m must have an outcome in E. The value state is what makes value attributions true, while the dynamic state determines which state-attributions are true. In the modal interpretation, each system and subsystem has at all times both a dynamic state and a value state. The Born rule is understood this way:

> if a measurement of A is made on system X in state W then the expected value of observable A at the end of measurement equals Tr(AW).

Note that the "if" concerns the initial dynamic state of X, while the "then" gives information about its final value state. The measurement result therefore does not, contrary to von Neumann, directly indicate the final dynamic state (Projection Postulate). But measurement outcomes give relevant information about what state the system was in. From information about what values the observables have, we can -- by looking for a statistical fit -- infer backward to the state. Using statistical inference it is possible to conclude, that a certain source or preparation procedure produces systems in state W -- or to attribute state W to the systems in an ensemble of which we have inspected a sample -- but not to determine that a single given system is in this state.

The values of observables play no direct predictive role. In that sense they are "empirically superfluous". The character of the initial values of the observables could at best be a symptom or clue to what the initial state is. The expectation and indeed character of the future is determined, to the partial extent that it is determined at all, by the dynamic state W (of the whole system) alone. The value state must be allowed to change unpredictably, within the limits set by the dynamic state, in order to allow for indeterminism in the phenomena.

2. PRINCIPLES OF THE MODAL ACCOUNT

Following Jauch[5], I shall use the word "event" to signify what is distinct from the state, but may be predicted on the basis of the state. Events are described by statements of form

> (2) Observable B pertaining to system X (actually) has value b

and the Born interpretation rule sometimes tells us the probability of this event actually occurring (this statement actually being true) at the end of an appropriate measurement. Statement (1) cannot be equated with any of the form "The system is in a state of type....", which can at best give us a probability of (1) being true. What is the logical connection between value attributions (form (1)) and state attributions? At the end of a von Neumann measurement of non-degenerate observable A on system X by apparatus Y with pointer observable B, we have the following description of total and reduced states:

(a) X is in state $W_X = \Sigma c_i^2 \, I_{|a(i)>}$

(b) Y is in state $W_Y = \Sigma c_i^2 \, I_{|b(i)>}$

(c) X + Y is in state $\phi_t = \Sigma c_i |a_i> \otimes |b_i>$

Now the Born rule, given our proposed modal alternative to von Neumann's interpretation rule leads here to

(d) for some index k, pointer observable B has value b_k; and the probability k = i, equals c_i^2.

Rule (d) only establishes a link between the state at the end of an A-measurement and the statements attributing values to B. Could there be a larger principle which connects states more generally with values of observables?

The interpretational question facing us is exactly: in general, which value attributions are true? The Copenhagen point of view introduced great conservatism in this respect. Copenhagen scientists appeared to doubt or deny that observables even have values, unless their state forces us to say so. I accordingly call the following very cautious answer the _Copenhagen variant_ of the modal interpretation. It says that if system X has dynamic state W at t, then the state attributions [M,E] which are true, are those such that $Tr(WI_E^M)=1$. About the value attributions, it says that they cannot be deduced from the dynamic state, but are constrained in three ways:

(i) if [M,E] is true then so is the value attribution <M,E>: observable M has value in E;

(ii) all the true value attributions could have Born probability 1 together;

(iii) the set of true value attributions is maximal with respect to feature (ii).

In order to discuss the import of these constraints, let us use the following terminology:

state W _makes_ [M,E] _true_ exactly if $Tr(WI_E^M)=1$.

state W' _is possible relative_ to W exactly if for all subspaces S, if $Tr(WI_S) = 1$ then $Tr(W'I_S) = 1$ also.

It is now easy to see that dynamic and value states admit the same mathematical representation. For (i)-(iii) entail that if the system is in dynamic state W then there is a pure state W' which is possible relative to W, and such that the set of true value attributions includes exactly those <M,E> such that W' makes [M,E] true.

That dynamic state W' is the bookkeeping device which identifies the true value attributions correctly. Hence it can be used to _represent_ the value state. That does not mean that value states are dynamic states, but only that each admits the same sort of mathematical representation. We can sum all this up in a single postulate, which describes our family of models for physical situations governed by quantum mechanics:

(e) Given that system X is in dynamic state W at time t, there is a certain pure state x (to be called the value state) which is possible relative to W, and such that for all observables M pertaining to X:

(e1) a state-attribution [M,E] is true if and only if W makes it true;

(e2) a value attribution <M,E> is true if and only if x makes [M,E] true.

The first consequence we deduce from this postulate concerns the identification of observables. My terminology and notation "Hermitean operator M represents observable m" allows for the possibility that two distinct observables be represented by the same operator. If that could happen, principle (e) above would have to be read as applying to all observables represented by given operator M. But so read, it implies that if two observables are represented by M, it is not only their measurement outcome probabilities that are always the same, but also their values. This is because of the "if and only if" in (e2). Hence there is then no difference at all between these observables; they are the same:

<u>Principle of identity of observables</u>. If observables m and m' are represented by the same Hermitean operator then m=m'.

This principle entails that the same name can designate an observable and its representing operator.

The second consequence is that <u>it is as if the ignorance interpretation of mixtures were correct</u>. For if system X is in mixed state W, then the actual values of observables pertaining to X are exactly those it would have had, if it had been in a certain pure state in the image space of W. The ignorance interpretation is not tenable, for a number of reasons, but some of what it entails for values of observables is right. The third consequence is the falsity of the classical principle that each observable always has one of its possible sharp values. We can derivatively attribute "unsharp" values. To observable M there corresponds a large set of observables, via spectral decomposition, which have eigenvalues 1 and 0 only:

$$P_W^M(E)=1 \text{ if and only if } \mathrm{Tr}(WI_E^M)=1$$

Therefore we have the following result concerning state attributing propositions: $[M,E]=[I_E^M,1]$. The adoption of (e2) -- or equivalently, of (i)-(iii) -- entails that

$\langle M,E\rangle = \langle I_E^M, 1\rangle$ too. Hence $\langle M,E\rangle$ is <u>not</u> the classical disjunction of the value attributions $\langle M,r\rangle : r\varepsilon E$. Indeed:

> If $\langle I_E^M, 1\rangle$ is true if and only if Borel set E_0 is included in E, then E_0 is also the smallest Borel set such that $\langle M,E_0\rangle$ is true

and we should say that M has <u>unsharp value</u> E_0.

4. WHAT HAPPENS IN A MEASUREMENT?

To be a candidate for designation as a measurement at all, a physical process must satisfy certain requirements. But (a) what is the significance of measurement?, and (b) what is really going on in a measurement? Philosophers tend to worry that the appearance of the term "measurement" in the Born rule bears its anthropocentric connotations essentially. Does that mean that we cannot think of quantum theory as a putative autonomous description of the world in neutral physical terms and prospectively complete? In the jargon: if that were so we could not be <u>realists</u> with respect to the theory, but only <u>instrumentalists</u>. This worry can be laid to rest: the requirements upon physical correlates of measurement involved no reference to us, to persons or consciousness, and not even to the macro/micro distinction.

But then, what is the significance of measurement? I believe it is the following. Every interpretation of quantum mechanics begins qualitatively, by speaking out on whether an observable always [sometimes, never] has sharp values, while the system is [is not] in one of its eigenstates, and so forth. But then must come a next step: assignment of probabilities. Human acts being nothing very special among the physical categories, the probabilities announced when Stern and Gerlach, or Aspect, or Compton, or Leggett designs an experiment to be actually carried out in a laboratory, must be a special case. Fine; but how do we draw out consistent, self-coherent probabilities in general?

Officially, quantum mechanics allows for only one way to assign probabilities -- via Born's rule. We can <u>interpret</u> this to extend to all processes, even on the microscopic level in the ionosphere, which meet the minimal requirements for a physical correlate of measurement. Can we go a little further? Von Neumann went a lot further. Can we go a little way along with him, exploiting for example the correlations set up in those special von Neumann, or von Neumann-Lueders measurements? The answer is <u>yes</u>, but we must proceed as if barefooted in a field strewn with glass. This problem -- the general problem of drawing a consistent general recipe for probability assignments from quantum theory -- is what is _really_ behind the seemingly disproportionate attention to measurement.

The situation of system X at given time t is characterized according to (e) by two states: its _dynamic state_ W and its _value state_ x. What does this look like when X is a compound system? If Z = X + Y, we have both a dynamic state and a value state for each of Z, X, Y. I shall here discuss only the case in which Z has a pure dynamic state ϕ. Let us designate the mixed states assigned to X and Y, by "reduction of the density-matrix" as $\#\phi$ and $\phi\#$. Then the situation is this:

(f) (X + Y), X, Y have dynamic states ϕ, $\#\phi$, $\phi\#$ respectively

(g) (X + Y), X, Y have as value states ϕ, x, y respectively, where x is possible relative to $\#\phi$ and y possible relative to $\phi\#$.

We cannot specify what x and y are: there are a number of possibilities. We only know that they must be possible relative to $\#\phi$ and $\phi\#$. This means mathematically that they are vectors in the image spaces of $\#\phi$ and $\phi\#$.

What we would like next is a probability distribution on these possibilities. The non-unique decomposability of mixtures stands in the way of a general rule for probability assignments which is at once simple and consistent. But the Born rule is meant to assign probabilities coherently if the situation comes _at the end of a measurement_. The rule presupposes that measurement is a process which is so structured that it singles out certain observables, and tells us the probabilities for their possible values -- without the inconsistencies that plague the ignorance interpretation. That means of course that the observables singled out are mutually compatible.

The major recognized classes of measurement in the literature satisfy the meta-criterion hereby suggested: roughly, that if a process measures various observables on an object system jointly, then it ends with the apparatus in a mixture of joint eigenstates of the pointer observables. It is to these, in the first instance, that the Born rule assigns probabilities. As usual, the paradigm illustration is the von Neumann measurement of an observable with non-degenerate spectrum. We can consistently add to our previous principles:

(h) If (f) and (g) hold with $\phi = \Sigma c_i \ |a_i\rangle \otimes |b_i\rangle$ at the end of an A measurement, with B designated as pointer observable, then the probability equals c_k^2 that $y = |b_k\rangle$.

This has Born's rule translated into our present representation of this physical situation, as corollary, since value state $y = |b_k\rangle$ exactly if B has value b_k. It is consistent with the preceding because x and y do, as they must, lie in the image spaces of $\#\phi$ and $\phi\#$. But of course a von Neumann measurement has a special feature. It does not just allow statistical inference backward to the initial state of the object system. In addition it effects a perfect correlation between measurement and pointer observables. Hence we can go one step further, and add

> (h') If (f) and (g) hold with $\phi = \Sigma c_i |a_i> \otimes |b_i>$ at the end of an A measurement, with B designated as pointer observable, then the probability equals c_k^2 that <u>both</u> y = $|b_k>$ <u>and</u> x = $|a_k>$.

The extension of (h) to (h') is important, because it entails that a pointer-reading result b_k implies with certainty that observable A had value a_k at that point:

> (i) If the situation is as in (h'), then the probability that value attribution $<A, a_k>$ is true, given that value attribution $<B, b_k>$ is true, equals 1.

The proof of this is simple: the probabilities of the combinations (x = $|a_i>$, and y = $|b_i>$ sum to <u>one</u>, therefore the probability of any combination (x = $|a_i>$ and y = $|b_j>$) with $i \neq j$ must equal <u>zero</u>. Notice also however, that a possibility which has probability <u>zero</u> may still really occur. Probability <u>zero</u> does not imply impossibility. The addition of (h) and (h') to our interpretation assigns probabilities; it does nothing else.

We are tempted to exclaim: <u>it is as if the Projection Postulate were correct</u>. For at the end of the measurement of A on system X, it is indeed true that A has the actual value which is the measurement outcome. But of course the Projection Postulate is not really correct. There has been a transition from possible to actual value, so what it entailed about values of observables is correct, but that is all. There has been no acausal state transition.

5. WHEN ONE, TWO OR NO MEASUREMENTS ARE BEING MADE

Quantum mechanics gives us probabilities only for measurement outcomes, conditional on the hypothesis that a measurement is made. That is very stingy. It has also to philosophers a very empiricist or even positivist sound, but that is deceptive. At least on our present interpretation, any process at all is a measurement if it satisfies certain purely physical conditions, which make no reference to conscious observers, and need not be macroscopic. But we are still left with the question: what can happen when no measurement is being made?

The answer on our present interpretation is: <u>anything is possible</u>. Of course, "anything" does not mean anything logically conceivable, for according to the distinctive Copenhagen principles laid down above, no value attributing propositions can be true together unless they <u>can</u> be certain together (in the sense of receiving Born probability 1). This is a very weak constraint on what actually happens, and it still allows things to happen in actuality, which there have <u>zero</u> probability of ever appearing as measurement outcomes. That is partly because <u>zero</u> probability is not strict impossibility, and partly because something which has Born probability 1 under <u>some</u> circumstances, need not have Born probability 1 under other circumstances. In a given particular case, the dynamic state may have correlations in it which are not to be found in all dynamic states of which the system is capable.

Should we not think (analogously to classical physics) that in quantum physics the process of interaction designated as measurement <u>changes</u> the probabilities? What, for instance, if I am the measurement apparatus, and I decide beforehand that if I detect a red light, I shall shoot the piano player? Does it not follow then that the occurrence of an actual value (and not just the total quantum mechanical state) affects the probability of what happens next?

There is a fallacy here. The interaction affects the measured system if we think of it <u>by itself</u>, apart from the total isolated system of which it is part. But if we have before us the description of the total evolution of the state of the entire system, then the description of some internal episode <u>as</u> a measurement <u>adds nothing</u> -- it is simply a classification. If we want to reflect on for example von Neumann's graphic "immediate repetition" demand, we do it as follows: we ask, what about the total evolution of a system in which a certain episode qualifies as an immediate succession of two measurements of the same observable, plus a measurement which checks on whether the two outcomes agreed? What are the probabilistic predictions for outcomes of <u>that</u>? To get the answer we look again to the wave function of a total system in which that happens, and the Born rule allows us to deduce: we can expect the outcome <u>the two agreed</u>! with probabilistic certainty.

It is exactly at this point that our intuitions tend to declare war on our deductions. "We don't want to know", they insist, "what probabilities are derived for the outcome of a third measurement, if performed -- we want to know purely and simply whether if the first two measurements occur, their outcomes do agree!" But this question has a presupposition -- namely that quantum mechanics, if it is to be complete and accurate, should give us such information. In other words, it is assumed that the theory must give us more than the probabilities of measurement outcomes. And this is an assumption that does not come from physics, but from somewhere outside. If it had some independent justification, it would be a desideratum which might be imposed as a criterion of adequacy for interpretation. But it does not.

This war of intuitions lies very deep, and appears nowhere so clearly as in the thought-experiments in which we imagine ourselves involved as observers and manipulators. In this area even subtle thinkers leave philosophical clarity behind, and regress to a primitive mind-body dualism or naive materialism. In <u>any</u> supposed total description of the universe and its entire history (let us call that a Universal Story), our actions and observations must appear no differently from other processes. The initial supposition of having a Universal Story entails that their designation as free actions or conscious observations can only be a <u>classification</u> of certain episodes already entirely encompassed by other supposed descriptions. If I then additionally suppose myself to be somehow outside the process and capable of reporting directly on the state of the system -- as if I had an 'ontological telescope' -- I am supposing that there is something not encompassed in the Universal Story. But by supposition it encompasses everything! The result is incoherence.

Here is a more down to earth issue. If the measurement outcomes' probabilities are set by the evolving state, but the outcomes do not provide a "feedback" input to that evolution, should we not expect a very chaotic life? Imagine the following measurement (observation) outcomes. Looking out of my window I see two cars collide; next the drivers emerge gesticulating toward each other; a crowd gathers; a police car shows up and stops at the accident. If at each moment the probabilities for possible events were given by the total state at that time, shouldn't we expect to see very often some 'unconnected series'? For example, a car stops skew in the middle of the intersection, a police car appears, a crowd gathers, and then another car whose speed has been fluctuating noticeably, hits the first one. Stranger sequences yet can be imagined. Also in the case of Schroedinger's cat, the account I gave above is logically comparable with the cat being sometimes alive and sometimes dead, between the triggering of the device and the opening of the box.

This is only a more complicated version of points about repeated measurement. Here we have a correlation over time; we could ask a similar question about a simultaneous correlation. In Aspect's experiment, each photon has a 50% probability of passing, if the filters' orientation is parallel: why should there be a correlation observed? The answer is the same to both questions: given the total state of the system, the uncorrelated measurement outcome combinations are given a zero or negligible probability.

6. MODAL INTERPRETATION OF COMPOSITION AND MEASUREMENT

The touchstone for the interpretation will be the quantum theoretical treatment of compound states. That is where the probabilities and possibilities are tangled, and where the holistic character of the quantum world becomes manifest. In general we must consider an N-body system S $= X_1 + \ldots + X_N$. The order in which the components are listed has no counterpart in reality; but when we describe the states and observables, some such order creeps into our representation. The system $X_1 + X_2 + X_3$ can also be mutually divided in various ways: into a 2-component system if thought of as $X_1 + (X_2 + X_3)$, or as $(X_1 + X_2) + X_3$. But $X_1 + X_3$ is also a subsystem; all these divisions are equally good as alternative representations.

The total system S has a dynamic state; so does each subsystem. If S is in state W, and S' is a subsystem, then S' is in reduced state #W[S']. #W[$X_1 + X_2$] can be reduced again to give a state to X_2, namely #(#W[$X_1 + X_2$])[X_2]. But, as consistency requires, that is just the state #W[X_2]. The reduced states are in general mixtures, even if the total state W of whole system S is pure. According to the modal interpretation, the actual values of observables correspond, for a system in mixed state W', to those made certain by some pure state x in the image space of W'. Thus we have for each subsystem S' a dynamic state #W[S'] and a value state λ[S']. If W itself was pure, let us represent it by vector ϕ, we arrive then at the following picture for a 3-component system:

$$\phi;\quad \lambda_{123}$$

$$\#\phi[1,2];\ \lambda_{12} \qquad \#\phi[2,3];\ \lambda_{23} \qquad \#\phi[1,3];\ \lambda_{13}$$

$$\#\phi[1];\ \lambda_1 \qquad \#\phi[2];\ \lambda_2 \qquad \#\phi[3];\ \lambda_3$$

where I indicate the subsystems of S = ($X_1 + X_2 + X_3$) just by indices -- e.g. #ϕ[1,2] and λ_{12} are respectively the dynamic state #ϕ[$X_1 + X_2$] and the value state λ[$X_1 + X_2$].

Both the ignorance interpretation of mixtures and the modal interpretation require this picture to remain consistent subject to the following requirements:

(a) #ϕ[S'] is the reduction of state ϕ of S with respect to subsystem S'

(b) $\lambda[S']$ is a pure state in the image space of $\#\phi[S']$ -- i.e. possible relative to the latter.

But it is clear that there is an additional requirement, which comes into play exactly when the compound has more than two components. The situation would not be as if the ignorance interpretation were true, if we could not <u>pretend</u> that the dynamic state of $X_1 + X_2$ is <u>really</u> $\lambda[X_1 + X_2]$! For that interpretation said that $\#\phi[1,2]$ just represents incomplete knowledge. So the following alteration of part of the picture:

$$\lambda_{12};\ \lambda_{12}$$

$$\#\ \lambda_{12}[X_1];\ \lambda_1 \qquad \#\ \lambda_{12}[X_2];\ \lambda_2$$

must still be consistent, subject to demands (a) and (b) above. That is, λ_1 should be possible not only relative to $\#\phi[X_1]$, or equivalently $\#(\#\phi[12])[X_1]$, but also possible relative to the reduction $\#\lambda_{12}[X_1]$ of the value state λ_{12}. This extra criterion therefore amounts to

(c) If S" is included in S' then $\lambda[S"]$ is a pure state also in the image space of $\#(\lambda[S'])[S"]$

Whenever several overlapping constraints are imposed jointly, inconsistency threatens. What about here?

We can settle part of the question at once, by reasoning from definitions. The consistency of (b) and (c) requires:

Theorem. If ϕ is in the image space of W, then the image space of $\#\phi[S']$ is part of the image space of $\#W[S']$.

which is easily proved. But this little result, though reassuring, is not enough. For the first alteration of the picture which had to remain consistent, is matched by another one:

$$\lambda_{23};\ \lambda_{23}$$

$$\#\ \lambda_{23}[X_2];\ \lambda_2 \qquad \#\ \lambda_{23}[X_3];\ \lambda_3$$

and we see λ_2 appearing in both altered pictures. To put it differently: criterion (c) demands that $\lambda[X_2]$ be possible relative to the reductions of the value states of <u>both</u> $(X_1 + X_2)$ and $(X_2 + X_3)$. Those value states were, as it were, random selections from two image spaces (of $\#\phi[12]$ and of $\#\phi[23]$). It is not obvious, and certainly not entailed by our little theorem, that consistency allows this sort of overlap, i.e. that there will exist a state (a candidate for being $\lambda[X_2]$) which is possible relative to the two reductions of those randomly selected elements of the two image spaces.

Though not obvious, this can be proved; that is the major consistency theorem for the modal interpretation.

7. IDENTICAL PARTICLES: AGGREGATE BEHAVIOUR

> An atom...possesses two kinds of symmetry properties: (1) the laws governing it are spherically symmetric, i.e. invariant under an arbitrary rotation about [its center]; (2) it is invariant under permutation of its...electrons.
>
> Herman Weyl (1929)[6]

An elementary particle is characterized first of all by certain constant features, which serve to classify it. Mass is such a constant: baryons have large mass, mesons intermediate, and leptons small. Other constants, such as charge, subdivide these classes. Particles characterized by all the same constants are often called "identical". In that sense, two particles can be identical, and yet be in different states of motion -- so the identity is not strict numerical identity, or even strict qualitative identity. If two identical particles are in the same state, they are "indistinguishable" in a sense going beyond that of "identical" as used above.

Given two systems X and Y which are capable of the same states, we represent the composite system X + Y by means of a tensor product $H \otimes H$ and in general, any state of X+Y by a statistical operator on $H \otimes H$. Any given vector in $H \otimes H$ takes the form $\phi = \Sigma c_{ij}\, x_i \otimes y_j$ where $\{x_i\}$ and $\{y_j\}$ are bases. Intuitively we describe this as a superposition of states $x_i \otimes y_j$ in which X has state x_i and Y has state y_j. What about the permuted situation, in which X has y_j and Y has x_i? Mathematically, we can see that the vector $y \otimes x \neq x \otimes y$. Similarly the general state ϕ has permutation

$$\phi' = \Sigma c_{ij}(y_j \otimes x_i) \neq \phi \text{ in general.}$$

In the case of N particles there are N! permutations, forming a larger group. Every permutation s can be represented by a unitary operator P_s which transforms the states in the obvious way, illustrated here by the change of ϕ into ϕ'.

Does the mathematical change in the vector by such a permutation, lead to the representation of a new state? Or is a permutation like rotation through 360 degrees? For identical particles X and Y, the assertion is that indeed there is no "real" difference in states represented by permuted vectors such as $x \otimes y$ and $y \otimes x$.

> Permutation Invariance. If ϕ is the state of a composite system whose components are identical particles then the expectation value of any observable A is the same for all permutations of ϕ.

The Permutation Invariance Requirement requires a limitation either on the states (perhaps not all vectors represent pure states that can really be found in nature) or else on the observables (perhaps not all Hermitean operators represent measurable physical quantities). If we regard ourselves as already knowing what the observables are we can try to state it in form

> (Perm. Inv. 1). If some observable A does not have the same expectation value in ϕ and its permutation ϕ', then ϕ does not represent a (physically possible) pure state of an aggregate of identical particles.

In this formulation the requirement has bite if we add for example that all Hermitean operators or a certain subclass thereof, represent (real, measurable) observables. If instead we regard ourselves as knowing already what states there are, we can attempt to formulate Permutation Invariance as

> (Perm. Inv. 2). If some state ϕ is such that Hermitean operator A does not have the same expectation value in state ϕ as in its permutation ϕ', then A does not represent a (real, measurable) observable.

This formulation acquires bite if for example we add that all vectors in $H \otimes H$ represent states (even if not uniquely, when superselection rules are in force).

In the case N = 2, there is only one non-trivial permutation, which replaces $x \otimes y$ by $y \otimes x$; extended by linearity, it replaces $\Sigma c_{ij}(x_i \otimes y_j)$ by $\Sigma c_{ij}(y_j \otimes x_i)$. If ϕ is the original vector, let ϕ' be its replacement by this non-trivial permutation. Then we have two special cases:

$\phi' = \phi$: ϕ is a <u>symmetric</u> vector
$\phi' = -\phi$: ϕ is an <u>anti-symmetric</u> vector

In both cases, the expectation value of any Hermitean operator A remains the same, i.e. $(\phi.A\phi) = (\phi'.A\phi')$. Hence we are sure that such a vector ϕ can be admitted as representing a state which satisfies (Perm. Inv. 1). The definition can be extended in general to: ϕ in H^N is <u>symmetric</u> if every two-particle permutation leaves it the same, and <u>anti-symmetric</u> if every such permutation turns it into $-\phi$. For the two-particle case, these form two subspaces:

<u>Symmetric states</u> $\phi = \Sigma c_{ij}(x_i \otimes y_j + y_j \otimes x_i)$

<u>Anti-symmetric states</u> $\phi = \Sigma c_{ij}(x_i \otimes y_j - y_j \otimes x_i)$

Having noted this, we can postulate the

> <u>Dichotomy Postulate</u>: Every type of particle is such that its aggregate can take only <u>symmetric</u> states (<u>boson</u>), or else such that its aggregates can only take <u>anti-symmetric</u> states (<u>fermion</u>).

The given definitions do indeed imply the applicability of Fermi-Dirac statistics to fermions, and of Bose-Einstein statistics to bosons.

We must however sharply distinguish the Dichotomy Postulate I have just described from the preceding weaker postulate of Permutation Invariance. In the 2-particle case, two sorts of states automatically satisfy Permutation Invariance for all Hermitean operators. The postulate of Dichotomy is a non-trivial generalization because the general N-particle case is not a simple extrapolation of the 2-particle case.

8. THE EXCLUSION PRINCIPLE

In 1925 Wolfgang Pauli[7] formulated his principle for orbital electrons in the atom:

> In an atom there cannot be two or more equivalent electrons for which the values of all four quantum numbers coincide. If an electron exists in an atom for which all of these numbers have definite values, then this state is "occupied".

In general terms, if states are specified completely, one would say then that two fermions of the same type cannot be in the same state. But this, though intuitively satisfying, is not accurate in relation to the general treatment. Suppose an aggregate of two fermions, X + Y, is in the anti-symmetric state:

$$\phi = \frac{1}{\sqrt{2}} [(x \otimes y) - (y \otimes x)] \tag{1}$$

It is indeed true that x = y is excluded, for that would make this the zero vector. But in what sense do the two particles "occupy" different states? Could we say that one is in state x and the other in state y? That does not sit well at all with our previous discussion of composition and reduction. The two reduced states are the same:

$$\#\phi[1] = \#\phi[2] = \frac{1}{2} I_x + \frac{1}{2} I_y \tag{2}$$

So Reduction assigns the same state to both.

To capture the distinction between fermions and bosons which the literature keeps alluding to in terms of exclusion, define an <u>exclusive product state</u> to be any tensor product state $x_1 \otimes \ldots \otimes x_n$ in which $x_i = x_j$ only if $i = j$. Then we have the following

> <u>QM Exclusion Principle</u>. All anti-symmetric states are superpositions of exclusive product states only; not all symmetric states are.

This distinguishes the anti-symmetric and symmetric subspaces from each other, but not the individual states.

Perhaps the appeal of the ignorance interpretation has helped to keep the slogan formulation of Exclusion alive. There is something intuitively feasible about a picture of identical fermions all occupying mutually orthogonal states, only we don't know (or it makes no sense to ask) which particle is in which state. The precise truth lies perhaps in the

<u>QM Exclusion Corollary</u>. If X_1 +..+ X_N is in anti-symmetric state ϕ, then the reduced states $\#\phi[X_j]$ are all the same, and have an image space of at least N dimensions.

In other words, each particle is in the same reduced state, which is a mixture of at least N mutually orthogonal pure states.

9. THE MODAL INTERPRETATION FOR IDENTICAL PARTICLES

In the modal interpretation permutation invariance for the dynamic states places constraints also on the values observables can have in actuality. But these constraints are not entirely rigid. It is possible to utilize the Exclusion Corollary for fermion states of section 8 above, to produce models in which fermions <u>are</u> individuated, and the Identity of Indiscernibles is not violated. In the case of bosons, however, that is not possible, at least not if we take the quantum mechanical description of reality, as understood in the modal interpretation, to be complete.

9.1 The Exclusion Principle

The idea of a connection with Leibniz' principle of the identity of indiscernibles was perhaps first suggested by Herman Weyl, who referred to the Exclusion Principle as "the Pauli-Leibniz principle of exclusion". The nomenclature suggests the connection. Consider two distinct orbital electrons in an atom. Each is characterized by certain constants, definitive of electrons, plus a state of motion. The quantum-mechanical description admits nothing further, so if we assume that description to be complete, then the identity of indiscernibles requires their states to be different. This is the content of PII: any entities which are numerically distinct, must differ in some significant respect. The assumption of completeness here may have a "metaphysical" air, but it is involved in the very application to atomic structure for which the Exclusion Principle was introduced. This can be seen in its use in the theory of atomic structure, which I described in section 1 above. PII plus a list of quantum numbers, determines the maximum number of atoms for each orbit. It appears therefore that the Exclusion Principle can appear simply as a corollary, if we assume that we have, in this list, a <u>complete</u> set of characteristics that could individuate the particles.

Margenau[8] cites the electron as violating Leibniz' principle, in writings specifically devoted to the

Exclusion Principle. As we saw in section 8, the individual components of an aggregate of identical fermions are all assigned the same state by reduction of the density matrix. The individual particles are therefore indiscernible, as far as the quantum-mechanical state goes. The violation of identity of indiscernibles is a disturbing idea.

At first sight we have only two options. The first is to insist on the ignorance interpretation. With some amendments it could perhaps allow us to say in the above case, that one particle is in state x and the other in state y, but we don't know which, and that this truth is in any case irrelevant when we predict their joint behavior. The second option is Margenau's, which insists that PII is violated, if the description is complete. However, the modal interpretation allows us a reconciliation. In its models the dynamic states are assigned exactly as by Margenau. But in addition, each subsystem S' (including the whole system) has a value state $\lambda[S']$ which assigns actual values to the observables. In some of the possible worlds, PII will be violated if we let $\lambda[S']$ be whatever it can be, subject only to the general constraints. But if we eliminate, or classify as unreal, those worlds in which this happens, the system of possible worlds is not rendered defective. All the same dynamic states can still occur, so the basis of prediction is left untouched.

As example, let us take again the simplest 2-particle fermion aggregate X + Y in dynamic state $\phi = (1/\sqrt{2})(x \otimes y - y \otimes x)$. The dynamic state of X and Y is the same, namely $W = (1/2)(I_x + I_y)$. If we ignore the fact that this is a fermion aggregate, we associate (at least) four distinct possible worlds with this dynamic state:

value states	X + Y	X	Y
in world 1	ϕ	x	y
in world 2	ϕ	y	x
in world 3	ϕ	x	x
in world 4	ϕ	y	y

But to represent a fermion aggregate, this is a "bad" system of possible worlds, according to the suggestion that fermions are individuated by their value states. So we must rule out the last two worlds. Happily the image space of W has dimension 2, so there is no need to assign the same states to X and Y; indeed we can assign them orthogonal states. In section 8 we saw that in any fermion aggregate state, the reduction to the single particle space has a dimension at least as great as the cardinality of the aggregate. Therefore, as far as that is concerned, there are indeed enough possible states for all the individual particles.

These remarks do not yet establish that the modal interpretation can be carried through in this fashion. As we saw above, that interpretation needs to meet certain special criteria for many-particle systems. The proof that these can be met here too, builds on the general consistency theorem. For the aggregate states of fermions, we need to fulfill three desiderata for a possible choice of value states:

(1) all three general criteria (a)-(c) for the attribution of component states, listed in section 6 are satisfied;
(2) the value state assigned to each subaggregate is itself also anti-symmetric;
(3) the value states assigned to distinct individual particles are different.

The tenability of the interpretation requires that some such choice be possible, for each anti-symmetric total state of a fermion aggregate. This can indeed be proved.

9.2 Bosons and genidentity

Boson aggregates are capable of symmetric states, and not of anti-symmetric ones. Margenau's argument can be given for such a symmetric state

$$\phi = \Sigma a_r (P_1 \phi_r + \ldots + P_{N!} \phi_r)$$

where each ϕ_r has form $(x_1^r \otimes \ldots \otimes x_N^r)$, in which repetitions are now allowed (no restriction to exclusive product states). We can assign reduced states to the N particles, but they will all be the same. Hence the assignment of states does not individuate bosons. If this description is complete, then PII is violated.

The modal interpretation cannot help if $\phi = x \otimes x$; for then both particles are assigned x as dynamic state, and hence also as value state. However, we have restricted our discussion so far to the state of the aggregate at a given moment. Should we not, now that we are stymied, bring in the time-variable, which certainly has a role in QM as well? Then we can follow Aquinas' course with respect to disembodied souls and say that each boson may be individuated by its history. Historical individuation would have to be regarded as empirically superfluous, since it goes beyond the description of (dynamic and value) states. For we can imagine a boson aggregate created in state $x \otimes x$ and annihilated before any change of state. Nothing in the theory forbids that.

It has been emphasized and clearly explained by Mittelstaedt[9] that re-identification over time has no empirical significance in quantum mechanics. He illustrates this specifically with a boson aggregate: a class of free He^4-atoms with a conserved momentum. When Mittelstaedt constructs possible world models, he interprets only definite descriptions, and does not allow singular terms (such as names) without descriptive content. The same issue is examined by Dalla Chiara and Toraldo di Francia[10], and graphically described for both electrons and photons. They propose the even more radical departure of a theory of "quasets" which are abstract objects which have a cardinality but no order type. I am more sanguine about the applicability of classical logic and set theory without presupposing PII. But one main conclusion we must accept: identity through time -- in Reichenbach's terminology, "genidentity" -- loses at least its empirical significance in quantum mechanics.

It was Reichenbach who tied the discussion of individuation of fermions and bosons to the question of identity through time. Reichenbach reformulated the traditional questions concerning genidentity (his term for identity across time), first in connection with relativity, and then for quantum mechanics[11]. The classical particle/wave distinction is used as illustration. A floating cork bobs up and down when a wave reaches it; thus we see that no water moves laterally, although the wave moves across the surface. If the individual water droplets, or better its molecules are entities persisting in time, the wave is merely a changing configuration of these entities. In Reichenbach, a particle has material identity (its temporal stages, or the events involving it, are genidentical with each other) and the wave does not. Hence questions of individuation or identification of waves (which may form superpositions) are either misplaced, or settlable by convention.

The suggestion may fit photons (which are in various ways atypical even among bosons) especially well. If bosons are "not genidentical", then instead of an assembly of N photons each persisting in time, we really have only at each moment N photon-stages (temporal slices), and there is no objectivity of any sort to the classification of one of these photon-stages at time t belonging to the same photon as one or other of the stages at time t + d. A photon-stage at a certain time is really no more than an event -- the being-occupied of a certain photon-state. All N events may have exactly the same character -- there are n being-occupied-s of the same photon state. If these events are not individuated by their historical connections to previous such events, nor in any other way, then we have a clear violation of PII.

Reichenbach did not draw this conclusion, but he gave an argument why we should regard bosons as not genidentical. If we think of bosons as entities persisting through time, then the correlations evident in

Bose-Einstein statistics mean that these particles tend to go into the same states. They exhibit this tendency even though there is no non-negligible interaction between them in the past. Hence this is correlation without common cause. Reichenbach had exactly the attitude toward interpretation as I espouse. Hence he writes[12]:

> We see now how the thesis concerning indistinguishable particles is to be qualified. In precise language we cannot simply say: The particles are indistinguishable. We must say: Either the particles are indistinguishable, or their behavior displays causal anomalies. We are left the choice of selecting the one or the other interpretation. Neither interpretation is "more true" than the other; the two are equivalent descriptions. However, only one of the two descriptions supplies a normal system, that is, a system free from causal anomalies; this is the description according to which the particles are indistinguishable. When we follow the usual rule of employing a normal system whenever it is possible, we may therefore say, without hesitation, that the particles are indistinguishable.

But Reichenbach was writing a decade or so before Bell's Inequalities were published, and well before the common cause principle had been clearly seen to contradict the empirical predictions of quantum mechanics. No normal interpretation, in his sense, is possible anyway.

In conclusion, I would maintain only the logical tenability of PII. Individuation of bosons, if it is insisted on, is by characteristics not describable in QM terms, as well as empirically superfluous. That is indeed one possible option for interpretation. The intelligibility an interpretation brings derives from the way it leads us in detail from one aspect of the story to another, always guided by the same thread -- that is all; but it is still worth making explicit.

ACKNOWLEDGEMENT. This paper summarizes parts of [1], which built on the author's papers in [2] (see also [13]); research support by the National Science Foundation is gratefully acknowledged.

REFERENCES

1. B. C. van Fraassen, Quantum Mechanics: An Empiricist View, (Oxford University Press, Oxford, 1991).
2. E. Beltrametti and B. van Fraassen, (eds.), Current Issues in Quantum Logic, (Plenum, New York, 1981.
3. J. von Neumann, Mathematical Foundations of Quantum Mechanics, (Princeton University Press, Princeton, 1955), tr. from the German (1932).
4. J. A. Wheeler, Reviews of Modern Physics 29 (1957),463-465.
5. J. M. Jauch, Foundations of Quantum Mechanics, (Addison-Wesley, New York, 1968), 173.
6. H. Weyl, Gesammelte Abhandlungen, (Springer-Verlag, New York, 1968), 268.
7. W. Pauli, Zeitschrift fuer Physik 31 (1925), 965.

8. H. Margenau, Philosophy of Science 11 (1944), 187-208; The Nature of Physical Reality, (McGraw-Hill, New York, 1950).
9. P. Mittelstaedt, in P. Mittelstaedt and E. W. Stachow, eds. Recent Developments in Quantum Logic (Bibliographisches Instituet, 1985).
10. M. L. Dalla Chiara and G. Toraldo di Francia, Versus 40 (1985), 29-50.
11. H. Reichenbach, The Philosophy of Space and Time, (Dover, New York, 1957), tr. from the German (1928), section 43; The Direction of Time, (University of California Press, 1956), section 26.
12. Reichenbach (1956), 234-234. 13. B. C. van Fraassen, in R. Colodny, ed. Paradigms and Paradoxes: the Challenge of the Quantum Domain, (University of Pittsburgh Press, Pittsburgh, 1972); in C. A. Hooker, ed. Contemporary Research in the Foundations and Philosophy of Quantum Theory, (Reidel, Dordrecht, 1973); Synthese 29 (1974), 291-309.

QUANTUM THEORY AS A THEORY OF HUMAN KNOWLEDGE

C. F. v. WEIZSÄCKER, TH. GÖRNITZ

Maximilianstr. 15, D-8000 Starnberg, Germany

1. The Question

For a mathematical educated reader, the basic assumptions of quantum theory can be expressed in four theses, on less than a half page of print. We propose these four:

1. The possible states of an object can be represented by vectors or linear subspaces of a Hilbert space.
2. Let x and y be two normalized vectors in a Hilbert space H, then the conditional probability of finding y in an experiment when x is known to be present, is $|\langle x|y\rangle|^2$.
3. Let H_1 and H_2 be the Hilbert spaces of two co-existing objects X_1 and X_2, then the Hilbert space of the total object is the tensor product $H_1 \otimes H_2$.
4. The change of states in time is such that their conditional probabilities stay invariant.

Quantum theory is pragmatically fully accepted by the scientific community; no convincing experiment contradicting it has ever been found. Yet, since more than sixty years, the debate about its interpretation has continued. What is the reason?

This is our question.

We give the tentative answer first in a parable, then directly.

The parable: Columbus knew that the earth is a sphere. He went to the West, to find a way to India. He discovered a new continent which then was conquered by the Europeans. Modern physicists, on the other hand, have discovered a new continent and conquered it. But they have not yet understood, how there could be a new continent there. They have still to learn that possible knowledge is a sphere.

The direct answer: Quantum theory is science, hence it is a piece of human knowledge. But it is also a theory <u>of</u> or <u>on</u> human knowledge. You can reach it from both sides, speaking on objects or on knowledge. This is why it has puzzled traditional physicists and even some philosophers of science.

What does this answer mean in detail?

Quantum theory is a theory on the states of objects in nature and their change with time. But in thesis 2 it is a theory on probabilities for the outcome of experiments on these objects and their states. An experiment is a human action, and a probability expresses limited human knowledge on its outcome. Therefore the debate on the meaning of quantum theory is mainly a debate on the theory of the measuring process.

Let me condense the debate into a fictious dialogue between Mr. Copenhagen (called **C.**) and Mr. Anti-Copenhagen (called **A.**).

C. says: "The Copenhagen interpretation (C.I.) is the attempt to formulate the pragmatically sufficient minimal semantics of quantum theory. We can condense it into the 'Golden Copenhagen Rule' (G.C.R.): 'No contradiction will arise if we accept that the probability for a result of a measurement is given by the absolute square of the inner product of the respective vectors.' But this means explicitly to accept that quantum theory contains statements on human knowledge. To say it in the parable: It means to approach quantum theory from both sides, from the known object as well as from the knowing subject. This was Bohr's philosophy."

A. replies: "But what do you mean by 'knowledge'? Knowledge means to know a reality; else it is not knowledge but fantasy. As Einstein said: to accept the probabilities as final, means to accept an incomplete picture of reality. In the theory of measurement, this is seen by the fact, that C.I. admits two completely different changes of the wave function with time:

1. According to the Schrödinger equation between the observations

2. by the collapse of the wave function in that act of observation."

C. interjects: "This is no paradox. According to the G. C. R., the wave function is nothing but the means to calculate probabilities. When no observation is made, it changes, as classical probabilities do, too, continuously in time; the probability of rain changes with the motion of the atmosphere. But probabilities are also changed by observation, else you would not be able to confirm an observation by repeating it."

A. continues: "True, you are not contradicting yourself, but only because you are too modest. Tell me: In which moment of time is the wave reduced? Is it in the moment when the observer becomes aware of the measuring result? Will you admit such an interference of consciousness with material events?"

C. answers: "This is why Bohr insisted that the measuring instrument must be classically described. In the classical approximation it does no harm to assume that the wave function was changed when an irreversible event of interaction happened in the instrument. Only under this condition can you speak of an objective result of the measurement, independent of the questions, whether it is later on entering into the mind of an observer, and who is the observer and when he observes."

A. makes an offer: "Then you will certainly be satisfied by papers offered in our symposium which say: Irreversibility is an approximation which becomes satisfactory for the measuring process in an infinite-dimensional Hilbert space of the instrument. In this approximation a quantum-mechanical pure state is transformed into a density matrix with classical probabilities."

C. replies: "These studies are very useful to describing irreversibility. But they do not at all eliminate the so-called collapse of the wave. Let the initial quantum probability in a yes-no-alternative, say in the measurement of a spin-1/2 component, be 50% for each of the two possible outcomes. Then it will still be 50% in the resulting density matrix. But the final observer finds one of the possible results and not the other. This reduction must, in our present picture, have happened when the irreversible event took place in the instrument. Thus your objection against quantum theory as being incomplete is by no means weekend by these calculations."

A. says: "I made an offer, and you rejected it. You were right. But this means that quantum theory stays incomplete. Some quite different processes may happen beyond it. Tell me, why do you so firmly adhere to it?"

C. says: "Let me now tell you how we try to reconstruct quantum theory explicitly, accepting it is a theory of human knowledge. Then, if you still do not like it, please come with further objections."

2. Reconstruction of "abstract" quantum theory

2.1 Introductory remarks

First we want to make some remarks about the starting points of our attempt.

In our view a classical theory of systems with infinitely many degrees of freedom can not be made into a consistent theory in connection with thermodynamics. This is well known since the days of Planck, who was forced by this circumstance to invent the quantum idea. Therefore we do not believe in the attempts to "improve" quantum theory by an underlying classical-like framework. Instead we understand quantum theory as the most fundamental concept for all the parts of physics.

With respect to this concept our point of view is beyond

Copenhagen, possibly it could be called a **Neo-Copenhagen** one.

The people who assume a universal validity of quantum theory may be divided into two groups. They can be distinguished by their understanding of systems with an infinite number of degrees of freedom. To say it in other words - in their understanding of the connection between quantum theory and cosmology.

On one side it is assumed that quantum theory is relevant for the problems of micro- and mesophysics, but not for cosmological questions, and, that there are systems, e.g. electromagnetic fields, with really actually infinitely many degrees of freedom. If such a decision is seen as fundamental (and not only as a practical arrangement) then it is in itself a decision about a special cosmology, of course of a possible one.

We, on the other hand, take not for granted that position space is infinite. Therefore also for electromagnetic fields it is not beyond any doubt, that there are not actually infinitely many degrees of freedom for it. In a compact position space perhaps no physical system could have actually infinitely many degrees of freedom. From this point of view the methods of algebraic quantum theory are seen as some very powerful tools to classify and to handle the idealized limiting behaviour of large systems. But from this point of view as well we are not forced to throw away quantum mechanics in Hilbert space as to insufficient and to old-fashioned. So we have taken quantum theory in it is historically given form in Hilbert space for the short-hand description at the beginning of our talk.

It is clear, that from such a point of view in a principal sense objects do not exist. Then it remains to explain, why - for mankind in the daily praxis - there are objects and nothing else! And further on, not making the fundamental difference between endo- and exophysics it is also obvious that the problem of self-reference cannot be avoided. Of course we are unable to solve it in a rigorous manner. But it seems also impossible to avoid the problem just by cleverly circumventing it.

2.2. Reconstruction of "abstract" quantum theory

By "abstract" quantum theory we designate the general frame of quantum theory, without reference to a three-dimensional position space, to concepts like particles or fields, or to special laws of dynamics.

"Reconstruction" means the attempt to formulate simple and plausible postulates on prediction and to derive the basic concepts of abstract quantum theory from them. Our reconstruction differs from most traditional descriptions or interpretations of quantum theory in a central point. It does not presuppose any set of laws of "classical" physics which would then have to be "quantized". The early papers in this field have started from some concepts of quantum logic [1,2]. Now the inftention is to formulate - as far as it is possible - preconditions of experience and to see to which degree it is possible to deduce the structure of quantum theory from them.

The level of abstraction on which we propose to start is indicated by the "terms of physics" contained in our description of abstract quantum theory. "Object" (also called "system" in physics), "state" and "composition of objects" are terms of some "abstract ontology of physics"; in our own procedure we shall even replace "object" by the purely logical concept of "alternative", and refer "state" to the logical concept of "statement". "Time" is the only intuitive concept without which, in our view, even the most abstract set of laws cannot be physics. "Dynamics" means the laws under which states change in time. "Probability" we consider as the prediction of relative frequency, thus as an abstract concept referring to the future hence to time.

The following subchapter is based to a large extend on the paper "Reconstruction of abstract quantum theory" of Drieschner, v. W., G. [3], in which the assertions are arranged in four groups: Heuristic principles, verbal definitions, the basic postulates, consequences.

2.2.1. Heuristic principles

Preconditions of experience:

As far as possible our postulates ought to express conditions without which we cannot expect experience to be possible at all.

Simplicity:

Without precisely defining simplicity, we wish for simple postulates rather than complicated ones.

Innocuous generality.

General rules are usually simpler than specialized ones. We shall confine ourselves to general rules as far as they give the hope of being "innocuous"; e.g. claiming the general existence of a set of states while under special conditions (like a dynamics implying a super-selection-rule) some of those might not actually come into being.

2.2.2.Definitions.

Experience

Experience means to learn from the past for the future.

Facticity of the past.

We speak of past events as of objective facts, independently of our actually knowing them.

Possibility of the future.

We are aware of future events only as possibilities.

Probability.

Probability is a quantification of possibility. We define it as the prediction (mathematically: the expectation value) of a relative frequency.

Temporal statements

A temporal statement (briefly "statement") is a verbal proposition (or a mathematical proposition with a physical meaning) referring to a moment in time.

States

States are recognizable events. A state is what is the case when some temporal statement is true. States at different times can be identical: it is meaningful to ask whether we observe now the same states as at a certain time before.

Conditional probability

Let x and y to be two states. Then $p(x,y)$ is the probability that, if x is a present state, y will be found as the state if searched for.

Alternatives

An n-fold alternative is a set of n mutually exclusive states, exactly one of which will turn out to be present if and when an empirical test of this alternative is made.

Connection

Two states x and y are called connected if there is a law of nature determining their conditional probabilities $p(x,y)$ and $p(y,x)$. If the connection is transitive, i.e. if the existence (by law of nature) of probabilities $p(x,y)$ and $p(y,z)$ implies the existence of a $p(x,z)$, then connection is an equivalence relation, defining a partition of the class of all states into subclasses of mutually connected states.

Separability

Two states are called separable if they are not connected. (This concept of separability should not be confused with the mathematical concept of separability of e.g. a Hilbert space.)

2.2.3. Postulates.

The wholeness of reality

Any part of reality is interacting with the other parts. From the existing quantum theory, which we want to reconstruct, we already know, that interacting objects constitute a total object, which almost never can be divided into its parts conceptually. Therefore in the strong sense of the theory we have a "zeroth" postulate:

Reality is a wholeness (in German: Ganzheit)

Separable alternatives.

In a very good approximation with only few exceptions we can divide reality into different parts and the parts of reality and not the whole of it are the entities which can only be caught by conceptual thinking.

Therefore as the first postulate we have:

There are empirically decidable finite alternatives whose states are separable from nearly all other states. If an n-fold alternative will be empirically decided, exactly one state will be found as being present and n-1 mutually exclusive other states as being not present.

"Nearly" will be defined as meaning all states not connected with the states of the alternative by the postulate of indeterminism. Infinite alternatives are useful idealizations the set of alternatives which are finite but without an upper bound.

Kinematics.

As second postulate we have:

The possible futuric states of alternatives change continuously in time. Thereby the conditional probabilities between connected states are not altered:

$$p[(x,t),(z,t)] = p[(x,0),(z,0)].$$

This is also called the "Darwinism of states". Hypothetical states not fulfilling this postulate are hardly able to be observable.

Indeterminism.

To correct and connect the two foregoing postulates we need a third one. This has to bridge the discontinuity of the discrete alternatives with the continuous time development.

Let x and y be two connected mutually exclusive states {p(x,y)=p(y,x)=0}, then one can find states z according to postulate one, different from x and y but not constructed from x and y by mere logical operations and which possess conditional probabilities p(z,x) and p(z,y) none of which are equal to zero or to one. It is postulated that any state z can be produced by an adequate kinematical law from the x's or y's and vice versa.

2.2.4. Consequences.

In our common work with Drieschner [3] we have derived as consequences a space of mutually equivalent states. This is a representation space for a symmetry group G(n) which preserves the probabilities between the states and further on a probability metric on this space. A possible dynamics will be a represented by a one-parameter subgroup D(t) of G(n). We wont to get observable states for this dynamics, therefore the one-parameter subgroup D(t) has to be diagonalizable. Then the state space must be a unitary one, i.e. a complex Hilbert space.

2.3. Concrete quantum theory

In our reconstruction we only have used time from the outset, the other concepts like space, particle, field and so on have to be derived. All these concepts, including the theories referring to them, like relativity and particle theory, should be developed as a <u>consequence</u> of abstract quantum theory. This is essentially to be done by adding to abstract quantum theory one single (and simple) idea: the reduction of all alternatives to the successive decision of <u>binary alternatives</u> (yes-no decisions, bits or "urs"). Some work in this field is already done (see e. g. [4] - [12]), much more is yet to do.

3. Remaining questions

The dialogue will now be resumed by <u>A</u>. and <u>C</u>., but a third partner, Mr. Between-and-Beyond (called <u>B</u>.), will participate.

<u>A</u>. begins: "My position is now stronger than before. I have seen that on your premise that quantum theory is a theory of human knowledge your friend G. and you accept quantum theory as it stands. That includes the Schrödinger equation. You must then apply it to the events in the measuring instrument, too. Then you are not permitted to accept a collapse of the wave function even in the approximately irreversible interaction."

<u>C</u>. answers: "But observations are facts. On your argument you will have to abandon quantum theory."

<u>A</u>. counters: "No. I shall deny your hypothesis on human knowledge as the final interpretation of quantum theory."

<u>C</u>. asks: "But which alternative interpretation you will offer?"

<u>A</u>. replies: "For instance, Everett's many-world theory: We should never reduce the wave function. But there is an ever increasing number of worlds. At every irreversible process like a measurement, the known world splits into as many worlds as there are possible results of the measure-

ment. You live in one of these worlds, not knowing that you also continuing do live in all other worlds which follow from other results than the one you have observed in this world of which you are aware."

C. counters: "This is a statement which cannot possibly be tested by experience. Do you call that physics?"

A. says: "It is physics as soon you see that it is the only logically possible interpretation of quantum theory."

C. says: "I offer an other one. I have a dictionary, translating Everett's language into mine by changing one single word."

A.: "Which word?"

C.: "The word 'world'. Replace 'many worlds' by 'many futuric possibilities'. The wave function is never reduced as far as it reefers to possible events in the future. It is reduced when possibilities become present and turn into facts."

A. says: "But then you stay with the impossibility to apply the Schrödinger equation to real, observable events. What do you about that?"

C. admits: "I cannot answer your question. There are limits to our knowledge."

B. enters the scene, saying: "Permit me to be between and beyond your two positions! Let me first ask a question. You described two material objects. Let me call the measuring object X_1, and the measuring instrument X_2. You seemed to agree that both of them, and hence the total object $X_1 \otimes X_2$, must obey quantum theory."

A. and C.: "We agree, at least tentatively."

B. continues: "Now we have a living observer. Let us call him Y. He observes the instrument X_2 and draws correct conclusions on X_1. Do you describe him by quantum theory, too?"

C.: "Bohr in his famous paper 'Light and life' [13], did not believe that quantum theory would be sufficient to explain the phenomenon of life."

A.: "But most biologist's today believe it. Max Delbrück, who started from Bohr's ideas, finally believed it, too."

B. resumes: "Let me accept the biologist's view, at least for arguments sake."

A.: "But of course only the body of the observer will be described by quantum theory. What about his mind?"

B.: "According to the reconstruction which we have heard, I see no difficulty in applying its probabilistic predictions also to empirically decidable alternatives on the observer's state of mind."

C.: "How do you observe the observer?"

B.: "We may introduce 'Wigner's friend', i.e. a second observer, called Z, who talks to Y and asks him about his knowledge. This description is methodologically easier to handle. In principle, Y can predict himself what he will probably know at a later moment, and then test it in the predicted moment of time by introspection."

A.: "But then the collapse of the wave function is only transferred in the observation of mental states. It is not eliminated."

B.: "Now, finally, we have reached the remaining question. In the Copenhagen Interpretation, new knowledge necessarily means a new wave function. If you want not to reduce the wave function - and I sympathize with this wish -, you must go beyond Copenhagen, perhaps to Trieste."

Here one Author (vW) interferes. Mr. **B**. here has referred to my remarks made in Trieste [6], better described in my book [7] Aufbau der Physik, p. 606-612.

B. continues: "There is a problem of description and a problem of philosophy. I beginn with the description.

The 3. thesis in the introduction of this paper is the law of composition. In the tensor product of the Hilbert spaces of two objects X_1 and X_2, the states which are products of well-defined states of the two single objects are a set of measure zero. Thus, strictly speaking, there are no separate objects in the world."

A. interjects: "And, if you describe minds by quantum theory, no separate minds, either!"

B. replies: "You jump into the philosophical question!"

C. interferes: "How do you define philosophy as distinct from science?"

B.: "Normal science, as Th. Kuhn describes it, is puzzel-solving under a successful paradigm, that is in a tradition of pragmatic success. Philosophy is the attempt to understand what we do in this success, and to express this understanding in words. But philosophy is just a little bit beyond human power. We try our best."

A.: "And what is your philosophy on mind and matter?"

B.: "You know the pun:

'What is mind? No matter. What is matter? Never mind.'

But my philosophy is the opposite. Quantum theory does not need any distinction of mind and matter. In evolution, the conscious mind slowly grows out of the immense sea of unconscious or half-conscious behaviour. Why should not my brain just be the form in which my mind appears to an outward observer?"

A.: "A nice fairy-tale. I accept it for argument's sake. But what then is your philosophy of measurement? How do you describe the event in which the wave function is reduced?"

B.: "Sokrates, and his re-incarnation Niels Bohr, taught us the philosophy of questioning the meaning of our basic concepts. I shall question the meaning of the term 'event'. In quantum theory we have the nice expression of a 'virtual event' or 'virtual state'. A particle of given momentum is virtually in all position states which have non-vanishing probability of being fund when we measure position. New actual facts are produced only by irreversible processes, and irreversibility is always only an approximation. Thus a fact is, strictly speaking, though with very small probability - and always virtually - not this fact."

C.: "And the fact, that I know something?"

B.: "Bohr sometimes quoted William James: Consciousness is an unconscious act.' If you have read Freud, you know that you never know precisely what is actually in your mind. A psychologist once said to me:

When I ask you: What are you actually thinking? and you give an answer, you are already lying.'

A. with emotion: "And where is reality, then? Einstein said, he believed in the God of Spinoza. Would not an all-embracing intellect know all things, all events?"

B.: "Spinozas God is the God of Greek metaphysics. This is a wonderful human idea on the most profound truth. But in the Platonic tradition, following Plato's dialogue 'Parmenides', so-called negative theology was developed, saying that whatever predicate of human thinking you ascribe to God will make a mistake. The Form of the Good is even beyond of Being, says Sokrates in Plato's Republik, and his partner Glaucon bursts into laughter, saying 'What an exceeding Goodness!'"

C.: "Why do you quote this philosophy?"

B.: "Return to our interpretation problem! Do we know what we mean by a separate event? In truth it is part of a far larger process. My nerves and my little mental emotions are necessary for my consciousness, but neither the nerve alone nor the emotion alone knows in conscious reflection to what consciousness it belongs. If my consciousness is just part of a larger Spirit, as Spinoza and Plato believed, can I know what my awareness of events means for the all-knowing God? Does God believe in events?"

A. and C.: "And so you intend to resolve the puzzle of the reduction of the wave function?"

B. makes an end of the debate by saying: "Friend C., you said that quantum theory is a theory of human knowledge. I fully agree. You will never run into trouble, if you stick to that modest wisdom. Friend A., you want to know more. I fully sympathize. But we have so far seen that you got no defensible answer if you search for it within the concepts which are made to describe separable objects. Your question seems to transcend what we have so far learnt to call physics'. Is this not an inspiring answer for future meditation?

Sokrates used to end his dialogues, saying to his partner: I confessed from the outset that I did not knew the answer to our common question. Now you have to admit that you do not know it either. Too bad! But to-morrow we shall continue our research!'"

Acknowledgement

One of us (Th. G.) is grateful to the Stifterverband der Wissenschaft for financial support for part of this work.

References

1. Weizsäcker, C. F. v.: Komplementarität und Logik I and II, *Naturwiss.* **42** (1955) 521-529 and 545-555
2. Drieschner, M.: *Quantum Mechanics as a General Theory of Objective Prediction* (in German), Thesis, Hamburg (1970)
3. Drieschner, M., Görnitz, Th., Weizsäcker, C. F. v.: Reconstruction of Abstract Quantum Theory, *Intern. Journ. Theoret. Phys.* **27** (1987) 289 - 306
4. Scheibe, E., Süssmann, G., Weizsäcker, C. F. v.: Mehrfache Quantelung, Komplementarität und Logik III, *Z. Naturforschung* **13a** (1958) 705
5. Weizsäcker, C. F. v.: *Die Einheit der Natur*, Hanser, München (1971), (engl. Ed.: *The Unity of Nature*, Farrar, Straus, Giroux, New York 1980)
6. Weizsäcker, C. F. v.: Classical and Quantum Descriptions, in: *The Physicist's Conception of Nature*, J. Mehra, ed. Reidel, Dordrecht (1973)
7. Weizsäcker, C. F. v.: *Aufbau der Physik*, Hanser, München (1985)
8. Weizsäcker, C. F. v.: Quantum Theory and space-time, in P. Lathi and P. Mittelstaedt (Eds.): *Symp. on the foundations of modern physics*, Joensuu ; World Scientific (1985)
9. Görnitz, Th., Weizsäcker, C. F. v.: De-Sitter Representations and the Particle Concept in an Ur-Theoretical Cosmological Model, in Barut, A. O., Doebner, H.-D. (Eds.): *Conformal Groups and Related Symmetries, Physical Results and Mathematical Background*, Lect. Notes in Physics 261, Springer Verl. Berlin ect. (1986)
10. Görnitz, Th.: Abstract Quantum Theory and Space-Time-Structure, Part I: Ur-Theory, Space Time Continuum and Bekenstein-Hawking-Entropy, *Intern. Journ. Theoret. Phys.* **27** (1988) 527 - 542
11. Görnitz, Th.: On Connections between Abstract Quantum Theory and Space-Time-Structure, Part II: A Model of cosmological evolution, *Intern. Journ. Theoret. Phys.* **27** (1988) 659 - 666
12. Görnitz, Th., Ruhnau, E.: Connections between Abstract Quantum Theory and Space-Time-Structure, Part III: Vacuum Structure and Black Holes, *Intern. Journ. Theoret. Phys.* **28** (1989) 651 - 657
13. Bohr, N.: Licht und Leben, *Naturwiss.* **21** (1933) 245-250

LOCAL VALUES OF QUANTUM OBSERVABLES AND LOCAL POSITION MEASUREMENTS

K KONG WAN
Department of Physics & Astronomy, University of St. Andrews, St. Andrews, Fife KY16 9SS, Scotland, UK.

ABSTRACT

There have been many schemes to assign numerical values to quantum observables at each point in the phase space ever since Wigner started doing it with the help of what is now known as the Wigner distribution function. Bohm's quantum potential theory further points to the possibility and usefulness of assigning numerical values to quantum observables at each point in the spatial space. However, these schemes to produce local values are applicable only to a limited number of quantum observables. We shall present a general scheme here together with an analysis of the meaning of such local values and their implications on quantum measurement.

1. Motivation

Classical mechanics is a local theory; everything is fixed right down to the neighbourhood of a point in the spatial space $M=\{x\}$ or in the phase space $\Gamma_c=\{p,x\}$. Quantum mechanics is nonlocal. Intuitively we can see that firstly states as wave functions are spread out in M, secondly observables as operators are of a global nature, ie, generally they do not vanish outside any bounded region in M, and thirdly expectation values are integrals over M and are not associated with any spatial region.

The question then arises as to whether it is possible to localise quantum mechanics in some sense for a variety of reasons, eg, mathematical curiosity, philosophical belief in things being local or physical reasoning that measuring devices are all of finite size and therefore should not be able to measure global quantities. We can try to localise any one of the global quantities mentioned:

(1) *Localisation of states.*
The most topical scheme is that of Ghirardi, Rimini and Weber[1];

(2) *Localisation of Observables.*
Local observables are well-established in the algebraic approach to quantum field theory[2,3]. We have introduced local observables into quantum mechanics[4,5]. Orthodox quantum mechanical observables are represented by operators of a

global nature, eg the momentum $\hat{p} = -i\hbar\,\partial/\partial x$ and the position x, which do not vanish outside any bounded spatial region. In sharp contrast, physical devices used to measure these global quantities are invariably of finite size. To reconcile this we have introduced operators of a local nature to correspond to finite measuring devices. For example a Geiger counter can ascertain whether a particle is in a certain region J or not. The counter carries out a yes-no experiment; it measures the proposition "the measured value of position x is contained in J" [6]. The proposition is represented by the position spectral projector E(x;J) which, in a less fancy language, can be identified with the characteristic function $\chi_J(x)$ of J [6]. Let us call this proposition a local position observable; $\chi_J(x)$ is clearly a local operator. Many results and analyses relating to this endeavour are reviewed in a recent paper by Wan[7]. The use of local observables can lead to a theory of quantum mechanics which is asymptotically separable.

(3) *Localisation of Expectation Values.*
In quantum theory it is the expectation values which are directly related to the physical world; it is these expectation values, not any particular operator representations of observables, which are obtained in a measurement. So, if quantum mechanics has any "local" meaning it should show up in expectation values. There are two conventional approaches to produce a localisation of expectation values:

(i) *Classical Phase Space Approach.*
This approach originated by Wigner[8] is an attempt to express quantum expectation values as the average of a classical observable $\mathcal{A}$ under a distribution function $P_c(p,x)$ on the classical phase space Γ_c. A classical observable $\mathcal{A}$ is a function defined on the phase space Γ_c, ie, $\mathcal{A} = \mathcal{A}(p,x)$. Let A be the quantum counterpart to $\mathcal{A}$ and ϕ be the state of the quantised system. The idea then is to express the quantum expectation value $\langle \phi | A \phi \rangle$ as an integral over Γ_c:

$$\langle\phi|A\,\phi\rangle = \iint \mathcal{A}(p,x)\,P_c(p,x)\,dp\,dx .$$

There is a large choice of the distribution functions corresponding to different quantisation rules to produce a quantised observable A from $\mathcal{A}(p,x)$ [9,10,11].

(ii) *Spatial Space Approach.*
In his quantum potential theory Bohm[12] showed that it was possible to assign a momentum value at each point in M.

It is the problem of localisation of expectation values we shall examine in this paper. Let us begin by explaining why the existing approaches are incomplete as well as problematical[13]. These schemes to produce local values of quantum observables are mathematically rather limited, capable of assigning local values to a limited number of quantum observables only, namely those which can be constructed from $\hat{p}$ and x in some well-defined manner. Wigner's approach depends on the existence of a proper quantisation rule to relate the classical observable $\mathcal{A}(p,x)$ to its quantum counterpart A. But there are no generally accepted quantisation rule at the moment capable of quantising even very simple classical observables[14,15,16]. Take for example $\mathcal{A}(p,x) = x^n p$. The operator

$$\bar{A} = -i\hbar\,(x^n\, d/dx + \tfrac{1}{2} n\, x^{n-1})$$

is obtained by the three most often used quantisation rules[16], ie the Weyl rule, the symmetrisation rule and the Born-Jordan rule [9,10]. A close examination reveals that for an odd value of n bigger than one the operator $\bar{A}$ is not selfadjoint and has no selfadjoint extention[16]. It follows that $\bar{A}$ cannot serve as a quantum observable. On the other hand there are quantum observables like the parity operator which have no classical counterpart; Wigner's scheme simply does not apply here. Moreover there are other mathematical difficulties in trying to assign values to points in Γ_c which will become apparent later. All these demonstrate the lack of generality of Wigner's approach. Similar remarks can also be said about Bohm's scheme[13,16]. The problem is not just a mathematical one. Wigner's distribution function is not positive definite; this gives rise to a severe difficulty of its interpretation unless a notion of "negative probability" is brought in[11].

A general approach should be based on quantum observables right from the start rather than on classical observables. It would then be able to deal with all quantum observables. A scheme based on such a general approach does exist capable of assigning local values to all quantum observables in the spatial space[13,16,17,18] which can be extended to generalised phase space[16]. Furthermore these local values themselves have a well-defined quantum mechanical meaning. Let us spell out the problems first.

2. The Problems and Their Chosen Solutions

For simplicity confine our attention to a one-dimensional quantum system. The Hilbert space is $L^2(\mathbb{R})$ and observables are represented by selfadjoint operators A in $L^2(\mathbb{R})$. Given a normalised wave function ϕ in the domain of selfadjoint operator A we want to solve the following problems:

Problem 1: Is there a real-valued function $f(A, \phi; x)$ on the spatial space $\mathbb{R}$ such that

$$<\phi| A \phi> = \int_{\mathbb{R}} f(A, \phi; x)\, dx \quad ?$$

Problem 2: Let S denote the spectrum of A. The generalised phase space is defined to be the space $\Gamma_A = S \times \mathbb{R}$. In the case of A being the momentum observable Γ_A is simply the conventional phase space in Wigner's theory. Is there a real-valued function on Γ_A, $W_\phi{}^A(\lambda,x)$, such that

$$<\phi| A \phi> = \int_{\mathbb{R}}\int_{S} W_\phi^A (\lambda, x)\, d\lambda dx \quad ?$$

Since a wave function is defined only up to a set of Lebesgue measure zero the above problems are not mathematically well-posed. To get round this mathematical difficulty we can rephrase our problems as follows:

Problem 1: Let $\{J_i\}$ be a partition of the spatial space $\mathbb{R}$ in terms of nonvanishing and bounded intervals J_i. Is there a sequence of real numbers $f(A, \phi; J_i)$, one for each interval J_i, such that

$$<\phi| A \phi> = \sum_i f(A, \phi; J_i) \quad ?$$

Problem 2: Let $\{\Lambda_j \times J_i\}$ be a similar partition of the phase space Γ_A. Is there a sequence of real numbers $W_\phi{}^A(\Lambda_j \times J_i)$, one for each bounded phase space region $\Lambda_j \times J_i$, such that

$$<\phi|A\,\phi> = \sum_i \sum_j W_\phi^A(\Lambda_j \times J_i) \quad ?$$

As it turns out there are many solutions to both above problems. One pair of solutions is as follows[13,16]:

$$f(A, \phi; J_i) = \mathrm{Re} \int_{J_i} \phi^*(x)\,(A\,\phi)(x)\,dx\,.$$

$$W_\phi^A(\Lambda_j \times J_i) = \mathrm{Re} \int_{J_i} \Big\{ \Big(\int_{\Lambda_j} \lambda \; dE(A;\lambda)\,\phi \Big)^*(x) \Big\} \phi(x)\,dx.$$

Here Re stands for the real part and $E(A;\lambda)$ is the spectral function of the operator A. Note that in the case when A possesses a discrete set of eigenvalues in the region Λ_j the integration over Λ_j reduces to a sum over the set of eigenvalues.

Now, it is natural to advance the interpretation that

$$f(A, \phi; J_i) \quad \text{and} \quad W_\phi{}^A(\Lambda_j \times J_i)$$

are respectively the local values of A in state ϕ in the spatial region J_i and in the generalised phase space region $\Lambda_j \times J_i$ in that the expectation value $<\phi|\,A\,\phi>$ is a sum of these local values over the entire spatial space or phase space.

3. Distinctive Features

The present scheme on local values in phase space has two distinctive features:

(1) The present theory does not require a probability distribution function. In contrast to Wigner's approach we do not seperate a local value into a product of a "value" and a "probability". So, the difficulty of negative probability encountered by Wigner's approach does not arise in our scheme. A negative local value presents no conceptual difficulty.

(2) Another feature of the present approach is that the local values themselves have a well defined quantum mechanical meaning.

Let us now examine in some detail how a local value could have a quantum mechanical meaning. Consider the value $W_\phi{}^A(\Lambda \times J)$ in region $\Lambda \times J$ in Γ_A. This turns out to relate to a restriction of A spatially to J and spectrally to Λ in the following sense. Let $\chi_\Lambda(A)$ be the characteristic function (of the operator A) of the set Λ; $\chi_\Lambda(A)$ is equal to the spectral projector $E(A;\Lambda)$ of A [19]. Introduce the operator $A_{\Lambda \times J}$ by

$$A_{\Lambda \times J} = \tfrac{1}{2}\,\{\,\chi_\Lambda(A)\;A\,\chi_J(x) + \chi_J(x)\,A\,\chi_\Lambda(A)\}.$$

Clearly the operator $A_{\Lambda\times J}$ is selfadjoint. It is easily seen that

$$W_\phi{}^A(\Lambda\times J) = \langle\phi|\, A_{\Lambda\times J}\,\phi\rangle.$$

So, the local value of A in region $\Lambda\times J$ in Γ_A is the quantum expectation value of the quantum observable represented by the operator $A_{\Lambda\times J}$.

Similar remarks can be made on spatial values. For a bounded observable A the operator

$$A_J = A_{\mathbb{R}\times J} = \tfrac{1}{2}\,\{A\,\chi_J(x) + \chi_J(x)\,A\}$$

is selfadjoint, and one can show easily that

$$f(A, \phi; J) = \langle\phi|\, A_J\,\phi\rangle.$$

So, the local value in J is the expectation value of the quantum observable represented by the operator A_J which may be regarded as a restriction of A to the spatial region J. Note that the value f(A, φ; J) is zero in region where the wave function is zero, ie no local value in region where the particle is never present. For the momentum observable our values agree with Bohm's values [11,13]. As with Bohm's theory it is possible to attach a hidden variables interpretation to the values f(A, φ; J) except that the values f(A, φ; J) in the present theory themselves are expectation values and are therefore not "hidden". A more detailed discussion can be found in reference 16.

4. Local Values and Local Position Measurements

Since every selfadjoint operator is decomposable in terms of projectors through the spectral theorem and the notion of a proposition-valued measure[6] we can confine our analysis on projectors P. The local value f(P, φ; J) of P in state φ in the region J is

$$f(P,\phi;J) = \langle\phi|\, P_J\,\phi\rangle,$$

where

$$P_J = \tfrac{1}{2}\,\{P\,\chi_J(x) + \chi_J(x)\,P\}$$

is clearly selfadjoint, and therefore represents an observable. We can go a step further to express the local value f(P,φ;J) in terms of expectation values of the local position observable $\chi_J(x)$ by

$$f(P,\phi;J) = \tfrac{1}{2}\{\langle\phi\,|\,\chi_J\,\phi\rangle + \langle P\phi\,|\,\chi_J\,P\phi\rangle - \langle P_\perp\phi\,|\,\chi_J\,P_\perp\phi\rangle\},$$

where $P_\perp$ is the orthogonal complement of P.

The purpose of a measurement is to obtain an expectation value. Since the expectation value of P, which equals the sum of local values, is now seen to be obtainable from expectation values of local position observables we are led to a very interesting, albeit familiar, question as to whether all quantum mechanical measurements are executed in the end by local position measurements. For example in a Stern-Gerlach type experiment we do perform local position measurements in the end[20]. While the present results may be open to many other interpretations, for example a hidden variables interpretation mentioned earlier, they do lend support to the contention that all measurements are executed in the end by local position measurements. Note that this does not mean that quantum mechanics is a local theory. To execute local position measurements at the end we need a process of filtering[21,22] to produce $P\phi$ and $P_\perp\phi$; this filtering process generally involves nonlocal interactions. A more formal analysis of the role played by local position measurements in the general context of quantum measurement theory will be discussed elsewhere[22].

5. Concluding Remarks

Finally we remark that the theory presented so far is capable of further generalisations including extentions to many-particle systems; relevant technical details are available in reference 16. We should also mention that progress is being made to extend the present theory to quantum field theory ; the results will be discussed elsewhere.

References

1. G Ghirardi, A Rimini and T Weber, *Nuovo Cim* **B102** (1988) 383.
2. R Haag and D Kastler, *J Math Phys* **7** (1964) 848.
3. N Bogolubov, A Logunov and I Todorov, *Introduction to Axiomatic Quantum Field Theory,* (W.A.Benjamin, Reading Mass.,1975) , Part Six.
   G Sewell, *Quantum Theory of Collective Phenomena*, (Clarendon Press, Oxford, 1986).
4. K Wan, T Jackson and I McKenna, *Nouvo Cim*, **B8** (1984) 165.
5. K Wan and T Jackson, *Phys Lett*, **A106** (1984) 219.
6. E Beltrametti and G Cassinelli, *The Logic of Quantum Mechanics*, (Addison-Wesley, Reading Mass.,1981),p38, p141, p147, p150.
7. K Wan, *Found Phys,* **18** (1988) 887.
8. E Wigner, *Phys Rev*, **40** (1932) 749.
9. N Cohen, *J Math Phys* **7** (1966) 781.
10. M Hillery, R O'Connell, M Scully, E Wigner, *Physics Reports* **106** (1984) 121.
11. W Muckenheim, *Physics Reports* **133** (1986) 337.
12. D Bohm, *Phys Rev* **85** (1952) 166.
13. K Wan and P Sumner, *Phys Lett* , **A128** (1988) 458.
14. K Wan and C Viazminsky, *Prog Theor Phys*, **58** (1977) 1030.
15. N Woodhouse, *Qeometric Quantization,* (Clarendon Press, Oxford,1986).
16. K Wan and P Sumner, preprint (1990) *Spatial and Phase Space Distributions of Observable Values in Quantum Mechanics and Local Quantization.*
17. L Cohen, *J Chem Phys*, **70** (1979) 788.
18. R Bader and H Essen, in *Local Density Approximations in Quantum Chemistry and Solid Sate Physics*, ed. J Dahl and A Avery, (Plenum, New York, 1984).
19. E Prugovecki, *Quantum Mechanics in Hilbert Space,* 2nd edition, (Academic Press, New York, 1981), p277.
20. K Gottfried, *Quantum Mechanics,* (W A Benjamin, New York, 1966), pp173-190.
21. R Feynmen, *Lectures on Physics Vol 3,* (Addison-Wesley, Reading Mass., 1965), Chapt 5.
22. K Wan and D McLean, preprint (1990) *Are All Quantum Measurements Reducible to Local Position Measurements?*

TOWARDS A DERIVATION OF A NON-LINEAR STOCHASTIC SCHRÖDINGER EQUATION FOR THE MEASUREMENT PROCESS FROM ALGEBRAIC QUANTUM MECHANICS

WOLFGANG ZAORAL

Laboratory of Physical Chemistry, ETH
ETH-Zentrum, CH-8092 Zürich, Switzerland

ABSTRACT

A non-phenomenological and individual description of the measurement process is given in the framework of algebraic quantum mechanics. The reduced dynamics of the measured system is governed by a non-linear, stochastic equation. The measurement problem may be regarded as solved, if non-linear, stochastic equations of motion are derived from first principles of quantum mechanics, and if the final pointer states are disjoint. A heuristic derivation of a non-linear Schrödinger equation, which seems to be a good candidate for an example of a model for the measurement process, is presented and discussed.

1. Outline of an Algebraic Description of the Measurement Process

In algebraic quantum mechanics, the starting point for an individual description is given by a C*-algebra $\mathscr{A}$ of intrinsic observables. The center of this algebra is trivial, there are no intrinsic classical observables. When an experiment is performed, a context is implicitly chosen, this means in algebraic language that one considers the W*-algebra $\mathscr{M}_\pi$ of contextual observables, which is obtained by choosing a suitable representation π of the algebra $\mathscr{A}$ and closing the represented C*-algebra $\pi(\mathscr{A})$ with respect to the weak topology. The algebra $\mathscr{M}_\pi$ of contextual observables must have a non-trivial center in order to allow an individual interpretation. In simplest examples, the von Neumann algebra $\mathscr{M}_\pi$ is decomposable into a W*- tensor product $\mathscr{M}_\pi = \mathscr{M}_d \overline{\otimes} \mathscr{M}_e$, where $\mathscr{M}_d$ is purely non-commutative, representing the *dressed object* and $\mathscr{M}_e$ is purely commutative, containing the pointer observables. The elements of the center are called classical observables and they have dispersionfree values for any pure state of the algebra.

In a two level approximation the dressed quantal object is described by a spin $\frac{1}{2}$ system, the observables of which are given by the complex 2×2 matrices, i.e. $\mathscr{M}_d = \mathscr{B}(\mathbb{C}^2)$. As an atomfree commutative W*-algebra is isomorphic to the algebra of essentially bounded, measurable functions over a classical phase space Ω,

$\mathcal{M}_e \cong L^\infty(\Omega)$, we may assume that $\mathcal{M}_\pi = \mathcal{B}(\mathbb{C}^2)\bar{\otimes}L^\infty(\Omega)$. The corresponding algebra of individual observables is $\mathcal{B}(\mathbb{C}^2)\otimes C^\infty(\Omega)$ (note the difference between the algebra of *intrinsic* observables and the algebra of *individual* observables). Now we describe the measurement process of the observable $\sigma_3 \in \mathcal{B}(\mathbb{C}^2)$, characterized by the following two equations:

$$\sigma_3 |\alpha\rangle = |\alpha\rangle, \qquad \sigma_3 |\beta\rangle = -|\beta\rangle, \qquad |\alpha\rangle, |\beta\rangle \in \mathbb{C}^2.$$

In an individual interpretation the system is always in a pure state, therefore, any candidate for a measurement process dynamics has to transform pure states into pure states. Since one factor of the algebra of individual observables is commutative, the pure states are of product form and we can set for the initial state of the whole system the following expression:

$$\rho(0) = \rho_d(0)\otimes\rho_e(0), \qquad \rho_d(0) \in \mathcal{B}(\mathbb{C}^2)^*, \rho_e(0) \in C^\infty(\Omega)^*$$

with

$$\rho_d(0)(A) = \langle\Psi(0)|A|\Psi(0)\rangle,$$

where the initial state vector $\Psi(0) \in \mathbb{C}^2$ is given by

$$\Psi(0) = c_\alpha |\alpha\rangle + c_\beta |\beta\rangle \qquad \text{with} \quad |c_\alpha|^2 + |c_\beta|^2 = 1,$$

and $\rho_e(0) \in C^\infty(\Omega)^*$ is the *pure* initial state of the measuring instrument. As the pure states of the C*-algebra $C^\infty(\Omega)$ are in one-to-one correspondence with the points of the phase space $\omega \in \Omega$, we henceforth identify these two sets.

A dynamics of measurement process type maps the initial state above into a pure final state $\rho(T)$, whereby in most experiments, *but not always*, $\rho(T)$ is given either by $\rho(T,\omega) = \rho_d^{(\alpha)}\otimes \rho_e^{(\alpha)}$ or by $\rho(T,\omega) = \rho_d^{(\beta)}\otimes \rho_e^{(\beta)}$. In order to describe a measurement process, $\rho_e^{(\alpha)}$ and $\rho_e^{(\beta)}$ have to be disjoint states that means the pointer observable $P \in C^\infty(\Omega)$ assumes different values on $\rho_e^{(\alpha)}$ and $\rho_e^{(\beta)}$. Moreover, σ_3 has to assume the values $\approx +1$ on $\rho_d^{(\alpha)}$ and ≈ -1 on $\rho_d^{(\beta)}$, respectively. Note that *the final state of a single measurement differs from experiment to experiment*, it all depends on the initial state of the environment $\omega \in \Omega$, which is not under control of the experimentalist. He is capable only of reproducing the initial state of the object, but not that of the environment. Therefore any dynamics for the measurement process is given by a family of equations indexed by a non-controlable parameter ω.

As the pure states of a spin $\frac{1}{2}$ system are in one-to-one correspondence with the points of the unit sphere, one can get a clear idea of the facts just dicussed by the following picture:

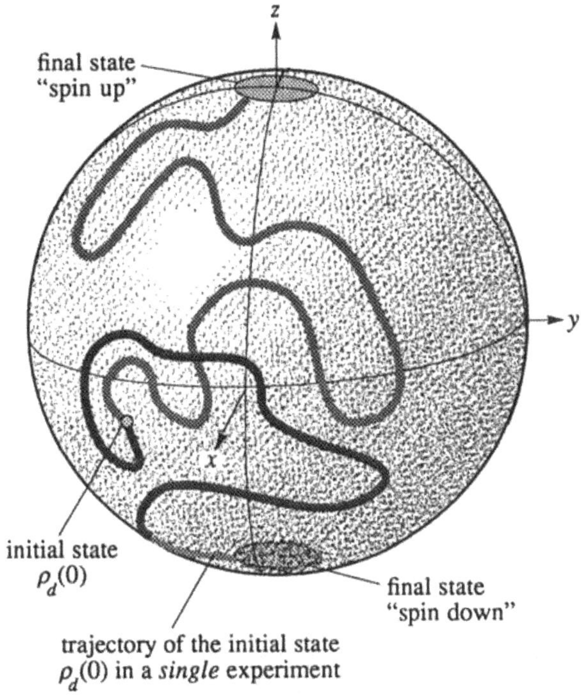

The pure initial state $\rho_d(0)$ of the object is given by a point of the sphere and since the dynamics has to transform pure states into pure states, the trajectory of the initial state is a curve on the surface of the sphere, which finally reaches a suitable neighborhood of one of the poles. In this picture, the course of two different measurements with the same initial state $\rho_d(0)$ is shown, in one experiment $\rho_d(0)$ is transformed into a neighborhood of the state $|\alpha\rangle\langle\alpha|$, that means the event "spin up" occurs, in the second experiment the event "spin down" occurs. There are also experiments in which neither of the results "spin up" or "spin down" is obtained. That is why one has to give information about the frequency of the occurence of the two events "spin up" and "spin down", respectively. This is done in the statistical interpretation, being the link between theory and experiment. For spin $\frac{1}{2}$ sytems, the result "spin up" is obtained with relative frequency of about $|c_\alpha|^2$ and the result "spin down" with relative frequency of about $|c_\beta|^2$. One realizes that the dynamics of the measurement process in a statistical interpretation is given by stochastic equations, which have to supply the correct transition probabilities.

Summing up we may state:

- No deterministic dynamics is able to transform the very same initial state $\rho_d(0)$ sometimes into the final state $\rho_d^{(\alpha)}$ and sometimes into a different final state $\rho_d^{(\beta)}$. *Time evolutions for measurement process models in an individual interpretation show stochastic behavior.*
- Linear stochastic equations conserve superpositions with probability one. Since linear individual time evolutions give rise to linear stochastic equations, *time evolutions of individual objects have to be non-linear.*

2. Spin $\frac{1}{2}$ System in Interaction with a Harmonic Environment as Measuring Instrument

In order to get an idea of a suitable dynamics, we start with a Hamiltonian model. The formal Hamilton operator is given by the following expression:

$$H = \tfrac{1}{2}\hbar\,\boldsymbol{B}\,\boldsymbol{\sigma}\otimes 1 + 1\otimes\int_0^\infty \tfrac{1}{2}\{P_k^2 + \nu_k^2\,Q_k^2 - \hbar\nu_k\}\,dk - \tfrac{1}{2}\hbar\,\sigma_3\otimes A \quad ,$$

with

$$A = \int_0^\infty g_k Q_k\,dk \quad ,$$

where $\sigma_1, \sigma_2, \sigma_3$ are Pauli matrices, and P_k and Q_k are canonical operators fulfilling Weyl's canonical commutation relations. The parameters $\boldsymbol{B} = (B_1, 0, B_3) \in \mathbb{R}^3$, $\nu_k > 0$, $g_k \in \mathbb{R}$ have to be chosen in such a way that the system can mimic a measurement process.

The formal Hamiltonian generates the following Hamiltonian equations of motion:

$$\dot{\boldsymbol{\sigma}}(t) = \boldsymbol{\sigma}(t)\times\boldsymbol{B} + \boldsymbol{\sigma}(t)\times\boldsymbol{A}(t) \quad ,$$

with $\boldsymbol{A}(t) = \{0, 0, A(t)\}$, where the function $t \to A(t)$ can be found by exact integration of the equations of motion for the harmonic oscillators.

From this equation, one deduces the dynamics of the quantum mechanical expectation value $\boldsymbol{M}(t)$ of $\boldsymbol{\sigma}(t)$:

$$\dot{M}(t,\omega) = M(t,\omega) \times B + M(t,\omega) \times a(t,\omega) + C\,, \quad \omega \in \Omega.$$

This equation depends on a function $t \to a(t|\omega) = \{0, 0, a(t|\omega)\}$ the explicit expression of which is given for $t \geq 0$ as follows:

$$a(t|\omega) := \int_k \{\lambda_k^* \omega(a_k(t)) + \lambda_k \omega(a_k^*(t))\}\, dk$$

$$= \omega(g(t)) + K(0)\, M_3(t,\omega) - K(t)\, M_3(0) - \int_0^t K(s)\, \dot{M}_3(t-s|\omega)\, ds\ ,$$

where $t \to \omega(g(t))$ is an external force, which depends only on the initial conditions of the environment

$$g(t) := \int_k \{\lambda_k^* e^{-i\nu_k t} a_k(0) + \lambda_k e^{+i\nu_k t} a_k^*(0)\}\ .$$

The function $t \to K(t)$ is a memory kernel, describing the interaction between object system and environment. The Einstein–Podolsky–Rosen correlations between object and environment are given by the functional $C := \{C_1, C_2, C_3\}$ with $C_i := \rho(\sigma_i \otimes A(t)) - \rho(\sigma_i \otimes \mathbf{1})\, \rho(\mathbf{1} \otimes A(t))$ and the dependence on the initial state of the free environment is indicated by the parameter $\omega \in \Omega$.

If A is a classical observable and ρ a pure state, then there are no Einstein–Podolsky–Rosen correlations, i.e. $C = 0$ and the equation above simplifies to

$$\dot{M}(t,\omega) = M(t,\omega) \times B + M(t,\omega) \times a(t,\omega)\ .$$

The validity of the equation above is proved, if one is able to find a representation of the C*-algebra of intrinsic observables, such that $A \in C^{\infty}(\Omega)$. There are good heuristic arguments but no rigorous proof for the existence of such a representation.

Since the pure initial state of the environment can in principle not be known to the experimentalist, it is necessary to use a statistical description, i.e. one assumes the set Ω to be equipped with a probability space structure (Ω, Σ, μ), where Σ and μ denote a suitable σ-algebra and probability measure, respectively. In this case, the function

$$a\colon \mathbb{R} \times \Omega \to \mathbb{R}$$

$$(t, \omega) \mapsto a(t, \omega)$$

represents a real valued stochastic process and the equations of motion for the expectation values of σ are stochastic differential equations, just as it is the case in classical statistical mechanics. In order to characterize this stochastic process, it is necessary to specify the probability measure μ. In most cases only the first two moments of μ are accessible empirically, that is why the assumption of Gaussian distributed initial states is optimal from an information-theoretical point of view. If in the mean the environment does not exert a force, then we can assume that the ensemble expectation values for all initial states of every degree of freedom vanish, so that under the assumption of stationarily distributed initial states of the free environment, we obtain the following relations

$$\begin{aligned} \mathcal{E}\{\omega(g(t))\} &= 0 \ , \\ \mathcal{E}\{\omega(g(t))\omega(g(s))\} &= D\tau^{-1}\exp(-|t-s|/\tau) \ . \end{aligned}$$

That is, the external force $t \mapsto \omega(g(t))$ is an Ornstein-Uhlenbeck process,

$$\omega(g(t)) = \frac{1}{\tau} U(t) \ ,$$

where the normalized dimensionless Ornstein-Uhlenbeck process $t \mapsto U(t)$ is defined as solution of the Itô differential equation

$$dU(t) = -\gamma U(t)dt + \sqrt{2D}\, dW(t) \ .$$

The one-dimensional standard Wiener process $\{W(t) \mid t > 0\}$ fulfills the relations

$$\mathcal{E}\{W(t)\} = 0 \ , \ \mathcal{E}\{[W(t)\, W(s)]^2\} = |t-s| \ .$$

In order to get a closed system of equations, we have to specify the interaction. The simplest equations are obtained if we choose the free parameters g_k and ν_k of the Hamiltonian in such a way that the memory kernel $K(t)$ is of exponential type, i. e. $K(t) = \gamma \exp(-|t|/\tau)$. In this case, the introduction of one additional variable $t \mapsto \boldsymbol{R}(t)$ is sufficient to obtain a closed system of stochastic differential equations for the expectation values of $\boldsymbol{\sigma}(t)$

$$\begin{aligned} \dot{\boldsymbol{M}}(t) &= \boldsymbol{M}(t) \times \{\boldsymbol{B} + \gamma \boldsymbol{M}(t) - \gamma \boldsymbol{R}(t) + \tau^{-1}\boldsymbol{U}(t)\}, \\ \dot{\boldsymbol{R}}(t) &= -\tau^{-1}\boldsymbol{R}(t) + \boldsymbol{M}(t) \ , \\ d\boldsymbol{U}(t) &= -\gamma\, d\boldsymbol{U}(t)dt + \sqrt{2D}\, d\boldsymbol{W}(t). \end{aligned}$$

For times $t \gg \tau$ the system above simplifies to:

$$\begin{aligned}
dM_1(t) &= -\Omega_3 M_2 dt - \eta\Omega_1 M_2^2(t)dt + \gamma M_3(t)M_2(t)dt + \sqrt{2D}\, M_2(t)\circ dW(t) \\
dM_2(t) &= +\Omega_3 M_1(t)dt - \Omega_1 M_3(t)dt + \eta\,\Omega M_1(t)M_2(t)dt - \gamma M_3(t)M_1(t)dt - \\
&\quad -\sqrt{2D}\, M_1(t)\circ dW(t) \\
dM_3(t) &= +\Omega_1 M_2(t)dt\ .
\end{aligned}$$

It can easily be seen that $M^2 = \boldsymbol{M}(t)\boldsymbol{M}(t)$ is a constant of motion, so that these equations in fact transform pure states into pure states. To the equations of motion for the quantum mechanical expectation values, there exist Schrödinger equations for the dynamics of the state vector, the simplest expression of which is given by

$$id\Psi_t = \tfrac{1}{2}\{\boldsymbol{B}\boldsymbol{\sigma} + \eta\Omega_1 M_2(t)\sigma_3 - \gamma M_3\sigma_3\}\Psi_t\, dt - \sqrt{\tfrac{1}{2}D}\,\sigma_3\Psi_t\circ dW(t) \quad ,$$

where γ, D, $\eta{:=}\gamma\tau$ are three independent parameters, which have to be chosen in such a way that the equation actually describes a measurement process and the symbol $\circ$ denotes Stratonovich multiplication.

3. Discussion of the Non-linear Schrödinger Equation

Ideal measurements of the first kind in the sense of von Neumann do not exist. In practice a measurement result, say "spin up", does not mean that the final state $\rho_d(T)$ equals $|\alpha\rangle\langle\alpha|$, but that $\rho_d(T)$ is an element of a neighborhood of $|\alpha\rangle\langle\alpha|$, which is essentially given by the sensitivity of the measuring device used in the experiment. The choice of some measuring device introduces a kind of classification on the set of pure states, because not all pure states can be distinguished by the measuring instrument used.

Since no experiment is performed for infinite time intervalls, the dynamics of measurement process models has to fulfil the following conditions:

There should be a choice of the relevant parameters γ, D, η for which there exist two times $T_1 \ll T_2$ and non-trivial neighborhoods N_α and N_β of the states $|\alpha\rangle\langle\alpha|$ and $|\beta\rangle\langle\beta|$, such that an initial state $|\Psi\rangle\langle\Psi|_i$, given by the vector $|\Psi\rangle_i = c_\alpha|\alpha\rangle + c_\beta|\beta\rangle$ is transformed at the instant T_1 into a state $\rho_d(T_1)$ where $\rho_d(T_1)$ lies *either* in the neighborhood N_α with a statistical probability of about $|c_\alpha|^2$ or in the neighborhood N_β with a statistical probability of about $|c_\beta|^2$. If $\rho_d(T_1)$ is an element of N_α or N_β, then $\rho_d(t)$ has to remain in N_α or N_β for all $t \in (T_1, T_2)$ with probability almost one.

These conditions represent the expectation value postulate, only taking into account that experimental data refer to classes of states and that no experiment is performed for infinite times.

Known properties of the non-linear Schrödinger equation:

i) For any neighborhood of the states $|\alpha\rangle\langle\alpha|$ and $|\beta\rangle\langle\beta|$ and any time intervals (T_1,T_2), there exists a D such that an initial state, being an element of one of these neighborhoods, remains therein for all $t \in (T_1,T_2)$ with probability almost one.

ii) There exist suitable parameter sets, so that for large times t, $\rho_d(t)$ is in a neighborhood of $|\alpha\rangle\langle\alpha|$ and $|\beta\rangle\langle\beta|$ with overwhelming probability.

iii) For $B_1 \neq 0$ there exist no subsets invariant for the reduced dynamics, i.e. the initial conditions are lost for infinite time intervals.

Conditions i) and ii) are necessary conditions for a caricature of the measurement process. They state in some sense that the result for the measurement of the observable σ_3 is ± 1 within experimetal error and that the expectation value postulate is fulfilled for initial states lying near the states $|\alpha\rangle\langle\alpha|$ and $|\beta\rangle\langle\beta|$. But the most important question, whether the transition probabilities induced by the stochastic Schrödinger equation approximately fulfil the expectation value postulate at time T_1 for arbitrary initial state is still open.

Acknowledgement

I wish to thank Prof. H. Primas who in long discussions patiently taught me all I know about measurement processes in algebraic quantum mechanics, my colleague P. Funck for drawing the beautiful sphere, and Mrs. M. Schiessl for reading the manuscript carefully.

QUANTUM REALITY AND HIGHER-ORDER CORRELATIONS: TWO REMARKS ON ENTANGLEMENT

ANTON ZEILINGER
Institut für Experimentalphysik, Universität Innsbruck
Technikerstrasse 25, A-6020 Innsbruck, Austria
and Atominstitut der Österreichischen Universitäten
Schüttelstraße 115, A-1020 Wien, Austria

HERBERT J. BERNSTEIN
Hampshire College
Amherst, Massachusetts 01002, U.S.A.

DANIEL. M. GREENBERGER
Department of Physics
City College of the City University of New York
New York, New York 10031, U.S.A.

MICHAEL A. HORNE
Department of Physics, Stonehill College
North Easton, Massachusetts 02356, U.S.A.

We propose an interferometric experiment on an ion-photon entangled state. There an Aharonov-Bohm phase shift is predicted only in the correlations between ion and photon fringes. We also analyze the role of rotational phases in three-particle correlations.

1. Introduction

The debate triggered by the Einstein-Podolsky-Rosen paper[1] in 1935 as to the relation between some common-sense ideas about physical reality on the one hand and quantum mechanics on the other hand is still gaining momentum as evidenced by the increasing number of citations of the original EPR-paper. Recently some interesting new aspects have been added by experiments exploiting momentum-correlated photons and by the analysis of correlations between more than two particles.

At the first Joenssuu conference two of us[2] pointed out that in principle experiments could be contemplated where Bell's inequalities apply in situations not involving spin but states of two particles with momentum entanglement. Since then it was realized that the production of correlated photons in the process of parametric down-conversion[3] can actually act as an appropriate source and the two-particle interferometry experiments proposed accordingly[4] have recently found a first experimental realization[5]. While the proposals and experiments just mentioned exploit momentum-position correlations, in a parallel development a proposal[6] to exploit energy-time correlations in parametric down-conversion has also recently found experimental verification[7]. For more detailed information on the two-particle case and specifically on quantum mechanically entangled states produced by parametric down-conversion the reader is referred to two recent review-type conference contributions[8].

To cover the whole field of theory and experiment on higher-order correlations and to discuss the meaning of reality in that context would by far be too extensive for the present paper. Therefore we wish to discuss here just two specific aspects connected with entangled states and possible related experiments. The first one concerns a proposal for future experiments where an Aharonov-Bohm phase shift would appear in a novel nonlocal way in an EPR experiment. The second remark concerns the role of rotational phases in higher-order correlations. Both points, we suggest, may indicate some connections between entanglement and geometry.

2. Proposal of an Einstein-Podolsky-Rosen Experiment Involving an Aharonov-Bohm Phase

Probably the great majority of physicists today attributes the fact that quantum mechanics and Nature both violate Bell's inequalities to some kind of breakdown of our pre-quantum conceptions of locality or separability. Without assuming any position in that debate and without going into any details of that intricate discussion we would like to point out that another class of experiments exists where most physicists also see evidence of nonlocality in quantum mechanics. These are experiments involving the Aharonov-Bohm effect[9]. In discussions the question has been raised to us by a number of colleagues as to whether there is some kind of connection between these two types of nonlocality. While we are not in a position to answer that probably rather profound question in a general way we would like to propose here some possible experiments where both aspects would show up in an intimately connected way.

Consider an ion interferometer[10] experiment. To contemplate the development of interferometers for beams of ions is quite reasonable in view of the development of atomic beam optics and ion optics. Suppose that the ions are in an excited state when they leave the beam splitter. Suppose furthermore that the lifetime of the excited state is such that the ion has a high probability to return to its ground state by emitting a photon before recombination of the two ionic beams and that an Aharonov-Bohm solenoid is arranged between the two ionic beams.

It is evident that, since it is undefined which path through the interferometer the ions take, it is also undefined from which of the two paths the photon is emitted when the ion returns to the ground state. Therefore after decay the photon and the ion together form an entangled state

$$|\psi\rangle = |\text{ion in beam path A}\rangle|\text{photon emerging from beam path A}\rangle$$
$$+ |\text{ion in beam path B}\rangle|\text{photon emerging from beam path B}\rangle. \quad (1)$$

Thus by measuring the photon in such a way that we obtain information about which path it came from, we also get information about the ion's path. This, by complementarity, excludes single-particle interference. Again, as is customary with entangled states, we will only obtain conditional interference fringes requiring recombination of the photon amplitudes emerging from the two beam paths as well. We therefore arrive at the well-known two-particle interferometry situation, this time with interferometers for two different kinds of particles, one for a photon the other one for an ion.

In such an experiment the relative position of the interference fringes is then a function of the difference of the phases in the two interferometers. Thus the introduction of an Aharonov-Bohm phase shift cannot anymore be said to cause a shift of the ion interference fringes, but, since these fringes in a two-particle interferometer are only defined conditionally, such a phase only causes a shift of the relative position of the interference fringes. Thus, finally, if we contemplate holding the ion detector fixed, we will note a shift of the photon fringes due to a change of the solenoid current. It is certainly of interest to note that this shift happens to the fringes of an electrically neutral particle. Clearly, that shift still is a function of the total flux enclosed and thus gauge-invariant.

For completeness we may simply note that a precursor to the type of experiments just discussed would be a two-particle interferometer where an excited atom, rather than an ion, decays by emitting a photon. Clearly, in such an experiment one would not see the original Aharonov-Bohm effect, but some generalized variants of it as they have been proposed and in part even been measured already for neutrons[11].

3. Rotational Phases in Three-Particle Correlations

It has recently been pointed out[12] that by analyzing the quantum correlations in three- or four-particle systems, more stringent limitations for local hidden variable theories arise than in the two-particle situation originally employed by Bell[13] in the derivation of his inequalities. In particular, it is not even possible to assign definite elements of reality in those experimental situations where one can make definite predictions for the experimental outcomes; in the standard Bell approach these were those cases where the two analyzers were oriented parallel. The original three- and four-particle derivation has been improved upon by Mermin[14] and we would like to refer the reader for more details to a comprehensive review[15]. Here we would like to give a brief demonstration how the strong correlations in the three-particle case arise as a consequence of interference. This has been pointed out previously by Stapp[16] for the specific choice of directions analyzed by Mermin, but the role of interference is more readily seen by generalizing his approach.

We start with the correlated state for three fermions

$$|\psi\rangle = \sqrt{1/2}\,[|z+\rangle|z+\rangle|z+\rangle + e^{i\chi}|z-\rangle|z-\rangle|z-\rangle] \tag{2}$$

where $|z+\rangle$ and $|z-\rangle$ mean states with spin up and down respectively along the z-direction. It is understood that in a triple product of kets the first ket describes particle 1, the second ket describes particle 2 and the third ket describes particle 3. Suppose we measure now the spin of particle one along some direction within the x-y-plane. We denote this direction by ϕ_1, the angle of that direction relative to the x-axis. Likewise we assume that the spin of particles 2 and 3 is measured along some direction within the x-y-plane denoted by ϕ_2 and ϕ_3 respectively. To obtain some information on the possible results we write down the state of Eq. 2 in terms of eigenstates along the measurement directions. Particularly, the state

$$|\phi+\rangle = R_z(\phi)|x+\rangle = \sqrt{1/2}\,[e^{-i\phi/2}|z+\rangle + e^{+i\phi/2}|z-\rangle] \tag{3}$$

describes a particle with spin up along the direction parametrized by the rotation angle ϕ. Here, $R_z(\phi)$ is the operator for a rotation through the angle ϕ around the z-axis. We also need the state

$$|\phi-\rangle = R_z(\pi)|\phi+\rangle = \sqrt{1/2}\,[-ie^{-i\phi/2}|z+\rangle + ie^{+i\phi/2}|z-\rangle] \tag{4}$$

which is orthogonal to $|\phi+\rangle$. We might remark that the state orthogonal to $|\phi+\rangle$ is only determined up to a sign because of the dichotomy of the spinor group. This fact, though interesting in other connections does not alter the result we wish to derive here. We thus can represent the states $|z+\rangle$ and $|z-\rangle$ as

$$|z+\rangle = \sqrt{1/2}\,e^{+i\phi/2}[|\phi+\rangle + i|\phi-\rangle]$$

$$|z-\rangle = \sqrt{1/2}\,e^{-i\phi/2}[|\phi+\rangle - i|\phi-\rangle] \tag{5}$$

for any possible direction within the x-y-plane. Inserting these relations separately for each particle into the initial expression for the three-particle state we obtain for the two terms in that state

$$\begin{aligned}\sqrt{1/2}\,|z+\rangle|z+\rangle|z+\rangle &= \\ &= (1/4)\,e^{i(\phi_1+\phi_2+\phi_3)/2}\,. \\ &\quad\cdot\,[|\phi_1+\rangle + i|\phi_1-\rangle][|\phi_2+\rangle + i|\phi_2-\rangle][|\phi_3+\rangle + i|\phi_3-\rangle]\end{aligned}$$

and

$$\begin{aligned}\sqrt{1/2}\,e^{i\chi}|z-\rangle|z-\rangle|z-\rangle &= \\ &= (1/4)\,e^{i[\chi-(\phi_1+\phi_2+\phi_3)/2]}\,. \\ &\quad\cdot\,[|\phi_1+\rangle - i|\phi_1-\rangle][|\phi_2+\rangle - i|\phi_2-\rangle][|\phi_3+\rangle - i|\phi_3-\rangle].\end{aligned} \tag{6}$$

Thus the final expression for the three-particle state becomes

$$
\begin{aligned}
|\psi\rangle = (1/2) &\Big\{\Big[e^{i(\phi_1+\phi_2+\phi_3)/2} - e^{i[\chi-(\phi_1+\phi_2+\phi_3)/2]}\Big] \\
&\cdot \Big[\ |\phi_1+\rangle|\phi_2+\rangle|\phi_3+\rangle - |\phi_1+\rangle|\phi_2-\rangle|\phi_3-\rangle - \\
&\quad - |\phi_1-\rangle|\phi_2+\rangle|\phi_3-\rangle - |\phi_1-\rangle|\phi_2-\rangle|\phi_3+\rangle\Big]\Big\} + \\
&+ (i/2) \Big\{\Big[e^{i(\phi_1+\phi_2+\phi_3)/2} + e^{i[\chi-(\phi_1+\phi_2+\phi_3)/2]}\Big] \cdot \\
&\cdot \Big[\ |\phi_1+\rangle|\phi_2+\rangle|\phi_3-\rangle + |\phi_1+\rangle|\phi_2-\rangle|\phi_3+\rangle + \\
&\quad + |\phi_1-\rangle|\phi_2+\rangle|\phi_3+\rangle - |\phi_1-\rangle|\phi_2-\rangle|\phi_3-\rangle\Big]\Big\}. \qquad (7)
\end{aligned}
$$

At first sight this new representation of the three-particle state may look a little confusing but upon closer inspection we can get some insights into the nature of the state. We therefore will analyze now the possible experimental results of spin measurements in the x-y-plane characterized by the three angles ϕ_i. Without loss of generality we restrict ourselves to the cases where we obtain, e.g., the result + both for the measurement result on the first particle and for that on the second particle. The terms in the equation above describing that case are (dropping common factors)

$$
\begin{aligned}
&\Big[e^{i(\phi_1+\phi_2+\phi_3)/2} - e^{i[\chi-(\phi_1+\phi_2+\phi_3)/2]}\Big]\ |\phi_1+\rangle|\phi_2+\rangle|\phi_3+\rangle + \\
&+ i\ \Big[e^{i(\phi_1+\phi_2+\phi_3)/2} + e^{i[\chi-(\phi_1+\phi_2+\phi_3)/2]}\Big]\ |\phi_1+\rangle|\phi_2+\rangle|\phi_3-\rangle \qquad (8)
\end{aligned}
$$

implying that for a general orientation ϕ_3 of the spin analyzer for the third particle both the result + and the result - may occur, the relative frequencies being given by the phase factor expressions in the square brackets. Yet, if either square bracket vanishes, only one result is possible for the spin measurement of the third particle. The condition, therefore, for that to happen is either

$$\phi_1+\phi_2+\phi_3-\chi = 2n\pi$$

or

$$\phi_1+\phi_2+\phi_3-\chi = (2n+1)\pi \qquad (9)$$

where n is an integer. In the first case the measurement result for the third particle will always be - and in the second case it will always be +. Clearly, a proper joint variation of the three angles will preserve this feature of a definite result for the measurement on the third particle (given the results of the measurements on the other two particles). It is exactly this flexibility in the choice of directions which, as shown in our earlier publications, rules out any possibility to attribute the measurement results to some intrinsic local properties of the particles. For completeness we might remark that the same analysis applies to all terms of the general expression for the three-particle state above. We have thus analyzed how the existence of the deterministic cases can be traced to an interference effect. These cases occur only if the phases associated with the eigenstates along the directions of measurement lead to destructive interference of possible alternative measurement results. Through equations (3) and (4) these phases can be considered to be resulting from rotations. We propose that the thus suggested connection between interference in Einstein-Podolsky-Rosen experiments and geometry deserves further investigation.

Acknowledgements

A. Zeilinger would like to appreciate useful discussions with Professor Henry Stapp. This work was supported by the Austrian Fonds zur Förderung der wissenschaftlichen Forschung under project No. P 6635 T and by the U.S. National Science Foundation (grants No. DMR 87-13559 and INT 87-13341).

References

1. A. Einstein, B. Podolsky and N. Rosen, *Phys. Rev.* 47 (1935) 777-780.
2. M.A. Horne and A. Zeilinger, "A Possible Spin-less Experimental Test of Bell's Inequality" in *Symposium on the Foundations of Modern Physics*, P. Lahti and P. Mittelstaedt (Eds.), World Scientific Publishing Co., Singapore 1985, pp. 435-9.
3. R. Gosh and L. Mandel, *Phys. Rev. Lett.* 59 (1987) 1903.
4. M.A. Horne, A. Shimony, A. Zeilinger, *Phys. Rev. Lett.* 62 (1989) 2209 and *Nature* 347 (1990) 429.
5. J.G. Rarity and P.R. Tapster, *Phys. Rev. Lett.* 64 (1990) 2498.
6. J.D. Franson, *Phys. Rev. Lett.* 62 (1989) 2205.
7. Z.Y. Ou, X.Y. Zou, L.J. Wang, and L. Mandel, to be published; P.G. Kwiat, W.A. Vareka, C.K. Hong, H. Natheel, and R.Y. Chiao, *Phys. Rev.* **A** 41 (1990) 2910.
8. M.A. Horne, A. Shimony, and A. Zeilinger, "Introduction to Two-Particle Interferometry" in A. Miller (Ed.) *"Sixty-Two Years of Uncertainty"* (Plenum, N.Y. 1990); M.A. Horne, A. Shimony, and A. Zeilinger, "Down-Conversion Photon Pairs: A New Chapter in the History of Quantum Mechanical Entanglement", paper presented at the Aharonov-Bohm Conference, North Carolina 1989 (in press).
9. Y. Aharonov and D. Bohm, *Phys. Rev.* 115 (1959) 485; see also W. Ehrenberg and R.E. Siday, *Proc. Roy. Soc. (London)* B 62 (1949) 8.

10. D.W. Keith, M.L. Schattenburg, W.I. Smith, and D.E. Ritchard, *Phys. Rev. Lett.* **61** (1988) 1589; H.W. Fink, *Physica Scripta* **38** (1988) 260; J. Summhammer and J. Schmiedmayer, *Physica Scripta* **42** (1990) 124.
11. A. Zeilinger, *J. de physique*, **45** (1984) C3-213; Y. Aharonov and A. Casher, *Phys. Rev. Lett.* **53** (1984) 379; A. Cimmino, A.G. Klein, G.I. Opat, H. Kaiser, and S.A. Werner, *Phys. Rev. Lett.* **55** (1988) 1309.
12. D.M. Greenberger, M.A. Horne, and A. Zeilinger, in *Bell's Theorem, Quantum Theory, and Conceptions of the Universe,* ed. M. Kafatos, (Kluwer, Dordrecht 1989), p. 73.
13. J.S. Bell, "On the Einstein-Podolsky-Rosen Paradox", *Physics* **1**, 195-200 (1964).
14. N.D. Mermin, *Phys. today* **43** (6), (1990) 9, and *Am. J. Phys.* **58** (1990) in press.
15. D.M. Greenberger, M.A. Horne, A. Shimony, and A. Zeilinger, "Bell's Theorem without Inequalities" in *Am. J. Phys.* **58** (1990) in print.
16. H.P. Stapp, Interpretation of an Experiment of the Greenberger-Horne-Zeilinger Kind, *Lawrence Berkeley Laboratory Report LBL-29377* (1990).

Closing Session

THE PROBLEM OF MEASUREMENT IN QUANTUM MECHANICS

Bas van Fraassen

Department of Philosophy
Princeton University, Princeton NJ 08544, USA

When Pekka Lahti asked me to give a summary of the discussions of measurement in this conference, I realized of course that there is no way I can do justice to all of them. In the panel on measurement David Albert said there are only two kinds of interpretations of quantum mechanics left that are feasible; according to the one quantum mechanics is wrong and according to the other one perhaps it is true but it's incomplete. For this summary, the best I can hope is that it is not in an even worse predicament.

I would like to divide the views that have been presented into two broad categories, and subdivide each of those into two subcategories. The first division is into those which *begin* with the usual Hilbert space formalism with unitary evolution operators as given, and those who do not want to take that as a starting point. In the first category I include the ones who wish to broaden the formalism in relatively standard ways, for example by allowing for superselection rules, or observables which are represented by positive operator valued measures. The second category is more radical.

That second category I am less familiar with than the first. But in this category we can still see another important divergence. There are those who are officially neutral to begin, and think they could possibly end up with the Hilbert space formalism, if that turns out to be justified. Here I think especially of the papers of Prof. von Weizsäcker and Prof. Peres. They try to develop very general principles which are so to say phenomenological; they concern what is epistemologically closest to us, the structure of experiment and observation. The program is then to give eventually what mathematicians call a *representation theorem.* In the sixties there were very large scale programs of that sort, such as the quantum logic program of Jauch, Piron, and others. It is not precluded from the beginning that the representation theorem comes up with Hilbert space and unitary groups, and, indeed, Prof Peres gave us reasons to think that it *will*, at least in part.

But we have also had examples of more radical departure. Especially I think here the GRW models and also the algebraic quantum theory as developed by professors Primas and Zaoral. These are both very fascinating departures.

Prof. Ghirardi has been showing us how his approach can make peace with relativity theory, that the GRW models can be generalized to yield relativistic models of wave function collapse. As before we see in GRW something that is truly a *rival* of standard quantum mechanics, but with empirical divergence so small that it can claim all the same experimental support. Still I wish to express some reservation. The GRW models are produced as modifications of more standard quantum mechanical models. I have to confess that I do not have confidence that I can answer the usual questions in the usual way, when I go and position myself in the GRW world. This is perhaps not so much a criticism as an appeal. We have had much working out in detail of the foundations of standard quantum mechanics – now please I appeal for a similar rigorous foundations for the GRW, too. Maybe it would not be totally unfair any way to say this: the GRW models has now had its Heisenberg, its Schrödinger, and its Pauli, I won't say which is which, but not yet its von Neumann, to my knowledge.

In the case of algebraic quantum mechanics we also have a genuine rival to quantum theory *now*. That is not how it began I think. The mathematics is a generalization which has standard quantum mechanics as a special case. But as Prof. Primas advocated it, this is seen as the *wrong* special case. The right special case does not have a locally compact phase space, and it admits the coexistence of classical and non-classical systems. There the evolution equations are not linear, we have quite a different situation from the usual quantum mechanics. Again, I find this fascinating as a theory.

But I think that the paper of Prof. Zaoral shows how much more work needs still to be done. Even the most basic questions are answered in detail only for a simplified case, which he himself actually called a "carricature". I think that perhaps we have as yet only seen the tip of the iceberg.

But I have a more important problem to ask about. As I listened to Prof. Primas' lecture I kept asking myself: what would be the *analogue* of the usual measurement problem in this theory? And I think that analogue began to emerge for me during the second half of the lecture, which was about endophysics and exophysics. Endophysics is the physical description of closed, isolated systems – so to speak from a God's eye point of view. In philosophy we now use the phrase "the view from nowhere". Here Prof. Primas offers us the C^* -algebra formalism which describes what he calls the *intrinsic observables* of the system. Exophysical description, on the other hand, is the description from the point of view of an observer, of the *rest* of the world – which is therefore an open system. We have this distinction also in ordinary quantum mechanics and of course we are reminded of that book which Prof. Peres mentioned yesterday in his reply to David Albert, *The Quantum Theory of Open Systems* by Davies.

Now how is this done in algebraic quantum mechanics? Well, the observer determines a certain context which is characterized by a topology on the space of intrinsic states – that is a *contextually selected* topology imposed on the space of states of the C^* -algebra model.

So far we have a parallel between the ordinary quantum mechanics and the algebraic quantum mechanics. We have in both cases two sorts of descriptions, one sort is perspectival, and determined by contextual elements, and the other sort is non-perspectival and non-contextual. Now let us raise the same problem for both. Suppose the observer is not a conscious human being but an automaton. The same distinction applies. But now the contextual selection is not by conscious free choice, in fact it must be determined by physical factors which characterize this automaton. So take the closed system which includes both automaton and the system it measures, and discuss its endophysical description. This must include exactly how the physical features of the automaton fix the contextual selection, so we can see how the correct exophysical description is related to the endophysical one.

I have raised the problem in Prof. Primas' terminology, but you can see *a*) that as applied to ordinary quantum mechanics, it is just the measurement problem in outline, and *b*) that Prof. Primas faces this problem too, *mutatis mutandis* and so he will need to address it also, just like his colleagues who favour a more ordinary approach.

Looking back now to the very beginning of Prof Primas' lecture I see that he is perfectly aware of this, and actually mentioned this problem at the outset and tried to get rid of it, to pre-empt it. What he said was that if we posit hypothetically the universal validity of the physics, we seem to be asking for a description of nature which involves self-reference. And he says that this is in any case impossible, by Tarski's theorem about truth, so if we try to give that description we try to do something impossible. Obviously we cannot be asked to do that.

But I do not think that this gets rid of the problem I raised. When we confront that very same problem for ordinary quantum mechanics, we do not think that we are in conflict with Tarski's theorem. No one is asking us to produce a theory of truth that takes into account self-reference – a project which is of course severely limited by that theorem. We are asked to do something much less ambitious, and I think Prof. Primas will have to try and do that too, and give an endophysical derivation of a closed system inside of which the contextual selections are determined by purely physical factors.

But I at least have been convinced that I must make much closer study of C^* and W^* -algebras now. However, I should emphasize again that I am not in a position to say very much about these various rivals to quantum mechanics except that they are fascinating and exciting proposals.

Perhaps I am in a little better position to discuss the first category, which is not so radical, and which is willing to *begin* with the usual Hilbert space formalism, and the unitary evolution operators, (that is, the usual sort of Schrödinger equation), and attempts to stay close to that point of departure. I do not claim that there has been a clear majority in favour of this policy, so if I spend more time on this, that reflects my own perspective.

To consider those standard approaches, I shall again divide them into two sorts. But the division is now not by starting point, but by tasks; most of the authors work on both those tasks to some extent.

The two tasks I have in mind are these: *i*) to give a purely quantum mechanical description of measurement, to the extent that is possible; *ii*) to address the problem of *interpretation* in which the idea of measurement appears.

I will begin with the first, the standard quantum mechanical theory of measurement. To my mind this work is of great value to *all* participants exactly because it maps out in a neutral way our common meeting ground. There has been great progress in this area. Such progress has been reported here in a number of papers, notably by Busch, Cassinelli, Mittelstaedt, Schroeck, Imoto, Kobayashi, Grabowski, Matsumoto, and Lahti. It is also exactly this year that two substantial surveys of the results were prepared, one finished recently by Beltrametti, Cassinelli, and Lahti, the other in progress by Busch, Lahti, and Mittelstaedt. Though we have all heard the crucial details already several times, I will give a quick summary of the basic results.

The question for quantum theory of measurement has to be phrased very carefully. We know that the *word* "measurement" carries certain connotations – it is a word from ordinary language and carries the connotation of human activity and intentions. But there is a core question which is purely physical: what does a process have to be *like* in order to be used for measurements? What kind of *physical process* is needed? In answer, certain conditions are laid down and these conditions define certain classes of physical interactions which are at least candidates for the role of measurement.

The first condition is the requirement that a certain probability distribution is reproduced. To express this condition I follow the usual picture of a measurement interaction where the object system starts with a state W and the apparatus with a state T, and they come out in the reduced states W' and T', respectively. Then the condition is that the initial Born probability distribution $(E \mapsto Tr(WI_E^A))$ for the measured observable A must be reproduced by the final Born probability distribution $(E \mapsto Tr(T'I_E^B))$ for the pointer observable B. Here, e.g., I_E^A is the spectral projection of A assosiated with the real number set E. In short:

Condition 1. $Tr(WI_E^A) = Tr(T'I_E^B)$.

There are two motivational reasons for this condition. The first is the form of the Born rule — it seems to say at least that much. But to me it seems that we also need it if we allow for chains of measurements.

There is now a very strong result, that sums up a number of previous results, namely the one reported by Prof. Cassinelli:

Result 1. For all discrete observables A condition 1 is met if and only if the process can be produced by a unitary operator U such that

$$U(\varphi_{ij} \otimes \Phi) = \Psi_{ij} \otimes \Phi_i.$$

where φ_{ij} are (a complete set of) eigenvectors of A corresponding to (disjoint) eigenvalues a_i, Ψ_{ij} are arbitrary unit vectors orthogonal with respect to the second index, and Φ_i are (a complete set of) eigenvectors of the (nondegenerate) pointer observable B.

There is no reason to think that Condition 1 is enough for a measurement process. One objection is this: The final state of the apparatus is *not* in general a mixture of pointer observable eigenstates. So we can't even begin to think that the Born probability is the probability of pointer observable values. I follow now a proposal of Prof. Beltrametti, given in a recent paper of his, to put down a second condition which makes the process look more like a measurement.

Condition 2. The reduced states of the system and the apparatus at end of the process are mixtures of eigenstates of the measured observable and the pointer observable.

There is again a result that answers all questions about the class of processes satisfying both conditions.

Result 2. For the processes of Result 1 Condition 2 holds if and only if the process sets up a strong correlation between the measured observable and the pointer observable, of the type characteristic of the von Neumann-Lüders measurement.

These are the basic results, but we have a very rich family of auxillary results. As Prof. Cassinelli showed, we now know how to classify these interactions and how to construct them – the pure abstract existence proofs have been replaced by constructive ones – and in addition we have other important classifications of all possible kind of interactions in terms of *probability* aspects. But these two results are I think the most basic.

There is a real concern here (I will discuss a very big problem in a moment), but yesterday in the panel people already complained that this sort of thing is too idealized. Real laboratory measurements seem to escape this idealized scheme. It seems to me however that there is already a good deal of work that addresses this concern to some extent; I mean work on *unsharp* measurements. The slogan should be something like this: we must characterize *crude* measurements as well, but a crude measurement of any observable is a sharp measurement of crude observable. Still, I think we all agree that more is needed here.

Now I come to the very last thing I want to talk about and that concerns interpretation in connection with this category of work that begins with the usual formalism. This is connected with a very serious problem. The people who are involved in the standard quantum theory of measurement pose this problem in a particular form: the *requirement of objectification.* Before saying exactly what that is, let me just mention what happened at the initial stage of the past decade, I mean in the book by Beltrametti and Cassinelli (1981) — a book which, if it has not been my bible, has at least been my Kalevala.

Is it possible (within standard quantum mechanics) that at the end of the measurement we could see the superposition change more or less in the way von Neumann postulated, into a *mixture*? The answer was *yes* if you allow for superselection rules. If the pointer observable eigenstates are separated by a superselection rule then in fact this final state, which we write as a single vector, i.e. as a superposition, really does represent a mixture, because it has components taken from different coherent subspaces. That is terrific in one way, because it means that you keep the Schrödinger equation, and you still have collapse of the wave packet. But there were problems. One was that the Hamiltonian could not be an observable. This was also pointed out by Kay Kong Wan, who discussed the same solution around that time. There is another problem: where do the superselection rules come from? And the suggestion was that they derive somehow from the macroscopic character of the measuring instrument. But when does the object get big enough, when are enough particles involved, for some aspects to become classical I have to say that this is the one point where my Kalevala disappoints me and in fact in a way still puzzles me (perhaps that is just typical of sacred scriptures). I don't just mean that I can't see how increased complexity can lead to such superselection rules – I can't see how, but who knows?

No, my real disappointment is this. Suppose we solve the measurement problem with an assumption like that – that the apparatus is *macroscopic*. In that case what do we have?

I want to remind you that physics is supposed to give us information about what happens in the world. Now if you ask a quantum physisist to give you some empirical information, he always does the same thing: *he calculates probabilities for measurement outcomes.*

That is fine, provided there are lots of measurements. I want to ask him what happened in the time of the dinosaurs, and also what happens on the other side of the moon. If measurements can be processes that happen without human agency, then that is O.K. . For example, a big rock fell on a dinosaur, and it got a lump on it's head, and this can be regarded as a measurement of the position of that rock at that time.

To be serious: if any criteria for measurement are purely physical criteria, then it is quite possible that measurement interactions are happening spontaneously in nature very often. So in that case those measurement outcome probabilities apply to inanimate nature and give a lot of information about it.

But now what about the microlevel? Does quantum mechanics give us any empirical information about what is happening there? By the same token, it gives us information about that only if some microprocesses too qualify as measurements.

Remember the two conditions I wrote down. It is quite possible that in some isolated region of space there are just a few particles, but they have interactions with each other which satisfy those two conditions. In that case, quantum mechanics tells us a lot about what happens there *provided* those interactions are measurements too.

So now you can see why I was disappointed with my Kalevala, and indeed with every solution to the measurement problem which concentrates on the macrolevel.

Anyway, since 1981, things have gotten a lot worse for this idea that a measurement ends with objectification.

Disaster of objectification. For this I shall briefly recapture the theorem described by Mittelstaedt and Busch.

a) The definition of objectification as given by Mittelstaedt: B has an objective value if and only if the state is a mixture of B-eigenstates which admits the ignorance interpretation (i.e. the system is really in some B-eigenstate).
b) The postulate of objectification: at the end of measurement, the pointer observable has an objective value.
c) Result: in that case the B-eigenstates are separated by a superselection rule, B commutes with all observables that pertain to the apparatus, i.e. the apparatus is a discrete classical system. Hence condition 1 can no longer be met!

Remarks. I take the liberty of closing with some comments on this situation, which bear reference to my own paper in this conference. The definition of objective value incorporates von Neumann's semantics, i.e. identifies value attributions with state attributions.

The problem of objectification so posed presupposes von Neumann's semantics for value attribution. And indeed, with that presupposition there can be only one sense to the idea of objectification.

Conclusion. We live now in an exciting time: we have learned so much, we know so much more now about what can and cannot be modelled in quantum mechanics with and without superselection rules, and also what sorts of alternatives can escape the NOGO hidden variable theorems. And yet, we keep encountering very fundamental problems; it is as if some single mystical dragon is slain and always again resurrects itself in a new form. But in each new life it is forced to take on a more precise form and in that way it becomes more vulnerable to rigorous, formal methods – and that, I guess, is why in the end I share the optimism that David Albert expressed: I think maybe this dragon does not have many more lives to go before he is dead for good.

REMARKS ON OUR CONCEPTION OF REALITY

Hans Primas

Laboratory of Physical Chemistry, ETH
ETH-Zentrum, CH-8092 Zürich, Switzerland

I have been asked to give a summary of our discussions *on the influence of the unconscious on our conception of reality.* I hope you can understand that I am not able to fulfill this task. All I can do is to give you some impressions and add some remarks which hopefully may soften some controversial positions.

At this conference, there has been a strong tendency to consider the open problems in quantum theory as purely physical or mathematical ones. Many speakers have stressed that these questions should not be mixed up with philosophical or psychological issues. I myself have agreed with this view, *but this should not be misunderstood* as denial of the fact that quantum mechanics has brought a deep change in our conception of reality, and has clearly put in evidence the *limitations* and the *narrowness* of today's scientific approach to understand reality. In particular, the modern developments in experimental science and in molecular biology make it necessary to discuss *individual events* and the associated *ontological* questions. These developments in science itself do not allow us any longer to adopt a cheap positivistic view, since a purely instrumentalistic view is simply insufficient to grasp modern science.

The genuinely probabilistic character of *individual quantum events* is an irreducible aspect of reality – according to Wolfgang Pauli, *an irrational aspect of reality.* The possibility of *free choice* of the experimental set-up implies that observations acquire the character of a unique irrational actuality. But Pauli also stressed that "the psychic state of the observer enters the laws of nature of quantum mechanics no more than those of classical physics", and that "the old question whether the psychic state of the observer may influence the external material evolution of Nature, has no place in contemporary physics". I think we simply *have* to acknowledge that there is a gap in the reasonings of present-day physics.

The fact that important features of reality cannot be reached by rational theories has been discussed at this conference, for example in the context of Pauli's dreams. Pauli considered the Cartesian separation of spirit and matter to be misconcieved and preferred a vision where spirit and matter are considered as two complementary aspects of reality. There can be no question that the dreams and the ideas of Wolfgang Pauli are most unconventional and most remarkable, to say the least. Not all participants of this conference have accepted Pauli's view of the utmost importance of the unconscious and the irrationality for our conception of reality. Nevertheless, I think all of us should pounder on this issue because it is a matter of importance for all who are engaged in the foundations of physics and science. Pauli's claim that reality contains something that rational theories are unable to describe is so important that we cannot reject it because perhaps we do not like this idea.

For example, during this conference many speakers have discussed the idea that quantum mechanics (or some variant of it) is a most promising candidate for a truly universal theory of matter. If we accept this view, we have to add that the reality encompassed by quantum mechanics or any scientific theory is certainly not the whole reality. It was, therefore, most appropriate that a conference devoted to the foundations of modern physics, the complementary aspects were not excluded.

The central problem is to bridge the dichotomy of psyche and matter. There are, indeed, remarkable correspondences which can be expressed in terms of quaternity. Quaternity is taken as a symbolic expression of all concepts of unbroken totality, both in physics and psychology. The "quaternian attitude" – an expression often used by Pauli – is the search for a new conception of reality, a reality which includes both its rational and irrational aspects. According to Pauli, quantum mechanics is indeed an *incomplete theory*, but *not* in the sense of Einstein but in the sense that it requires complementary descriptions which cannot be associated to quantum theory as hidden variables or any rational constructs. Pauli thinks of physics and psychology as complementary sciences. That is, both physics and psychology are regarded as incomplete descriptions of reality, both are needed for a more complete concept of reality. In order to express this idea of an unbroken reality, Pauli was looking for a *neutral language*, a language neutral with respect to the distinction between psychic and physical phenomena.

I think there is a growing part in the scientific community which agrees with Pauli that the *repression of the irrational* in our Western culture is dangerous, and that we have to re-evaluate fundamental questions in natural science. That, however, is a most difficult problem. As Pauli himself wrote, the problem is that we must not thereby sacrifice the positive values of the trinitarian attitude. At this conference, we have seen that there are at least two different strategies. One possibility is to take Pauli's version of the Copenhagen view as an indicaton for the proper road to proceed. Another possibilty is represented by a group of scientists who agree with the deep insights and concern of Pauli, but nevertheless do not accept the idea that the Copenhagen view is the last word for questions of interpretations of quantum mechanics, but who stress the necessity to develop quantum mechanics in all directions as far as possible.

I personally prefer the second approach, but not because I have the slightest intention to repress the existing genuinely irrational factors. On the contrary, I think that Pauli's message of the role of the unconscious in exact natural science and the role of the irrational in our conception of reality is of such an importance that I cannot take the risk to declare something as irrational which I simply do not understand because I did not try hard enough to understand it rationally.

It is really great that there are conferences like this one, where one can not only discuss problems of science in the narrow sense, but also problems like the role of the irrational in our scientific approach. That this is possible is due to the ardent efforts of our chairman, Professor Laurikainen. We have to be deeply grateful to Professor Laurikainen that he has opened the door for this, not only at the Joensuu conferences but also via his book "Beyond the Atom", through which the revolutionary ideas of Wolfgang Pauli became known to a large audience for the first time. Thanks to Professor Laurikainen, we can today afford to have the courage to speak openly about formerly repressed ideas. For this opening of a new dimension in discussion, Professor Laurikainen deserves our deepest thanks.

I am the last speaker at this conference and I take this opportunity to thank Professor Laurikainen and his whole staff in the name of all participants for the really outstanding organization of this congress. I should perhaps give explicitly the names of his co-organisators because they have done a difficult job in an excellent manner. I will not do so, because I know that I would forget some very important members of the staff who have worked efficiently and so quietly in the background that I just failed to notice it. So I would like to ask Professor Laurikainen to transmit our thanks to all of his co-workers. Thank you very much for this excellent conference, and let us hope that the tradition will continue to include in scientific conferences also aspects which are not covered by science in the old-fashioned narrow sense. Thank you.

Closing remarks

K.V.Laurikainen

I wish to thank Professors Van Fraassen and Primas for these interesting reviews, and especially Professor Primas for the kind words he addressed to me. Perhaps I may still add some remarks from the "Joensuu point of view".

In this series of conferences concerning the foundations of modern physics we have tried to clarify, in particular, *philosophical* problems. This reflects the local interest in natural philosophy which I mentioned in my opening address.

In order to really understand the criticism concerning the traditional Copenhagen interpretation, we have included various approaches to quantum mechanics in our programs, and in this symposium the problems of the theory of measurements have been one of the main themes. My personal view, however, remains very conservative: I do not understand why other approaches than the Copenhagen one are needed. I have found that the ontological views opened by Pauli and based essentially on a *psychological* analysis of observations are far more important than the arguments in favour of the various later approaches to quantum theory.

Thus, I find that the old principles of quantum mechanics, *complementarity and correspondence*, are still useful and as important as ever. This presupposes, however, that we are ready to change our conception of reality, which is not the case in the new approaches and in the theory of measurements. There the objectivity of the physical world is presupposed, and this is in conflict with the philosophy based on the idea of complementarity.

The following figure might illustrate the different basic attitudes concerned:

Logical analysis of the "paradoxes". Psychological view points are neglected. | *Results*: New approaches to quantum mechanics. Theory of measurements.

⇑

Quantum mechanics = formalism + the Copenhagen interpretation

⇓

Metaphysical analysis of the "paradoxes". Strong emphasis on psychology, incl. *the unconscious*. | *Results*: A new ontological view on *one world* containing both "material" and "psychic" aspects.

As I understand the situation, it cannot be resolved through logical analysis but instead requires a new conception of reality, which contains *irrational* aspects. The basic lesson of atomic research is that there are aspects in reality which cannot be described using rational theories.

* * * * * *

One characteristic remark by Pauli is: "Scientists went a little too far in the 17th century."

Pauli meant by this that the direction of modern science is *too rational*. We should remember that all aspects of nature cannot be described by using an exact language and mathematics. Empirical knowledge is, in fact, never purely rational but is always based on certain unconscious beliefs which influence the direction of our endeavours.

If one believes that all aspects of reality can be reached through the methods of exact sciences, the direction of research becomes one-sided and dangerous. Pauli wrote in a letter that "matter" can react in an unexpected way, destroying human beings who do not understand the very nature of "matter". This reminds me of the ecological problems of today and the threat of nuclear war.

It takes time before one can really understand Pauli's philosophy. In the Bohr-Einstein dispute Pauli was definitely on Bohr's side. However, he understood the implications of complementarity more profoundly than even Bohr himself.

* * * * * *

I wish to thank all of the participants for their contributions to this Symposium. Perhaps I may add that the next conference in Joensuu will most likely be held in two or three years.

Thank you.

Posters

Posters.

QUANTUM NONREPRODUCIBILITY AND THE DESCRIPTION OF NATURE

W. De Baere

Laboratory for Theoretical Physics, State University of Ghent, Proeftuinstraat 86, B-9000 Ghent, Belgium

HEISENBERG'S CUT AND THE LIMITATIONS OF ITS MOVABILITY

H.-H. von Borzeszkowski

Einstein-Laboratorium für Theoretische Physik der Akademie der Wissenschaften der DDR, Potsdam 1590, DDR

PEIRCEAN LOCAL REALISM DOES NOT IMPLY BELL'S INEQUALITIES

Peder Voetmann Christiansen

Institute of Mathematics and Physics, Roskilde University Center, DK 4000 Roskilde, Denmark

ON THE NATURE OF QUANTUM CORRELATION

Domenico Costantini, Ubaldo Garibaldi

Institute of Statistics, University of Genoa, 16146 Genova, Italy

CNR-GNSM, Department of Physics, University of Genoa, 16146 Genova, Italy

PHYSICAL REALITY IN CONTEMPORARY PHYSICS

Mehdi Golshani

Department of Physics, Sharif University of Technology, Tehran, Iran

CORRELATIONS-AT-A-DISTANCE IN QUANTUM THEORY: A NEW INTERPRETATION

Martin Hoffman

Institute for Physical and Theoretical Chemistry, University of Tübingen, 7400 Tübingen, West Germany

STABILITY OF PARTICLE TRAJECTORIES IN GENERALIZED COHERENT STATE FORMALISM

Kiyotaka Kakazu

Department of Physics, University of the Ryukyus, Okinawa 903-01, Japan

WHITEHEAD'S COSMOLOGY AND THE COLLAPSE OF QUANTUM STATES

Shimon Malin

Department of Physics, Colgate University, Hamilton, NY 13346, USA

THE ENTROPY CREATION NECESSARY TO PERFORM A MEASUREMENT AND TO RESET IT

Théo Muller

Center de Recherches Nucléaires, IN2P3-CNRS/Université Louis Pasteur, BP 20, F-67037 STRASBOURG CEDEX

COMPLEMENTARITY IN THE DOUBLE-SLIT EXPERIMENT REVISITED

Hiromichi Nakazato

Department of Physics, University of the Ryukyus, Okinawa 903-01, Japan

SOME REMARKS ON THE EPR PROBLEM

Izumi Ojima and Sho Tanaka

Research Institute for Mathematical Sciences, Kyoto University, Kyoto 606
Department of Physics, Kyoto University, Kyoto 606

FUZZY SET THEORY – NEW APPROACH TO FOUNDATIONS OF QUANTUM MECHANICS

Jarosław Pykacz

Insitute of Mathematics, University of Gdańsk, ul. Wita Stwosza 57, 80-952 Gdańsk, Poland

THE PRINCIPLE OF DECREASING MIXING DISTANCE IN CLASSICAL STATISTICAL PHYSICS AND FURTHER OUTLOOK

Ralf Quadt

University of Cologne, West Germany

QUANTUM MECHANICS IN TERMS OF MAPPING PROCESSES ONTO SPACE-TIME CURVES

Patrick Sibelius

Åbo Akademy, Department of Mathematics, Fänriksgatan 3, 20500 Turku, FInland

GAINING KNOWLEDGE FROM QUANTUM EXPERIMENTS

Johann Summhammer

Atominstitut der österreichischen Universitäten, Schüttelstraße 115, A-1020 Wien, Austria

SYMPLETIC HIDDEN - VARIABLES THEORIES - THE MISSING LINK IN ALGEBRAIC CONTEXTUAL APPROACHES

Milan Vujičič, Mićo Djurdjevič and Fedor Herbut

Faculty of Physics, University of Belgrade, P.O. Box 550, 11001 BEOGRAD, Yugoslavia

LOGICAL ASPECTES OF NON-LOCALITY IN MICROPHYSICS

Claes Åberg

Institute of Theoretical Physics, S-412 68 Göteborg, Sweden

LIST OF PARTICIPANTS

Alanen P.
TYHL
Lemminkäisenk. 2 A
20520 Turku
Finland

Albert D.
Department of Philosophy
Columbia University
Morningside Heights
New York
NY 10027 USA

Andrews T.B.
3828 Atlantic Avenue
Brooklyn
NY 11224-1209
USA

Awaya K.
Akatsuki Gakuen
Junior College
238 Shiroyama,Kayou-cho
Yokkaichi-city Mie 512
Japan

Barut A.O.
Department of Physics
University of Colorado
Boulder CO 80309
USA

De Baere W.
Laboratory of Theoretical Physics
State University of Ghent
9000 Gent
Belgium

Bedford D.
University of Natal
Department of Physics
King George V Avenue
Durban 4001
South Africa

Beltrametti E.
Department of Physics
University of Genova
Via Dodecaneso 33
16146 Genova
Italy

Ben-Dov Y.
Institute for History and Phil.of Science
Tel-Aviv University
Tel-Aviv 69978
Israel

Bitbol M.
Institut de Biologie
Physico-Chimique
13, Rue P.et M.Curie
5005 Paris
France

von Borzeszkowski H.H.
Einstein-Laboratorium
Akademie der Wissenschaften der DDR
Rosa-Luxemburg-Straße 17a
1590 Potsdam
DDR

Borg F.G.
Palo 15
67400 Karleby
Finland

Broekaert J.
Theoretische Natuurkunde
Vrije Universiteit Brussel
Pleinlaan 2
1050 Brussels
Belgium

Burinskii A.
USSR Academy of Sciences
Institute of Nuclear Safety
B. Tulskaya 52
Moscow 113191
USSR

Busch P.
Institut für Theoretische Physik
Universität zu Köln
5000 Köln 41
BRD

Cassinelli G.
Department of Physics
University of Genova
Via Dodecaneso 33
16146 Genova
Italy

Cattaneo G.
Department of Information Sciences
Via Moretto da Brescia 9
20133 Milano
Italy

Christiansen P.
Institute of Mathematics and Physics
Roskilde University Center
P.O.Box 260
4000 Roskilde
Denmark

Clifton R.K.
Department of History and Phil. Sci.
Cambridge University
Cambridge CB2 3RH
England

Costantini D.
Institute of Statistics
University of Genova
16125 Genova
Italy

Cushing J.
University of Notre Dame
Department of Physics
Notre Dame, IN 46556
USA

van Erkelens H.
Faculty of Physics and Astronomy
Free University
De Boelelaan 1083
1081 HV Amsterdam
The Netherlands

Fennema J.
Marterlaan 42
1216 EX Hilversum
The Netherlands

Ferrero M.
University of Oviedo
Department of Physics
Avd. Calvo Sotelo s/r
33007 Oviedo
Spain

Folse J.H.
Department of Philosophy
Loyola University
New Orleans
Louisiana 70100
USA

Frodl P.
Gutenberg-Universität
Institut für Physik
6500 Mainz
BRD

Garbaczewski P.
Institute of Theoretical Physics
University of Wroclaw
Cybuskiego 36
50205 Wroclaw
Poland

Garola C.
Dipartimento di Fisica
Universita de Lecce
Via Arnesano
73100 Lecce
Italy

Gervat A.
DS/EM
B.P.27
94195 Villeneuve-Saint-Georges-Cedex
France

Ghirardi G.
Department of Theoretical Physics
Strada Costiera 11
Miramare - Grignano
34014 Trieste
Italy

Gisin N.
GAP Universita de Genve
20, rue de l'Ecole-de-Médecine
1211 Genève 4
Suisse

Golshani M.
Sharif University of Technology
Physics Department
P.O.Box 11365-8639
Tehran
Iran

Graber J.S.
407 Seward Square SE
Washington, DC 20003
USA

Grabowski M.
Institute of Physics
N. Copernicus University
Torun
Poland

Hall E.
Department of Philosophy
1879 Hall
Princeton University
Princeton NJ 08544
USA

Hegstrom R.A.
Department of Chemistry
Wake Forest University
Winston-Salem,
North Carolina 27109
USA

Herbut F.
Faculty of Physics
University of Belgrad
11001 Belgrad
Yugoslavia

Hoffmann M.
Institute of Physical and Theoretical Chemistry
University of Tübingen
7400 Tübingen
BRD

Hokkanen Veli-Matti
Käyräkatu 9 B 23
40600 Jyväskylä
Finland

Huhmar E.
Rantalaivurintie 15
00850 Helsinki
Finland

Hämäläinen R.
Department of Physics
University of Joensuu
80101 Joensuu
Finland

Imoto N.
NTT Basic Research Laboratories
Midori-cho 3-9-11
Musashino-shi
Tokyo 180
Japan

Jones M.
NOAA Environment Research Laboratories
U.S.Department of Commerce
325 Broadway
Boulder CO 80303
USA

Jämsä H.
Taskisentie 3
90800 Oulu
Finland

Kakazu K.
Department of Physics
University of Ryukyus
Okinawa 903-01
Japan

Kamefuchi S.
Institute of Physics
University of Tsukuba
Ibaraki 305
Japan

Kanatchikov I.
Department of Theoretical Physics
Lepedev Physics Institute
Leninsky prospect 53
Moscow 117924
USSR

Ketolainen P.
Department of Physics
University of Joensuu
80101 Joensuu
Finland

Kobayashi T.
University of Tsukuba
Institute of Physics
Tsukuba, Ibaraki 305
Japan

Koh Y.
Department of Physics
Ibaraki University
Mito 310
Japan

Krüger J.
Institute for Theoretical Physics
State University of Ghent
9000 Ghent
Belgium

Lahti P.
Department of Physical Sciences
University of Turku
20500 Turku
Finland

Lam M.M.
Physics Department
Portsmouth Polytechnic
Portsmouth
United Kingdom

Laurikainen E.
Satamakatu 4 A 5
00160 Helsinki
Finland

Laurikainen K.V.
Research Institute for Theoretical Physics
University of Helsinki
00170 Helsinki
Finland

Lévy-Leblond J.-M.
Physique Théorique
Université de Nice
Parc Valrose
06034 Nice Cedex
France

Lierheimer M.
Röthelbachweg 40
8609 Bischberg
BRD

Liukkonen S.
Department of Chemistry
Laboratory of Physical Chemistry
Helsinki University of Technology
02150 Espoo
Finland

Malin, S.
Department of Physics
Colgate University
Hamilton NY 13346
USA

Mansfield V.
Department of Physics and Astronomy
Colgate University
Hamilton NY 13346
USA

Matsumoto S.
Shino-machi Miharu 2-5-4
Kameoka, Kyoto
Japan

McLean R.G.
Department of Mathematics
City University
Northampton Square
London EC1V 0HB
England

Mittelstaedt P.
Institut für Theoretische Physik
Universität zu Köln
5000 Köln 41
BRD

Muller Th.
Centre de Recherces Nuclaires
IN2P3-CNRS/Université
BP 20
67037 Strasbourg Cedex
France

Nagel B.
Theoretical Physics
Royal Institute of Technology
100 44 Stockholm
Sweden

Nakazato H.
Department of Physics
University of the Ryukyus
Okinawa 903-01
Japan

Nimtz G.
II Physikalisches Institut
Universität zu Köln
5000 Köln 41
BRD

Noyes H.P.
SLAC
Stanford University
Stanford CA 94309
USA

Ollila M.
Friisinniityntie 15
02240 Espoo
Finland

Peres A.
Department of Physics
Technion - Israel Institute of Technology
32000 Haifa
Israel

Pietilä H.
Department of Physical Sciences
University of Turku
20500 Turku
Finland

Primas H.
Laboratory of Physical Chemistry
ETH-Zentrum
8092 Zürich
Switzerland

Pusa K.
Ukonvaaja 2 D 88
00130 Espoo 13
Finland

Pykacz J.
Institute of Mathematics
University of Gdansk
80-952 Gdansk
Poland

Pylkkänen P.
Kylänevantie 16 D 58
00320 Helsinki
Finland

Quadt R.
Institut für Theoretische Physik
Universität zu Köln
5000 Köln
BRD

Rahnama M.
Department of Physics
Kerman University
Kerman
Iran

Raczynski A.
Institute of Physics
N. Copernicus University
87-100 Torun
Poland

Rauch H.
Atominstitut der Österreichischen Universitäten
1020 Wien
Austria

Rimini A.
Department of Theoretical and Nuclear Physics
University of Pavia
27100 Pavia
Italy

Ropolyi L.
Department of Natural Philosophy
Eötvös University
1088 Budapest
Hungary

Rosen J.
Department of Physics
The Catholic University of America
Washington DC 20064
USA

Roussel P.
Institut de Physique Nucleaire
Universite Paris XI
91406 Orsay Cedex
France

Saito S.
Advanced Research Laboratory
Hitachi Ltd.
P.O.Box 2
Kokubunji Tokyo 185
Japan

Santos E.
Department of Modern Physics
University of Cantabria
39005 Santander
Spain

Schleich W.
Max-Planck-Institut für Quantenoptik
8046 Garching
BRD

Schroeck F.E., Jr.
Department of Mathematics
Florida Atlantic University
Boca Raton
Florida 33431
USA

Shield P.A.
Department of Physics
University of St. Andrews
North Haugh
St. Andrews Fife
KY16 9AS Scotland
Great Britain

Sibelius P.
Matematisk Institut
Åbo Akademi
20500 Åbo
Finland

Sigman J.
1181 East 49 Street
Brooklyn
New York 11234
USA

Simpson M.
University of Adelaide
Box 498, G.P.O.
Adelaide
South Australia 5001

Smulter Leif
Fysik Institut
Åbo Akademi
20500 Åbo
Finland

Squires E.J.
Department of Mathematical Sciences
University of Durham
Durham DH1 3LE
England

Stapp H.
Lawrence Berkeley Laboratory
University of California
Berkeley CA 94720
USA

Summhammer J.
Atominstitut der Österreichischen Universitäten
1020 Wien
Austria

Sunnarborg K.
Kulorastaantie 3 T 155
01450 Vantaa
Finland

Tanaka S.
Department of Physics
Kyoto University
Kyoto 606
Japan

Tikochinsky Y.
Racah Institute of Physics
The Hebrew University
Jerusalem 91904
Israel

Unnerstall T.
Laboratorie de Physique Théor.
Institut Henri Poincare
75231 Paris Cedex 05
France

van Fraassen B.C.
Department of Philosophy
1879 Hall
Princeton University
Princeton N.J. 08544
USA

Venkula J.
Tehtaankatu 10 A 9
00140 Helsinki
Finland

Viiri J.
Department of Physics
University of Joensuu
Joensuu
Finland

Vujicic M.
Faculty of Physics
University of Belgrade
11001 Belgrade
Yugoslavia

Wan K.K.
Physics Department
St. Andrews University
St. Andrews
Fife KY 16 9SS
Scotland
United Kingdom

von Weizsäcker C.F.
Bahnhofplatz 4
8130 Starnberg
BRD

Zapatrin R.R.
Department of Mathematics
Finance and Economics Institute
191023 Leningrad
USSR

Zaoral W.
Laboratory of Physical Chemistry
ETH-Zentrum
8092 Zürich
Switzerland

Zeilinger A.
Atominstitut der Österreichischen Universitäten
1020 Wien
Austria

Zukowski M.
Institute of Theoretical Physics and Astronomy
University of Gdansk
80952 Gdansk
Poland

Åberg C.
Institute of Theoretical Physics
41296 Göteborg
Sweden

www.ingramcontent.com/pod-product-compliance
Lightning Source LLC
Chambersburg PA
CBHW041305210926
51851CB00002B/9
9789810203870